D0992479

Applied Mathematical Sciences

Volume 150

Springer
New York
Berlin
Heidelberg
Barcelona
Hong Kong
London
Milan
Paris
Singapore
Tokyo

Applied Mathematical Sciences

1. *John:* Partial Differential Equations, 4th ed.
2. *Sirovich:* Techniques of Asymptotic Analysis.
3. *Hale:* Theory of Functional Differential Equations, 2nd ed.
4. *Percus:* Combinatorial Methods.
5. *von Mises/Friedrichs:* Fluid Dynamics.
6. *Freiberger/Grenander:* A Short Course in Computational Probability and Statistics.
7. *Pipkin:* Lectures on Viscoelasticity Theory.
8. *Giacaglia:* Perturbation Methods in Non-linear Systems.
9. *Friedrichs:* Spectral Theory of Operators in Hilbert Space.
10. *Stroud:* Numerical Quadrature and Solution of Ordinary Differential Equations.
11. *Wolovich:* Linear Multivariable Systems.
12. *Berkovitz:* Optimal Control Theory.
13. *Bluman/Cole:* Similarity Methods for Differential Equations.
14. *Yoshizawa:* Stability Theory and the Existence of Periodic Solution and Almost Periodic Solutions.
15. *Braun:* Differential Equations and Their Applications, 3rd ed.
16. *Lefschetz:* Applications of Algebraic Topology.
17. *Collatz/Wetterling:* Optimization Problems.
18. *Grenander:* Pattern Synthesis: Lectures in Pattern Theory, Vol. I.
19. *Marsden/McCracken:* Hopf Bifurcation and Its Applications.
20. *Driver:* Ordinary and Delay Differential Equations.
21. *Courant/Friedrichs:* Supersonic Flow and Shock Waves.
22. *Rouche/Habets/Laloy:* Stability Theory by Liapunov's Direct Method.
23. *Lamperti:* Stochastic Processes: A Survey of the Mathematical Theory.
24. *Grenander:* Pattern Analysis: Lectures in Pattern Theory, Vol. II.
25. *Davies:* Integral Transforms and Their Applications, 2nd ed.
26. *Kushner/Clark:* Stochastic Approximation Methods for Constrained and Unconstrained Systems.
27. *de Boor:* A Practical Guide to Splines: Revised Edition.
28. *Keilson:* Markov Chain Models—Rarity and Exponentiality.
29. *de Veubeke:* A Course in Elasticity.
30. *Sniatycki:* Geometric Quantization and Quantum Mechanics.
31. *Reid:* Sturmian Theory for Ordinary Differential Equations.
32. *Meis/Markowitz:* Numerical Solution of Partial Differential Equations.
33. *Grenander:* Regular Structures: Lectures in Pattern Theory, Vol. III.
34. *Kevorkian/Cole:* Perturbation Methods in Applied Mathematics.
35. *Carr:* Applications of Centre Manifold Theory.
36. *Bengtsson/Ghil/Källén:* Dynamic Meteorology: Data Assimilation Methods.
37. *Saperstone:* Semidynamical Systems in Infinite Dimensional Spaces.
38. *Lichtenberg/Lieberman:* Regular and Chaotic Dynamics, 2nd ed.
39. *Piccini/Stampacchia/Vidossich:* Ordinary Differential Equations in $\mathbf{R}^n$.
40. *Naylor/Sell:* Linear Operator Theory in Engineering and Science.
41. *Sparrow:* The Lorenz Equations: Bifurcations, Chaos, and Strange Attractors.
42. *Guckenheimer/Holmes:* Nonlinear Oscillations, Dynamical Systems, and Bifurcations of Vector Fields.
43. *Ockendon/Taylor:* Inviscid Fluid Flows.
44. *Pazy:* Semigroups of Linear Operators and Applications to Partial Differential Equations.
45. *Glashoff/Gustafson:* Linear Operations and Approximation: An Introduction to the Theoretical Analysis and Numerical Treatment of Semi-Infinite Programs.
46. *Wilcox:* Scattering Theory for Diffraction Gratings.
47. *Hale/Magalhães/Oliva:* Dynamics in Infinite Dimensions, 2nd ed.
48. *Murray:* Asymptotic Analysis.
49. *Ladyzhenskaya:* The Boundary-Value Problems of Mathematical Physics.
50. *Wilcox:* Sound Propagation in Stratified Fluids.
51. *Golubitsky/Schaeffer:* Bifurcation and Groups in Bifurcation Theory, Vol. I.
52. *Chipot:* Variational Inequalities and Flow in Porous Media.
53. *Majda:* Compressible Fluid Flow and System of Conservation Laws in Several Space Variables.
54. *Wasow:* Linear Turning Point Theory.
55. *Yosida:* Operational Calculus: A Theory of Hyperfunctions.
56. *Chang/Howes:* Nonlinear Singular Perturbation Phenomena: Theory and Applications.
57. *Reinhardt:* Analysis of Approximation Methods for Differential and Integral Equations.
58. *Dwoyer/Hussaini/Voigt (eds):* Theoretical Approaches to Turbulence.
59. *Sanders/Verhulst:* Averaging Methods in Nonlinear Dynamical Systems.

(continued following index)

Anatoli V. Skorokhod Frank C. Hoppensteadt
Habib Salehi

Random Perturbation Methods with Applications in Science and Engineering

With 31 Figures

Anatoli V. Skorokhod
Institute of Mathematics
Ukrainian Academy of Science
3 Tereshchenkivska Street
Kiev 01601, Ukraine
and
Department of Statistics and Probability
Michigan State University
East Lansing, MI 48824
USA
skorokhod@stt.msu.edu

Frank C. Hoppensteadt
Systems Science and Engineering Research Center
Arizona State University
Tempe, AZ 85287-7606
USA
fchoppen@asu.edu

Habib Salehi
Department of Statistics and Probability
Michigan State University
East Lansing, MI 48824
USA
salehi@stt.msu.edu

Mathematics Subject Classification (2000): 37-02, 60JXX, 70K50

Library of Congress Cataloging-in-Publication Data
Skorokhod, A.V. (Anatolii Vladimirovich), 1930–
Random perturbation methods with applications in science and engineering / Anatoli V. Skorokhod, Frank C. Hoppensteadt, Habib Salehi.
p. cm. — (Applied mathematical sciences ; 150)
Includes bibliographical references and index.
ISBN 0-387-95427-9 (acid-free paper)
1. Perturbation (Mathematics) 2. Differentiable dynamical systems. I. Hoppensteadt, F.C. II. Salehi, Habib. III. Title. IV. Applied mathematical sciences (Springer-Verlag New York, Inc.) ; v. 150.
QA1 .A647 vol. 150b
[QA871]
510 s—dc21
[515′.35] 2001059799

ISBN 0-387-95427-9 Printed on acid-free paper.

Printed in the United States of America.

9 8 7 6 5 4 3 2 1 SPIN 10864032

Typesetting: Pages created by authors using a Springer TEX macro package.

www.springer-ny.com

Springer-Verlag New York Berlin Heidelberg
A member of BertelsmannSpringer Science+Business Media GmbH

Preface

This book has its roots in two different areas of mathematics: pure mathematics, where structures are discovered in the context of other mathematical structures and investigated, and applications of mathematics, where mathematical structures are suggested by real–world problems arising in science and engineering, investigated, and then used to address the motivating problem. While there are philosophical differences between applied and pure mathematical scientists, it is often difficult to sort them out. The authors of this book reflect these different approaches.

This project began when Professor Skorokhod of the Ukranian Academy of Sciences joined the faculty in the Department of Statistics and Probability at Michigan State University in 1993. After that, the authors collaborated on numerous joint publications that have culminated in the production of this book.

The structure of the book is roughly in two parts: The first part (Chapters 1–7) presents a careful development of mathematical methods needed to study random perturbations of dynamical systems. The second part (Chapters 8–12) presents nonrandom problems in a variety of important applications, reformulations of them that account for both external and system random noise, and applications of the results from the first part to analyze, simulate, and visualize these problems perturbed by noise. In most cases, we identify which results are novel and which are closely related to developments by others. We have tried to acknowledge the primary sources on which our work is based, but we apologize for those we have inadvertently omitted.

The authors are grateful for the support in these projects from the Department of Statistics and Probability and the College of Natural Science at Michigan State University, the Systems Science and Engineering Research Center at Arizona State University, and the National Science Foundation.

Kiev, Ukraine — Anatoli V. Skorokhod
Paradise Valley, Arizona, USA — Frank C. Hoppensteadt
East Lansing, Michigan, USA — Habib Salehi

December 2001

Contents

Introduction

Answers to the following questions describe the problems that we study in this book and how we approach them.

What Are Dynamical Systems?

Many physical and biological systems are described using mathematical models in terms of iterations of functions, differential equations, or integral equations. Respectively, these models have the form of a discrete–time iteration equation

$$x_{n+1} = f(n, x_n, b),$$

a differential equation

$$\dot{x} = f(t, x, b),$$

(here and below we write $\dot{x} = dx/dt$), or an integral equation

$$x(t) = \int_0^t f(t, s, x(s), b)\, ds.$$

The vector x in each case describes the state of the system, f describes the system itself, and the vector b describes a collection of parameters that characterizes the system f. These are usually observable or designable quantities in applications. For example, the forced harmonic oscillator

$$\dot{x}_1 = \omega\, x_2,$$
$$\dot{x}_2 = -\omega\, x_1 + A \cos \mu t,$$

has

$$x = \begin{pmatrix} x_1 \\ x_2 \end{pmatrix},$$

where

$$f_1 = b_1\, x_2,$$
$$f_2 = -b_1\, x_1 + b_2 \cos b_3 t,$$

and the parameters $b_1 = \omega, b_2 = A, b_3 = \mu$ describe the natural frequency of oscillation, the forcing amplitude, and the forcing frequency, respectively.

There are many problems in the physical and life sciences that require more sophisticated mathematical models, such as partial differential equations. While some of the methods that we present and develop here can be applied to them, we restrict attention in this book to dynamical systems represented by iterations, differential equations, and Volterra integral equations.

The problems of dynamical systems revolve around determining:

1. When solutions exist.

2. How many solutions there are.

3. Which solutions, or parts of them, are observable.

4. What are critical points for the system, for example, bifurcation values of b where the solution behavior changes dramatically, say from being at rest to oscillating. (Such bifurcation phenomena are often observable in experiments.)

5. What properties of the solutions do and do not persist when the system is disturbed.

6. How do solutions behave? Are they stable? Are they periodic or asymptotically periodic? What are their domains of attraction?

Once these questions are answered, we can begin to pursue further investigations of the system's qualitative and quantitative behaviors.

In particular, in this book we develop and apply methods for modeling and analyzing dynamical systems whose parameters (b here) involve random noise.

What Is Random Noise?

Engineering systems are described in terms of inputs, state variables, and outputs. There are inherently many sources of noise in each of these. Some are random influences on inputs, such as ionospheric fluctuations affecting radio signals; some are fluctuations in the system's components affecting the state variables, such as thermal fluctuations in electronic components, wear in mechanical parts, and manufacturing imperfections; and some are measurement errors.

Measurement errors illustrate the problem: We cannot measure things to infinite precision; we always make some kinds of errors. In fact, if we perform many supposedly identical measurements of a physical attribute, such as the thickness of a machined part or the voltage output of a semiconductor, we get many different answers that depend on the resolution of our measuring device. The collection of all our measurements forms a data base that we can analyze. To do this, we need a mathematical structure that will enable us to carry out precise calculations and, eventually, to quantify randomness in useful ways. Applications of this are, among many others, to quality control, system identification, signal processing, and system design.

Models are usually not known exactly, but only (at best) up to some mathematical processes, such as differentiation, iteration, or integration. Seeing how a model responds in the presence of noise gives some insight to its usefulness in reality. For example, one of the strongest attributes of dynamical systems is that "hyperbolic points persist under smooth perturbations of the system," but this is not necessarily true for systems with random perturbations. So more robust stability concepts are needed, like stability under persistent disturbances and stochastic stability.

Describing Noise Using Mathematics

A brief description of principal definitions and results in the theory of probability that we use is given in Appendix A. As described there, the underlying mathematical structure is a probability space, which we denote by $(\Omega, \mathcal{F}, P)$, where Ω is the set of samples, and $\mathcal{F}$ is a designated collection of subsets of Ω, called the *events*, for which the probability measure P is defined.

Solutions of dynamical systems are functions, say of time. If they are, in addition, random, we must describe both randomness and time dependence simultaneously. We refer to the result as either a *random process*, a *stochastic process*, or a *random signal.*

Random processes are written as $y(t, \omega)$, where t is a timelike variable and $\omega \in \Omega$ is a sample in a probability space $(\Omega, \mathcal{F}, P)$, and they are functions mapping a time interval and the sample space into the real numbers. Usually, ω is dropped from the notation, and we simply write $y(t)$ for this random process. The process is described in terms of its distribution function, which we write as

$$F_{y(t)}(a) = P\{\omega : y(t,\omega) \le a\},$$

where the probability P is defined in the sample space of ω. Obviously, for this to make sense we must require that this set of ω's be in $\mathcal{F}$ for each real number a, so that its probability can be calculated using P. It follows that F is a nondecreasing function of a, and we define the density function of $y(t)$ to be

$$f_y(a,t) = \frac{\partial F_{y(t)}}{\partial a}(a).$$

So, formally we have that

$$F_{y(t)}(a) = \int_{-\infty}^{a} f_y(a',t)\,da'.$$

We usually make some additional natural assumptions on how f_y depends on a and t, but not at this point. These ideas carry over directly for vectors of random processes, where each component is a random process, etc.

Our approach to modeling randomness in dynamical systems is through allowing the parameters b in a system to be random processes. For example, in the case of a differential equation we write

$$\dot{x} = f(t, x, b(y(t,\omega))),$$

indicating that the parameters can change with some underlying random process $y(t)$. The resulting solutions $(x(t,\omega))$ will also be random processes, and the methods developed in this book describe x in terms of f, b, and y.

Our approach using perturbation theory is based on assuming that the underlying noise processes operate on a faster time scale than the system. So, we write $y(t/\varepsilon)$, where we drop the ω and include ε, the ratio of the system time scale to the noise time scale. Consider y to be an ergodic random process in a space Y with ergodic measure ρ in Y (precise definitions and conditions for the following calculation are given later). Consider the integral

$$x(t) = \int_0^t y(t'/\varepsilon)\,dt',$$

which solves the initial value problem $\dot{x} = y(t/\varepsilon), x(0) = 0$.

First, the ergodic theorem (Chapter 1) states that

$$\int_0^t y(t'/\varepsilon)\,dt' = \varepsilon \int_0^{t/\varepsilon} y(s)\,ds \approx t \int_Y y\,\rho(dy) = t\,\bar{y} \tag{1}$$

for $t > 0$ as $\varepsilon \to 0$, where $\bar{y}$ is the average value of $y(t)$ over Y. Second, the central limit theorem (Chapter 2) states that

$$\int_0^t (y(t'/\varepsilon) - \bar{y})\,dt' = \sqrt{\varepsilon}\sqrt{\varepsilon} \int_0^{t/\varepsilon} (y(s) - \bar{y})\,ds \approx \sqrt{\varepsilon}\,W(t)$$

for $t > 0$ as $\varepsilon \to 0$, where W is a known random process, called Wiener's process. The result is that

$$x(t) = t\,\bar{y} + \sqrt{\varepsilon}\,W(t) + \text{ Error.}$$

The first term in this approximation essentially comes from the law of large numbers, and the second from the central limit theorem. Estimating the size of the Error (Chapters 3, 5)depends on particular properties of y. For the most part, we consider y to be an ergodic Markov process or an ergodic stationary process. This example illustrates our approach to continuous–time differential equations. We also develop a similar approach for continuous–time Volterra equations and for discrete–time models.

A key element in using this procedure is determining what ρ is. Often, this can be avoided by using methods of signal processing to estimate $\bar{y}$ and W directly. For example, a useful approximation to $\bar{y}$ can usually be found by evaluating the first integral in equation (1) over a single observation of y.

Simulating Random Noise Using MATLAB

A computer simulation illustrates these facts. Suppose that y is defined to be constant except at a series of random stopping times, where it can jump to another value. For example, the stopping times might be exponentially distributed so that the time between jumps is described by the probability

$$P\left(t_{n+1} - t_n \geq \delta\right) = \exp\left(-\lambda\delta\right),$$

where λ is a positive constant. Further, suppose that at each stopping time y is a random variable whose probability density function is $f_y(y) = \mathbf{1}_{[0,1]}(y)$, which indicates the function whose value is 1 if $y \in [0,1]$ and zero otherwise. This random variable is generated in MATLAB using the function `rand`, and the stopping time increments are generated using $-\lambda *$ $\log(\texttt{rand}(\texttt{nstop}, 1))$, where `nstop` is the number of stopping times (e.g., we use for `nstop` the final time divided by ε). Such a process $y(t)$ is called a jump process, and Figure 1 shows one sample path of this jump process that is generated using MATLAB.

The integral of y is shown in Figure 2.

Wiener's process is related to one of the most important random processes that arises in science and engineering, namely *Brownian motion.* The physical phenomenon of erratic movement of microscopic particles was observed by the botanist R. Brown in 1827, and the theory that eventually described it rests on work by Laplace (the normal distribution), Einstein (the movement can be described by a diffusion process), and Wiener (there is a probability space and on it a random process whose sample paths describe particle motions).

Brownian motion is characterized by three properties:

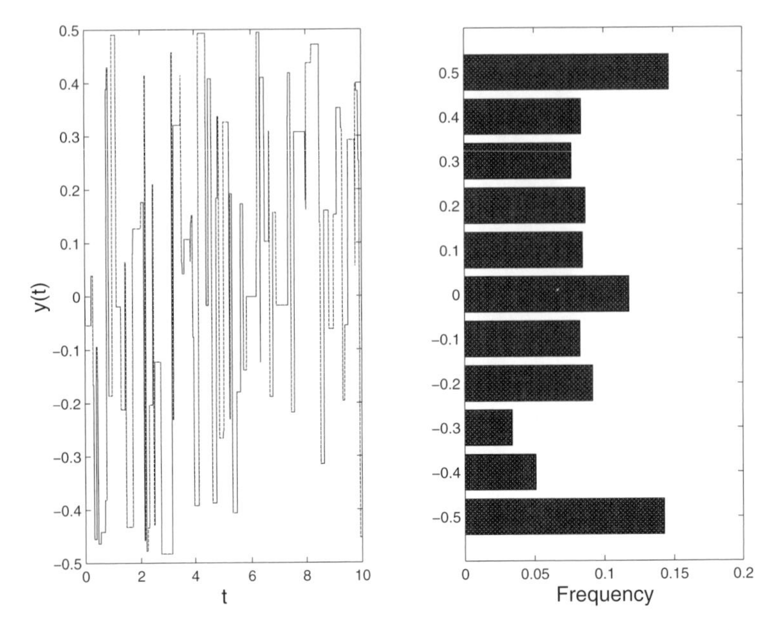

Figure 1. Left: A sample path of the jump process $y(t/\varepsilon)$ for $\varepsilon = 0.1, \lambda = 3.0$. Note that the vertical lines are not part of y, but they help visualize it. Right: The distribution of y–values for this sample path.

1. The sample paths, say $B(t)$, are continuous functions of time.
2. The nonoverlapping increments $B(t_{i+1}) - B(t_i)$ are random variables that are independent for any choice of $t_0 < t_1 < \cdots < t_n$.
3. The distribution of the increments $B(t+h) - B(t)$ for $h > 0$ is normal; that is,

$$P\{B(t+h) - B(t) \leq \Delta\} = \frac{1}{\sqrt{2\pi\sigma^2 h}} \int_{-\infty}^{\Delta} \exp\left(-\frac{x^2}{2\sigma^2 h}\right) dx.$$

The process B is called Brownian motion or *Wiener's process* with diffusivity $\sigma^2/2$, and its derivative, which exists in some sense, is called white noise. White noise is an engineering term that describes the generalized function dB/dt, although $B(t)$ is not differentiable anywhere in the ordinary sense.

There is a close connection between Brownian motion and diffusion equations. This is suggested by the facts that the increments are Gaussian and that the Gaussian function

$$u(x,t) = \frac{1}{\sqrt{2\pi\sigma^2 t}} \exp\left(\frac{-x^2}{2\sigma^2 t}\right)$$

is a solution of the diffusion equation

$$\frac{\partial u}{\partial t} = \frac{\sigma^2}{2} \frac{\partial^2 u}{\partial x^2}.$$

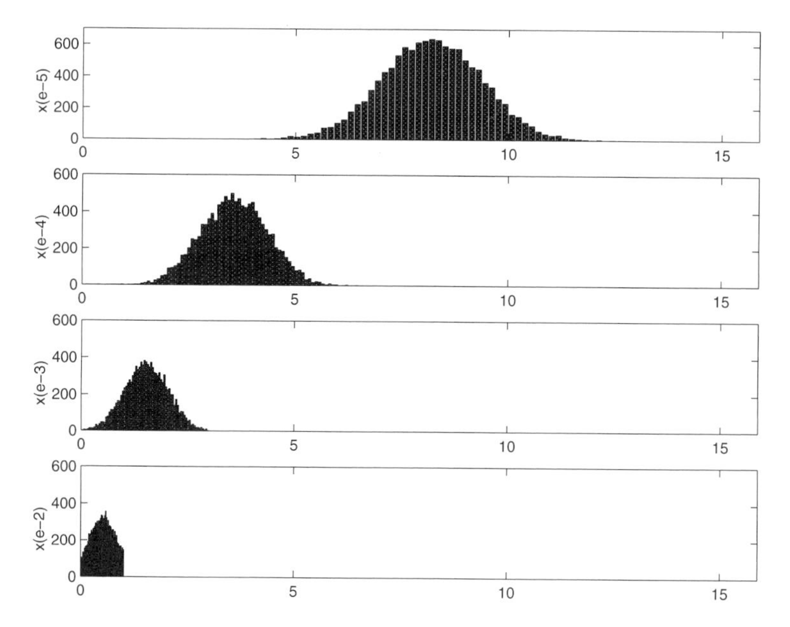

Figure 2. Histograms of 10,000 evaluations of the integral of $x(\varepsilon^n) = \int_0^1 y(t/\varepsilon^n)dt$ for $n = 1, 2, 3, 4$. This shows that the values of this integral take the form of a normal distribution whose mean and variance are as described in Figure 1 and the text.

This connection makes possible the derivation of many useful results about Brownian motion and more general diffusion processes. In particular, we can use this approach to describe the shape of the density functions appearing in Figure 2.

Estimating the Impact of Noise on a System

In some cases, noise simply passes through the system, and the random process $x(t)$ will have (essentially) the same distribution as that of the parameters $b(y)$. In other cases, noise can have dramatic impact. For instance, it can stabilize an unstable system, and it can destabilize a stable system.

For example, the (non–random) system

$$\dot{x} = \alpha\, x$$

has the solution $x(t) = \exp(\alpha t)\, x(0)$. If $\alpha < 0$ and $x(0) \neq 0$, this approaches 0 as $t \to \infty$, and if $\alpha > 0$, this grows without bound. We say in the first case that $x = 0$ is stable and in the second that it is unstable.

The ideas of stability are more complicated in noisy systems. Let us consider a linear system that is driven by white noise:

$$\dot{x} = \alpha\, x + \sigma\, x\, \dot{W},$$

where W is a Wiener process. The solution of this problem is

$$x(t) = \exp\left((\alpha - \sigma^2/2)t + \sigma W(t)\right) x(0).$$

This surprising formula results from Ito's formula for geometric Brownian motion.

We define three useful kinds of stability for the solution (Chapter 6):

1. Almost surely stable: $P\{x(t) \to 0 \text{ as } t \to \infty\} = 1$.
2. Stable in probability: $\lim_{\varepsilon \to 0} P\{|x(t)| \geq \varepsilon\} = 0$ as $t \to \infty$.
3. p-stability: $E(|x(t)|^p) \to 0$ as $t \to \infty$.

In this example, $x = 0$ is almost surely stable if $\alpha - \sigma^2/2 < 0$, so α can be positive, *and* the solution $x = 0$ can be stable in a probabilistic sense! In this sense the noise stabilizes the system. At the same time, the expected value of $x(t)$ grows: $Ex(t) = \exp(\alpha t)x(0)$. This apparent contradiction is resolved by observing that with very small probability a sample path grows very quickly. Such large deviations are rare, but they can be catastrophic or beneficial for a system. For example, this model has been used to describe financial wealth [141], and while most sample paths approach zero, infrequently one will grow dramatically.

But for p-stability, we must have $\alpha + \sigma^2(p-1)/2 < 0$, so α must be sufficiently negative to overcome noise in this sense. This example also shows that the perturbation $(\sigma\, x\, \dot{W})$ might be small in the sense that its mean is zero, but its deviations, which are scaled by σ, can be large and disrupt the system.

What Are Ergodic Theorems?

The word *ergodic* was used by Boltzmann in referring to mechanical systems that operate under the constraint of constant energy (*erg* means "work" in Greek). An important problem in statistical mechanics is that any experimental measurement of such a system requires time to perform, so it is an average of some observable of the system over a time interval. Which path is measured and how much variation is possible in such a measurement were addressed by physicists using the ergodic hypothesis, which states that the time average of an observation of a particle system is the same as the average over the phase space in which the system operates. A rigorous mathematical result due to George Birkhoff resolved this issue (for mathematicians at least) and opened the way for other uses of the result by mathematicians, physicists, and engineers.

Ergodic theorems are considered here for nonrandom dynamical systems and for randomly perturbed ones. In the nonrandom case, they are based on the assumption that there exists a measure in the phase space of the system that is preserved by the flow of the system. For example, consider a discrete–time system: Let its phase space be R^d and its dynamics be described by iteration of a function $\Phi(x) : R^d \to R^d$, so

$$x_{n+1} = \Phi(x_n), \tag{2}$$

where $x_{n+1} \in R^d$ is the state of the system at time $n+1$. Assume that there is a probability measure m on R^d for which

$$m\left(\{x : \Phi^{-1}(x) \in A\}\right) = m(A)$$

for events A in R^d. In this case, we say that the measure m is invariant under Φ. (We take the events here to be the Borel sets in R^d.) The ergodic theorem states that for all continuous bounded functions $\phi(x) : R^d \to R$ (i.e., the observables) and for almost all initial values x_0, the time average of ϕ exists:

$$\lim_{n\to\infty} \frac{1}{n+1} \sum_{k=0}^{n} \phi(x_k).$$

The dynamical system is called *ergodic* if this limit is the same as the integral of ϕ over the phase space, weighted by the measure m; that is, if

$$\lim_{n\to\infty} \frac{1}{n+1} \sum_{k=0}^{n} \phi(x_k) = \int_{R^d} \phi(x)\, m(dx).$$

If this is the case, the time average of an observable (ϕ) along almost any path is the same as its average over the whole phase space.

For example, the iteration $x_{n+1} = 4x_n(1 - x_n)$ operates in the unit interval $x_n \in [0, 1]$. An invariant measure for this is $m(x) = 1/\sqrt{\pi x(1-x)}$, and a computer simulation of this mapping reflects the invariant measure in the sense that for almost any initial point x_0, a histogram of the iterates approximates the graph of the measure's density function, as shown in Figure 3.

An analogous ergodic theorem for stochastic processes can be formulated: A discrete–time stochastic process $\{X_n, n = 1, 2, 3, \dots\}$ is called *stationary* if

$$P(X_0 \in A_0, \dots, X_n \in A_n) = P(X_1 \in A_0, \dots, X_{n+1} \in A_n)$$

for any events $A_0, \dots, A_n$. If $E|X_k| < \infty$ for all k, then with probability 1 there exists the limit

$$\lim_{n\to\infty} \frac{1}{n} \sum_{k=1}^{n} X_k,$$

and the distribution of this limit (which is also a random variable) can be calculated. The ergodic theorem establishes stability of the system in the sense of the strong law of large numbers [90].

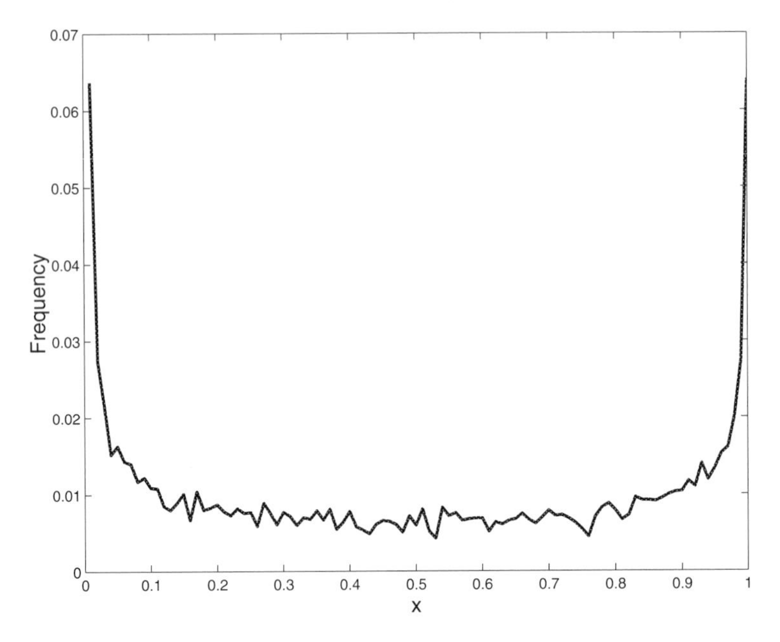

Figure 3. Histogram of iterates of $x_{n+1} = 4\,x_n\,(1 - x_n)$ beginning at $x_0 = 0.2475$.

The simulation in Figure 4 shows the distribution of the values of the jump process $y(t/\varepsilon)$ described earlier. Its values reflect the invariant measure $\mathbf{1}_{[0,1]}(y)$.

It should be noted that, as a rule, systems that are perturbed by stationary processes are not stationary, and investigation of their stability (in the general sense) is not a simple problem.

Estimating Errors

Our procedures for estimating errors in approximation methodologies involve the use of the ergodic theorem and other limit theorems for stochastic processes. For example, consider a sequence of stochastic processes $\{x_n(t)\}$. If there is a stochastic process $x_0(t)$ such that for any times $t_1, t_2, \ldots, t_k$, the joint distribution of the k random variables

$$x_n(t_1), \ldots, x_n(t_k)$$

converges to the joint distribution of

$$x_0(t_1), \ldots, x_0(t_k),$$

then the sequence $x_n(t)$ is said to be *weakly convergent* to the process x_0. Special kinds of convergence are also considered if the processes $x_n(t)$ and $x_0(t)$ are continuous functions of t: For example, let $\Phi(x(\cdot))$ be a continuous function that is defined on the space C of all continuous functions. Suppose

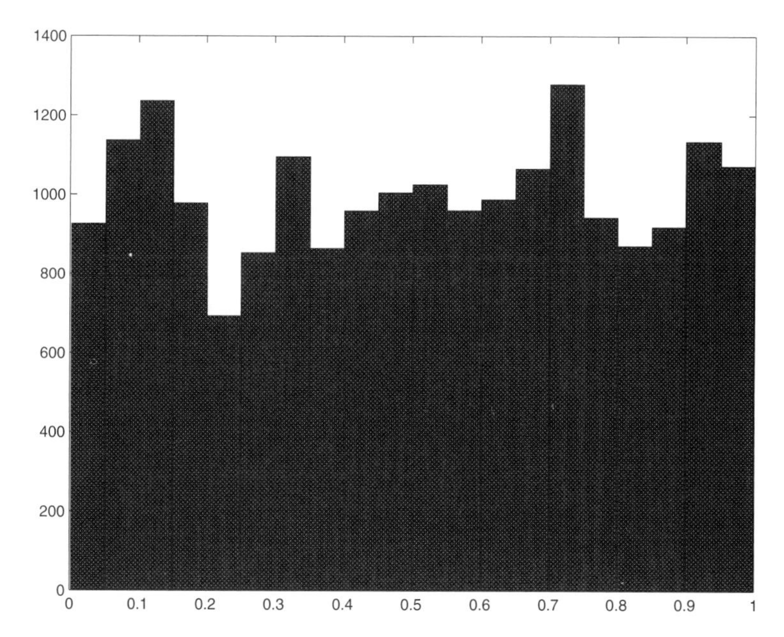

Figure 4. Distribution of values of the jump process $y(t/\varepsilon)$ that is generated by $\mathbf{1}_{[0,1]}(y)$. Here $t/\varepsilon \leq 100$ and the iterates are shown using 20 cells.

that $E\Phi(x_n(t_1), \ldots, x_n(t_k)) \to E\Phi(x_0(t_1), \ldots, x_0(t_k))$ as $n \to \infty$. Then the sequence $x_n(t)$ is said to be *weakly convergent in* C to the process $x_0(t)$. Weak convergence in C implies, for example, that

$$\lim_{n\to\infty} E \int_a^b f(s, x_n(s))\, ds \to E \int_a^b f(s, x_0(s))\, ds$$

if the function $f(s, x)$ is bounded and continuous.

We consider three classes of limit theorems, which will be useful for investigating the behavior of randomly perturbed systems:

I. *Convergence to a Wiener process*
   There are various theorems that extend the central limit theorem to more general classes of random variables. Let $y(t)$ be a random process taking values in a set Y that satisfies certain natural conditions (e.g., it is homogeneous and Markovian). We describe how y changes from one time to another by the probability of it hitting certain target sets: Let $P(t, y, A) = P\{y(t) \in A/y(0) = y\}$, which denotes the probability that $y(t) \in A$ given that $y(0) = y$. This function is called the *transition probability* for y, and we use it to describe what ergodic means for y.

We say that the process $y(t)$ is *ergodic* with an ergodic distribution $\rho(A)$ if

$$\rho(A) = \lim_{T \to \infty} \frac{1}{T} \int_0^T P(t, y, A)\, dt,$$

so $\rho(A)$ is the average probability of beginning at y and landing in the set A. A technical condition, akin to a mixing condition, is that the integral

$$R(y, A) = \int_0^\infty |P(t, y, A) - \rho(A)|\, dt$$

converge for any subset $A \subset Y$. We suppose that these two conditions are satisfied. If $g(y)$ is a measurable, bounded, real-valued function for which $\int g(y)\rho(dy) = 0$, then the distribution function of the random variable

$$\frac{1}{\sqrt{T}} \int_0^T g(y(t))\, dt$$

converges to a Gaussian distribution with mean value 0 and variance

$$\beta = 2 \iint g(y)g(y')R(y, dy')\, \rho(dy').$$

(Here and below, integrals without limits are assumed to be over Y, the noise space.) Thus, knowing R enables us to calculate the variance of the limiting process. In this case we say may that the stochastic process

$$X_T(t) = \frac{1}{\sqrt{\beta T}} \int_0^{tT} g(y(s))\, ds$$

converges weakly to a Wiener process in C as $T \to \infty$. The integral here is scaled by $\sqrt{T}$, which is suggested by the lines of constant probability in a Gaussian distribution where $x^2/(2\sigma^2 t) =$ constant, so $x \propto \sqrt{t}$.

Theorems of this kind will be considered for vector-valued functions g and for stationary processes $y(t)$.

II. *Convergence to a diffusion process*

In many important applications we can actually find a density function for the transition probability. We call this the *transition probability density*, and we write it as $f(s, t, x_0, x)$, which is the conditional density of the random variable $x(t)$ at the point x under the condition $x(s) = x_0$. The function f satisfies the second–order partial differential equation

$$\frac{\partial f}{\partial s}(s, t, x_0, x) = L_{s,x_0} f(s, t, x_0, x) \tag{3}$$

as a function of s and x_0. The differential operator

$$L_{s,x_0}u(s,x_0) = (u_{x_0}(s,x_0), a(s,x_0)) + \frac{1}{2}\mathrm{Tr}(u_{x_0x_0}(s,x_0))B(s,x_0),$$

where $a(s,x_0)$ is a vector–valued function, $B(s,x_0)$ is a matrix–valued function, (a,b) indicates the dot product, or inner product, of two vectors a and b, and Tr is the trace of the matrix. (The subscripts $u_{x_0} = \partial u/\partial x_0$, etc., indicate partial derivatives.) The operator $L_{t,x}$ is called the generator of the diffusion process $x(t)$.

It can be proved that a diffusion process $x(t)$ with the generator $L_{t,x}$ satisfies the equation that for all $t_1 < t_2 < \cdots < t_{k+1}$,

$$E\Phi(x(t_1),\ldots,x(t_k)) \times \left[f(x_n(t_{k+1})) - f(x_n(t_k)) - \int_{t_k}^{t_{k+1}} f_{xx}(x_n(s))\,ds\right] = 0 \tag{4}$$

for any continuous bounded function $\Phi(x_1,\ldots,x_k)$ and any continuous function f with continuous bounded derivatives f_x and f_{xx}.

The following statement is based on this property: If the sequence of stochastic processes $x_n(t)$ satisfies the equation

$$\lim_{n\to\infty} E(\Phi(x_n(t_1),\ldots,x_n(t_k))) \times \left[f(x_n(t_{k+1})) - f(x_n(t_k)) - \int_{t_k}^{t_{k+1}} L_{s,x}(f(x_n(s))\,ds\right] = 0$$

for all $t_1, t_2, \ldots, t_{k+1}$, Φ, and f being the same as in equation (4), then the processes $x_n(t)$ are said to converge weakly in C to the diffusion process whose generator is

$$Lu(x) = \frac{1}{2}\mathrm{Tr}\; u_{xx}(x) = \frac{1}{2}\nabla^2 u.$$

So, if the sequence of stochastic processes $x_n(t)$ satisfies

$$\lim_{n\to\infty} E(\Phi(x_n(t_1),\ldots,x_n(t_k))) \times \left[f(x_n(t_{k+1}) - f(x_n(t_k)) - \int_{t_k}^{t_{k+1}} f_{xx}(x_n(s))\frac{ds}{2}\right] = 0,$$

then the sequence $x_n(t)$ converges weakly in C to a Wiener process.

III. *Large deviation theorems*

Large deviation theorems deal with estimation of the difference between the time average of a process and its average over the noise space. We consider theorems on large deviations for Markov processes in Chapters 2, 3, and 6, in terms of which random noise will be described later.

For instance, let $y(t)$ be a homogeneous, ergodic Markov process in a compact space Y with ergodic distribution ρ. The empirical distribution of $y(t)$ is defined by the equation

$$\nu_t(A) = \frac{1}{t}\int_0^t \mathbf{1}_{\{y(s)\in A\}}(s)\,ds$$

for any set A that is a measurable subset of Y. The ergodic theorem states that the measure ν_t converges to the ergodic distribution ρ as $t \to \infty$.

Denote by $\mathcal{M}$ the set of probability measures on Y. It is known that for $m \in \mathcal{M}$ there is a nonnegative function $I(m)$ such that if the probability that the distance in $\mathcal{M}$ between ν_t and m is small enough, then it is of order $\exp(-t\,I(m))$. In fact, the function $I(m)$ is given by the formula

$$I(m) = -\inf\left\{\int_Y \frac{L\phi(y)}{\phi(y)}\,m(dy) : \phi \in D_L, \text{ and } \phi(y) > 0\,\forall y \in Y\right\},$$

where L is the generator of the Markov process y and D_A is its domain of definition.

For example, let $Y = \{0, 1\}$, and consider a process y whose generator is given by the matrix

$$\begin{pmatrix} -a_0 & a_0 \\ a_1 & 1 - a_1 \end{pmatrix}.$$

Then

$$\rho_0 = \frac{a_1}{a_0 + a_1}, \qquad \rho_1 = \frac{a_0}{a_0 + a_1},$$

and

$$I(m) = \left(\sqrt{m_0 a_0} - \sqrt{m_1 a_1}\right)^2, \; m_i = m(\{i\}),\; i = 0, 1.$$

This formula enables us to estimate the probability of a large deviation occurring during convergence to ρ. We develop these ideas further and use them to estimate errors in perturbation methods in Chapters 2, 3, and 6.

Averaging Dynamical Systems over Random Noise

Most of our averaging statements compare the behavior of a system containing ergodic random perturbations with an associated nonrandom system. For example, a system defined by a differential equation in R^d has the form

$$\dot{x}_\varepsilon(t) = a(t, x_\varepsilon(t), y(t/\varepsilon)), \tag{5}$$

where $x_\varepsilon(t)$ is the state of the system at time t and $y(t)$ is a noise process (supposed here to be a Markov or stationary process) in some space Y. The function $a(t, x, y)$ is an vector valued function that is regular enough to ensure that the solution of (5) exists and is unique when the initial value $x_\varepsilon(0)$ is given. Note that we can make no assumption about smoothness of the system with respect to y, since there is no topology given for Y. An appropriate restriction with respect to the noise variables is that the data (in this case the function a) be measurable as a function of y. We investigate the behavior of $x_\varepsilon(t)$ as $\varepsilon \to 0$.

Assume that $y(t)$ is an ergodic process with an ergodic distribution ρ; namely,

$$\lim_{T\to\infty} \frac{1}{T} \int_0^T g(y(t))\, dt = \int_Y g(y)\rho(dy)$$

for all functions g for which

$$\int_Y |g(y)|\rho(dy) < \infty.$$

Define

$$\bar{a}(t,x) = \int_Y a(t,x,y)\rho(dy). \tag{6}$$

The averaged system is defined by the differential equation

$$\dot{\bar{x}} = \bar{a}(t, \bar{x}(t)). \tag{7}$$

In this case, the averaging theorem claims that if $x_\varepsilon(0) = \bar{x}(0)$, then $x_\varepsilon(t) \to \bar{x}(t)$ uniformly on any fixed finite time interval as $\varepsilon \to 0$ with probability 1.

Following are two examples of averaging.

First, consider the system to be time–invariant (i.e., $a(t,x,y) = a(x,y)$), and let Y be a finite set. The noise $y(t)$ is a homogeneous Markov process that is ergodic, and ρ is its ergodic probability measure. A stochastic process $x_\varepsilon(t)$ is defined by the differential equation

$$\dot{x}_\varepsilon = a(x_\varepsilon, y(t/\varepsilon)). \tag{8}$$

Further, suppose that y is a jump process for which there is a sequence of random stopping (or jumping) times

$$0 = \tau_0 < \tau_1 < \cdots < \tau_n < \cdots$$

and y is constant between jumps:

$$y(t) \equiv y(\tau_k)$$

if $\tau_k \le t < \tau_{k+1}$. Then the sequence $\{y_k = y(\tau_k)\}$ is a finite Markov chain. We suppose that the jump increments $\tau_{k+1} - \tau_k$ have an exponential distribution

$$P(\tau_{k+1} - \tau_k \ge t \,/\, y(s), s \le \tau_k) = \exp(-\lambda t).$$

Equation (8) on the interval $[\varepsilon\tau_k, \varepsilon\tau_{k+1})$ can be rewritten as

$$\dot{x}_\varepsilon = a(x_\varepsilon, y_k). \tag{9}$$

So, to find the solution of equation (8) on the interval $[0, T]$, we must solve equation (9) for each k for which $\tau_k < T/\varepsilon$. As $\varepsilon \to 0$, the number of equations that must be solved to determine a sample path tends to infinity like $1/\varepsilon$. On the other hand, if the ergodic probabilities of the process $y(t)$ are ρ_y, we can average a over Y:

$$\bar{a}(x) = \sum_{y \in Y} a(x, y)\rho_y.$$

Then instead of finding the solution to a large collection of equations to determine a sample path, we need solve only one equation:

$$\dot{\bar{x}} = \bar{a}(\bar{x}).$$

The solution of this equation gives an approximation to $x_\varepsilon(t)$ if ε is sufficiently small. Figure 5 illustrates this calculation for an inertialess pendulum having a random applied torque at the support point: The model is

$$\dot{x} = 1.0 + y(t/\varepsilon) + \sin x, \tag{10}$$

where y is a jump process having exponentially distributed stopping times and having a uniform transition probability given by $\mathbf{1}_{[0,1]}(y)$.

Second, let

$$a(t, x, y) = A(t)x + f(t, x, y), \tag{11}$$

where $\int f(t, x, y)\,\rho(dy) = 0$ and $A(t)$ is a continuous matrix–valued function. Then the averaged equation is a linear one:

$$\dot{\bar{x}} = A(t)\bar{x}. \tag{12}$$

The solution to this linear problem can be found using methods of ordinary differential equations (e.g., [16]).

Constructing the First–Order (Order $\sqrt{\varepsilon}$) Correction

The next step in the investigation of randomly perturbed systems is analysis of the deviations of the perturbed system from the averaged one. We consider

$$z(t) = x_\varepsilon(t) - \bar{x}(t).$$

We illustrate results for this using equation (8). This difference satisfies the equation

$$z(t) = \int_0^t \bar{a}_x(s, \bar{x}(s))z(s)\,ds + \int_0^t (a(s, \bar{x}(s), y(s/\varepsilon)) - \bar{a}(s, \bar{x}(s)))\,ds + r_\varepsilon(t),$$

where $r_\varepsilon(t)$ is a stochastic process that can be calculated by applying Taylor's formula to the difference $a(s, x_\varepsilon, y) - a(s, \bar{x}, y)$.

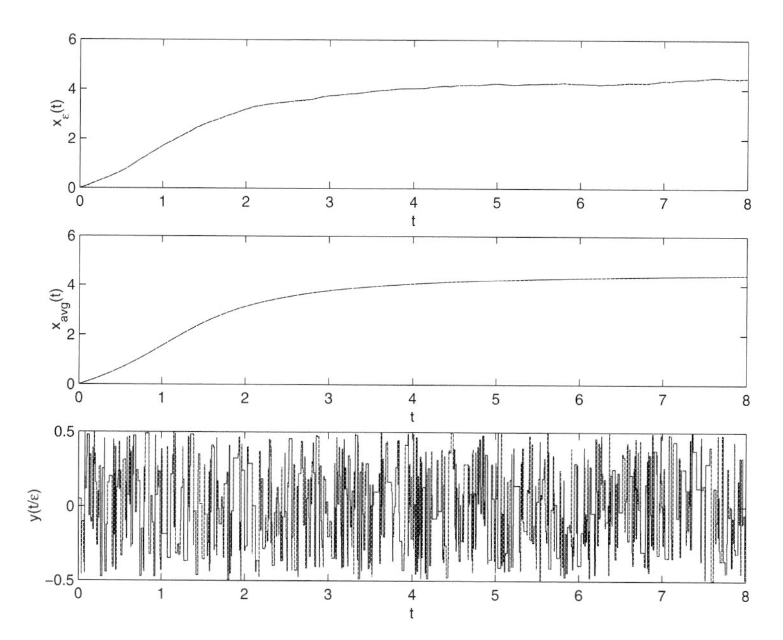

Figure 5. Comparison of the averaged system to the perturbed system (10). The top figure shows the perturbed solution; the middle figure shows the solution of the averaged equation; and the bottom figure shows the noise function $y(t/\varepsilon)$. In this simulation, the elapsed time for solving the averaged system was 0.002 times that for solving the perturbed system. Here $\varepsilon = 0.01$, and the same fixed time steps were used in both simulations.

It can be shown that $E|r_\varepsilon|^2 = o(\varepsilon)$. It follows from a limit theorem for stochastic processes that with an appropriate mixing condition satisfied, the process

$$\frac{1}{\sqrt{\varepsilon}} \int_0^t [a(s, \bar{x}(s), y(s/\varepsilon)) - \bar{a}(s, \bar{x}(s))]\, ds,$$

which is like an external forcing on the system, converges weakly in C to a Gaussian process $\gamma(t)$ for which for any vector ζ, $E(\gamma(t), \zeta) = 0$ and

$$E(\gamma(t), \zeta)^2 = \int_0^t \int \int (a(s, \bar{x}(s), y), \zeta)(a(s, \bar{x}(s), y'), \zeta) R(y, dy') \rho(dy')\, ds.$$

Therefore, the stochastic process

$$\tilde{x}_\varepsilon(t) = \frac{x_\varepsilon(t) - \bar{x}(t)}{\sqrt{\varepsilon}}$$

converges weakly in C to a stochastic process $\tilde{x}(t)$ that is the solution to the linear integral equation

$$\tilde{x}(t) = \int_0^t \bar{a}_x(s, \bar{x}(s)) \tilde{x}(s)\, ds + \gamma(t). \tag{13}$$

This result has the following interpretation: The distribution of x_ε is close to the distribution of the stochastic process $\bar{x}(t) + \sqrt{\varepsilon}\,\tilde{x}(t)$ in the sense that for any function Φ that is sufficiently smooth,

$$E\Phi(x_\varepsilon(t)) = E\Phi\left(\bar{x}(t) + \sqrt{\varepsilon}\tilde{x}(t)\right) + o(\sqrt{\varepsilon}). \quad (14)$$

This gives rigorous meaning to the expression

$$x_\varepsilon(t) \approx \bar{x}(t) + \sqrt{\varepsilon}\,\tilde{x}(t).$$

For example, consider equation (11). Denote by $K(s,t)$ the solution of equation (12) on the interval $t \in [s,\infty)$ for which $K(s,s) = 1$. Then

$$\tilde{x}(t) = \int_0^t K(s,t)d\gamma(s),$$

where $\gamma(t)$ is a Gaussian stochastic process with independent increments for which $E(\gamma(t),\zeta) = 0$ and

$$E(\gamma(t),\zeta)^2 = \int_0^t \int\int (f(s,\bar{x}(s),y),\zeta)(f(s,\bar{x}(s),y'),\zeta)\, R(y,dy')\,\rho(dy)\,ds,$$

which enables us to calculate the correlation between the solution and γ.

A computer simulation of the distribution of the first–order correction that is based on equation (10) is shown in Figure 6; also shown there is a simulation of a Gaussian process.

What Happens for t Large?

If we consider the differential equation (8) under the assumption that $\bar{a}(t,x) = 0$, then $\bar{x}(t) = \bar{x}(0)$ is a constant. The averaging theorem then implies that $x_\varepsilon(t)$ converges to $\bar{x}(0)$ for all t as $\varepsilon \to 0$. We investigate next the behavior of the random variable $x_\varepsilon(t)$ as $\varepsilon \to 0$ and $t \to \infty$ simultaneously.

To illustrate some possibilities of the evolution of such a system, we consider a linear problem in R^1:

Let

$$a(t,x,y) = a_0(t,y) + a_1(t,y)x,$$

and let $x_\varepsilon(t)$ be the solution of the equation

$$\dot{x}_\varepsilon(t) = a_0(t,y(t/\varepsilon)) + a_1(t,y(t/\varepsilon))x_\varepsilon(t), \quad (15)$$

where the continuous, bounded, real-valued functions a_k, $k = 0,1$, satisfy the condition

$$\int a_k(t,y)\rho(dy) = 0.$$

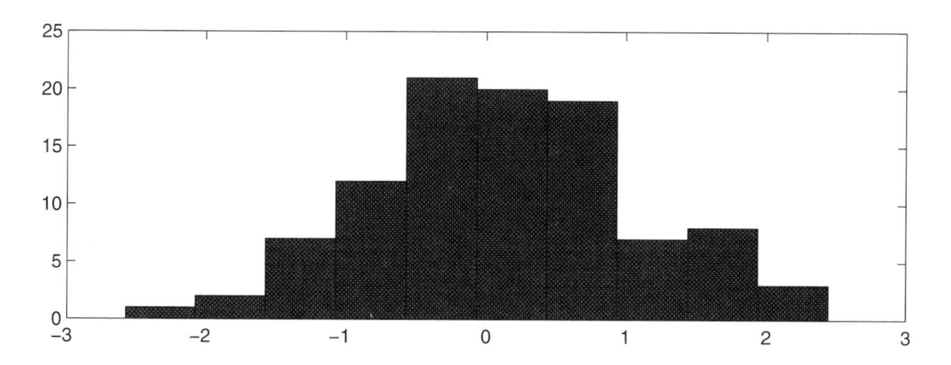

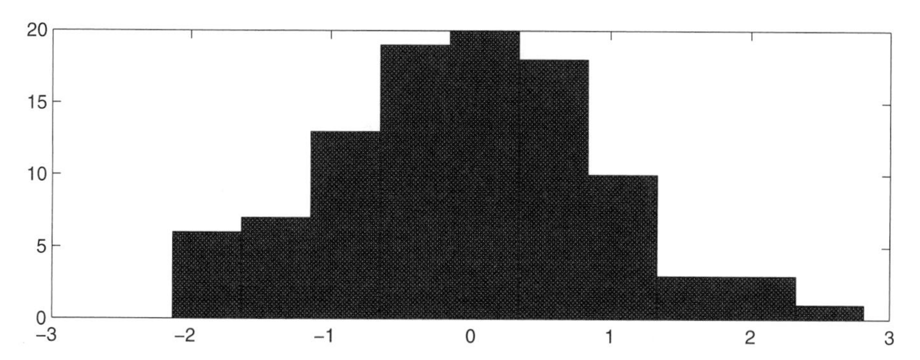

Figure 6. Top: $\tilde{x}_\varepsilon(8) = (x_\varepsilon(8) - \bar{x}(8))\varepsilon^{-1/2}$ for 100 sample paths of equation (10). In this case, $\varepsilon = 0.01$, $0 \le t \le 8$, and y is generated as above but with $\bar{y} = 0.9$. The final value of the solution is shown on the horizontal axis, and the histogram of 100 samples is plotted in the top figure. Bottom: A histogram of 100 samples of `randn`, the MATLAB function for normal random variables, is plotted for comparison.

The solution to equation (15) can be found using the variation–of–constants formula as

$$x_\varepsilon(t) = x_\varepsilon(0)e^{\int_0^t a_1(s,y(s/\varepsilon))\,ds} + \int_0^t e^{\int_u^t a_1(s,y(s/\varepsilon))\,ds} a_0(u, y(u/\varepsilon))\,du. \quad (16)$$

Consider the stochastic processes

$$\gamma_1^\varepsilon(t) = \int_0^t a_1(s, y(s/\varepsilon))\,ds, \ \gamma_2^\varepsilon(t) = \int_0^t a_2(s, y(s/\varepsilon))\,ds.$$

If the limits

$$\lim_{T\to\infty} \frac{1}{T} \int_0^T \iint a_k(s,y)a_k(s,y')R(y,dy')\rho(dy)\,ds = b_{kk},$$

for $k = 0, 1$, and

$$\lim_{T\to\infty} \frac{1}{T} \int_0^T \iint [a_0(s,y)a_1(s,y') + a_1(s,y)a_0(s,y')] R(y, dy')\rho(dy)\, ds = b_{01}$$

exist, then the two–dimensional process

$$(\gamma_0^\varepsilon(t), \gamma_1^\varepsilon(t))$$

converges weakly in C to a two–dimensional Gaussian process with independent increments, say $(\gamma_0(t), \gamma_1(t))$, for which $E\gamma_k = 0$, $E\gamma_k^2(t) = 2\,b_{kk}$, and $E\gamma_0(t)\gamma_1(t) = 2\,b_{01}$.

Under these conditions, we can use formula (16) to prove that the stochastic process $x_\varepsilon(t/\varepsilon)$ converges weakly to the stochastic process

$$\tilde{x}(t) = \exp\left(\gamma_1(t)\right) \left[\tilde{x}(0) + \int_0^t \exp\left(-\gamma_1(u)\right) d\gamma_0(u)\right] \tag{17}$$

as $\varepsilon \to 0$ if additionally $x_\varepsilon(0) \to \tilde{x}(0)$.

It follows from formula (17) that $\tilde{x}(t)$ satisfies the stochastic differential equation

$$d\tilde{x}(t) = (b_{01} + b_{11}\tilde{x}(t))\, dt + d\gamma_0(t) + \tilde{x}(t)\, d\gamma_1(t).$$

Equivalently, $\tilde{x}(t)$ is a diffusion process with the generator

$$Lu(x) = (b_{01} + b_{11}x)u_x(x) + (b_{00} + b_{01}x + b_{11}x^2)u_{xx}.$$

Using the law of the iterated logarithm [41], we see that the solution is approximately constant until time T/ε, after which the noise terms begin to dominate the approximation.

For example, consider a similar linear differential equation in R^d,

$$\dot{x}_\varepsilon = A(t, y(t/\varepsilon))x_\varepsilon + a_0(t, y(t/\varepsilon)), \tag{18}$$

where $A(t,y)$ is a $d \times d$ matrix–valued function and $a_0(t,y)$ is a vector–valued function. We assume that they are bounded, continuous functions, and that

$$\int A(t,y)\rho(dy) = 0, \int a_0(t,y)\rho(dy) = 0.$$

To analyze this problem, we introduce the two stochastic processes

$$A^\varepsilon(t) = \int_0^t A(s, y(s/\varepsilon))\, ds, \quad a^\varepsilon(t) = \int_0^t a(s, y(s/\varepsilon))\, ds.$$

With some additional conditions that ensure that these two processes have Gaussian limits, say

$$(A^\varepsilon(t), a^\varepsilon(t)) \to (\Gamma(t), \gamma(t))$$

as $\varepsilon \to 0$, we rewrite (18) as an equivalent integral equation:

$$x_\varepsilon(t) = x_\varepsilon(0) + \int_0^t dA^\varepsilon(s)\, x_\varepsilon(s) + a^\varepsilon(t). \tag{19}$$

Then we show that the stochastic process $\tilde{x}_\varepsilon(t) \equiv x_\varepsilon(t/\varepsilon)$ converges weakly in C to a stochastic process $\tilde{x}(t)$ that is a solution to the stochastic integral equation

$$d\tilde{x}(t) = \int_0^t d\Gamma(s)\tilde{x}(s)\,ds + d\gamma(t) + (B\tilde{x}(t) + a_0)\,dt. \tag{20}$$

This equation has an additional term $(B\tilde{x} + a_0)$, whose appearance is due to the difference between ordinary and stochastic analysis; namely, this results from Ito's formula [141].

So the stochastic processes $x_\varepsilon(t/\varepsilon)$ converge weakly in C to a diffusion process. This result can be extended to equations of the form given by formula (5) if the function $a(t, x, y)$ satisfies the relation

$$\int a(t, x, y)\rho(dy) = 0$$

and is sufficiently smooth in x, and if there exists the limit

$$\lim_{T\to\infty} \frac{1}{T}\int_0^T \int\int ((a_x(t, x, y'), f_x(x)), a(t, x, y))R(y, dy')\rho(dy)\,dt \tag{21}$$

for all twice continuously differentiable functions f having bounded derivatives.

If $y(t)$ is an ergodic process for which the central limit theorem holds, then Lf is the generator of the diffusion process $\tilde{x}(t)$ to which the process $x_\varepsilon(t/\varepsilon)$ converges weakly in C.

What Is in This Book?

The book is in three parts. The first part (Chapters 1 and 2) presents the mathematical tools that we use later to investigate randomly perturbed systems. These include an ergodic theorem and various limit theorems for stochastic processes.

The second part (Chapters 3–7) develops the theory of randomly perturbed systems. We consider only systems that are in some sense close to averaged ones, as measured by a small parameter $\varepsilon > 0$. For continuous–time systems this parameter ε is the ratio of the response time of the system to the (relatively fast) time scale of noise. So, we write the noisy parameters in the problem in the form $b = y(t/\varepsilon)$. For discrete–time systems we assume that the variance of the noise is small, and so ε characterizes this attribute of a random perturbation. In each case, we investigate the asymptotic behavior of systems as $\varepsilon \to 0$ and $t \to \infty$. Note that randomly perturbed systems with *fixed* ε can represent any stochastic process, so the asymptotic properties of these systems as $t \to \infty$ is a problem of too general interest for us here.

The third part (Chapters 8–12) presents applications of the theory of random perturbations to problems in mechanics, engineering, and the life sciences.

Dynamical Systems

Topics in dynamical systems of special interest to us are listed next:

Diffusion Approximations of a First Integral

Consider the differential equation

$$\dot{x}_\varepsilon = a(x_\varepsilon(t), y(t/\varepsilon)) \tag{22}$$

and suppose the averaged equation

$$\dot{\bar{x}} = \bar{a}(\bar{x}(t)) \tag{23}$$

has a first integral; that is, there is a function $\phi(x)$ such that

$$(\nabla\phi(x), \bar{a}(x)) = 0$$

for all x. Then $\phi(\bar{x}(t)) \equiv$ constant along any solution of equation (23).

Suppose that the set $\Gamma_c = \{x : \phi(x) = c\}$, the level set for $\phi = c$, is a bounded smooth surface in R^d and the solution of equation (23) is ergodic on the set Γ_c. That is, there is a probability measure m_c on Γ_c such that

$$\lim_{T\to\infty} \frac{1}{T}\int_0^T g(\bar{x}(t))\,dt = \int_{\Gamma_c} g(x) m_c(dx) \tag{24}$$

for all continuous bounded functions $g : R^d \to R$. Then under some general conditions (including ergodicity and mixing conditions for $y(t)$ and smoothness of the integral on the right-hand side of equation (24)), the stochastic process $z_\varepsilon(t) = \phi(x_\varepsilon(t/\varepsilon))$ converges weakly to a diffusion process $\tilde{z}(t)$ in R. We also calculate the generator of $\tilde{z}$.

For example, consider the system of differential equations in R^2

$$\begin{aligned} \dot{x}^1_\varepsilon &= -a_1(y(t/\varepsilon))x^2_\varepsilon, \\ \dot{x}^2_\varepsilon &= a_2(y(t/\varepsilon))x^1_\varepsilon, \end{aligned} \tag{25}$$

where $\bar{a}_1 = \bar{a}_2 = a$. The averaged system is

$$\begin{aligned} \dot{\bar{x}}^1 &= -a\,\bar{x}^2, \\ \dot{\bar{x}}^2 &= a\,\bar{x}^1. \end{aligned} \tag{26}$$

This has the first integral $\phi(u, v) = u^2 + v^2$. We show that the stochastic process

$$z_\varepsilon(t) = \phi(x^1_\varepsilon(t/\varepsilon), x^2_\varepsilon(t/\varepsilon))$$

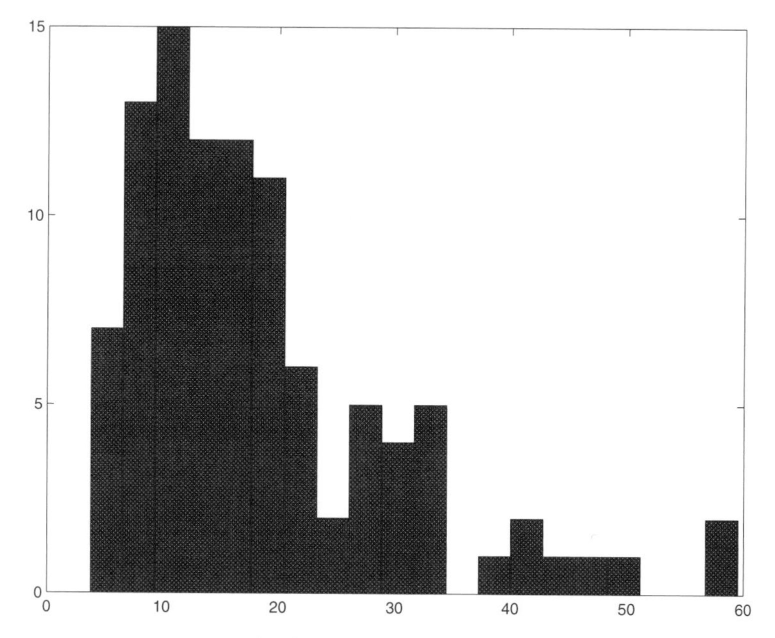

Figure 7. Simulation of $\phi(x_\varepsilon^1, x_\varepsilon^2)$ for 100 sample paths of the harmonic oscillator $\dot{x}_\varepsilon^1 = -y_1\, x_\varepsilon^2$ and $\dot{x}_\varepsilon^2 = y_2\, x_\varepsilon^1$ for $0 \le t \le 6\pi$ and $\varepsilon = 0.05$. Plotted here is the histogram of values of ϕ at $t = 6\pi$. Here y is a jump process as described earlier.

converges to a diffusion process $\tilde{z}$ whose generator we can compute. In this case, m_c is the uniform measure (Haar measure) on the circle having center 0 and radius $\sqrt{c}$.

A simulation of the harmonic oscillator with noisy frequencies, $y = (y_1, y_2)$, is shown in Figure 7.

Stability of Linear Systems

Stability of nonlinear systems is often investigated by considering certain associated linear problems. For linear systems the stability of an unperturbed system is equivalent to boundedness of its solutions for all initial values.

Consider the linear perturbed system

$$\dot{x}_\varepsilon = A(y(t/\varepsilon))x_\varepsilon,\ x_\varepsilon(0) = x_0. \tag{27}$$

Now, x_ε is an R^d–valued stochastic process, $A(y)$ is a $d \times d$ matrix–valued function whose components are bounded measurable functions, and y is an ergodic process. Set $\bar{A} = \int A(y)\rho(dy)$. The averaged equation with initial value x_0 has the solution

$$\bar{x}(t) = e^{t\bar{A}}\, x_0.$$

This system is asymptotically stable if $\lim_{t\to\infty} |e^{t\bar{A}}| = 0$. We show in this case that for sufficiently small ε there is a positive number δ_1 such that the solution of the perturbed system satisfies

$$P\left\{\sup_{t>0} |x_\varepsilon(t)| e^{\delta_1 t} < \infty\right\} = 1.$$

This implies the asymptotic stability of the noisy system with probability 1! In particular,

$$P\left\{\lim_{t\to\infty} x_\varepsilon(t) = 0\right\} = 1,$$

so it is almost surely stable. In addition,

$$\lim_{t\to\infty} P(|x_\varepsilon(t)| > \delta) = 0$$

for all $\delta > 0$, so $x_\varepsilon(t) \to 0$ in probability as $t \to \infty$.

Asymptotic Behavior of Gradient Systems

A gradient system in R^d is determined by a differential equation of the form

$$\dot{x}(t) = -\nabla F(x(t)),$$

where the function $F : R^d \to R^1$ is sufficiently smooth. Note that any local minimum of the function F is a stable static state for this system. We consider random perturbations of this system in the form

$$\dot{x}_\varepsilon(t) = -\nabla F(x_\varepsilon(t)) + B(x_\varepsilon(t))v(y(t/\varepsilon)), \tag{28}$$

where $v : Y \to R^d$ is a bounded measurable function satisfying the relation

$$\int v(y)\,\rho(dy) = 0.$$

The asymptotic behavior of solutions to such systems can be investigated using the theory of large deviations.

This is of the following nature: Set

$$I^*(b) = \inf\left[I(m) : \int v(y)m(dy) = b\right].$$

where I was defined earlier. Assume that $\bar{x}_1, \ldots, \bar{x}_N$ are minima of the function F, and that $F(x) \to \infty$ as $|x| \to \infty$. Then the system spends almost all of its time moving in some small neighborhood of the set of points $\{\bar{x}_k : k = 1, \ldots, N\}$. Moving from a neighborhood of a point $\bar{x}_i$ to a neighborhood of $\bar{x}_k$ requires a random time that has an exponential distribution with a parameter that is of order

$$\exp\left[\frac{G(\bar{x}_i, \bar{x}_k)}{\varepsilon}\right],$$

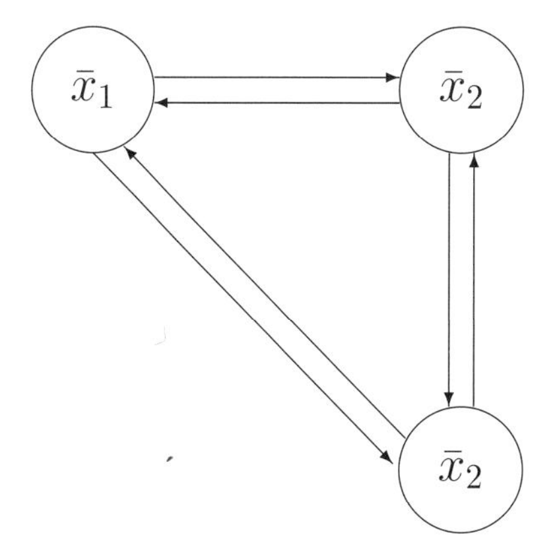

Figure 8. Transitions between minima of the gradient function.

where

$$G(\bar{x}_i, \bar{x}_j) = \inf \int_0^T I^* \left(B^{-1}(f(t))(\dot{f}(t) + \nabla F(f(t)) \right) dt$$

and the infimum is taken over all functions $f(t)$ for which $f(0) = \bar{x}_i$ and $f(T) = \bar{x}_k$, for any $T > 0$. While this is a complicated construction, it does provide an algorithm for estimating the occurrence of large deviations.

As an example, suppose that $m = 3$, $G(\bar{x}_1, \bar{x}_2) = 1$, $G(\bar{x}_2, \bar{x}_1) = 2$, $G(\bar{x}_1, \bar{x}_3) = 3$, $G(\bar{x}_3, \bar{x}_1) = 1$, $G(\bar{x}_2, \bar{x}_3) = 3$, and $G(\bar{x}_3, \bar{x}_2) = 1$. Then the system changes its position as depicted in the graph in Figure 8, and it is almost always moving in a neighborhood of the point $\bar{x}_2$.

We note that many exchange–of–stability problems, for example those described by the Ginzburg–Landau equation of fluid mechanics, fall into this class of problems, and so our methods enable us to analyze exchange–of–stability problems in gradient systems perturbed by random noise. We do not pursue here fluid mechanics problems specifically, but describe general methods for analysis of multistable systems.

As an example, consider the system

$$\dot{x} = -\nabla F(x, y(t/\varepsilon)),$$

where

$$F = y_3(t/\varepsilon) \left(\exp(y_1(t/\varepsilon)(x_1 - 1)^2 + x_2^2) + \exp(y_2(t/\varepsilon)(x_1^2 + (x_2 - 1)^2) \right).$$

Sample trajectories of this and the averaged system are shown in Figure 9.

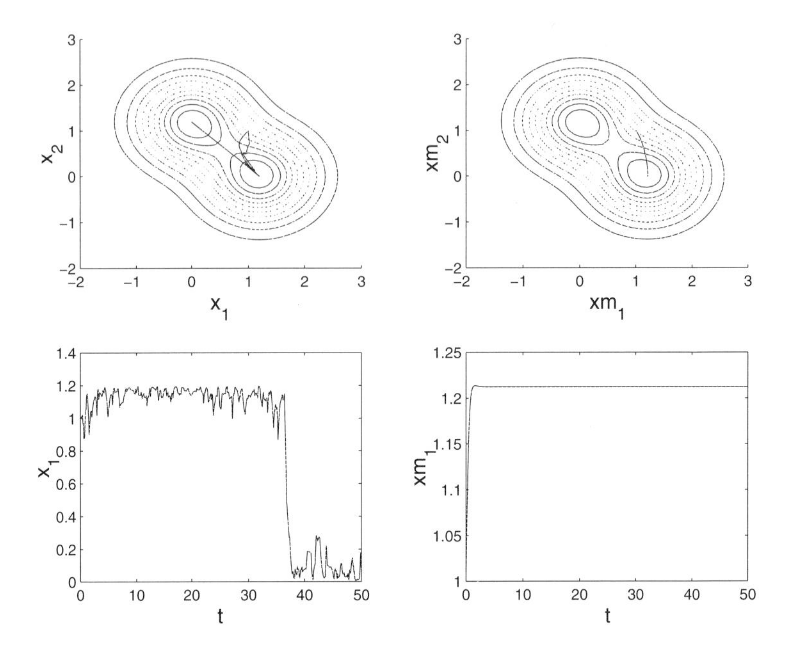

Figure 9. A contour map of the surface defined by F is shown at the top, and on the bottom is the first component of the solution of the system perturbed by noise (left) and the averaged system (right). The solution that starts at $(1, 1)$ is shown in each case. For the averaged system on the right, the solution goes into the energy well near $(1, 0)$. However, the random perturbation of the system moves the solution from one energy well to the other. Here $0 \leq t \leq 50$ and $\varepsilon = 0.1$. Note that the transition between energy wells occurs along a trajectory that takes the shortest path from an equilibrium to the boundary of its basin of attraction. This happens with high probability and suggests how transitions between equilibria occur.

Stochastic Resonance

Stochastic resonance is a closely related phenomenon. Applying noise can uncover a great deal of information about the underlying system. As we've seen, all possible static states will (probably) be visited by any trajectory of a gradient system perturbed by noise, but only one will be visited otherwise.

Moreover, the time it will take to visit these minima can be estimated: For example, suppose that the potential function F in equation (28) depends on an additional parameter $\alpha \in R$, so $F = F(x, \alpha)$, and suppose that it is periodic in α with period 1; i.e., $F(x, \alpha + 1) \equiv F(x, \alpha)$ for all α. We consider the case where the parameter α is slowly changing in time, say $\alpha = t/T(\varepsilon)$, where $\log T(\varepsilon) = O(1/\varepsilon)$ and $\varepsilon \ll 1$. Finally, assume that the function $F(x, \alpha)$ has two minima, say $\bar{x}_1(\alpha)$ and $\bar{x}_2(\alpha)$, for all α, and the functions

$$g_1(\alpha) \equiv \bar{G}(\bar{x}_1(\alpha), \bar{x}_2(\alpha)), \quad g_2(\alpha) \equiv \bar{G}(\bar{x}_2(\alpha), \bar{x}_1(\alpha)),$$

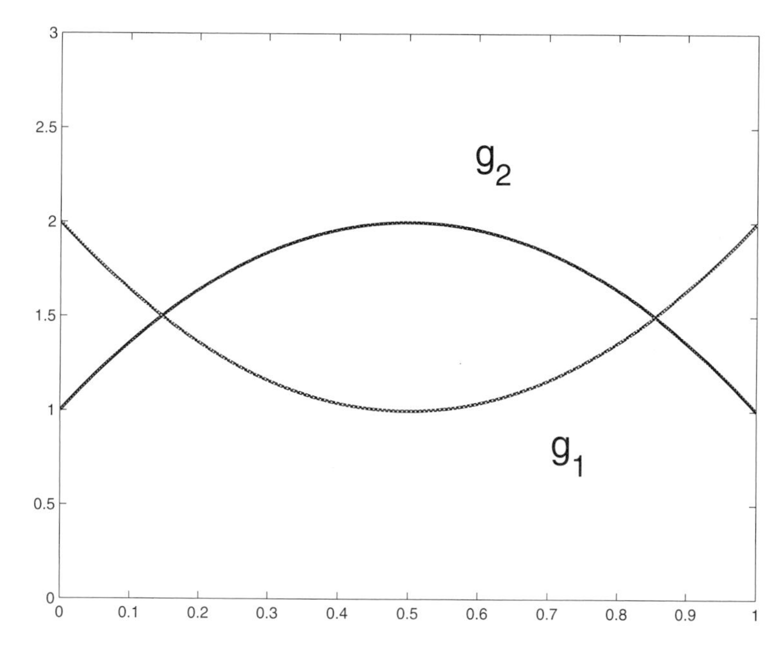

Figure 10. g_1 and g_2 as functions of α. Note that $g_1(\frac{1}{2}) = g_2(0) = g$.

are of the form shown in Figure 10. If $T(\varepsilon) = \exp\{g/\varepsilon\}$, then the system at times $(n + \frac{1}{2})T(\varepsilon)$ is expected to pass from a neighborhood of $x_1(\alpha)$ to a neighborhood of $x_2(\alpha)$, and at times $nT(\varepsilon)$ in the reverse direction.

Markov Chain in a Random Environment

We consider various discrete–time systems. In addition to considering difference equations of the form

$$x_{n+1} = x_n + \varepsilon\phi(x_n, y_n),$$

we describe random perturbations of discrete–time, discrete–state Markov chains. A Markov chain is determined by its transition probabilities, which are summarized in a matrix

$$P_n = (p_{jk}(n))$$

for time steps $n = 0, 1, 2, \ldots$. The indices $j, k \in I$ describe the states, a finite set that is called the phase space of the Markov chain. The component $p_{j,k}(n)$ is the probability that the system, being at j at time n, jumps to state k. Interesting problems arise in biology, where the transition probabilities have the form

$$P_n = P + \varepsilon Q_n,$$

where P is a fixed matrix, $\varepsilon > 0$ is a small parameter, and Q_n is a sequence of random matrices. Our main problem is to investigate the behavior of the product

$$(P + \varepsilon Q_0)(P + \varepsilon Q_1) \cdots (P + \varepsilon Q_n)$$

as $n \to \infty$ and $\varepsilon \to 0$. For stationary ergodic sequences $\{Q_n\}$ we investigate the difference between this product and the nonrandom matrix product

$$(P + \varepsilon \bar{Q})^{n+1},$$

where $\bar{Q} = EQ_n$.

The following simulation for $\varepsilon = 0.1$ illustrates the result.

```
 0.0746    0.0087    0.1439    0.4599    0.3129
 0.0920    0.0484    0.3289    0.2832    0.2474
 0.2072    0.3154    0.0623    0.2159    0.1992 = P
 0.1084    0.3805    0.0746    0.0624    0.3740
 0.2619    0.2573    0.1803    0.2673    0.0331

-0.0227    0.0002    0.0084    0.0086    0.0056
 0.0020   -0.0141    0.0018    0.0025    0.0078
 0.0014    0.0074   -0.0206    0.0036    0.0082 = Qbar
 0.0019    0.0079    0.0049   -0.0177    0.0030
 0.0044    0.0091    0.0047    0.0066   -0.0249

 0.1502    0.2163    0.1627    0.2418    0.2291
 0.1502    0.2163    0.1627    0.2418    0.2291
 0.1502    0.2163    0.1627    0.2418    0.2291 = C
 0.1502    0.2163    0.1627    0.2418    0.2291
 0.1502    0.2163    0.1627    0.2418    0.2291

 0.1499    0.2162    0.1628    0.2414    0.2296
 0.1499    0.2162    0.1628    0.2414    0.2296
 0.1499    0.2162    0.1628    0.2414    0.2296 = D
 0.1499    0.2162    0.1628    0.2414    0.2296
 0.1499    0.2162    0.1628    0.2414    0.2296
```

The the exact result, namely the matrix $C = \prod_{n=1}^{100}(P + \varepsilon Q_n)$, is approximated to order 0.001 by the matrix $D = (P + \varepsilon \bar{Q})^{100}$. Note that the matrix C is ergodic (all states are accessible), 1 is a unique eigenvalue, and the rows are identical; they form its left eigenvector, which describes the ergodic measure for this limiting matrix.

Mechanical and Electrical Applications

As mentioned at the start of this introduction, random noise is inherent to engineering systems in what they receive and what they deliver. The following examples illustrate how we apply random perturbation theory to various problems in engineering and science.

Randomly Perturbed Conservative Mechanical Systems

We first consider conservative systems with two degrees of freedom. The state of the system is determined by the pair $(x, \dot{x})$: position and velocity. We suppose that the unperturbed system has a first integral, namely the energy of the system:

$$E = (\dot{x}^2)/2 + U(x),$$

where U is the potential energy. An orbit is determined by its potential energy U. Under typical conditions the orbits of such a system with fixed energy are periodic or are saddle–saddle connections.

We consider a random perturbations of this system in which the potential energy depends on a fast ergodic Markov process satisfying some strong mixing conditions, as we described earlier. For such systems, we can use the theorems on averaging and on normal deviations.

It follows from the averaging theorem that the randomly perturbed system moves near an orbit of the corresponding averaged system up to a time of order $1/\varepsilon$, but with small normal random deviations; these transfer the system to near another orbit. The transitions from one orbit to another involves changing the total energy of the system. Using the theorem on diffusion approximations for the first integral, we can describe the evolution of the total energy of the system (it is a stochastic process) as being a slow diffusion process on a graph. This graph is determined by the set of the orbits of the averaged system. So, the motion of the perturbed system can be described as the motion of the averaged system with total energy that is a slowly changing diffusion process on the graph.

Let the potential energy for the system, say $U(x)$, be a smooth function that has a finite set of minima and maxima, and $U(x) \to \infty$ as $|x| \to \infty$. The graph of orbits for the system is determined by its local extrema.

For example, suppose that $U(x)$ has no local maxima and only one minimum: $U(x) \geq 0$, $U(0) = 0$, $U''(0) > 0$. Then any orbit is determined by a nonnegative constant c. Namely, $\{(x, \dot{x}) : \dot{x}^2/2 + U(x) = c\}$. In this case, the graph has one edge ($[0, \infty)$) and one vertex (0). The total energy of the system is $E_\varepsilon(t) = \dot{x}_\varepsilon^2(t)/2 + U(x_\varepsilon(t))$, and the function $u_\varepsilon(t) = E_\varepsilon(t/\varepsilon)$ converges weakly in C to a diffusion process in the interval $[0, \infty)$ for which the point 0 is the repelling boundary.

Now let $U(x)$ have one local maximum, say at $x = 0$, separating two local minima, say at points $x = \pm 1$, as in Duffing's equation

$$\ddot{x} + x - x^3 = 0. \tag{29}$$

More generally, if $c_1 > U(0)$, then there exists only one orbit

$$\{(x, \dot{x}) : \dot{x}^2/2 + U(x) = c_1\}.$$

For $c_2 \in (\max[U(-1), U(1)], U(0))$ we have two orbits for which $E = c_2$. For $c_3 \in (\min[U(-1), U(1)], \max[U(-1), U(1)])$ there exists only one orbit

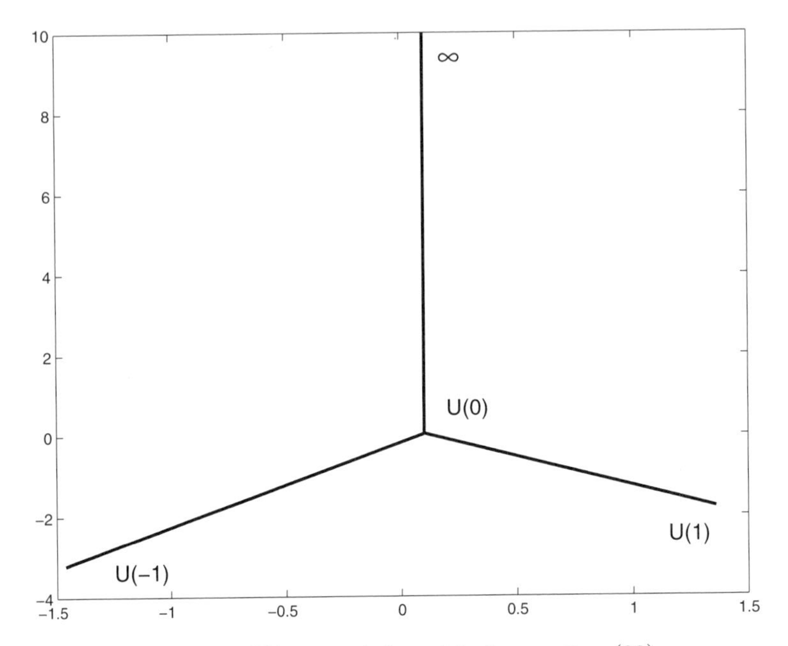

Figure 11. Edge–graph for orbits in equation (29).

with $E = c_3$. The graph of the orbits for this system can be represented as in Figure 11.

Now, there are three edges, $[U(0), \infty)$, $[U(1), U(0))$, and $[U(-1), U(0))$, and three vertices, $U(-1), U(1), U(0)$, where the first two vertices are endpoints. If the initial energy of the system is greater than $U(0)$, then the process $u_\varepsilon(t)$ converges as $\varepsilon \to 0$ weakly to a diffusion process on the interval $(U(0), \infty)$. This is true unless the process u_ε hits the point $U(0)$. In this case the process begins to move in the interval $(U(1), U(0))$ with some probability, say P_1, and in $(U(-1), U(0))$ with probability P_{-1}. It stays in the interval $(U(0), \infty)$, reflecting if it hits the boundary. If the initial energy is less than $U(0)$, and $x_\varepsilon(0) < 0$, then u_ε converges to a diffusion process in the interval $(U(-1), U(0))$ with a repelling boundary at $U(-1)$. When it hits the point $U(0)$, its behavior is as described before. The same occurs if $E < U(0)$ and $x_\varepsilon(0) > 0$, but with $(U(-1), U(0))$ replaced by $(U(1), U(0))$.

A second problem is to investigate the asymptotic behavior of a perturbed linear conservative system. Consider an N–dimensional system with potential energy

$$U(x) = (\Lambda x, x)/2,$$

where $x \in R^N$ and Λ is a positive symmetric matrix of order N. The kinetic energy of the system is

$$T(\dot{x}) = (\dot{x}, \dot{x})/2.$$

We can choose a coordinate system in which Λ is a diagonal matrix, and then the equations for the system are of the form

$$\ddot{x}_k = -\lambda_k x_k$$

for $k = 1, \ldots, N$. The solutions are

$$x_k(t) = a_k \cos(\lambda_k t + \phi_k).$$

Here the constants λ_k are the eigenvalues of Λ, and the amplitude and phase deviation of the solutions are determined by the initial conditions. The functions

$$E_k(x, \dot{x}) = \dot{x}_k^2/2 + \lambda_k^2 x_k^2$$

are first integrals for the system.

A perturbed system is described by the system of differential equations

$$\ddot{x}_{\varepsilon k} = -\lambda_k x_{\varepsilon k} + f_k(x_\varepsilon, \dot{x}_\varepsilon, y(t/\varepsilon))$$

for $k = 1, \ldots, N$, where y is a (vector) Markov process. The solution of this system can be represented in the form

$$x_{\varepsilon k} = z_k^\varepsilon \cos(\lambda_k t + \phi_k^\varepsilon(t)), \ \dot{x}_{\varepsilon k} = -\lambda_k z_k^\varepsilon \sin(\lambda_k t + \phi_k^\varepsilon(t)).$$

We prove that the stochastic process $(z^\varepsilon, \tilde{\phi}^\varepsilon)$, where $z_k^\varepsilon(t) = z_{\varepsilon k}(t/\varepsilon)$ for $k = 1 \ldots, N$, and $\tilde{\phi}_k^\varepsilon(t) = \phi_{k+1}^\varepsilon(t/\varepsilon) - \phi_k^\varepsilon(t/\varepsilon)$ converge weakly in C to a $(2N-1)$–dimensional diffusion process.

Randomly Perturbed Dynamical Systems on a Torus

Dynamical systems of these kinds arise in numerous applications in engineering, such as for describing rotating machinery and frequency based communication devices. Consider a system of differential equations of the form

$$\begin{aligned} \dot{x}_1 &= a(x_1, x_2), \\ \dot{x}_2 &= b(x_1, x_2), \end{aligned} \tag{30}$$

where x_1, x_2 are real–valued functions, and $a > 0$ and b are real–valued, doubly periodic functions of (x_1, x_2):

$$a(x_1 + 1, x_2) = a(x_1, x_2 + 1) = a(x_1, x_2),$$

etc. These functions are sufficiently smooth so that this system has a unique solution for any initial values. It is known that there exists the limit

$$\lim_{t \to \infty} x_2(t)/x_1(t) = \mathbf{r},$$

called the rotation number, and $\mathbf{r}$ does not depend on the initial conditions of the solution defining it. If $\mathbf{r}$ is irrational, then the system is ergodic on the torus $\mathcal{T}$; if $\mathbf{r}$ is rational, then each solution of the system approaches a periodic solution on the torus. Such periodic solutions are referred to as torus knots, since a closed trajectory defines a knotted curve in R^3.

We consider a perturbation of this system in the form

$$\dot{x}_1 = a(x_1, x_2, y(t/\varepsilon)),$$
$$\dot{x}_2 = b(x_1, x_2, y(t/\varepsilon)),$$

where y is an ergodic Markov process. We assume that the averaged system is the original one (30). First, we prove that

$$\lim_{\varepsilon \to 0} P\left\{ \limsup_{t \to \infty} |x_2(t)/x_1(t) - \mathbf{r}| < \delta \right\} = 1$$

for any $\delta > 0$. The system (30) is called purely periodic if all solutions are periodic (a fortiori, $\mathbf{r}$ is rational). Using the diffusion approximation theorem for first integrals, we prove that

$$x_{2,\varepsilon}(t) = \mathbf{r}\, x_{1,\varepsilon}(t) + \varepsilon t A + o(\varepsilon t),$$

where A is the coefficient of the generator of the diffusion process on the circle to which the first integral converges. Systems that are not purely periodic are also investigated.

A computer simulation of the rotation number for the system

$$\dot{x} = 1 + y_1(t/\varepsilon) + \sin(x_1 - x_2), \dot{x}_2 = 1 + y_2(t/\varepsilon) - \sin(x_1 - x_2), \qquad (31)$$

where $\varepsilon = 0.5$ and $0 \le t \le 50$, is shown in Figure 12. In the absence of noise, the rotation number is 1 for this system.

Pendulums and Phase–Locked Loops

Pendulums play important roles in engineering, science, and mathematics. They arise in a number of surprising places, most recently in models of electronic devices, called Josephson junctions, and in other quantum mechanics applications. We study the impact of noise on a particularly important example of this: an electronic circuit called a phase–locked loop. This device, and variants of it, are central to timing devices in computers, radar signal demodulators, and FM radio, and they are useful analogues of circuits found in our brains [78].

The system is described by the equations

$$\begin{aligned} \tau \dot{z} &= -z + \cos\theta, \\ \dot{\theta} &= \omega + z, \end{aligned} \qquad (32)$$

where z is the output voltage of a low–pass filter whose time constant is τ, and ω is the center frequency of a voltage–controlled oscillator whose phase is θ. The state of the system at time t is $(z(t), \theta(t))$, and $z(t)$ and $\cos\theta(t)$ are observables.

The problem of interest to us here is the behavior of the system as $t \to \infty$, in particular that of the phase $\theta(t)$.

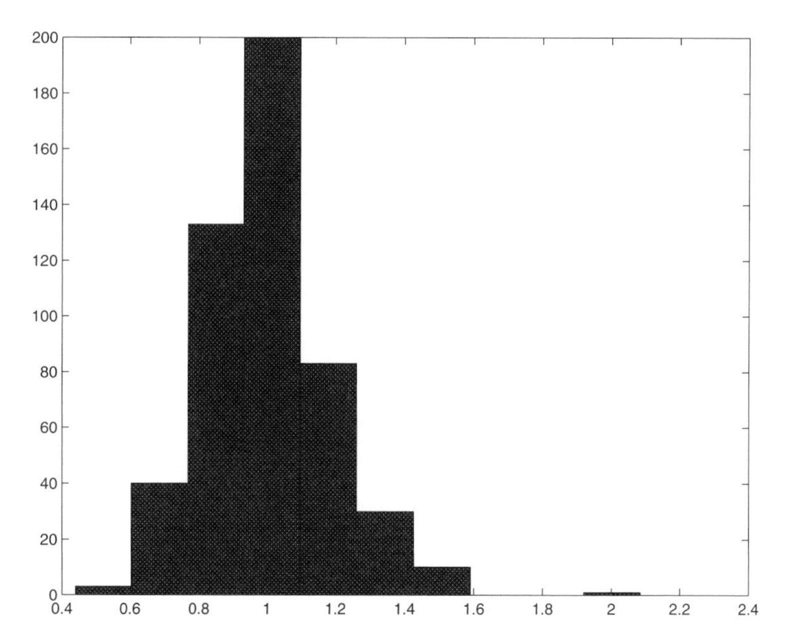

Figure 12. Simulation of the rotation number of the model system (31): Shown here is a histogram of the values of $\mathbf{r}(50)$ for $\varepsilon = 0.5$.

The trajectories of system (32) can be considered on the cylinder $z \in (-\infty, \infty)$, $\theta \in [0, 2\pi)$, and they are determined by the scalar differential equation

$$\tau \frac{dz}{d\theta} = \frac{-z + \cos\theta}{\omega + z}. \tag{33}$$

The behavior of equilibrium solutions to this system depends on ω, which we suppose is positive:

1. For $\omega = 1$ equation (33) has a singular point $(-1, \pi)$, which is an unstable equilibrium for the system (32). There is one trajectory moving from $-\infty$ making an infinite number of rotations around the cylinder and the ending in this singular point.

2. For $\omega < 1$, equation (33) has two singular points, $(-\omega, \theta_1^*), (-\omega, \theta_2^*)$, where $\cos\theta_k^* = -\omega$ for $k = 1, 2$, and $0 < \theta_1^* < \theta_2^* < 2\pi$. The point (ω, θ_1^*) is a stable equilibrium for the system, and the other is unstable. There are two trajectories ending at (ω, θ_2^*) that form the boundary of a region G on the cylinder that contains the stable equilibrium.

3. There is a number $\omega^* \in [0, 1)$, which depends on τ, and for $\omega > \omega^*$ equation (33) has a periodic solution. This trajectory is closed on the

cylinder, it divides the cylinder into two parts, and it is called the running periodic solution, or running wave.

4. For $\omega = \omega^*$, the running periodic solution forms a saddle–saddle connection. It coincides with a trajectory emanating from the unstable state and ending at the same point after one rotation around the cylinder. It divides the cylinder into two parts, and we denote the upper part by C_+.

The behavior of transient solutions to system (32) is described next:

1. For $\omega > 1$, every solution of the system (32) tends to the running wave.

2. For $\omega = 1$, any solution but one with initial values not at the equilibrium $(-1, \pi)$ tends to the running wave.

3. For $\omega^* < \omega < 1$, we have three possibilities: If the initial values are in G, then the solution tends to the stable equilibrium. If the initial values lie outside $G \cup \partial G$ (i.e., either in G or its boundary ∂G), the solution tends to the running wave. If the initial value lies in ∂G, the solution tends to the unstable state.

4. If $\omega \leq \omega^*$, then the solution with initial values in G tends to the stable equilibrium $(-\omega, \theta_1^*)$, and solutions with initial values in ∂G tend to the unstable equilibrium.

We consider random perturbations of this system in the form

$$\begin{aligned} \tau(t/\varepsilon)\, \dot{z}_\varepsilon &= -z_\varepsilon(t) + \cos\theta_\varepsilon(t), \\ \dot{\theta}_\varepsilon(t) &= \omega(t/\varepsilon) + z_\varepsilon(t), \end{aligned} \tag{34}$$

where $(\tau(t), \omega(t))$ is a two–dimensional stationary (or Markov) ergodic random process. Set

$$\tau = E(\tau(t)),\ \omega = E(\omega(t)).$$

(In the case of Markov processes, we define the expectations using the ergodic distribution.) Then system (32) is the averaged system for (34). It follows from the averaging theorem that the solution to system (34) converges to the solution of (32) with the same initial values with probability 1 on any finite interval as $\varepsilon \to 0$.

This implies that the solutions to system (34) for large t are close to either the running wave or to the stable equilibrium state. The first is possible if $\omega > \omega^*$, and the second if $\omega < 1$. In addition, if $\omega \leq \omega^*$, $\theta_\varepsilon(t)$ converges in probability to θ_1^* as $t \to \infty$ and $\varepsilon \to 0$. Note that the statement "$\theta_\varepsilon(t) \to \theta_1^*$ with probability 1" is incorrect.

What is the difference between solutions of the unperturbed and perturbed systems? First, we note that under general conditions, a solution of the perturbed system is an ergodic stochastic process, and its ergodic distribution is positive on any open subset of the cylinder for all $\varepsilon > 0$. If $\varepsilon \to 0$, then the ergodic distribution of any ball that does not contain the stable equilibrium and has the no intersection with the running wave tends to zero. So the limiting ergodic distribution is concentrated on the trajectory of the running wave and the stable equilibrium.

The unperturbed system is not ergodic for $\omega \leq 1$, and for $\omega > 1$ there are two parts to the cylinder, one above the running wave, and one below it. A solution starting in one part never reaches the other. If an unperturbed solution is close to the running wave, it cannot reach some small neighborhood of the stable equilibrium state, nor can one reach from such a neighborhood to a neighborhood of the running wave. On the other hand, starting from any point on the cylinder the perturbed system can reach with probability 1 any ball about a state, but the time to reach such a ball will tend to ∞ as $\varepsilon \to 0$.

We also consider the forced problem for the first-order phase–locked loop. In this case, the filter is removed (i.e., $\tau = 0$), and we consider an input signal whose phase is $\eta(t) = \mu t + \phi$. The model for this system is

$$\dot{\theta} = \omega + F(\theta, \eta(t)),$$

where $F(\theta, \eta)$ is a smooth function that is doubly periodic,

$$F(\theta + 2\pi, \eta) = F(\theta, \eta + 2\pi) = F(\theta, \eta),$$

for all θ, η. Therefore, we consider the system of equations

$$\begin{aligned} \dot{\eta}(t) &= \mu, \\ \dot{\theta}(t) &= \omega + F(\theta(t), \eta(t)). \end{aligned} \tag{35}$$

Since this system is doubly periodic, its behavior can be studied using methods for flows on a torus.

A question of special interest in signal processing is the existence of a correlation function of the form

$$\lim_{T\to\infty} \frac{1}{T} \int_0^T g\left(\theta(t), \eta(t)\right) g\left(\theta(t+h), \eta(t+h)\right) \, dt \equiv \mathbf{R}_g(h)$$

for any smooth, doubly periodic function g. We prove here the existence of this limit and provide a method for calculating it.

A random perturbation of system (35) has the form

$$\begin{aligned} \dot{\eta}_\varepsilon(t) &= \mu(t/\varepsilon), \\ \dot{\theta}_\varepsilon(t) &= \omega + F(y(t/\varepsilon), \theta_\varepsilon(t), \eta_\varepsilon(t)), \end{aligned} \tag{36}$$

where $(\mu(t), y(t))$ is an ergodic stationary process in t, and F is 2π periodic in θ and η.

Denote by $\mathbf{R}_g^\varepsilon(h)$ the correlation function for $g(\theta_\varepsilon(t), \eta_\varepsilon(t))$. We prove that if $E(\mu(t)) = \mu$ and $E(F(y, \theta, \eta)) = F(\theta, \eta)$, so system (33) is the averaged system for (36), which we now suppose to be ergodic, then the function $\mathbf{R}_g^\varepsilon(h)$ converges uniformly for $|h| \leq c$, where $c > 0$ is an arbitrary constant, to the function $\mathbf{R}_g(h)$ in probability as $\varepsilon \to 0$. Our approximation to the solution of (33) involves an offset diffusion process whose variance can be approximated using $\mathbf{R}_g(0)$.

A particularly interesting application of these methodologies is to the problem of cycle slipping in phase–locked loops. Consider the system where noise enters only through the center frequency ω:

$$\tau\ddot{\theta} + \dot{\theta} + \cos\theta = \omega(y(t/\varepsilon)).$$

In the absence of damping, the averaged equation defines a conservative system with a potential function

$$\bar{\omega}\,\theta + \sin\theta.$$

If $\omega < 1$, there can be infinitely many wells, each approximately 2π units apart. With damping each of these is stable. However, with the addition of noise to this system through ω, the solution can be driven from one of these wells to another, and the system will execute a full oscillation during such a transition. An example is shown in the simulation in Figure 13.

Finally, one of the motivations for our investigation of Volterra dynamical systems is to study filters more complicated than the low pass filter just considered. Filters are described in the engineering literature using the notation of linear time–invariant (LTI) systems: For instance, if $x(t)$ is an input signal to such a system, $h(t)$ is the impulse response function, and $X(t)$ is the output, then we write

$$X(t) = \int_0^t h(t-s)x(s)\,ds.$$

Placing such a filter in the phase–locked loop circuits gives

$$\dot{x}(t) = \omega + \int_0^t h(t-s)\cos x(s)\,ds. \tag{37}$$

Our methods enable us to analyze such systems in the presence of noise. For example, consider the perturbed system

$$\dot{x}_\varepsilon(t) = \omega(y(t/\varepsilon)) + \int_0^t h(t-s, y(s/\varepsilon))\cos x_\varepsilon(s)\,ds, \tag{38}$$

where h and the other data are as above. If (37) is the averaged system for (38) and its solution is $\bar{x}(t)$, then we show that

$$x_\varepsilon(t) \approx \bar{x}(t) + \sqrt{\varepsilon}\,x_1(t),$$

where $x_1(t)$ is a Gaussian process that is determined by solving the linear equation

$$\dot{x}_1(t) = z(t) + \int_0^t K(t,s)\sin\bar{x}(s)\,x_1(s)\,ds,$$

where the kernel K is found by averaging h and z is a Gaussian process.

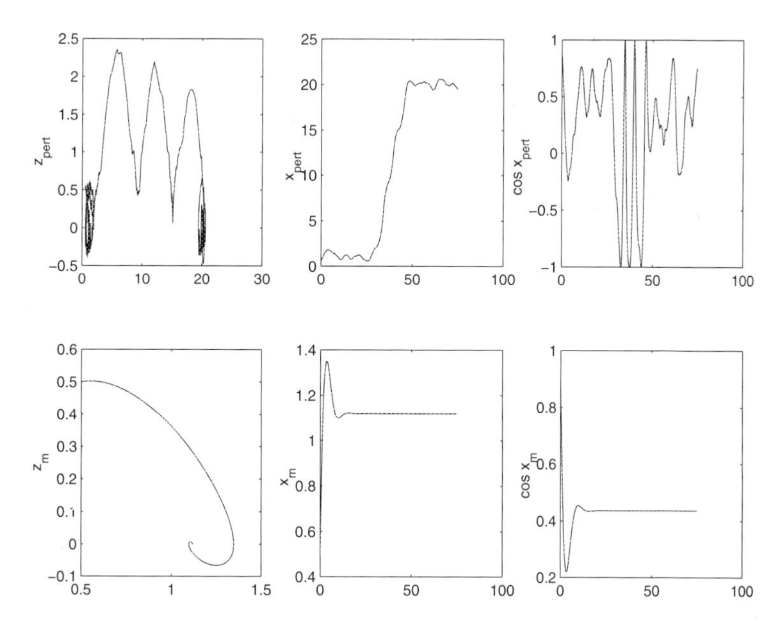

Figure 13. Phase–locked loop with noisy center frequency: The lower plots show the solution of the averaged system from various points of view. On the left is the phase plane of (θ, z), (we write x for $\theta(t)$ here) in the middle is θ vs. t, and on the right is $\cos\theta(t)$ vs. t. The top plots show the same thing for a sample path of the perturbed system: In the upper left is the phase plane of θ (horizontal axis) vs. z. This shows that θ has slipped over three energy wells. Each slip results in a jump in the voltage, as shown on the upper right, which would be heard in FM radio as a click. In this case $0 \le t \le 50$ and $\varepsilon = 0.1$.

Mathematical Population Biology

We list next several examples of how the methods in this book can be applied to various problems in population biology.

Ecology

Structures of ecological systems, ranging from bacteria in the gut to food chains in the sea, have been described using mathematical models, and these have been useful in understanding how these systems work. We investigate here several models that are widely used in studies of ecological systems when they are perturbed by random noise.

The simplest model of how several species interact is the prey–predator system introduced and studied by Lotka and Volterra. If $x_1(t)$ is the number of prey and $x_2(t)$ is the number of predators at time t, then the Lotka–Volterra model of interaction between the two is

$$\begin{aligned}\dot{x}_1 &= \alpha x_1 - \beta x_1 x_2,\\ \dot{x}_2 &= -\gamma x_2 + \delta x_1 x_2.\end{aligned}$$

In the absence of prey (i.e., $x_1 = 0$) the predators decrease, and in the absence of predators, the prey grow without bound. In addition, the growth rate of prey is $\alpha - \beta x_1$, which decreases as the predator population grows, and conversely for the predators.

The set $(x_1 \geq 0, x_2 \geq 0)$ is invariant for this system, and there are two equilibria: the coexistence equilibrium $(x_1^* = \gamma/\delta, x_2^* = \alpha/\beta)$ and the extinction equilibrium $(x_1 = 0, x_2 = 0)$.

There is a first integral for this system, namely,

$$\phi(x_1, x_2) = x_1^\gamma x_2^\alpha \exp(-\delta x_1 - \beta x_2).$$

Thus, any solution starting away from an equilibrium is periodic with a period $T(c)$, where $c = \phi(x_{1,0}, x_{2,0})$, and $T(c) = o(-\log c)$ as $c \to 0$.

Random perturbations of the data in this system take the form

$$\begin{aligned}
\dot{x}_{1,\varepsilon} &= \alpha(y(t/\varepsilon))x_{1,\varepsilon} - \beta(y(t/\varepsilon))x_{1,\varepsilon}x_{2,\varepsilon}, \\
\dot{x}_{2,\varepsilon} &= \gamma(y(t/\varepsilon))x_{2,\varepsilon} + \delta(y(t/\varepsilon))x_{1,\varepsilon}x_{2,\varepsilon},
\end{aligned}$$

where y is an ergodic Markov process in a measurable space (Y, C). We suppose that the functions $\alpha(y)$, etc., are nonnegative, bounded and measurable functions.

Set

$$\begin{aligned}
\alpha &= \int \alpha(z)\rho(dz), \quad & \beta &= \int \beta(z)\rho(dz), \\
\gamma &= \int \gamma(z)\rho(dz), \quad & \delta &= \int \delta(z)\rho(dz).
\end{aligned}$$

Then the original system is the averaged system. The averaging theorem in Chapter 3 implies that the solution of the perturbed system is close to the averaged solution on any finite interval $(t_0, t_0 + T)$ for ε sufficiently small. Thus, the solution is approximately periodic with period $T(\phi(x_{1,\varepsilon}(t_0), x_{2,\varepsilon}(t_0)))$ along the trajectory $\phi(x_1, x_2) = \phi(x_1(t_0), x_2(t_0))$. This function is slowly changing in t_0, since it is constant for the averaged system. Thus, to describe the behavior of the solution to the perturbed system, we analyze the process $\phi(x_{1,\varepsilon}(t_0), x_{2,\varepsilon}(t_0))$.

We show that the stochastic process

$$u_\varepsilon(t) = \log \phi(x_{1,\varepsilon}(t), x_{2,\varepsilon}(t))$$

converges weakly in C to a diffusion process $u(t)$ on the interval $(-\infty, \log \phi(a, b))$, and the point $\log \phi(a, b)$ is a natural boundary for the process $u_\varepsilon(t)$. That is, the population of predators vanishes by time T/ε with probability that tends to 1 as $T \to \infty$ and $\varepsilon \to 0$.

The simulation in Figure 14 shows the averaged system and a sample path of the perturbed system. It is interesting how dramatically different are the behavior of the two systems.

Other systems studied are for two species competing for a limited resource and generalizations of the Lotka–Volterra equations to a three

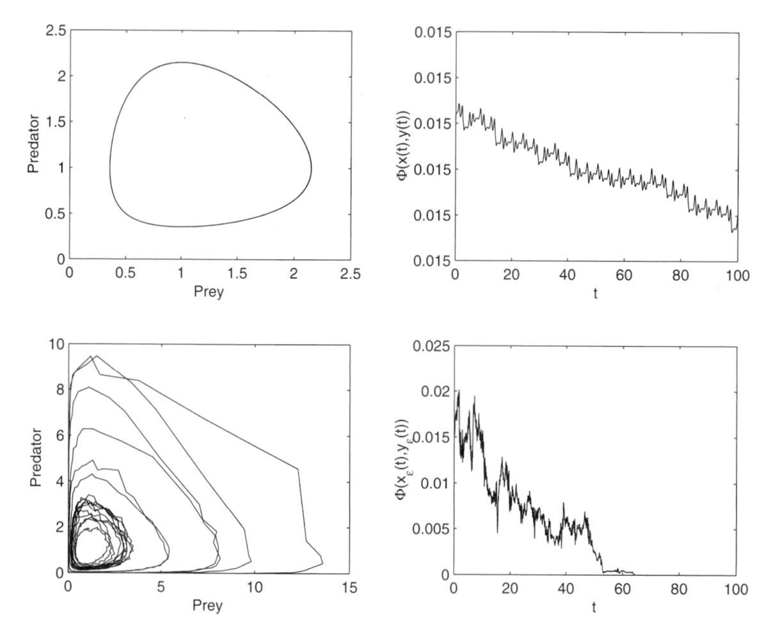

Figure 14. Simulation of the Lotka–Volterra system. The averaged system is shown at the upper left, and its "energy" function is shown at the upper right. Note the scale on the vertical axis in this figure, which indicates that the energy is approximately fixed at the value 0.015. The trajectory on the lower left is for the perturbed system, and its energy is shown on the lower right. In this case, the perturbed system spends a great deal of time near the origin and only occasionally sojourns. Thus, the noise can lead to a significant lengthening in the period of oscillation of the system. Here $\varepsilon = 0.01$ and $0 \leq t \leq 10$.

species food chain, where the first produces nutrient for the second, and the second produces nutrient for the third.

Epidemics

A problem in epidemiology is to predict whether an infection will propagate in a population. For diseases that impart permanent immunity, the Kermack–McKendrick model has been useful, which we describe next. Let S denote the number of susceptibles, I the number of infectives, and R the number of removals (those who are immune). The model is described by the system of differential equations

$$\dot{S} = -\gamma I\, S,$$
$$\dot{I} = \gamma I\, S - \lambda I,$$
$$\dot{R} = \lambda I,$$

where γ is the infection rate and λ is the rate at which infectives are removed from the process, for example through quarantine, death, or cure.

Denote the initial conditions by $S_0 = S(0), I_0 = I(0), R_0 = R(0)$, and let $\alpha = \lambda/\gamma, I_m = \max I(t)$, and $\tilde{S} = \lim_{t\to\infty} S(t)$. Let t_m denote the time at which I peaks: $I(t_m) = I_m$.

If $S_0 > \alpha$, then even one infective will cause an epidemic to propagate with severity and duration characterized by t_m, I_m, and $\tilde{S}$. If $S_0 \leq \alpha$, then $t_m = 0$, and $I(t)$ decreases, so the initial infectives do not replace themselves; that is, there is no increase in infectives for any starting value I_0.

We consider a random perturbation of this system with data $\gamma(t/\varepsilon)$ and $\lambda(t/\varepsilon)$ forming a two-dimensional process that is ergodic and stationary with nonnegative and bounded components. We denote the solution of the corresponding equations by

$$S_\varepsilon(t),\ I_\varepsilon(t),\ R_\varepsilon(t).$$

We suppose that $S_\varepsilon(0) = S_0, I_\varepsilon(0) = I_0$, and $R_\varepsilon(0) = 0$. If we set $E\gamma(t) = \gamma, E\lambda(t) = \lambda$, then the average equations coincide with the original system. The main result related to the behavior of the perturbed system is that

$$\sup_{t>0}(|S_\varepsilon(t) - S(t)| + |I_\varepsilon(t) - I(t)|) \to 0$$

in probability as $\varepsilon \to 0$. In particular, $I_m^\varepsilon \to I_m$ and $t_m^\varepsilon \to t_m$ in probability as $\varepsilon \to 0$. So random perturbations of the kind considered here do not change the threshold behavior of the Kermack–McKendrick epidemic process, except near $S_0 = \bar{\alpha}$, where there is a bifurcation of final sizes.

Demographics

Linear models in demographics are described by the renewal equation

$$x(t) = \phi(t) + \int_0^t M(t-s)\,x(s)\,ds,$$

where $x(t)$ is the population's birth rate at time t, $\phi(t)$ is a function with compact support, and $M(s)$ is the probability that a newborn at time zero produces an offspring at time s. The main result related to the behavior of the birth rate is formulated as follows: There is a number α (usually negative) such that the limit

$$\lim_{t\to\infty} x(t)e^{\alpha t} = B_0$$

exists. The number α is the unique real solution of the characteristic equation

$$\int_0^\infty e^{\alpha t} M(t)\,dt = 1.$$

Random perturbations of this system take the form

$$x_\varepsilon(t) = \phi(t) + \int_0^t M(t - s, y(s/\varepsilon))x_\varepsilon(s)\,ds,$$

where $y(s)$ is an ergodic, homogeneous Markov process in a measurable space (Y, C) satisfying a strong mixing condition. The function $M(t, y)$ is a positive, bounded and measurable function, and as before, we denote by M the mean of this random process:

$$M(t) = \int M(t, y)\,\rho(dy),$$

where ρ is the ergodic distribution of the process $y(t)$.

We prove that

$$\lim_{c\to\infty} \limsup_{\varepsilon\to 0} P\left\{ \sup_{c\le t\le T/\varepsilon} \left| \frac{\log x_\varepsilon(t)}{t} + \alpha \right| > \delta \right\} = 0$$

for all $T > 0$ and $\delta > 0$. The growth rate α is determined in the same way as before.

The last relation implies that

$$P\left\{ e^{(-\alpha-\delta)t} \le x_\varepsilon(t) \le e^{(-\alpha+\delta)t} \right\}$$

tends to 1 as $\varepsilon \to 0$ and $t \to \infty$ for all $\delta > 0$.

A related problem is that of Malthus's model with a random growth rate. In our setting, we consider the equation

$$\dot{x}_\varepsilon = (1.0 + y(t/\varepsilon))x_\varepsilon.$$

This problem has been considered extensively in the literature when y is white noise [141]. In Figure 15 is a simulation of this problem when y is a process like that used in the other simulations here (i.e., a jump process with jumps uniformly distributed on $[0, 1]$).

Diploid Genetics

Consider a population of diploids and a gene in it having two possible alleles, say A and B. Then all members of the population are either of type AA, AB, or BB with respect to this gene. The proportions of the population that describe these subpopulations satisfy a binomial Markov chain whose probabilities change with the population's distribution from one generation to the next. The chain is called the Fisher–Wright chain for its discoverers Ronald Fisher and Sewell Wright.

A deterministic version of this chain is a model for the proportion g_{n+1} of the gene pool that is of type A in the $(n + 1)$th generation:

$$g_{n+1} = \frac{r_n g_n^2 + s_n g_n(1 - g_n)}{r_n g_n^2 + 2s_n g_n(1 - g_n) + t_n(1 - g_n)^2},$$

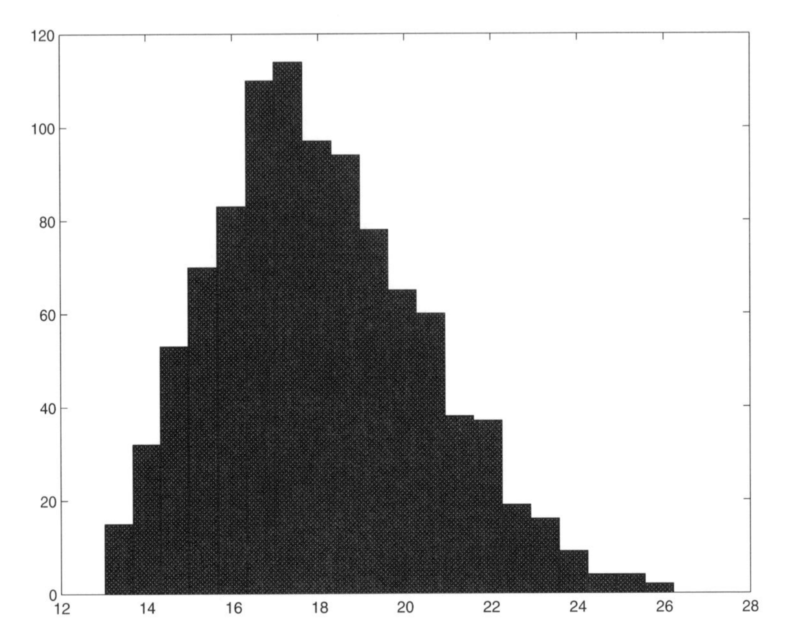

Figure 15. This figure shows the distribution of $x(1,\varepsilon)$ for 1000 sample paths of Malthus's model. Here $x(0) = 4.0$, $\bar{x}(1) = 17.93$, $0 \leq t \leq 1.0$, and $\varepsilon = 0.1$.

where the sequences r_n, s_n, and t_n describe the selection coefficients of the genotypes AA, AB, and BB, respectively, over the generations.

Selection is slow when the selection coefficients are nearly identical; for example,

$$r_n = r + \varepsilon \rho_n, \ s_n = r + \varepsilon \sigma_n, \ t_n = r + \varepsilon \tau_n,$$

where $\varepsilon \ll 1$. If $\rho_n = \rho$, $\sigma_n = \sigma$, and $\tau_n = \tau$, then the discrete-time sequence $\{g_n^\varepsilon\}$ describing the evolution of the gene pool satisfies

$$g_{n+1}^\varepsilon - g_n^\varepsilon = \varepsilon \frac{Q(g_n^\varepsilon)}{1 + \varepsilon P(g_n^\varepsilon)},$$

where Q and P are polynomials with

$$Q(x) = x(1-x)(ax+b), \ a = \rho + \tau - 2\sigma, \ b = \rho - \tau.$$

The asymptotic behavior of the sequence as $\varepsilon \to 0$ is determined through the following results, which we establish in Chapter 12:

1. If $b \geq 0, a + b \geq 0, |a| + |b| > 0$, then A dominates the gene pool:

$$\lim_{\varepsilon \to 0, \varepsilon n \to \infty} g_n^\varepsilon = 1.$$

2. If $b \leq 0, a + b \leq 0, |a| + |b| > 0$, then B dominates the gene pool:

$$\lim_{\varepsilon \to 0, \varepsilon n \to \infty} g_n^\varepsilon = 0.$$

3. If $b > 0, a + b < 0$, then there is a polymorphism in the gene pool:

$$\lim_{\varepsilon \to 0, \varepsilon n \to \infty} g_n^\varepsilon = -b/a.$$

4. If $b < 0, a + b > 0, |a| + |b| > 0$, then there is disruptive selection in the gene pool: If $g_0^\varepsilon \in (-b/a, 1)$, then

$$\lim_{\varepsilon \to 0, \varepsilon n \to \infty} g_n^\varepsilon = 1,$$

but if $g_0^\varepsilon \in (0, -b/a)$, then

$$\lim_{\varepsilon \to 0, \varepsilon n \to \infty} g_n^\varepsilon = 0.$$

We study random perturbations of this system in the form

$$\rho_n = \rho(y_n), \ \sigma_n = \sigma(y_n), \ \tau_n = \tau(y_n)$$

where the sequence $\{y_n\}$ is a homogeneous ergodic Markov chain in a measurable space (Y, C) with ergodic distribution $m(dy)$. We suppose that ρ, σ, τ are positive, bounded, and measurable functions. If we set

$$\rho = \int \rho(y)m(dy), \ \sigma = \int \sigma(y)m(dy), \ \tau = \int \tau(y)m(dy),$$

then we show that the solution of the differential equation

$$\dot{g}(t) = Q(g(t))$$

is closely related to the gene pool distribution: In particular, $g_n^\varepsilon \approx g(\varepsilon n)$, where the approximation is with probability 1. A version of the central limit theorem is also applicable to this. For example, in the case of selection favoring a polymorphism, the process

$$z_n^{\varepsilon,N} = (g_{N+n}^\varepsilon + b/a)/\sqrt{\varepsilon}$$

converges weakly to a stationary Gaussian process.

In Figure 16 is shown the averaged system's solution and a sample path of the perturbed system in the case of selection for a polymorphism.

Bacterial Genetics

A bacterium contains a chromosome, which is a large strand of DNA, and also extrachromosomal DNA elements called plasmids. Each new cell contains exactly one chromosome, but it might have many copies of particular plasmids. The chromosome is propagated mainly through replication and cell division (vertical propagation), but plasmids can be propagated vertically and horizontally, meaning that they can be passed from one cell to another before replication. We study here the distribution of vertically propagated plasmid types as cells grow and divide.

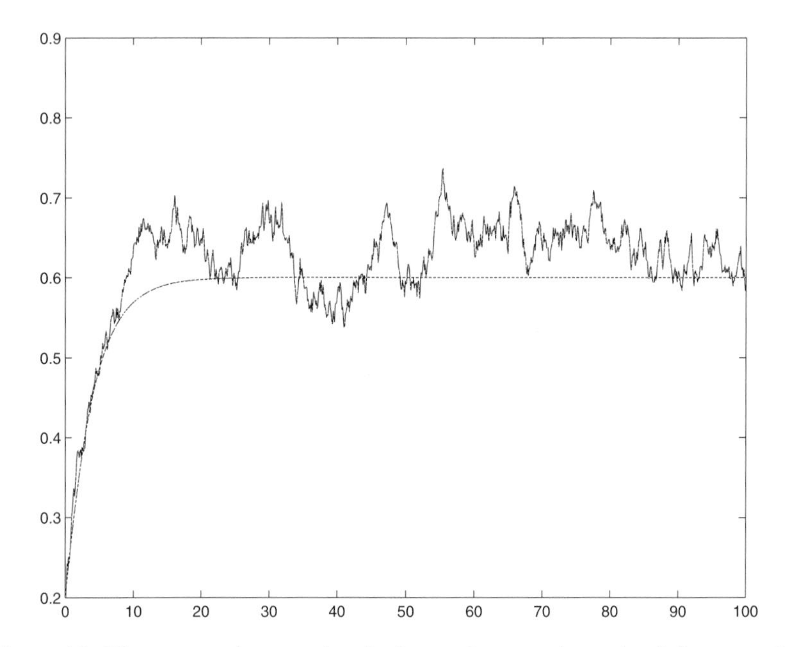

Figure 16. The averaged system's solution and a sample path of the perturbed system in the case of selection for a polymorphism. Here $\varepsilon = 0.1$ and the number of iterations is 1000.

Suppose that each cell has exactly N plasmids in each generation. These might be of r different types, say $T_1, \ldots, T_r$. Then the cells are of

$$m = \binom{N+r-1}{N}$$

different types, where $\binom{k}{j} = k!/(j!(k-j)!)$ is the binomial coefficient. The type of the cell is denoted by the vector $(n_1, \ldots, n_r)$, whose components are integers that add to N and describe the number of various plasmid types in the cell. Denote by $C_1, \ldots, C_m$, the different types of cells that are possible. A population of bacteria is characterized by the proportions of these various types. Let p^i denote the proportion of the population that is of type C_i. Then the population's distribution is described by the vector $\vec{p} = (p^1, \ldots, p^m)$.

We investigate the dynamics of the population distribution $\vec{p}_t$, for $t = 0, 1, 2, \ldots$, as the population evolves from generation to generation. We suppose that this sequence satisfies the relation

$$\vec{p}_{t+1} = \vec{p}_t \, A(t),$$

where $A(t)$ is a stochastic matrix of order $m \times m$, which is determined from the laws of genetics.

Random perturbations of such a system can have the form $A(t) = A + \varepsilon B(t)$, where the unperturbed system is described by the constant matrix A and random noise is described by the matrix $B(t)$, which we discuss later.

During the synthesis phase of the cell reproductive cycle, each plasmid is replicated, so that just before splitting, the plasmid pool has size $2N$ with $2n_i$ plasmids of type i for $i = 1, \ldots, m$. During mitosis, one daughter receives a distribution of plasmids, say described by the vector $\vec{k} = (k_1, \ldots, k_m)$ with probability

$$a_{\vec{n},\vec{k}} = \frac{\binom{2n_1}{k_1} \cdots \binom{2n_m}{k_m}}{\binom{2N}{N}}.$$

In this case, $\vec{p}_t = \vec{p}_0 A^t$, where A is the stochastic matrix with components $a_{\vec{n},\vec{k}}$. If we denote by $P_{\vec{n},\vec{k}}(t)$ the transition probability from the state $\vec{n}$ to the state $\vec{k}$ after t steps, then its elements are those of A^t. Our main result for this system is the following: There are absorbing states described by the vectors $\vec{n}^I = (\delta[I-1], \ldots, \delta[I-N])N$, where all N plasmids are from type C_I. (Here $\delta(0) = 1$ and $\delta(j) = 0$ if $j \neq 0$.) Then

$$\lim_{t \to \infty} P_{\vec{n},\vec{n}^I}(t) = \frac{n_I}{N},$$

but $P_{\vec{n},\vec{k}}(t) \to 0$ if $\vec{k} \neq \vec{n}^I$ for all I.

The randomly perturbed system has

$$A(t) = A + \varepsilon B(t),$$

where $B(t)$ is a stationary matrix–valued stochastic process. We assume:

1. $B(t)\,\vec{1} = 0$, where $\vec{1} = (1, \ldots, 1)^{\mathrm{T}}$. (i.e., the row sums of B are all zero.)
2. $\|B(t)\| \leq K$, where K is a constant.
3. $B(t)$ is ergodic and satisfies a mixing condition.

Then under some additional conditions the asymptotic behavior of the transition matrix for the perturbed problem can be determined. In fact, if $P^\varepsilon(k)$ is the transition probability matrix for k steps, then there exists, for sufficiently small ε, the limit

$$\lim_{t \to \infty} \frac{1}{t} \sum_{k=1}^{t} P^\varepsilon(k) = \tilde{\Pi}^\varepsilon,$$

where the matrix $\tilde{\Pi}^\varepsilon$ is a matrix having identical rows. Moreover, the limit

$$\lim_{\varepsilon \to 0} \tilde{\Pi}^\varepsilon = \tilde{\Pi}^0$$

can be calculated explicitly. Using this methodology, we can describe the asymptotic distributions of plasmids as $t \to \infty$ and $\varepsilon \to 0$.

Evolutionary Paths in Random Environments

A bacterial chromosome is described by a vector whose components are taken from among the symbols $\{A, C, G, T\}$. If the length of the chromosome is N base pairs, then the genetic structure of the organism is completely described by a corresponding chromosome vector, say $s \in S = \{A, C, G, T\}^N$, which denotes the set of all possible chromosomes, and s lists the DNA sequence of one in particular. We assume that all cells in the population reproduce at the same time, so reproduction is synchronized. A population of these organisms can be described by a vector whose components give the proportions of type s for all $s \in S$. Our interest is in how the vector of proportions changes from one generation to the next.

At reproduction, a cell produces two daughters, each having one chromosome that will generally be the same as the mother's, except for possible mutations, recombinations, and transcription errors. In our model, all changes of the type of a chromosome at the time of reproduction are random, although selective pressures can be accounted for as indicated earlier in the example of diploid genetics.

We describe the evolution of the population as a branching Markov process with the set of types S. This means that there exist probabilities $\pi(s, s_1, s_2)$ that a cell of type s splits into two daughters, one of type s_1 and the other of type s_2, so

$$\sum_{s_1, s_2 \in S} \pi(s, s_1, s_2) = 1.$$

We suppose that splitting of different cells is independent in the sense of probability. Let the matrix A have elements

$$A(s', s) = 2\pi(s', s, s) + \sum_{s_1 \in S \setminus \{s\}} \pi(s', s, s_1).$$

Denote by $\nu_s(t)$, for $s \in S$, the number of cells of type s at time t, and let $\vec{\nu}(t)$ be the vector of these. Then

$$E\vec{\nu}(t) = E\vec{\nu}(0)\ A^t.$$

Note that the row sums of A are identical,

$$\sum_{s \in S} A(s', s) = 2,$$

so $A/2$ is a stochastic matrix. If the matrix $\Pi = A/2$ is irreducible and aperiodic, then

$$\lim_{t \to \infty} \Pi^t = P,$$

where P has all rows the same as $\vec{p}$, the ergodic distribution for the Markov chain with transition probabilities describing one step of Π. Then

$$\lim_{t \to \infty} 2^{-t} E\vec{\nu}(t) = \left(E\vec{\nu}(0), \vec{1}\right)\ \vec{p}$$

and

$$\lim_{t\to\infty} 2^{-t}\vec{\nu}(t) = \left(\vec{\nu}(0), \vec{1}\right) \vec{p},$$

where $\left(\vec{a}, \vec{1}\right) = \sum_i a_i$ is the sum of the components of $\vec{a}$. We consider more general matrices Π as well.

For describing evolution in a random environment we use the same model, but with a random matrix A_n^* that describes reproduction at the nth time step. Assume that $\{A_n^*, n = 0, 1, 2, \dots\}$ is an ergodic stationary matrix–valued stochastic process satisfying some mixing conditions. (This sequence is called a random environment.) Denote by $\mathcal{E}$ the σ-algebra generated by the random environment. Then

$$E(\vec{\nu}(t+1)/\mathcal{E}) = \vec{\nu}(0) A_0^* \cdots A_t^*,$$

and the matrices $\Pi_n^* = A_n^*/2$ are stochastic matrices. Under certain natural conditions on the random environment, we prove the following statements:

1. $\lim_{n\to\infty} E\Pi_0^* \cdots \Pi_n^* = \Pi_0$.
2. With probability 1,
$$\lim_{n\to\infty} \frac{1}{n}(\Pi_0^* + \Pi_0^*\Pi_1^* + \cdots + \Pi_0^* \cdots \Pi_{n-1}^*) = \Pi_0.$$
3. The matrix Π_0 has identical rows.

For evolution in a random environment with small random perturbations, we consider matrices of the form

$$\Pi_n^* = \Pi + \varepsilon\hat{\Pi}_n^*,$$

where Π is a nonrandom stochastic matrix. The investigation of the asymptotic behavior of the products

$$\left(\Pi + \varepsilon\hat{\Pi}_0^*\right) \cdots \left(\Pi + \varepsilon\hat{\Pi}_{n-1}^*\right)$$

rests on the results in Chapter 7; in particular, we prove the existence of the limits

$$\lim_{n\to\infty} \left(\Pi + \varepsilon\hat{\Pi}_0^*\right) \cdots \left(\Pi + \varepsilon\hat{\Pi}_{n-1}^*\right) = \Pi_\varepsilon$$

and the existence with probability 1 of the limit

$$\lim_{n\to\infty} \frac{1}{n} \sum_{k=1}^{n} \left(\Pi + \varepsilon\hat{\Pi}_0^*\right) \dots \left(\Pi + \varepsilon\hat{\Pi}_{k-1}^*\right) = \Pi_\varepsilon.$$

The general inference that we make in this case is that if the averaging system is ergodic, then the behavior of the random system is likely to be similar to the behavior of the unperturbed system. If the averaging system is not ergodic, then the perturbed one can be ergodic with simpler behavior.

This completes our introduction to the topics, methods, and applications in this book.

1 Ergodic Theorems

Ergodic theorems have played major roles in the development of theoretical physics and in how we study random processes. In the physical setting, it is plausible that a collection of gas molecules in a confined volume will move in such a way that they will hit any sub–volume a proportion of time comparable to the size of the sub–volume. It was proposed by Boltzmann in the "ergodic surmise" that time averages and space averages agree in this sense, and mathematicians tried to prove this, but with only partial success. Notably, George Birkhoff, an expert in dynamical systems, and Norbert Wiener, a probabilist, found useful results that were applicable to certain cases in physics. These turned out to shape how we think of dynamical systems, randomness, and noise. Among the great many applications of these ideas are studies in genetics, population biology, financial planning, molecular diffusion, chemical reactions, medicine, and signal processing in engineering.

The purpose of this chapter is to describe ergodic processes from a mathematical point of view and to establish some interesting results that we will use throughout this book. We first review some of Birkhoff's work, and then demonstrate some of its consequences in modern studies of randomness.

1.1 Birkhoff's Classical Ergodic Theorem

The notation of modern probability involves a collection of possible samples; a collection of possible observable events; which are certain collections of samples; and a measure of the likelihood or probability of a given event occurring. The set of samples is denoted here by Ω; the events by a collection of subsets of Ω, which is denoted by $\mathcal{F}$; and the probability measure P, which associates with every event $A \in \mathcal{F}$ a number $P(A) \in [0,1]$. The collection $\mathcal{F}$ is a σ-algebra of sets, meaning that it is not empty and it is closed under complements and the formation of countable unions. The probability measure satisfies certain properties: For example, $0 \leq P(A) \leq 1$, $P(\Omega) = 1$, and $P(A \cup B) = P(A) + P(B)$ if $A \cap B \neq \phi$. We refer to the triple $(\Omega, \mathcal{F}, P)$ as a *probability space*. (These ideas are introduced more carefully in the Appendix.)

A function, say f, from Ω to the real numbers is said to be measurable for this probability space if the inverse images of intervals are events in $\mathcal{F}$, that is, $f^{-1}([a,b)) \in \mathcal{F}$ for any numbers $-\infty < a \leq b < \infty$. This is weaker than continuity, since there is no topology in Ω to describe "closeness", but it ensures that the mapping is consistent with the probability space's attributes; in particular, $P\{\omega : a \leq f(\omega) < b\}$ is defined for any real numbers $a \leq b$. A random variable, or measurement of samples, is a measurable function from Ω to the real numbers. If f is measurable, then its *distribution function*

$$F_f(a,b) \equiv P(a \leq f < b) = P\{\omega \in \Omega : a \leq f(\omega) < b\}$$

makes sense for any real numbers $a \leq b$. The distribution function plays a central role in studies of random variables. We write $F_f(x) = F_f(-\infty, x)$, and we denote the derivative of this function by $F_f'(x) = \partial F_f(x)/\partial x$. The existence of this derivative is discussed in the Appendix.

Let $\{\Omega,\ \mathcal{F},\ P\}$ be a probability space and let T be a measurable transformation $T : \Omega \to \Omega$. We say that T preserves the measure P if $P\{T^{-1}(A)\} = P(A)$ for $A \in \mathcal{F}$, where $T^{-1}(A) = \{\omega : T\omega \in A\}$.

For example, consider the probability space where $\Omega = [0,1]$ with measurable sets the Borel sets in Ω. These are the σ-algebra generated by the intervals $[a,b)$, where $0 \leq a \leq b \leq 1$. The probability measure is defined by Lebesgue measure, so $P[a,b) = b - a$. The mapping $Tx = 2x \pmod 1$ is a measure–preserving mapping of $[0,1]$ into itself, and Lebesgue measure is the invariant measure, as shown in Figure 1.1.

A set $B \in \mathcal{F}$ is called *invariant* if $T\omega \in B$ whenever $\omega \in B$ (i.e., $T(B) \subset B$). In this case, $B \subset T^{-1}(B)$, and the probability of the difference between $T^{-1}(B)$ and B is

$$P(T^{-1}(B) \setminus B) = P(T^{-1}(B)) - P(B) = 0.$$

More generally, a set $B' \in \mathcal{F}$ is called *quasi–invariant* if

$$P(T^{-1}(B') \setminus B') = 0.$$

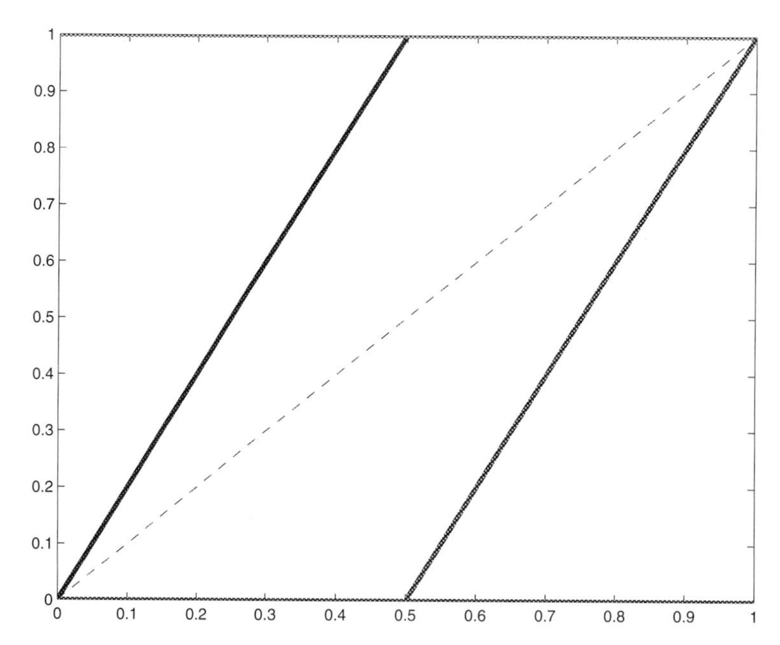

Figure 1.1. This figure depicts the mapping $Tx = 2x \pmod 1$ of $[0,1]$ into itself. Note that the preimage of an interval of length $b-a$ has length $2 \times (b-a)/2$.

We define $T^{-n}(C) = \{x : T^n x \in C\}$, where $T^0 x = x$ and for $n = 1, 2, \ldots$, $T^n x = T(T^{n-1}x)$. Then if B' is a quasi–invariant set, the sets

$$B_1 = \bigcap_n T^{-n}(B'), \quad B_2 = B' \cup \bigcup_n T^{-n}(B')$$

are invariant, and we have

$$B_1 \subset B' \subset B_2, \quad P(B_1) = P(B') = P(B_2).$$

A set is quasi-invariant if it is invariant as far as probability can determine.

Denote by $\mathcal{I}$ the collection of all invariant sets in $\mathcal{F}$ and denote by $Q\mathcal{I}$ the collection of all quasi–invariant sets in $\mathcal{F}$. Then we have the following properties:

(a) $Q\mathcal{I}$ is a σ–algebra;

(b) $\mathcal{I} \subset Q\mathcal{I}$;

(c) If $\mathcal{N}$ is the σ–algebra that is generated by the null sets of P, i.e., $\{C \in \mathcal{I} : P(C) = 0\}$, then $\mathcal{N} \subset Q\mathcal{I}$ and $Q\mathcal{I} = \mathcal{I} \vee \mathcal{N}$ is the smallest σ–algebra containing $\mathcal{I}$ and $\mathcal{N}$.

With these definitions we have the following result, which describes a relationship between *time* averages and averages over Ω:

Theorem 1 *Let $\xi(\omega) \in L_1(P)$ (that is, ξ is a function that is Lebesgue integrable over Ω with respect to the measure P), and let T be a transformation of Ω into itself that preserves the measure P. Then the limit*

$$\lim_{n\to\infty} \frac{1}{n} \sum_{k=1}^{n} \xi(T^k\omega) = \bar{\xi}(\omega) \tag{1.1}$$

exists for almost all (with respect to measure P) $\omega \in \Omega$, and

$$\bar{\xi}(\omega) = E\big(\xi(\omega)/Q\mathcal{I}\big).$$

The expectation E in the last formula is defined in the Appendix. This theorem is referred to as Birkhoff's ergodic theorem, and it is proved in [8]. The proof is not presented here.

In our example, the iterates of the mapping $Tx = 2x \pmod 1$ will take almost every $x_0 \in [0,1]$ all over the interval. The simulation in Figure 1.2 illustrates this. Select a point $x_0 \in [0,1]$ at random. Calculate the numbers $\{T^n x_0\}$ for $n = 1, \dots, 10^5$, and plot a histogram of the results. The outcome of this for $x_0 = 0.9218$ is shown in the figure.

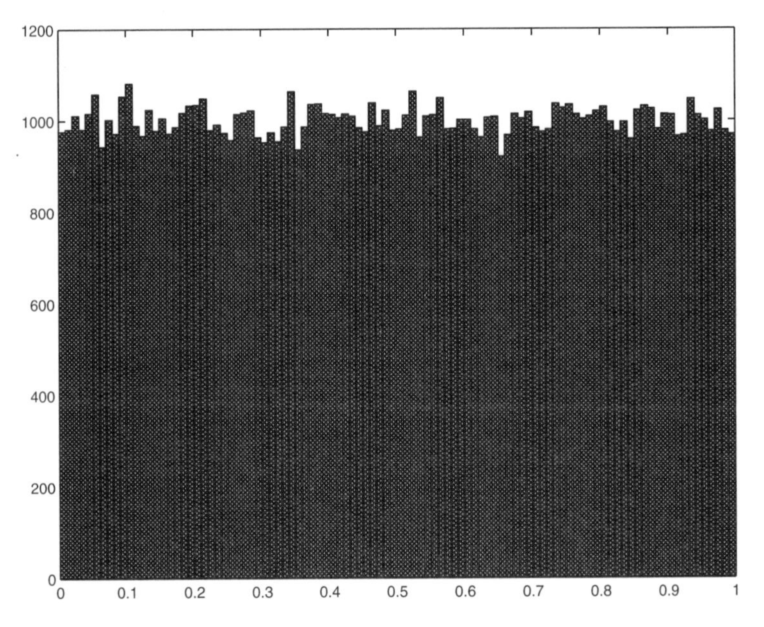

Figure 1.2. This figure depicts a histogram of iterates of the mapping $Tx = 1.99999x \pmod 1$ of $[0,1]$ into itself. This begins with the point $x_0 = 0.9218$ and plots a histogram of the results for $n = 1, \dots, 10^5$. This shows that the iterates are trying to reproduce the graph of the invariant measure's density function $F_T'(x) \equiv 1.0$.

The following definitions are useful:

Definitions.

(1) *A measure P is called* ergodic *with respect to a transformation T if $Q\mathcal{I} \subset \mathcal{N}$*, that is, if quasi–invariant sets have measure zero.

(2) *A transformation T is called* metrically transitive *with respect to the probability measure P if $\mathcal{I} \subset \mathcal{N}$, and so $Q\mathcal{I} \subset \mathcal{N}$.*

Remark 1 *If $Q\mathcal{I} \subset \mathcal{N}$, then in Theorem 1 $\bar{\xi}(\omega) = E\xi(\omega)$ with probability 1.*

Remark 2 *If $\bar{\xi}(\omega) = E\xi(\omega)$ with probability 1 for all $\xi(\omega) \in L_1(P)$, then $Q\mathcal{I} \subset \mathcal{N}$.*

Remark 3 *Suppose that $L \subset L_1(P)$ and L is dense in $L_1(P)$. If $\bar{\xi}(\omega) = E\xi(\omega)$ with probability 1 for $\xi \in L$, then $Q\mathcal{I} \subset \mathcal{N}$.*

The following lemma is useful to establish ergodicity of measure P.

Lemma 1 *Let $|\xi(\omega)| \le c$ with probability 1. Then $\bar{\xi}(\omega) = E\xi(\omega)$ if*

$$\lim_{n\to\infty} \frac{1}{n} \sum_{k=1}^{n} \left[E(\xi(\omega)\xi(T^k\omega)) - \left(E\xi(\omega)\right)^2 \right] = 0. \tag{1.2}$$

Proof of Lemma 1 Let $a = E\xi(\omega)$ and

$$\xi_n(\omega) = \frac{1}{n} \sum_{k=1}^{n} \xi(T^k\omega).$$

Equation (1.2) is equivalent to the relation

$$E\xi(\omega)\bar{\xi}(\omega) = a^2. \tag{1.3}$$

It is easy to see that $\bar{\xi}(T^k\omega) = \bar{\xi}(\omega) \pmod P$ for all k. The relation (1.1) implies that $E\xi_n(\omega)\bar{\xi}(\omega) = a^2$, so $E\bar{\xi}^2(\omega) = a^2$.

□

Theorem 2 *A measure P is ergodic with respect to a mapping T if there exists a linear set of bounded random variables L that is dense in $L_1(P)$ for which (1.2) is fulfilled.*

The proof of this theorem follows from Remark 3 and Lemma 1.

1.1.1 Mixing Conditions

We say that the mapping T satisfies the *mixing condition* if for any bounded random variables $\xi(\omega)$ and $\eta(\omega)$,

$$\lim_{k\to\infty} E\xi(\omega)\eta(T^k\omega) = E\xi(\omega)E\eta(\omega). \tag{1.4}$$

This relation implies that

$$\lim_{n\to\infty} E\xi(\omega)\frac{1}{n}\sum_{k=1}^{n}\eta(T^k\omega) = E\xi(\omega)E\eta(\omega). \tag{1.5}$$

If (1.5) is true for all bounded random variables ξ and η, then we say that T satisfies the *weak mixing condition.* Note that (1.3) and (1.5) are equivalent, so any metrically transitive mapping T satisfies the weak mixing condition.

1.1.2 Discrete–Time Stationary Processes

A measurable space $(X, \mathcal{B})$ is a set X of elements and a σ-algebra $\mathcal{B}$ of subsets of X. A discrete–time sequence of X-valued random variables has the form $\xi_n(\omega)$, $n \in N$, where N is either the positive integers $\mathbf{Z}_+$ or the set of all integers $\mathbf{Z}$. We assume that the joint distributions of $\vec{\xi}(\omega) \equiv \{\xi_n(\omega)\}$ satisfy the relation

$$P\{\xi_k \in A_0, \dots, \xi_{k+l} \in A_l\} = P\{\xi_{k+1} \in A_0, \dots, \xi_{k+l+1} \in A_l\}$$

for all $A_0, \dots, A_l \in \mathcal{B}$, and all relevant k, l. We will say the sequence $\{\xi_n(\omega)\}$ is a stationary random sequence.

In $(X^N, \mathcal{B}^N)$ we introduce a probability measure μ for which

$$\mu\big(\{\vec{x} : x^{(k)} \in A_0, \dots, x^{(k+l)} \in A_l\}\big) = P\{\xi_k \in A_0, \dots, \xi_{k+l} \in A_l\}, \tag{1.6}$$

where $\vec{x} \in X^N$ and $\vec{x} = \{x^{(n)}, n \in N\}$.

Let $S\vec{x} = \{x^{(n+1)}, n \in N\}$, so S is the shift operator in the space of sequences. It follows from relation (1.6) that S preserves the measure μ. Hence, we have the following theorem.

Theorem 1′ *Let g: $X^N \to R$ be a $\mathcal{B}^N$-measurable and μ-integrable function. Then with probability* 1 *there exists the limit*

$$\lim_{n\to\infty}\frac{1}{n}\sum_{k=1}^{n} g\big(S^k\vec{\xi}(\omega)\big) = \bar{g}(\omega), \tag{1.7}$$

and $\bar{g}(\omega) = E\big(g(\vec{\xi}(\omega))/Q\mathcal{I}\big)$, where $Q\mathcal{I}$ is the σ–algebra generated by the random variables

$$\Big\{\eta = g\big(\vec{\xi}(\omega)\big) : g\big(S\big(\vec{\xi}(\omega)\big)\big) = g\big(\vec{\xi}(\omega)\big)\Big\}.$$

(Note that two random variables are considered equal if they are equal for almost all $\omega \in \Omega$.)

We say that a stationary sequence is *ergodic* if we have in equality (1.7) that $\bar{g}(\omega) = Eg\big(\vec{\xi}(\omega)\big)$.

Theorem 2′ *A stationary sequence $\{\xi_n\}$ is ergodic if for every positive integer $l \in \mathbf{Z}_+$ and all sets $A_0, \dots, A_l \in \mathcal{B}$ and $B_0, \dots, B_l \in \mathcal{B}$, there exists* *the limit*

$$\lim_{n\to\infty} \frac{1}{n} \sum_{k=1}^{n} P\{\xi_0 \in A_0, \ldots, \xi_l \in A_l,\, \xi_k \in B_0, \ldots, \xi_{k+l} \in B_l\}$$

$$= P\{\xi_0 \in A_0, \ldots, \xi_l \in A_l\} P\{\xi_0 \in B_0, \ldots, \xi_l \in B_l\}. \tag{1.8}$$

Finally, we define the mixing condition:

Mixing Condition: A stationary sequence $\{\xi_n\}$ satisfies the mixing condition if for all sets $A_0, \ldots, A_l$ and $B_0, \ldots, B_l$ in $\mathcal{B}$, the following relation is fulfilled:

$$\lim_{n\to\infty} P\{\xi_0 \in A_0, \ldots, \xi_l \in A_l,\, \xi_n \in B_0, \ldots, \xi_{n+l} \in B_l\}$$

$$= P\{\xi_0 \in A_0, \ldots, \xi_l \in A_l\} P\{\xi_0 \in B_0, \ldots, \xi_l \in B_l\}. \tag{1.9}$$

For example, consider the sequence of independent identically distributed random variables $\{Z_n\}$. This is a stationary process. Moreover, suppose that these random variables have finite variance σ^2. Then $\chi_m \equiv E\, Z_n\, Z_{m+n} = 0$ if $m \neq 0$, but is σ^2 when $m = 0$. In addition, if EZ_n is finite, then by the law of large numbers there exists the limit

$$\lim_{n\to\infty} \frac{1}{n} \sum_{k=1}^{n} Z_k = m,$$

and $m = E(Z_n)$. Thus, in this case the process is stationary and ergodic. (Note, by the way, that the process

$$Y_n = \sum_{k=1}^{n} Z_k$$

is not stationary even if $m = 0$, since $EY_n^2 = n\sigma^2$.)
Next consider a random variable Z^* having finite mean (m) and variance (σ^2), and let the process $\{Z_n\}$ be such that each $Z_n = Z^*$. In this case, $\chi_m \equiv E\, Z_n\, Z_{m+n} = \sigma^2$ for any m, n, and we have the limit

$$\lim_{n\to\infty} \frac{1}{n} \sum_{k=1}^{n} Z_k = Z^*,$$

which is not the result of the ergodic theorem. The reason is that this process does not satisfy an appropriate mixing condition. (See [90] for further discussion of these processes.) In the first case, the sequence χ_m converges to zero quickly (in one step), but in the second case, it does not converge to zero.
Figure 1.3 shows a simulation of these two cases using 10^4 sample paths in each case. In the first case the independent variables Z_n are uniformly

distributed on the unit interval with mean $\frac{1}{2}$ and variance $\frac{1}{12}$. In the second case Z^* has the same distribution.

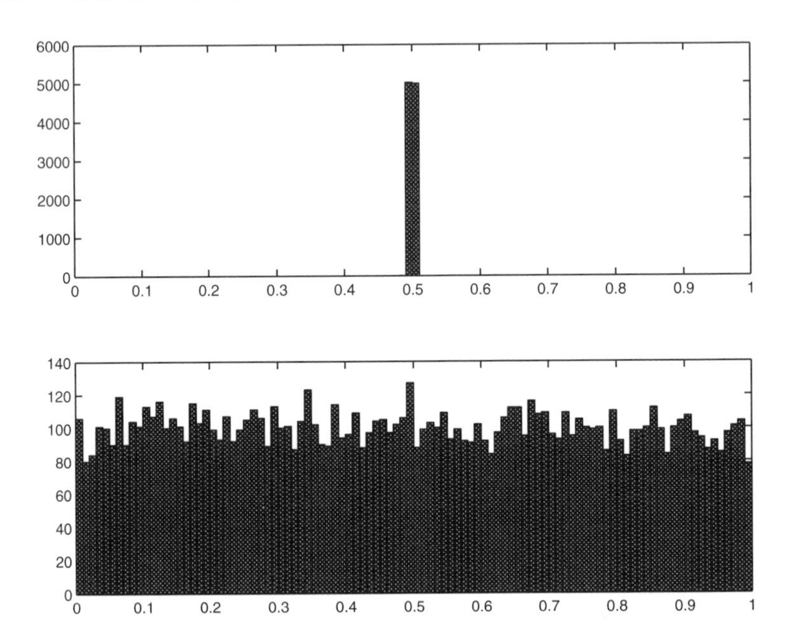

Figure 1.3. The top figure shows the convergence of the sample means in the case of an i.i.d. process. The law of large numbers tells us that the sample means converge to the common mean, $\frac{1}{2}$ in this case. In the bottom figure, the sample means are exactly samples of the random variable Z^*. In this simulation, 10^4 sample paths were used in each case.

1.2 Discrete–Time Markov Processes

We consider next a homogeneous Markov process in a measurable space $(X, \mathcal{B})$ with the transition probability for one step being $P(x, B)$, $x \in X$, $B \in \mathcal{B}$. That is, $P(x, B)$ gives the probability that starting at x the process arrives in the set B in one step. Let ρ be a probability measure on X defined for sets $B \in \mathcal{B}$. The measure ρ is called an *invariant distribution* for the transition probability $P(x, B)$ if

$$\rho(B) = \int \rho(dx) P(x, B), \tag{1.10}$$

for each set $B \in \mathcal{B}$. Let ξ_n for $n \in \mathbf{Z}_+$ be a Markov chain with transition probability $P(x, B)$ and initial distribution ρ, which means that $P\{\xi_0 \in B\} = \rho(B)$. Then $\{\xi_n\}$ is a stationary random sequence in X. We say that ρ is an *ergodic distribution* if this sequence is ergodic. A subset $B \in \mathcal{B}$ is

called ρ–invariant if $\rho\{x \in B : P(x, B) < 1\} = 0$. Then ρ is an ergodic distribution if $\rho(B) = 0$ or $\rho(B) = 1$ for every ρ–invariant set B.
We denote by P_x the measure on $X^{\mathbf{Z}_+}$ for which

$$P_x\{\vec{x} : x^0 \in A_0, \ldots, x^n \in A_n\}$$
$$= 1_{A_0}(x_0) \int P(x^0, dx^1) 1_{A_1}(x_1) \cdot P(x^1, dx^2) \cdots 1_{A_n}(x_n) P(x^{n-1}, dx^n).$$

The function P_x is the distribution of $\vec{\xi} = \{\xi_0, \xi_1, \ldots\}$ under the condition $\xi_0 = x$. (Here $1_A(x) = 1$ if $x \in A$, and $1_A(x) = 0$ if $x \notin A$.)

Theorem 3 *Let ρ be an ergodic distribution and $F : X^{r+1} \to R$ be a $\mathcal{B}^{r+1}$-measurable function for which*

$$\int |F(x_0, \ldots, x_r)| \rho(dx_0) P(x_0, dx_1) \cdots P(x_{r-1}, dx_r) < \infty.$$

Then

$$P_x\Bigg\{ \lim_{n\to\infty} \frac{1}{n} \sum_{k=1}^n F(\xi_k, \cdots, \xi_{k+r})$$
$$= \int F(x_0, \ldots, x_r) \rho(dx_0) P(x_0, dx_1) \cdots P(x_{r-1}, dx_r) \Bigg\} = 1 \qquad (1.11)$$

for almost all x with respect to the measure ρ.

Under what conditions is relation (1.11) true for all x? To answer this question we need some further ideas. A Markov process $\{\xi_n\}$ is called *Harris-recurrent* if there exists a nonzero measure m on $\mathcal{B}$ for which

$$P_x\Big\{ \sum_n 1_{\{\xi_n \in A\}} = +\infty \Big\} = 1$$

for all x if $m(A) > 0$. A measure m on $\mathcal{B}$ is called an *invariant measure* for the Markov process $\{\xi_n\}$ if

$$m(B) = \int P(x, B) m(dx)$$

for all $B \in \mathcal{B}$.
With these ideas, we have the following theorem:

Theorem 4 *Let the Markov process $\{\xi_n\}$ be Harris-recurrent. Then*
(i) *it has a unique (up to a constant multiplier) σ–finite invariant measure π, and*
(ii) *a π-integrable positive function $G : X \to R$ exists for which*

$$P_X\Big\{ \sum_n G(\xi_n) = +\infty \Big\} = 1$$

for all $x \in X$.
If a function F satisfies the conditions of Theorem 3, then for all $x \in X$,

(iii)

$$P_x\left\{\lim_{n\to\infty}\sum_{k=1}^{n} F(\xi_k,\ldots,\xi_{k+1})\Big/\sum_{k=1}^{n} G(\xi_k)=\hat{F}\right\}=1 \tag{1.12}$$

and

(iv)

$$\lim_{n\to\infty} E_x\sum_{k=1}^{n} F(\xi_k,\ldots,\xi_{k+1})\Big/E_x\sum_{k=1}^{n} G(\xi_k)=\hat{F}, \tag{1.13}$$

where

$$\hat{F}=\int F(x_0,\ldots,x_r)\pi(dx_0)P(x_0,dx_1)\cdots P(x_{r-1},dx_r)\Big/\int G(x)\pi(dx) \tag{1.14}$$

and E_x is the expectation with respect to the measure P_x.

For the proof, see [168, Chapter 6, Theorem 6.5].

Definition. A Markov chain $\{\xi_n\}$ is called uniformly ergodic if (1) it has a unique stationary distribution, (2) it is Harris-recurrent with respect to the ergodic distribution ρ, and (3)

$$\lim_{n\to\infty}\sup_x E_x\left(\frac{1}{n}\sum_{k=1}^{n}\psi(\xi_k)-\int\psi(x)\rho(dx)\right)^2=0$$

for every bounded measurable function $\psi: X\to R$.

Remark 4 *Let $P_n(x,B)$ be the transition probability of the Markov chain for n steps. Assume that the measure $P_n(x,\cdot)$ is absolutely continuous with respect to the ergodic distribution ρ and that*

$$\frac{dP_n(x,\cdot)}{d\rho}(x')=p_n(x,x'),$$

so

$$P_n(x,B)=\int_B p_n(x,x')\rho(dx').$$

Then the condition

$$\lim_{n\to\infty}\sup_x\int|p_n(x,x')-1|\rho(dx')=0$$

implies the uniform ergodicity of the Markov chain.

For example, consider a finite-state Markov chain, say $\{Z_n\}$, having states $\{1,2\}$ and transition probability matrix

$$P=\begin{pmatrix} a & 1-a\\ 1-b & b\end{pmatrix},$$

where $a, b \in (0, 1)$. The eigenvalues of this matrix are 1 and $a + b - 1$, for which $|a + b - 1| < 1$. The eigenvector corresponding to the larger eigenvalue is $(1 - b, 1 - a)/c$, where $c = 2 - a - b$. This Markov chain is stationary and ergodic, and the ergodic measure is defined by the table of values $\rho(1) = (1-b)/c$, $\rho(2) = (1-a)/c$. In fact, the spectral decomposition of P is

$$P = \begin{pmatrix} (1-b)/c & (1-a)/c \\ (1-b)/c & (1-a)/c \end{pmatrix} + (a+b-1)P_2,$$

where $P_2P_2 = P_2$ and $P_2P_1 = P_1P_2 = 0$, where P_1 is the first matrix in this formula. Thus, for large m, $P^m \approx P_1$.
The simulation illustrated in Figure 1.4 is used later.

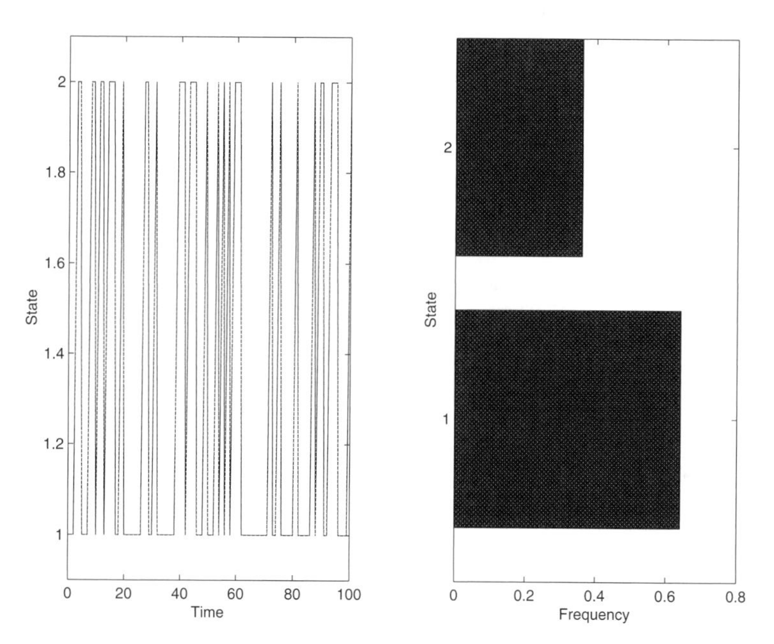

Figure 1.4. The figure on the left shows a sample path for the case where $a = 0.75, b = 0.4$. The figure on the right shows the proportion of times each of the two states was visited by this sample path. The ergodic measure indicates that the expected proportion of visits to state 1 is 0.706 and to state 2 is 0.294. The proportion of visits observed in this sample path of length 100 are 0.69 and 0.31, respectively.

Simulations of the kind illustrated in Figure 1.4 do not converge rapidly if many states are involved. For example, consider a Markov chain having 10 states and a transition probability matrix P whose first row is $(1 - 0.2, 0.1, \ldots, 0.1)$ and the rest of whose rows are determined as shift permutations of this row, so the resulting transition matrix is a circulant matrix. The simulation in Figure 1.5 shows a sample path of length 10000, and the histogram at the right shows the proportion of visits to each state by this single sample path.

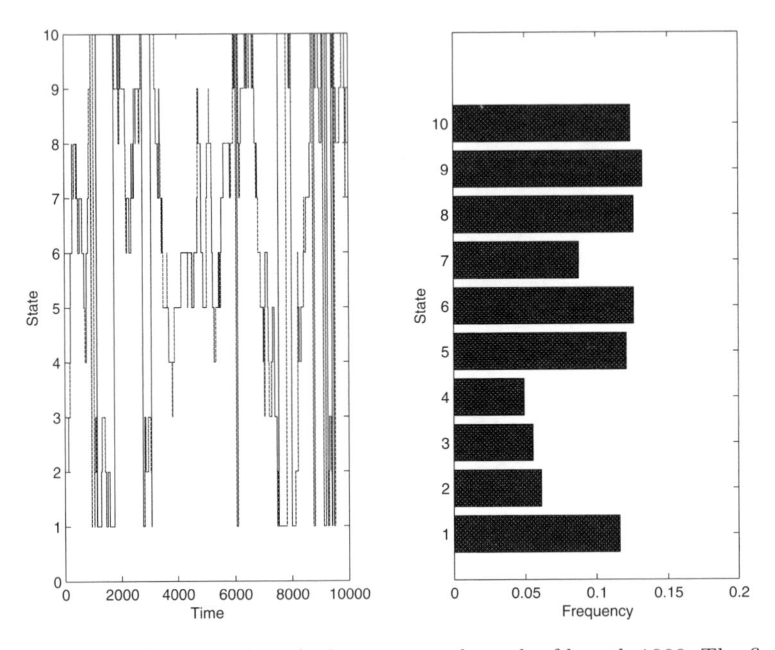

Figure 1.5. The figure on the left shows a sample path of length 1000. The figure on the right shows the proportion of visits to each of the ten states by this sample path. The ergodic measure indicates that the expected proportion of visits to each state is 0.1, but the histogram shows substantial deviation from this even at 1000 time steps. Things improve significantly at 10^4 time steps, but at greater computational expense.

1.3 Continuous–Time Stationary Processes

We consider an X-valued stationary stochastic process $\xi(t)$, $t \in \Delta$, where $\Delta = R_+$ or $\Delta = R$. This means that

$$P\{\xi(t_1) \in A_1, \ldots, \xi(t_k) \in A_k\} = P\{\xi(t_1 + s) \in A_1, \ldots, \xi(t_k + s) \in A_k\}$$

for all $t_1, \ldots, t_k, s \in \Delta$ and for all sets $A_1, \ldots, A_k \in \mathcal{B}$. We denote by $\mathcal{B}_\Delta$ the Borel σ–algebra in the set Δ.
Suppose that $\xi(t, \omega)$ is $\mathcal{B}_\Delta \otimes \mathcal{F}$-measurable. We denote by L_2^ξ the Hilbert space generated by the random variables of the form

$$\eta = \int \cdots \int f\big(t_1, \ldots, t_k, \xi(t_1), \ldots, \xi(t_k)\big)\, dt_1 \cdots dt_k, \tag{1.15}$$

where $f : R^k \times R^k \to R$ is a bounded $\mathcal{B}_{R^k} \otimes \mathcal{B}^k$-measurable function. Note that η_1 and η_2 are identical in L_2^ξ if $P\{\eta_1 = \eta_2\} = 1$. We introduce the operators in L_2^ξ by the following formula: If η is given by (1.15), then

$$S_\tau \eta = \int \cdots \int f\big(t_1 - \tau, \ldots, t_k - \tau, \xi(t_1), \ldots, \xi(t_k)\big)\, dt_1 \cdots dt_k, \quad \tau \in \Delta.$$

It is easy to see that $S_{\tau_1+\tau_2} = S_{\tau_1} S_{\tau_2}$ and that S_τ is an isometric linear operator. (If $\Delta = R$, then S_τ is a unitary operator.)
Let I^ξ be the subspace that consists of those elements ζ for which $S_\tau \zeta = \zeta$ for all $\tau \in \Delta$. Let $\mathcal{I}^\xi$ be the σ-algebra generated by $\{\zeta, \zeta \in I^\xi\}$.

The following theorem is a consequence of Theorem 1.

Theorem 5 *Suppose that a measurable function $F : X^k \to R$ satisfies the condition*

$$E\big|F\big(\xi(t_1), \ldots, \xi(t_k)\big)\big| < \infty.$$

Then with probability 1 there exists the limit

$$\bar{\zeta} = \lim_{T\to\infty} \frac{1}{T} \int_0^T F\big(\xi(t_1 + s), \ldots, \xi(t_k + s)\big) ds, \tag{1.16}$$

where $\bar{\zeta} \in \mathcal{I}^\xi$ and

$$\bar{\zeta} = E\Big(\frac{1}{h} \int_0^h F\big(\xi(t_1 + s), \ldots, \xi(t_k + s)\big) ds / \mathcal{I}^\xi\Big) \tag{1.17}$$

for any $h > 0$.

The process $\xi(t)$ is called *ergodic* if I^ξ contains the constants only. Then $\mathcal{I}^\xi$ is the trivial σ-algebra. In this case

$$\bar{\zeta} = EF\big(\xi(t_1), \ldots, \xi(t_k)\big). \tag{1.18}$$

1.4 Continuous–Time Markov Processes

Now consider a homogeneous Markov process in the space $(X, \mathcal{B})$ with transition probability $P(t, x, B)$ for $t \in R_+$ and $B \in \mathcal{B}$. Assume that $P(t, x, B)$ satisfies the following conditions:

I. P is a probability measure for $B \in \mathcal{B}$ for each fixed t and x;

II. P is measurable in t and x for each fixed B;

III. P satisfies the Chapman–Kolmogorov equation:

$$P(t+s, x, B) = \int P(t, x, dx') P(s, x', B), \quad t, s \in R_+,\ x \in X,\ B \in \mathcal{B}.$$

Finally, we suppose that the σ-algebra $\mathcal{B}$ is countably generated.
A σ-finite measure ρ is called an invariant measure for the Markov process if

$$\rho(A) = \int \rho(dx)P(t,x,A) \quad \text{for all } A \in \mathcal{B}. \tag{1.19}$$

Suppose that there exists an invariant probability measure π. The X-valued process $\xi(t)$ for which

$$P\{\xi(0) \in A_0, \dots, \xi(t_k) \in A_k\} = \iint \pi(dx_0)1_{A_0}(x_0)P(t_1, x_0, dx_1)1_{A_1}(x_1)$$
$$\cdots \times P(t_k - t_{k-1}, x_{k-1}, dx_k)1_{A_k}(x_k), \quad 0 < t_1 < \cdots < t_k, \tag{1.20}$$

is a stationary process in X. So Theorem 5 is true for this process. To formulate the next theorem we introduce the measure P_x that is the probability for the Markov process under the condition that its value at $t = 0$ is x.

Theorem 5′ *Let a measurable function $F : X^k \to R$ satisfy the condition*

$$\int \cdots \int |F(x_1, \dots, x_k)| \pi(dx_1) \prod_{j=2}^{k} P(t_j - t_{j-1}, x_{j-1}, dx_j) < \infty.$$

Then the limit (1.16) *exists with probability $P_x = 1$ for almost all x with respect to the measure π, and it can be represented by formula* (1.17).

We describe $\mathcal{I}^\xi$ in this case.

Definition A set $B \in \mathcal{B}$ is called π-invariant if for all $t > 0$,

$$\int_B P(t,x,X \setminus B)\pi(dx) + \int_{X\setminus B} P(t,x,B)\pi(dx) = 0. \tag{1.21}$$

Denote by $\mathcal{I}^\xi_\pi$ the collection of all π-invariant subsets of X. It is easy to show that $\mathcal{I}^\xi_\pi$ is a σ-algebra.

Theorem 6 *$\mathcal{I}^\xi$ is generated by the set of random variables*

$$\{1_B(\xi(0,\omega)),\ B \in \mathcal{I}^\xi_\pi\}.$$

For the proof, see [168, Section 2.3].

Definition *An invariant probability distribution π is* ergodic *if for all $B \in \mathcal{I}^\xi_\pi$ we have either that $\pi(B) = 0$ or $\pi(B) = 1$.*
In this case we say that the σ-algebra $\mathcal{I}^\xi_\pi$ is trivial with respect to the measure π. Note that if there exists only one invariant probability measure, then it is ergodic.

Definition *We will say that the Markov process is uniformly ergodic if it satisfies the following conditions:*

(1) *it has a unique invariant probability measure* π,

(2) *for all* $A \in \mathcal{B}$

$$P_x\left\{\int_0^\infty 1_A(\xi(t))dt = +\infty\right\} = 1$$

for all x *if* $\pi(A) > 0$,

(3) *The relation*

$$\lim_{T\to\infty} \sup_{x\in X} E_x\left(\frac{1}{T}\int_0^T \psi(\xi(s))ds - \int \psi(x)\pi(dx)\right)^2 = 0$$

holds for every bounded $\mathcal{B}$*-measurable function* $\psi : X \to B$.

Remark 5 *If the transition probability of the Markov process* $P(t, x, B)$ *is absolutely continuous with respect to the measure* π *and if*

$$p(t, x, x') = \frac{dP(t, x, \cdot)}{d\pi}(x'),$$

then the condition

$$\lim_{T\to\infty} \frac{1}{T} \sup_x \int_0^T |p(t, x, x') - 1|\pi(dx') = 0$$

implies uniform ergodicity of the Markov process.

This completes our introduction to the ergodic theory of stochastic processes.

2

Convergence Properties of Stochastic Processes

In this chapter we describe ways that sequences of stochastic processes can converge. Our general results about perturbation methods give the first two terms of an approximation: The leading term is based on convergence (as $\varepsilon \to 0$) to some nonrandom values, and the second term is based on convergence to a diffusion process. These are in the spirit of the law of large numbers and the central limit theorem, respectively. In this chapter we define the senses in which these limits are valid; we describe technical conditions that are sufficient for various kinds of convergence; and at the end, we describe a central limit theorem for ergodic stochastic processes. While the results in this chapter are developed mostly for continuous–time processes, there are analogous results for discrete–time processes that are left to the reader.

2.1 Weak Convergence of Stochastic Processes

The concept of weak convergence is important for our approximations.

Definition 1. *Let $\xi_n(t)$, $t \in R_+$, $n = 0, 1, 2, \dots,$ be a sequence of continuous–time R^d-valued stochastic processes. We say that $\xi_n(t)$ converges weakly to $\xi_0(t)$ as $n \to \infty$ if*

$$\lim_{n\to\infty} Ef\big(\xi_n(t_1), \dots, \xi_n(t_k)\big) = Ef\big(\xi_0(t_1), \dots, \xi_0(t_k)\big)$$

for every $k \in \mathbf{Z}_+$, for all times $t_1, \dots, t_k \in R_+$, and for all $f \in C((R^d)^k)$.

In this case, we say that all finite-dimensional distributions of the process $\xi_n(t)$ converge weakly to the corresponding finite-dimensional distributions of the process $\xi_0(t)$.

Definition 2. *A sequence* $\xi_n(t)$, $t \in R_+$, $n = 1, 2, \ldots,$ *is said to be* weakly compact *if each subsequence* $n_1, n_2, \ldots$ *admits a further subsequence* $n_{l_1}, n_{l_2}, \ldots$ *such that* $\xi_{n_{l_k}}(t)$ *converges weakly to a stochastic process* $\tilde{\xi}(t)$.

Remark 1 *Let* $\{\tilde{\xi}(\cdot)\}$ *be the set of stochastic processes that are the limits for weakly convergent subsequences* $\{\xi_{n_k}(t)\}$. *The entire sequence* $\xi_n(t)$ *converges weakly to some stochastic process if and only if all the processes* $\tilde{\xi}(\cdot)$ *have the same distribution.*

The proof of this is left to the reader. The following result gives sufficient conditions for a sequence of processes to be weakly compact:

Proposition 1 *The collection of stochastic processes* $\{\xi_n(t)\}$ *is weakly compact if it satisfies the conditions*

$$\text{(a)} \quad \lim_{r\to\infty} \limsup_{n\to\infty} P\big\{|\xi_n(0)| > r\big\} = 0$$

and

$$\text{(b)} \quad \lim_{h\to 0} \limsup_{n\to\infty} \sup_{t\le T} \sup_{|t-t'|\le h} P\big\{|\xi_n(t) - \xi_n(t')| > \delta\big\} = 0$$

for every tolerance $\delta > 0$ *and terminal time* $T \in R_+$.

Proof Conditions (a) and (b) imply that the finite–dimensional distributions of the processes $\xi_n(t)$ are weakly compact. Denote by Q_+ the set of nonnegative rational numbers. Let $\{n_l : l = 1, 2, \ldots\}$ be a sequence of natural numbers, $n_l \to \infty$ as $l \to \infty$. Using the diagonal method we can construct a subsequence $\{n_{l_k}, k = 1, 2, \ldots\}$ for which a limit exists,

$$\mathcal{L}(f, t_1, \ldots, t_r) = \lim_{k\to\infty} Ef(\xi_{n_{l_k}}(t_1), \ldots, \xi_{n_{l_k}}(t_r)),$$

for all $t_1, \ldots, t_r \in Q_+$ and $f \in C((R^d)^r)$. This implies the existence of a stochastic process $\tilde{\xi}(t)$, $t \in Q_+$, for which

$$\mathcal{L}(f, t_1, \ldots, t_r) = Ef(\tilde{\xi}(t_1), \ldots, \tilde{\xi}(t_r)),$$

for $t_1, \ldots, t_r \in Q$, $f \in C((R^d)^r)$. It follows from condition (b) that

$$\lim_{h\to 0} \sup \Big(P\{|\tilde{\xi}(t) - \tilde{\xi}(t')| > \varepsilon\} : t, t' \in Q_+, |t - t'| \le h \Big) = 0$$

for all $\varepsilon > 0$. So a limit in probability exists,

$$\lim_{t'\to t, t'\in Q_+} \tilde{\xi}(t'), \quad \text{for all } t \in R_+,$$

which we denote by $\tilde{\xi}(t)$. Then the sequence $\xi_{n_k}(t)$ converges weakly to the stochastic process $\tilde{\xi}(t)$.

□

2.1.1 Weak Convergence and Weak Compactness in C

Denote by $C_{[a,b]}(R^d)$, for $[a,b] \subset R_+$, the space of continuous functions $x(t)$: $[a,b] \to R^d$ with the norm of a function being defined by

$$||x(\cdot)||_{C_{[a,b]}} = \sup_{t\in[a,b]} |x(t)|.$$

Let $\xi(t)$ be a continuous R^d-valued stochastic process. Denote by μ_ξ^T the distribution of $\xi(t)$, $t \in [0,T]$, in the space $C_{[0,T]}(R^d)$. That is, $\mu_\xi^T(A) = P\{\xi(\cdot) \in A\}$ for all measurable sets A in $C_{[0,T]}(R^d)$.

Definition 3. *The sequence $\xi_n(t)$, $n = 1, 2, \ldots$, of continuous R^d-valued processes converges weakly in C to a continuous R^d-valued process $\xi(t)$ if $\mu_{\xi_n}^T$ converges weakly to μ_ξ^T for all $T \in R_+$. That is, for all $T \in R_+$,*

$$\lim_{n\to\infty} EF^T\left(\xi_n(\cdot)\right) = EF^T\left(\xi(\cdot)\right)$$

for any continuous function $F^T : C_{[0,T]}(R^d) \to R$.

Weak Compactness in C of a Sequence of Continuous Processes

A sequence of continuous processes $\xi_n(t)$, $t \in R_+$, $n = 1, 2, \ldots$, is said to be *weakly compact in C* if any subsequence $\{n_k,\ k = 1, 2, \ldots\}$ admits a further subsequence $\{n_{k_l}\}$, $l = 1, 2, \ldots$, such that $\xi_{n_{k_l}}(t)$ is weakly convergent in C.

Remark 2 *A process $\tilde{\xi}(t)$ is called a partial C-limit for the sequence $\{\xi_n(\cdot)\}$ if there exists a subsequence $\{n_k\}$ such that $\xi_{n_k}(t)$ converges weakly to $\tilde{\xi}(t)$ in C. A sequence of continuous functions is C-weakly convergent (i.e., there exists a process to which it converges weakly in C) if*
(1) the sequence is weakly compact in C,
and
(2) all partial limits have the same distributions.
In particular, $\xi_n(t)$ converges weakly in C if (1) is true and if $\xi_n(\cdot)$ converges weakly.

Proposition 2 *Let $\{\xi_n(t)\}$ for $t \in R$ or $t \in R_+$ be a sequence of continuous processes. It is weakly compact in C if and only if*

$$(a) \quad \lim_{r\to\infty} \limsup_{n\to\infty} P\big\{|\xi_n(0)| > r\big\} = 0$$

and

$$(b) \quad \lim_{h\to 0} \limsup_{n\to\infty} P\left\{\sup_{t\le T}\ \sup_{|t-t'|\le h} |\xi_n(t) - \xi_n(t')| > \epsilon\right\} = 0$$

for all $\epsilon > 0$ and $T \in R_+$.

For the proof see [7, Section 7].

Theorem 1 *Let a sequence of continuous processes* $\xi_n(t)$ *satisfy the following conditions:*
(a) *condition* (a) *of Proposition* 2*;*
(b) *there exists a sequence* $\delta_n \downarrow 0$ *for which*

$$\lim_{n\to\infty} P\left\{\sup_{t\le T,\,|t-s|\le\delta_n} |\xi_n(t)-\xi_n(s)| > \epsilon\right\} = 0$$

for all $\epsilon > 0$ *and* $T \in R_+$*;*
(c) *there exist* $\alpha > 0$, $\beta > 0$ *for which*

$$E\left(1 \wedge |\xi_n(t)-\xi_n(s)|^\alpha\right) \le C_T\left(|t-s|^{1+\beta} + \delta_n^{1+\beta}\right) \quad \textit{if } t \le T,$$

where δ_n *is the same as in* (b), $C_T < \infty$ *for every* $T > 0$.
Then the sequence $\{\xi_n(t)\}$ *is weakly compact in* C.

Proof Let

$$\rho_{nk} = \max_{i\le 2^k T}\left\{\left|\xi_n\left(\frac{i+1}{2^k}\right) - \xi_n\left(\frac{i}{2^k}\right)\right|\right\},$$

and let

$$\hat\rho_n = \max\{|\xi_n(t)-\xi_n(s)| : t\le T,\ |t-s|\le\delta_n\}.$$

Then

$$\max\{|\xi_n(t)-\xi_n(s)|,\ t\le T,\ |t-s|\le h\}$$

$$\le \hat\rho_n + \sum_{\log_2\frac{1}{\delta_n}\ge k>\log_2\frac{1}{h}} \rho_{nk},$$

If $2^{-k} \ge \delta_n$, then

$$P\{\rho_{nk} > \theta\} \le \sum_{i/2^k\le T} \frac{1}{\theta^\alpha}(C_T+1)\frac{1}{2^{k(1+\beta)}} \le (T+1)\frac{\theta^\alpha}{2^{k\beta}}.$$

Therefore,

$$P\{\sup_{t\le T,\,|t-s|\le\delta_n} |\xi_n(t)-\xi_n(s)| \ge \epsilon\} \le P\left\{\hat\rho_n + \sum_{\log_2 h^{-1}\le k\le\log_2\delta_n^{-1}} \delta_n^{-1}\rho_n k > \epsilon\right\}$$

$$\le P\left\{\hat\rho_n > \frac{\epsilon}{2}\right\} + \sum_{\log_2 h^{-1}<k} \frac{T+1}{(\theta_k\epsilon)^\alpha}2^{-k\beta},$$

where

$$\sum \theta_k = \frac{1}{2} \quad \text{and} \quad \sum \frac{1}{\theta_k^\alpha 2^{k\beta}} < \infty.$$

So relation (b) of Proposition 2 is satisfied, which implies the result.

□

We can apply this theorem to a sequence of random polygonal functions.

Corollary *Let*

$$\xi_n(t) = \xi_n\left(\frac{k}{n}\right) + (nt-k)\left(\xi_n\left(\frac{k+1}{n}\right) - \xi_n\left(\frac{k}{n}\right)\right)$$

for $t \in \left[\frac{k}{n}, \frac{k+1}{n}\right]$ *and let*

$$\sup_{k\le nT}\left|\xi_n\left(\frac{k+1}{n}\right) - \xi_n\left(\frac{k}{n}\right)\right| \to 0$$

in probability as $n \to \infty$ *for every* $T > 0$. *Suppose that for each* $T > 0$ *there exists a pair of* $\alpha > 0$ *and* $\beta > 0$ *for which*

$$E\big(1 - \exp\{-|\xi_n(t) - \xi_n(s)|^\alpha\}\big) \le C_T\left(|t-s|^{1+\beta} + \frac{1}{n^{1+\beta}}\right)$$

for $t \le T$. *Then the sequence* $\xi_n(t)$ *is weakly compact in* C.

2.2 Convergence to a Diffusion Process

2.2.1 Diffusion Processes

We consider a continuous Markov process $\xi(t)$ in R^d with transition probability $P(s,x,t,B)$ for $x \in R^d$, $0 \le s < t < \infty$ and for $B \in \mathcal{B}(R^d)$. This process is called a *diffusion process* if there exist continuous functions $a(t,x)\colon R_+ \times R^d \to R^d$ and $b(t,x)\colon R_+ \times R^d \to L_+(R^d)$, which is the set of nonnegative symmetric $d \times d$ matrices, such that

$$\int g(x')P(s,x,t,dx') - g(x) = \int_s^t \int L_u g(x')P(s,x,u,dx')\,du, \tag{2.1}$$

where $g \in C^2(R^d)$, $0 \le s < t$, and

$$L_u g(x) = \sum_{i=1}^d a^i(u,x)g_{x^i}(x) + \frac{1}{2}\sum_{i,j=1}^d b^{ij}(u,x)g_{x^ix^j}(x), \tag{2.2}$$

where $a^1, \ldots, a^d$ are the coordinates of the vector a and b^{ij}, $i,j \in \overline{1,d}$, are the elements of the matrix b. The operator L_u is called the *generator of the Markov process*.

Proposition 3 *Let* $a(t,\ x)$ *and* $b(t,\ x)$ *be continuous and satisfy a local Lipschitz condition: For every* $r > 0$ *there exists a constant* l_r *for which*

$$|a(t,x) - a(t,x')| + \|b(t,x) - b(t,x')\| \le l_r|x - x'|$$

if $|x| \le r$, $t \le r$, *and there is a constant* q *for which* $(a(t,x),x) + b(t,x) \le q(1+(x,x))$. *Then the transition probability is determined by the functions* $a(t,x)$ *and* $b(t,x)$ *through relation* (2.1).

This statement is a consequence of the following observation.

Remark 3 *Consider the stochastic differential equation*

$$d\xi(t) = a(t,\xi(t))dt + b^{1/2}(t,\xi(t))dw(t), \tag{2.3}$$

where $b^{1/2}(\cdot)$ *is the positive symmetric square root of the matrix* $b(\cdot)$ *and* $w(t)$ *is the Wiener process in* R^d *(i.e.,* $Ew(t) = 0$, $E(w(t),z)^2 = t|z|^2$, $z \in R^d$*). It follows from the theory of stochastic differential equations [108, p.166] that the conditions of Proposition* 3 *imply the uniqueness and the existence of the solution to equation* (2.3) *for a given initial condition. Denote by* $\xi_{x,s}(t)$ *the solution to equation* (2.3) *on the interval* $[s,\infty)$ *satisfying the initial condition*

$$\xi_{x,s}(s) = x.$$

Then

$$P(s,x,t,B) = P\{\xi_{x,s}(t) \in B\}.$$

Remark 4 *Let* $g \in C^{(2)}(R^d)$, *and let* $\xi(t)$ *be the diffusion process with a generator* L_t. *Then*

$$g(\xi(t)) - \int_0^t L_s g(\xi(s))ds$$

is a martingale with respect to the filtration $(\mathcal{F}_t^\xi, t \in R_+)$ *generated by the process* $\xi(t)$. (See Appendix A for definitions.)
This follows from relation (2.1).

Proposition 4 *Assume that* $a(t,x)$ *and* $b(t,x)$ *satisfy the conditions of Proposition* 3. *Let* $\tilde{\xi}(t)$ *be a measurable* R^d*-valued stochastic process on* R^+ *and let* $(\tilde{\mathcal{F}}_t, t \in R_+)$ *be the filtration generated by the stochastic process* $(\tilde{\xi}(t), t \in R_+)$. *If for all* $g \in C^{(2)}(R^d)$ *the stochastic process*

$$\mu_g(t) = g(\tilde{\xi}(t)) - \int_0^t L_s g(\tilde{\xi}(s))ds$$

is a local martingale with respect to the filtration $(\tilde{\mathcal{F}}_t, t \in R_+)$, *then* $\tilde{\xi}(t)$ *is the diffusion process with the generator* L_s.

A proof of Proposition 4 is presented in [188].
We also will consider diffusion processes in a region of R^d having an absorbing boundary. Let G be an open set in R^d and let G' be its boundary. Let $\xi(t)$ be a Markov process in $G \cup G'$ satisfying the following properties:

(1) If $\tau = \inf\{t : \xi(t) \in G'\}$, then $\xi(t) = \xi(\tau)$ for $t \geq \tau$ (i.e., the boundary is "sticky").

(2) For every $g \in C^{(2)}(R^d)$ for which $g(x) = 0$ if $x \in R^d \setminus G$ (the complement of G) the relation (1) is fulfilled. Then $\xi(t)$ is called the diffusion process in G with the generator L_t and absorption on G'.

Propositions 3 and 4 are true for diffusion processes in a region G with absorbing boundary conditions on G'.

Remark 5 *Assume that $\xi(t)$ is a diffusion process in R^d with the generator L_t and $G \subset R^d$ is an open set in R^d, G' its boundary. Set*

$$\tau = \inf\{t : \xi(t) \in G'\}, \quad \xi'(t) = \xi(t \wedge \tau).$$

Then $\xi'(t)$ is the diffusion process in G with the generator L_t and absorption on G'.

Remark 6 *Let $\tilde{\xi}(0) = 0$, and let $\tilde{\xi}(t)$ satisfy the conditions of Proposition 3 with*

$$L_u g = \frac{1}{2} \sum_{i=1}^{d} g_{x^i x^i}.$$

Then $\tilde{\xi}(t)$ is a Wiener process for which $E\tilde{\xi}(t) = 0$ and $E\big(\tilde{\xi}(t), z\big)^2 = t(z, z)$.

2.2.2 *Weak Convergence to a Diffusion Process*

Theorem 2 *Let $\{\xi_n(t)\}$ be a sequence of measurable R^d-valued random processes for $t \in R_+$. Suppose that*
(1) The distributions of $\xi_n(0)$ converge weakly to some distribution $m_0(dx)$ on $\mathcal{B}(R^d)$ as $n \to \infty$.
(2) There exists a generator L_t of a diffusion process with its coefficients satisfying the conditions of Proposition 3 for which

$$\lim_{n\to\infty} EG\Big(\xi_n(t_1),\dots,\xi_n(t_k)\Big)\Big[g\big(\xi_n(t+h)\big)-g\big(\xi_n(t)\big)-\int_t^{t+h} L_u g\big(\xi_n(u)\big)du\Big] = 0$$

for all $k \in \mathbf{Z}_+$, $0 \leq t_1 \leq \cdots t_k < t < t+h$, and all $G(x_1,\dots,x_k) \in C\big((R^d)^k\big)$ and $g(x) \in C^{(2)}(R^d)$, uniformly in $h \in [0,1]$.
Then $\xi_n(t)$ converges weakly to the diffusion process $\tilde{\xi}(t)$ with the generator L_t for which the distribution of $\tilde{\xi}(0)$ is $m_0(dx)$.

Proof First we show that the sequence $\{\xi_n(t)\}$ satisfies the conditions of Proposition 1. Set

$$\varphi_c(x) = \frac{(x,x)}{c + (x,x)}.$$

Then a constant Q exists that does not depend on c and for which

$$|L_t\varphi_c(x)| \le Q\left[\varphi_c(x) + \frac{1}{\sqrt{c}}\right].$$

As a result,

$$\limsup_{n\to\infty} E\varphi_c(\xi_n(t)) \le \int \varphi_c(x)m_0(dx) + Q\int_0^t \limsup_{n\to\infty} E\varphi_c(\xi_n(s))ds + \frac{1}{\sqrt{c}}Qt.$$

Using the limit relation in condition (2) of the theorem in the case $G = 1$, $t = 0$, $h = t$, $g = \varphi_c$, we have the relation

$$\limsup_{n\to\infty} E\varphi_c(\xi_n(t)) \le \left(\frac{1}{\sqrt{c}}Qt + \int \varphi_c(x)m_0(dx)\right)e^{Qt}.$$

So

$$\begin{aligned}\limsup_{n\to\infty} P\{|\xi_n(t)| > \sqrt{c}\} &\le \limsup_{n\to\infty} P\left\{\varphi_c(\xi_n(t)) > \frac{1}{2}\right\}\\ &\le 2\left(\frac{1}{\sqrt{c}}Qt + \int \varphi_c(x)m_0(dx)\right)e^{Qt}.\end{aligned}$$

The last expression tends to zero as $c \to \infty$. Thus, condition (a) of Proposition 1 is fulfilled. In the same way we can obtain the inequalities

$$\begin{aligned}&\limsup_{n\to\infty} E[1-\varphi_c(\xi_n(t))]\varphi_1(\xi_n(t+h)-\xi_n(t))\\ &\le \limsup_{n\to\infty} E[1-\varphi_c(\xi_n(t))]\int_t^{t+h} |L_s\varphi_1(\xi_n(s)-\xi_n(t))|ds\\ &\le Q_ch,\end{aligned}$$

where

$$Q_c = \sup_{t,x}(1-\varphi_c(x))L_t^y\varphi_1(y-x).$$

The operator L_t^y is the differential operator in y having coefficients $a^i(t,y)$, $b^{ij}(t,y)$. This implies (together with (a)) that condition (b) of Proposition 1 is satisfied.

Let n_l be a subsequence for which the sequence ξ_{n_l} converges weakly to some stochastic process $\tilde{\xi}(t)$. The distribution of $\tilde{\xi}(0)$ is $m_0(dx)$, and $\tilde{\xi}(t)$ is a stochastically continuous process, i.e., $P\{|\tilde{\xi}(t) - \tilde{\xi}(s)| > \varepsilon\} \to 0$ as $s \to t$ for all $\varepsilon > 0$. Using this property, it is easy to check that

$$\begin{aligned}&\lim_{l\to\infty} EG(\xi_{nl}(t_1),\ldots,\xi_{n_l}(t_k))\int_t^{t+h} f(\xi_{n_l}(s))ds\\ &= EG(\tilde{\xi}(t_1),\ldots,\tilde{\xi}(t_k))\int_t^{t+h} f(\tilde{\xi}(s))\,ds\end{aligned}$$

if G satisfies the conditions that are listed in (2) and if f is a bounded continuous function.

So, condition (2) implies that for $0 \le t_1 < \cdots < t_k \le t < t+h$,

$$EG(\tilde{\xi}(t_1), \ldots, \tilde{\xi}(t_k)) \left[g(\tilde{\xi}(t_h)) - g(\tilde{\xi}(t)) - \int_t^{t+h} L_s g(\tilde{\xi}(s)) ds \right] = 0.$$

Set

$$\tilde{\mu}_g(t) = g(\tilde{\xi}(t)) - \int_0^t L_s g(\tilde{\xi}(s)) ds.$$

It follows that

$$E[\tilde{\mu}_g(t+h) - \tilde{\mu}_g(t)] G\left(\tilde{\xi}(t_1), \ldots, \tilde{\xi}(t_k)\right) = 0$$

for all continuous bounded functions G and $t_1 < t_2 < \cdots < t_k \le t$. This means that $\tilde{\mu}_g(t)$ is a martingale. Proposition 4 implies that $\tilde{\xi}(t)$ is a diffusion process with generator L_t. The distribution of $\tilde{\xi}(t)$ is determined by the generator and by the initial distribution, which is $m_0(dx)$. The remainder of the proof follows from Remark 1.

□

Remark 7 *If, in addition, the sequence $\{\xi_n(t)\}$ in Theorem 2 is weakly compact in C, then $\xi_n(t)$ converges weakly in C to the diffusion process $\tilde{\xi}(t)$, which is described in the theorem.*

Remark 8 *The statement of Theorem 2 is true if condition (2) is fulfilled for all functions $g \in \mathcal{D}$, where $\mathcal{D}$ is a dense linear subset in $C^{(2)}(R^d)$. In particular, $\mathcal{D}$ can be chosen as $C^{(3)}(R^d)$.*

Remark 9 *Let V be an open set in R^d and $\tau^n = \inf(t : \xi_n(t) \in V')$, where V' is the boundary of the set V, $\{\xi_n(t)\}$ is a sequence for which condition (1) of Theorem 2 is fulfilled, and $P\{\xi_n(0) \in V\} = 1$. Suppose that L_t is a second-order differential operator that is defined in the set V and that is the generator of a diffusion process $\tilde{\xi}(t)$ in V with absorbing boundary conditions on the V' and*

$$\lim_{n\to\infty} EG(\xi_n(t_1), \ldots, \xi_n(t_k))$$
$$\times \left[g(\xi_n(t+h)) - g(\xi_n(t)) - \int_t^{t+h} L_u g(\xi_n(u))\, du \right] 1_{\{\tau_V^n \le t+h\}} = 0$$

for all $k \in$ mathbfZ$_+$, $0 \le t_1, \ldots, \le t_k \le t < t+h$, $G \in C((R^d)^k)$, $g(x) \in C^{(2)}(R^d)$ for which supp $g \subset V$, *where* supp g *is the closure of the set $\{x : g(x) > 0\}$.* Then the sequence $\xi_n(t \wedge \tau^n)$ converges weakly to the process $\tilde{\xi}(t)$ with initial value $\tilde{\xi}(0)$ having the distribution $m_0(dx)$.

2.3 Central Limit Theorems for Stochastic Processes

We consider central limit theorems for both continuous– and discrete–time processes in this section.

2.3.1 *Continuous–Time Markov Processes*

We consider a homogeneous Markov process $y(t)$ in a measurable space $(Y, \mathcal{C})$ along with its transition probability $P(t, y, C)$, where $t \in R_+$, $y \in Y$, $C \in \mathcal{C}$. We will make some assumptions about the process $y(t)$, which we refer to as *strong mixing conditions (SMC)*.

SMC I. *Suppose that the Markov process $y(t)$ is uniformly ergodic with ergodic distribution $\rho(dy)$. If g is a bounded $\mathcal{C}$-measurable function, for which $g : Y \to R$ and $\int g(y)\rho(dy) = 0$, then*

$$\sup_y \left| \int_{t_1}^{t_2} \int_Y g(y')P(t, y, dy')dt \right| \to 0 \quad as\ t_1, t_2 \to \infty.$$

In this case there exist the limits

$$\lim_{T\to\infty} \int_t^T \int g(y')P(s, y, dy')ds = R^t g(y)$$

and

$$\sup_y \left| R^t g(y) \right| \to 0 \quad as\ t \to \infty.$$

Let

$$R(y, B) = \int_0^\infty \int P(t, y, dy')\left[1_B(y') - \rho(B)\right]dt = \int_0^\infty \left[P(t, y, B) - \rho(B)\right] dt.$$

We note that R is sometimes referred to as a *quasi-potential.*

SMC II.
We assume **SMC I** *and that*

$$\int_0^\infty \sup_y \sup_B \left| P(t, y, B) - \rho(B) \right| dt < \infty.$$

In this case $R(y, A)$ is an additive function of bounded variation in A, and this variation is uniformly bounded in y.

Theorem 3 *Let $g(y)$ be a bounded measurable function for which $\int g(y)\rho(dy) = 0$. Set*

$$\xi_T(t) = \frac{1}{\sqrt{T}} \int_0^{Tt} g(y(s))\, ds.$$

Denote by $w(t)$ the Wiener process with $Ew(t)=0$, $Ew^2(t)=bt$, where $b=2\int g(y)R^0g(y)\rho(dy)$.

(1) *Suppose that $y(t)$ is a random process that satisfies the condition* SMC I. *Then $\xi_T(t)$ converges to $w(t)$ weakly as $T\to\infty$.*

(2) *Suppose that $y(t)$ satisfies the condition* SMC II.

Then $\xi_T(t)$ converges to $w(t)$ weakly in C as $T\to\infty$.

Proof For $f\in C^{(3)}(R)$ we have

$$\begin{aligned} f(\xi_T(t_2))-f(\xi_T(t_1)) &= \int_{t_1}^{t_2} f'(\xi_T(s))\sqrt{T}g(y(Ts))\,ds\\ &= \frac{1}{\sqrt{T}}\int_{t_1T}^{t_2T} f'\left(\xi_T\left(\frac{s}{T}\right)\right)g(y(s))\,ds\\ &= \frac{1}{\sqrt{T}}\int_{t_1T}^{t_2T} f'(\xi_T(t_1))g(y(s))\,ds\\ &+\frac{1}{T}\int_{t_1T}^{t_2T} f''\left(\xi_T\left(\frac{u}{T}\right)\right)\int_u^{t_2T} g(y(u))g(y(s))\,ds\,du. \end{aligned}$$

Denote by $\mathcal{F}_t^T$ the filtration generated by $\xi_T(t)$. Then

$$\begin{aligned} &E\bigg(f(\xi_T(t_2))-f(\xi_T(t_1))-\frac{1}{2}\int_{t_1}^{t_2} bf''(\xi_T(s))\,ds/\mathcal{F}_{t_1}^T\bigg)\\ &=E\bigg(\frac{1}{\sqrt{T}}f'(\xi_T(t_1))[R^0g(y(t_1T))-R^{t_2T}g(y(t_1,T))]\\ &+\frac{1}{T}\int_{t_1T}^{t_2T} f''(\xi_T(u))\left(g(y(u))R^0g(y(u))-\frac{1}{2}b\right)du\\ &-\frac{1}{T}\int_{t_1T}^{t_2T} f''(\xi_T(u))R^{t_2T-u}g(y(u))\,du/\mathcal{F}_{t_1}^T\bigg)\\ &=E\left(\frac{1}{T}\int_{t_1T}^{t_2T} f''(\xi_T(u))\left(g(y(u)R^0g(y(u)))-\frac{1}{2}b\right)du/\mathcal{F}_{t_1}^T\right)\\ &+O\left(\frac{1}{\sqrt{T}}+\frac{1}{T}\int_0^{T(t_2-t_1)}\sup_y|R^ug(y)|\,du\right). \end{aligned}$$

The expectation in the last expression tends to zero as $T\to\infty$ because of *SMC* I and the uniform ergodicity of $y(t)$, which implies that

$$E\left(\left|\frac{1}{L}\int_s^{s+L} g(y(u))R^0g(y(u))du-\frac{1}{2}b\right|\Big/ y(s)=y\right)\longrightarrow 0$$

uniformly in s,y as $L\to\infty$.
The remaining expression tends to zero as $T\to\infty$. With this, the proof of statement (1) follows from Theorem 2 and Remark 5.

Suppose that SMC II is true. We will calculate $E\big((\xi_T(t_2)-\xi_T(t_1))^4/\mathcal{F}_{t_1}^T\big)$. Let

$$q_h(T,y) = E\big(\xi_T^4(h)/y(0)=y\big).$$

Then

$$E\big((\xi_T(t_2)-\xi_T(t_1))^4/\mathcal{F}_{t_1}^T\big) = q_{t_2-t_1}\big(T, y(Tt_1)\big).$$

On the other hand, we have

$$q_h(T,y) = \frac{24}{T^2}\int_{0<t_1<t_2<t_3<t_4\le Th} \iiiint g(y_1)g(y_2)g(y_3)g(y_4)$$
$$\times P(t_1,y,dy_1)P(t_2-t_1,y_1,dy_2)P(t_3-t_2,y_2,dy_3)P(t_4-t_3,y_3,dy_4).$$

Set $Q(u,y,dy') = P(u,y,dy')-\rho(dy')$. Using the relation $\int g(y')\rho(dy')=0$, we can rewrite the expression for $q_h(T,y)$ in the form

$$q_h(T,y) = \frac{24}{T^2}\int_{0<t_1<t_2<t_3<t_4\le Th} \iiiint g(y_1)g(y_2)g(y_3)g(y_4)$$
$$\times\Big\{\rho(dy_1)Q(t_2-t_1,y_1,dy_2)\rho(dy_3)Q(t_4-t_3,y_3,dy_4)$$
$$+Q(t_1,y,dy_1)\rho(y_2)Q(t_3-t_2,y_2,dy_3)Q(t_4-t_3,y_3,dy_4)$$
$$+\rho(dy_1)Q(t_2-t_1,y_1,dy_2)Q(t_3-t_2,y_2,dy_3)Q(t_4-t_3,y_3,dy_4)$$
$$+Q(t_1,y,dy_1)Q(t_2-t_1,y_1,dy_2)\rho(dy_3)Q(t_4-t_3,y_3,dy_4)$$
$$+Q(t_1,y,dy_1)Q(t_2-t_1,y_1,dy_2)Q(t_3-t_2,y_2,dy_3)Q(t_4-t_3,y_3,dy_4)\Big\}$$
$$\le C\Big(h^2+\frac{1}{T^2}\Big).$$

The fact that $\int\int\big|Q(u,y,dy')g(y)\big|du<\infty$ implies

$$E\big|\xi_T(t_2)-\xi_T(t_1)\big|^4 \le C\Big(|t_1-t_2|^2+\frac{1}{T^2}\Big).$$

Since $\big|\xi_T(t_1)-\xi_T(t_2)\big|\le C_1\sqrt{T}|t_1-t_2|$, the sequence $\xi_T(t)$ is compact in C because of Theorem 1.

□

2.3.2 *Discrete–Time Markov Processes*

We consider a Y-valued homogeneous Markov process $\{y_n^*, n=0,1,\dots\}$ with transition probability for one step being $P(y,B)$ for $y\in Y$ and $B\in\mathcal{C}$. Denote by $P_n(y,B)$ the transition probability for n steps. The strong mixing conditions for discrete time are analogous to the continuous–time versions:

SMC I. *The process $\{y_n^*\}$ is ergodic with ergodic distribution $\rho(dy)$. If g is a bounded $\mathcal{C}$-measurable function $Y \to R$ and $\int g(y)\rho(dy) = 0$, then*

$$\lim_{n_1,n_2\to\infty} \sup_y \left| \sum_{n_1\le k\le n_2} \int g(y')P_k(y,dy')\right| = 0.$$

In this case there exists the sum

$$R^\circ g(y) = \sum_{k=1}^{\infty} \int P_k(y,dy')g(y'),$$

and $\sup_y \left|R^\circ g(y)\right| < \infty$.

SMC II. *Condition* **SMC I** *is fulfilled and*

$$\sum_{k=1}^{\infty} \sup_y \sup_{B\in\mathcal{C}} \left|P_k(y,B) - \rho(B)\right| < \infty.$$

The next theorem is a discrete–time analogue of Theorem 3.

Theorem 4 *Let $g(y)$ be a $\mathcal{C}$-measurable bounded function for which*

$$\int g(y)\rho(dy) = 0.$$

Set

$$\xi_n(t) = \frac{1}{\sqrt{n}} \int_0^{tn} g(y^*(s))\,ds,$$

where $y^*(s) = y_k^*$ for $s \in [k, k+1)$. Let $w(t)$ be the Wiener process for which $Ew(t) = 0$ and $Ew^2(t) = bt$, where

$$b = \int (g^2(y) + 2g(y)R^0 g(y))\rho(dy).$$

Then the following statements hold:

I. *If $\{y_n^*\}$ satisfies condition , then $\xi_n(t)$ converges weakly to $w(t)$.*

II. *If $\{y_n^*\}$ satisfies condition* **SMC II**, *then $\xi_n(t)$ converges to $w(t)$ weakly in C.*

The proof of this statement can be obtained in the same way as the proof of Theorem 3.

2.3.3 Discrete–Time Stationary Processes

We consider here an ergodic stationary process $\{\xi_k,\ k = \pm 1, \pm 2, \dots\}$. We suppose that $E\xi_k = 0$ and introduce the process

$$\xi_n(t) = \frac{1}{\sqrt{n}} \int_0^{tn} \xi^*(s)\,ds,$$

where $\xi^*(s) = \xi_k$ for $k \le s < k+1$.
We will find conditions under which $\xi_n(t)$ converges weakly in C to a Wiener process. Denote by $\mathcal{F}_k$ the σ-algebra generated by $\{\xi_l,\, l \le k\}$.

Theorem 5 *Let $\{\xi_k\}$ satisfy, in addition, the following two conditions:*
(1) there exists $\delta > 0$ for which $E|\xi_0|^{2+\delta} < \infty$,
(2) the series $\sum_{k=1}^{\infty} E(\xi_k/\mathcal{F}_0)$ converges in $L_2(P)$.
Set

$$\eta_0 = \sum_{k=1}^{\infty} E(\xi_k/\mathcal{F}_0), \quad b = E(\xi_0^2 + 2\xi_0\eta_0).$$

Then $\xi_n(t)$ converges weakly in C to a Wiener process $w(t)$ for which $Ew(t) = 0$, $Ew^2(t) = bt$.

Proof We introduce the notation: $S_k = \xi_1 + \cdots + \xi_k$, $\eta_k = \sum_{i=k+1}^{\infty} E(\xi_i/\mathcal{F}_k)$, and $\eta_k^l = E(S_l - S_k/\mathcal{F}_k)$. Let $f \in C^{(3)}(R)$. Then for all $t > 0$,

$$\lim_{n\to\infty} \sup_{m<nt} E\left|E\left(f\left(\frac{1}{\sqrt{n}}S_m\right) - f(0) - \frac{1}{2}b\frac{1}{n}\sum_{k=0}^{m-1} f''\left(\frac{1}{\sqrt{n}}S_k\right)\Big/\mathcal{F}_0\right)\right| = 0. \tag{2.4}$$

To prove this we define functions

$$\alpha(z,x) = \frac{1}{x^2}\left[f(z+x) - f(z) - f'(z)x - \frac{1}{2}f''(z)x^2\right],$$
$$\beta(z,x) = \frac{1}{x}\left[f'(z+x) - f'(z) - f''(z)x\right].$$

Then

$$f\left(\frac{1}{\sqrt{n}}S_k\right) - f(0) = \sum_{k=0}^{m-1}\left(f\left(\frac{1}{\sqrt{n}}S_{k+1}\right) - f\left(\frac{1}{\sqrt{n}}S_i\right)\right)$$
$$= \sum_{k=0}^{m-1}\left(f'\left(\frac{1}{\sqrt{n}}S_k\right)\frac{\xi_{k+1}}{\sqrt{n}} + \frac{1}{2}f''\left(\frac{1}{\sqrt{n}}S_k\right)\frac{\xi_{k+1}^2}{n} + \alpha\left(\frac{S_k}{\sqrt{n}}, \frac{\xi_{k+1}}{\sqrt{n}}\right)\frac{\xi_{k+1}^2}{n}\right)$$
$$= \sum_{k=0}^{m-1}\left[\frac{\xi_{k+1}}{\sqrt{n}}\left(f'(0) + \sum_{i=0}^{k-1}\left(f''\left(\frac{1}{\sqrt{n}}S_i\right)\frac{\xi_{i+1}}{\sqrt{n}} + \beta\left(\frac{1}{\sqrt{n}}S_i, \frac{\xi_{i+1}}{\sqrt{n}}\right)\right)\right)\right.$$
$$\left. + \frac{1}{2}f''\left(\frac{1}{\sqrt{n}}S_k\right)\frac{\xi_{k+1}^2}{n} + \alpha\left(\frac{1}{\sqrt{n}}S_k, \frac{\xi_{k+1}}{\sqrt{n}}\right)\frac{\xi_{k+1}^2}{n}\right].$$

Therefore,

$$\mathcal{D}_m = E\biggl(f\Bigl(\frac{1}{\sqrt{n}}S_m\Bigr) - f(0) - \frac{b}{2n}\sum_{k=0}^{m-1} f''\Bigl(\frac{1}{\sqrt{n}}S_k\Bigr)\Big/\mathcal{F}_0\biggr)$$

$$= \frac{1}{\sqrt{n}}f'(0)\eta_0^m + E\biggl(\sum_{k=0}^{m-1}\alpha\Bigl(\frac{1}{\sqrt{n}}S_k, \frac{\xi_{k+1}}{\sqrt{n}}\Bigr)\frac{\xi_{k+1}^2}{n}\Big/\mathcal{F}_0\biggr)$$

$$+E\biggl(\sum_{k=0}^{m-2}\beta\Bigl(\frac{1}{\sqrt{n}}S_k, \frac{\xi_{k+1}}{\sqrt{n}}\Bigr)\frac{1}{n}\xi_k\eta_k^m\Big/\mathcal{F}_0\biggr)$$

$$+E\biggl(\frac{1}{n}\sum_{k=0}^{m-1} f''\Bigl(\frac{1}{\sqrt{n}}S_k\Bigr)[\xi_k\eta_k^m + \frac{1}{2}\xi_k^2 - \frac{1}{2}b]\Big/\mathcal{F}_0\biggr)$$

and

$$E|\mathcal{D}_m| \le O\Bigl(\frac{1}{\sqrt{n}}\Bigr) + \frac{1}{n}\sum_{k=0}^{m-2} E\biggl(\Bigl|\alpha\Bigl(\frac{1}{\sqrt{n}}S_k, \frac{\xi_{k+1}}{\sqrt{n}}\Bigr)\Bigr|\xi_{k+1}^2$$

$$+\Bigl|\beta\Bigl(\frac{1}{\sqrt{n}}S_k, \frac{\xi_{k+1}}{\sqrt{n}}\Bigr)\Bigr||\xi_{k+1}||\eta_k^m|\biggr) + E|\mathcal{D}_m^*|,$$

where

$$\mathcal{D}_m^* = E\biggl(\frac{1}{n}\sum_{k=0}^{m-1} f''\Bigl(\frac{1}{n}S_k\Bigr)\Bigl[\frac{1}{2}\xi_k^2 + \xi_k\eta_k^m - \frac{1}{2}b\Bigr]\Big/\mathcal{F}_0^T\biggr).$$

Using the relation

$$|\alpha(z,x)| + |\beta(z,x)| \le C(|x| \wedge 1),$$

where $c > 0$, and using condition (1) of the theorem, we can prove that

$$\lim_{n\to\infty}\frac{1}{n}\sum_{k=0}^{m-2} E\biggl(\Bigl|\alpha\Bigl(\frac{1}{\sqrt{n}}S_k, \frac{\xi_{k+1}}{\sqrt{n}}\Bigr)\Bigr|\xi_k^2 + \Bigl|\beta\Bigl(\frac{1}{\sqrt{n}}S_k, \frac{\xi_{k+1}}{\sqrt{n}}\Bigr)\Bigr||\xi_k||\eta_k^h|\biggr) = 0$$

if $m/n \le t$. It follows from the ergodic theorem (Chapter 1, Theorem 1′) that

$$\frac{1}{l}E\biggl|\sum_{k=m}^{m+l}\Bigl(\frac{1}{2}\xi_k^2 + \xi_k\eta_k^m - \frac{1}{2}b\Bigr)\biggr| \longrightarrow 0 \quad \text{as } l \to \infty$$

uniformly in m (this expectation does not depend on m!). This implies that

$$E|\mathcal{D}_m^*| \longrightarrow 0 \quad \text{as } n \to \infty, \quad m \le nt.$$

So relation (2.4) is proved. This relation can be rewritten in the form

$$\lim_{T\to\infty}\sup_{t\le t_0} E\biggl|E\Bigl(f(\xi_T(t)) - f(\xi_T(0)) - \int_0^t \frac{1}{2}bf''(\xi_T(s))ds\Big/\mathcal{F}_0^T\Bigr)\biggr| = 0 \quad (2.5)$$

for $t_0 > 0$, $f \in C^{(3)}(R^d)$. Since $\{\xi_k\}$ is a stationary discrete–time stochastic process, relation (2.5) implies that for all $t_1 < t_2$,

$$\lim_{T\to\infty} E\left|E\left(f(\xi_T(t_2)) - f(\xi_T(t_1)) - \int_{t_1}^{t_2} \frac{1}{2} bf''(\xi_T(s))ds \Big/ \mathcal{F}_{t_1}^T\right)\right| = 0.$$

So Theorem 2 implies that the stochastic process $\xi_T(t)$ converges weakly to $w(t)$ as $T \to \infty$.

To prove the theorem we prove that the family of stochastic processes $\{\xi_T(t), T > 0\}$ is weakly compact in C. Set $\Psi(x) = |x|^\gamma$ where $2 < \gamma \le (2 + \delta/2) \wedge 3$. Then

$$\psi'(x) = \gamma|x|^{\gamma-1}x, \qquad \psi''(x) = \gamma(\gamma-1)|x|^{\gamma-2},$$

and

$$|\alpha_\psi(x,y)| \le c(|x|^{\gamma-2} + |y|^{\gamma-2}),$$
$$|\beta_\psi(x,y)| \le c(|x|^{\gamma-2} + |y|^{\gamma-2}),$$

where $c > 0$ is a constant, and the functions α_ψ and β_ψ are determined in the same way as α and β for the function Ψ. We can obtain the following for some positive constants c_1, c_2:

$$E\Psi\left(\frac{1}{\sqrt{n}}S_m\right) \le \frac{c_1}{n} E \sum_{k=0}^{m-1} \left(\left|\frac{S_k}{\sqrt{n}}\right|^{\gamma-2} (\xi_{k+1}^2 + |\xi_{k+1}\eta_k^m|) + \frac{E|\xi_{k+1}|^\gamma}{n^{\gamma/2-1}}\right)$$
$$\le \frac{c_2}{n} \sum_{k=1}^{m-1} E\Psi\left(\frac{1}{\sqrt{n}}S_k\right) + \frac{C_2}{n^{\gamma/2}} m.$$

This inequality implies that for $m \le nh$ and $h \le 1$,

$$E\Psi\left(\frac{1}{\sqrt{n}}S_m\right) \le C_3\left(h^{\gamma/2} + \frac{1}{n^{\gamma/2}}\right),$$

where C_3 is a constant. This implies the inequality

$$E|\xi_n(t_2) - \xi_n(t_1)|^\gamma \le C_3\left(h^{\gamma/2} + \frac{1}{n^{\gamma/2}}\right).$$

So the weak compactness of the sequence $\{\xi_n(t)\}$ follows from the corollary to Theorem 1.

□

2.3.4 Continuous–Time Stationary Processes

We now consider an ergodic stationary stochastic process $\{\xi(t),\ t \in R\}$ that satisfies the following properties:

(1) $E\xi(t) = 0$, $E\big|\xi(t)\big|^{2+\delta} < \infty$ for some $\delta > 0$;

(2) $\int_0^t E(\xi(s)/\mathcal{F}_0)ds$ converges in $L_2(P)$, as $t \to \infty$, to some random variable η_0, where $\mathcal{F}_0$ is the σ-algebra generated by $\{\xi(t), t \le 0\}$.

Set

$$\xi_T(t) = \frac{1}{\sqrt{T}} \int_0^{Tt} \xi(u)du, \quad T > 0.$$

Theorem 6 *$\xi_T(t)$ converges weakly in C, as $T \to \infty$, to a Wiener process $w(t)$ with $Ew(t) = 0$, $Ew^2(t) = bt$, where $b = E\eta_0\xi(0)$.*

Proof Set $\hat{\xi}_k = \int_k^{k+1} \xi(u)du$. It is easy to check that the $\hat{\xi}_k$ satisfy the conditions of Theorem 5. Let us define

$$\hat{\xi}_n(t) = \frac{1}{\sqrt{n}} \int_0^{nt} \hat{\xi}^*(u)du,$$

where $\hat{\xi}^*(u) = \hat{\xi}_k$ if $k \le u < k+1$. It follows from Theorem 5 that $\hat{\xi}_n(t)$ converges weakly in C to $w(t)$ as $n \to \infty$. We can check that for any t_0 and $\epsilon > 0$,

$$\lim_{T\to\infty} P\left\{ \sup_{s \le t_0} \left|\xi_T(s) - \hat{\xi}_{[T]}(s)\right| > \epsilon \right\} = 0,$$

where $[T]$ is the integer part of T.

□

2.4 Large Deviation Theorems

Large deviation theorems are related to the estimation of probability that the difference between the time–averaged value of some ergodic process and its expectation is significant. The first such theorem was proved by H. Cramer in 1938 [24]. Let $\xi_k, k = 1, 2, \dots$, be a sequence of independent identically distributed random variables for which

$$E \exp\{t|\xi_1|\} < \infty$$

for some $t > 0$. Then the relations

$$\lim_n \frac{1}{n} \log P\left\{ \sum_{k \le n} \xi_k > nx \right\} = -H(x),\ x > E\xi_1$$

and

$$\lim_n \frac{1}{n} \log P\left\{ \sum_{k \le n} \xi_k < nx \right\} = -H(x),\ x < E\xi_1$$

hold, where

$$H(x) = \sup_t \left(tx - \log Ee^{t\xi_1}\right).$$

2.4.1 Continuous–Time Markov Processes

Let X be a compact metric space with metric r, and let $\mathcal{B}(X)$ be the Borel σ–algebra in X. Denote by $\mathcal{M}(X)$ the space of probability measures on $\mathcal{B}(X)$, and let d be a distance in $\mathcal{M}(X)$ for which convergence in d is equivalent to the weak convergence of measures. Denote by $C(X)$ the space of continuous functions $f : X \to R$ with the norm

$$\|f\| = \sup_{x \in X} |f(x)|.$$

We consider a homogeneous Markov process $x(t) \in X, t \in R_+$, with the transition probability function

$$P(t, x, B), \;\; t \in R_+, \, x \in X, \, B \in \mathcal{B}(X).$$

We assume that the process satisfies Feller's condition:
(**FC**)
(a) For a ball $B_\delta(x) \subset X$, the formula

$$\lim_{t \downarrow 0} P(t, x, B_\delta(x)) = 1$$

holds,
and
(b) the function

$$T_t f(x) = \int f(x') P(t, x, dx') \tag{2.6}$$

is continuous in x for all $f \in C(X)$.

If condition (FC) holds, then formula (2.6) defines the family of the operators $T_t, t > 0$, in the space $C(X)$, which is called the semigroup associated with the Markov process. We define the generator A of the semigroup $T_t, t > 0$, by the relation

$$Af(x) = \lim_{t \downarrow 0} \frac{1}{t}(T_t f(x) - f(x)) \tag{2.7}$$

on all $f \in C(X)$ for which the limit on the right-hand side of formula (2.7) exists uniformly for $x \in X$. The set of such functions f is the domain of definition of the operator A, and it is denoted by $\mathcal{D}(A)$. It is easy to see that if $f \in \mathcal{D}(A)$, then $T_t f \in \mathcal{D}(A)$ for all $t > 0$, and we have the following formulas:

$$\frac{d}{dt} T_t f(x) = A T_t f(x) = T_t A f(x),$$

$$T_t f(x) - f(x) = \int_0^t T_s A f(x) ds.$$

This implies that the stochastic process given by the relation

$$\xi_t(f) = f(x(t)) - f(x(0)) - \int_0^t Af(x(s)) ds$$

is a martingale with respect to the filtration $\{\mathcal{F}_t, t > 0\}$ generated by the stochastic process $x(t)$ for all $f \in \mathcal{D}(A)$.

Lemma 2 *Let $\phi \in \mathcal{D}(A)$ be a nonnegative function. Set*

$$\mu_t(\phi) = \phi(x(t)) \exp\left\{-\int_0^t (\phi(x(s)))^{-1} A\phi(x(s))ds\right\}. \tag{2.8}$$

The function $\mu_t(\phi)$ is a martingale with respect to the filtration $\{\mathcal{F}_t, t > 0\}$.

Proof We can represent the process $\mu_t(\phi)$ in the form

$$\mu_t(\phi) = (\xi_t(\phi) + \int_0^t A\phi(x(s))ds) \exp\left\{-\int_0^t (\phi(x(s))^{-1} A\phi(x(s))ds\right\}.$$

So $\mu_t(\phi)$ is a semimartingale, and we can write the relation

$$d\mu_t(\phi) = \exp\{-\int_0^t (\phi(x(s))^{-1} A\phi(x(s))ds\} d\xi_t(\phi).$$

With this, the lemma is proved.

□

Denote by $CL(X)$ the space of all right–continuous functions $z(\cdot) : R_+ \to X$ having a left limit $z(t-)$ for all $t > 0$. Condition (FC) implies the existence of the measure Q on the σ–algebra $\mathcal{B}(CL(X))$ for which

$$P\{x(\cdot) \in \widetilde{C}\} = Q(\widetilde{C}), \quad \widetilde{C} \in \mathcal{C}(X),$$

where $\mathcal{C}(X) = \bigcup_t \mathcal{C}_t(X)$ and $\mathcal{C}_t(X)$ is the algebra generated by the sets

$$\mathcal{C}_s(B) = \{z(\cdot) \in X^{R_+} : z(s) \in B\}, \quad s \le t, \ B \in \mathcal{B}(X).$$

Let $\phi \in \mathcal{D}(A)$, $\phi > 0$. Introduce a probability measure on $\mathcal{B}(CL(X))$ defined by the relation

$$Q_\phi(\widetilde{C}) = E1_{\{x(\cdot)\in\widetilde{C}\}} \frac{1}{\phi(x(0))} \mu_t(\phi), \quad \widetilde{C} \in \mathcal{C}_t(X). \tag{2.9}$$

With these preliminaries, we have the following lemma:

Lemma 3 *The stochastic process $z(t)$ having trajectories in the space $CL(X)$ and distribution Q_ϕ is a homogeneous Markov process satisfying Feller's condition (FC) and having generator*

$$A_\phi f = \frac{1}{\phi}(A(\phi f) - fA\phi), \quad \mathcal{D}(A_\phi) = \{f \in C(X) : \phi f \in \mathcal{D}(A)\}.$$

Proof It suffices to prove that for all $f \in \mathcal{D}(A_\phi)$ the stochastic process

$$\xi_t^*(f) = (f(x(t)) - \int_0^t A_\phi f(x(s))ds)\mu_t(\phi)$$

is a martingale. We introduce the notation $g = \phi f$,

$$\Phi_t = (\phi(x(t)))^{-1} \mu_t(\phi),$$

and

$$\Psi_t = \int_0^t \frac{Ag(x(s)) - g(x(s))A\phi(x(s))}{\phi(x(s))} ds.$$

Then

$$\xi_t^*(f) = \Phi_t g(x(t)) - \mu_t \Psi_t.$$

So

$$d\xi_t^*(f) = \Phi_t d\xi_t(g) - \Psi_t d\mu_t(\phi).$$

With this, the lemma is proved.

□

We define a function $I : \mathcal{M}(X) \to [0, +\infty]$ by the relation

$$I(m) = -\inf\left\{\int (\phi(x))^{-1} A\phi(x) m(dx) : \phi \in \mathcal{D}(A), \phi > 0\right\}. \tag{2.10}$$

This function has the properties listed in the next lemma.

Lemma 4 *(i) $I(m) \geq 0$ for all $m \in \mathcal{M}(X)$, and $I(m) = 0$ if and only if m is an invariant measure for the Markov process,*
(ii) $I(m)$ is a convex function, i.e.,

$$I(sm_1+(1-s)m_2) \leq sI(m_1)+(1-s)I(m_2), \;\; m_i \in \mathcal{M}(X), i = 1, 2, \;\; 0 < s < 1.$$

(iii) $I(m)$ is a lower continuous function, i.e.,

$$\liminf_{m_n \to m} I(m_n) \geq I(m).$$

(iv) for any $b > 0$ the set

$$C_b(I) = \{m : I(m) \leq b\}$$

is compact, and the function $I(m)$ is continuous on this compact set.

Proof Note that the function $\phi =$ constant satisfies the relation $A\phi = 0$, so $I(m) \geq 0$. Let ρ be an invariant measure for the Markov process; this means that

$$\int P(t, x, B)\rho(dx) = \rho(B), \;\; B \in \mathcal{B}(X), \, t > 0.$$

Then

$$\int T_t f(x)\rho(dx) = \int f(x)\rho(dx), \;\; f \in C(X), \, t > 0.$$

It is easy to see that

$$\int (\phi(x))^{-1} A\phi(x) m(dx) = \lim_{t \downarrow 0} \frac{1}{t} \int \log \frac{T_t\phi(x)}{\phi(x)} m(dx).$$

Jensen's inequality can be used to show that

$$\int (\log T_t\phi(x) - \log \phi(x))\rho(dx) \geq \int (T_t \log \phi(x) - \log \phi(x))\rho(dx) = 0.$$

So

$$\int (\phi(x))^{-1} A\phi(x)\rho(dx) \geq 0, \quad \phi \in \mathcal{D}(A),\, \phi > 0.$$

This implies that $I(\rho) = 0$.
On the other hand, suppose that $I(m) = 0$. Then

$$I(m, \phi) = \int (\phi(x))^{-1} A\phi(x) m(dx) \geq 0$$

for all $\phi \in \mathcal{D}(A)$. This implies that

$$\int T_t \log \phi(x) m(dx) \geq \int \log \phi(x) m(dx), \quad \phi \in \mathcal{D}(A),\, t > 0.$$

So

$$\int T_t f(x) m(dx) \geq \int f(x) m(dx), \quad f \in C(X),$$

since $\mathcal{D}(X)$ is dense in $C(X)$, and m is an invariant measure. With this, statement (i) is proved.
Statement (ii) follows from the definition of the function $I(m)$.
Now we note that the function $I(m, \phi)$ is continuous in m for every $\phi > 0$ with $\phi \in \mathcal{D}(A)$, so the function

$$\inf\{I(m, \phi) : \phi > 0,\, \phi \in \mathcal{D}(A)\}$$

is upper semicontinuous. This implies statement (iii), from which follows that the set $C_b(I)$ is closed. The function $I(m)$ is continuous on the compact set $C_b(I)$ as a bounded convex function.
This completes the proof of the lemma.

□

Now we introduce a random measure in $\mathcal{B}(X)$ by the relation

$$\nu_t(B) = \frac{1}{t} \int_0^t 1_{\{x(s) \in B\}} ds, \quad B \in \mathcal{B}(X).$$

Theorem 7 *For any closed set $F \subset \mathcal{B}(X)$ we have that*

$$\limsup_{t \to \infty} \frac{1}{t} \log P\{\nu_t \in F\} \leq -\inf\{I(m) : m \in F\}.$$

Proof For any $\phi \in \mathcal{D}(A)$ with $\phi > 0$ we can write the inequality

$$\begin{aligned} E\phi(x(0)) = E\mu_t(\phi) &\geq \inf_{x \in X} \phi(x) E \exp\{-tI(\nu_t, \phi)\} \\ &\geq cE1_{\{\nu_t \in F\}} \inf_{m \in F} \exp\{-tI(m, \phi)\}, \end{aligned}$$

where $c > 0$ is a constant. This implies the inequality

$$\limsup_{t\to\infty} \frac{1}{t} \log P\{\nu_t \in F\} \le \sup_{m\in F} I(m,\phi).$$

Using the inequality

$$\inf\{\sup_{m\in F} I(m,\phi) : \phi \in \mathcal{D}(A), \phi > 0\}$$
$$\le \sup_{m\in F} \inf\{I(m,\phi) : \phi \in \mathcal{D}(A), \phi > 0\} = \sup_{m\in F}(-I(m))$$

we obtain the proof.

□

Note that condition (FC) implies the existence of invariant measures for the Markov process. The process is ergodic if there exists a unique invariant measure. Next, we present a sufficient condition of ergodicity.

Lemma 5 *Assume that there is a probability measure π for which $P(t,x,\cdot) \ll \pi$ for all $t > 0$ and $x \in X$, (i.e., this measure is absolutely continuous with respect to π and*

$$\inf_{x\in X} \int_0^1 \frac{dP(s,x,\cdot)}{d\pi}(x')ds > 0$$

for almost all x' with respect to the measure π). Then the Markov process is ergodic, and its ergodic distribution ρ satisfies the condition $\rho \ll \pi$.

The proof follows from the observation that the conditions of the lemma imply that the Markov process is Harris recurrent.
If the conditions of Lemma 5 are fulfilled, then the same condition holds for the Markov process whose distribution Q_ϕ is given by the formula (2.9). Denote by ρ_ϕ the ergodic distribution for the Markov process corresponding to the measure Q_ϕ. Then

$$I(\rho_\phi) + I(\rho_\phi, \phi) = 0. \tag{2.11}$$

The proof of the last formula follows from the relations

$$I_\phi(m) = -\inf\left\{\int \frac{A_\phi g(x)}{g(x)} m(dx) : g \in \mathcal{D}(A_\phi), g > 0\right\} = I(m) + I(m,\phi)$$

and $I_\phi(\rho_\phi) = 0$, which follows from statement (i) of Lemma 4.

Theorem 8 *Let the transition probability $P(t,x,B)$ satisfy the conditions of Lemma 5. Then for every open set $G \in \mathcal{B}(X)$ we have that*

$$\liminf_{t\to\infty} \frac{1}{t} \log P\{\nu_t \in G\} \ge -\inf_{m\in G} I(m).$$

Proof Let $\phi \in \mathcal{D}(A)$ and $\phi > 0$, where ρ_ϕ is the same as in formula (2.11). Assume that $\rho_\phi \in G$ and that $\delta > 0$ satisfies the relations $B_\delta(\rho_\phi) \subset G$.

The ergodic theorem implies that

$$\lim_{t\to\infty} Q_\phi\{\tilde{\nu}_t \in B_\delta(\rho_\phi)\} = 1,$$

where

$$\tilde{\nu}_t(B) = \frac{1}{t}\int_0^t 1_{\{z(s)\in B\}} ds, \ \ z(\cdot) \in CL(X), \ \ B \in \mathcal{B}(X).$$

It follows from formula (2.11) that

$$\begin{aligned} Q_\phi\{\tilde{\nu}_t \in B_\delta(\rho_\phi)\} &= E1_{\{\nu_t\in G\}}\frac{\phi(x(t))}{\phi(\phi(0))}\exp\{-tG(\nu_t,\phi)\} \\ &\le c\exp\{-t(G(\rho_\phi,\phi)+\theta_\delta)\}P\{\nu_t \in G\}, \end{aligned}$$

where $c > 0$ is a constant, and $\theta_\delta \to 0$ as $\delta \to 0$. Using formula (2.8) we can write the inequality

$$\liminf_{t\to\infty}\frac{1}{t}\log P\{\nu_t \in G\} \ge \lim_{\delta\to 0}\liminf_{t\to\infty}\frac{1}{t}\log P\{\nu_t \in B_\delta(\rho_\phi)\} \ge -I(\rho_\phi).$$

This completes the proof.

□

2.4.2 Discrete–Time Markov Processes

Now we consider a homogeneous Markov chain $\{\xi_n, n = 0, 1, \dots\}$ with transition probability for one step, $P(x, B), x \in X, B \in \mathcal{B}(X)$, satisfying the following conditions:

(1) $\int f(x')P(x, dx') \in C(X)$ for all $f \in C(X)$,

(2) the Markov chain is ergodic with ergodic distribution ρ and condition **SMCII** of Section 2.3(b) is satisfied.

For $\phi \in C(X), \phi > 0$, set

$$S\phi(x) = \log\int \phi(x')P(x, dx') - \log\phi(x).$$

Introduce the function

$$I(m) = -\inf\left\{\int S\phi(x)m(dx) : \phi \in C(X), \phi > 0\right\}, \ \ m \in \mathcal{M}(X).$$

We can verify that for this function all of the statements of Lemma 4 hold. Let the random measure ν_n be defined by the relation

$$\nu_n(B) = \frac{1}{n}\sum_{k<n} 1_{\{\xi_k\in B\}}, \ \ B \in \mathcal{B}(X).$$

The following theorem is a main result in large deviation theory.

Theorem 9 *Let $F \subset \mathcal{B}(X)$ be a closed set, and $G \subset \mathcal{B}(X)$ be an open set. Then the relations*

$$\limsup_n \frac{1}{n} \log P\{\nu_n \in F\} = -\inf_{m \in F} I(m)$$

and

$$\liminf_n \frac{1}{n} \log P\{\nu_n \in G\} = -\inf_{m \in G} I(m)$$

hold.

The first relation can be proved in the same way as Theorem 7. The second one we can prove in the same way as Theorem 8 using instead of the martingale $\mu_t(\phi)$, the martingale

$$\mu_n(\phi) = \phi(\xi_n) \exp\left\{-\sum_{k<n} S\phi(\xi_k)\right\}, \quad n > 0.$$

This completes our review of convergence results for stochastic processes.

3
Averaging

In this chapter we consider random perturbations of Volterra integral equations, of differential equations, and of difference equations, and we develop an averaging method for each. In each case there is an equation of the form $F_\varepsilon(x(\cdot),\omega)=0$, where F_ε is an operator acting on functions $x(t)$. This equation is to be solved for a function $x=x_\varepsilon(t,\omega)$, where ω is a sample from a probability space on which the random perturbations are defined and ε is a small positive parameter. We average this equation over the probability space by defining $\bar{F}(x(\cdot)) = EF(x(\cdot),\omega)$, and we use ergodic theorems to show how the limit as $\varepsilon\to 0$ of the perturbed problem is related to the averaged problem $\bar{F}(\bar{x}(\cdot))=0$. The results in this chapter show how to derive, under natural conditions, convergence properties of $x_\varepsilon(t,\omega)-\bar{x}(t)$ as $\varepsilon\to 0$. In general, this error will approach zero in a probabilistic sense that is made precise in each case.

3.1 Volterra Integral Equations

We consider first an equation of the form

$$x(t)=\phi(t)+\int_0^t K(t,x(s),s)\,ds, \tag{3.1}$$

where ϕ is a continuous, bounded function $R_+\to R^d$ and $K(t,x,s)$ is a continuous function from $R_+\times R^d\times R_+\to R^d$. We seek a solution $x(\cdot)\in C_{R_+}(R^d)$. This equation has a unique solution if the kernel $K(t,x,s)$

satisfies a Lipschitz condition, that is, for some constant $l > 0$,

$$|K(t,x,s) - K(t,x',s)| \leq l|x - x'|,$$

for all relevant t, s, x, and x'.

A stochastic Volterra integral equation has the form

$$x(t,\omega) = \phi(t,\omega) + \int_0^t K(t, x(s,\omega), s, \omega)\, ds, \tag{3.2}$$

where $\phi(t,\omega) : R_+ \times \Omega \to R^d$ is a stochastic process and $K(t,x,s,\omega) : R_+ \times R^d \times R_+ \times \Omega \to R^d$ is a random function. It is easy to see that if $\phi(t,\omega)$ is a continuous–time stochastic process and the random function $K(t,x,s,\omega)$ is continuous in t, s and satisfies the Lipschitz condition uniformly over time and over the sample space Ω, say

$$|K(t,x,s,\omega) - K(t,x',s,\omega)| \leq l|x - x'|, \tag{3.3}$$

then equation (3.2) has a unique solution, and this solution is a continuous–time stochastic process. This follows from the standard results of the theory of integral equations (see, e.g., [16]), since we can consider equation (3.2) for any fixed value of ω.

We focus here on equations having rapidly changing stationary kernels:

$$x_\varepsilon(t,\omega) = \phi(t) + \int_0^t K\big(t, x_\varepsilon(s,\omega), \frac{s}{\varepsilon}, \omega\big)\, ds, \tag{3.4}$$

where for fixed t, x, the kernel $K(t,x,s,\omega)$ is a stationary R^d-valued process. We assume that $E|K(t,x,s,\omega)| < \infty$, and we set

$$\bar{K}(t,x) = EK(t,x,s,\omega). \tag{3.5}$$

The expectation on the right-hand side of this equation does not depend on s, since $K(\cdot)$ is a stationary stochastic process.

Denote by $\bar{x}(t)$ the solution of the averaged integral equation

$$\bar{x}(t) = \phi(t) + \int_0^t \bar{K}\big(t, \bar{x}(s)\big)\, ds. \tag{3.6}$$

Theorem 1 *Let $\phi(t)$ be a continuous function, $\phi : R_+ \to R^d$, and let $K(t,x,s,\omega)$ satisfy the following conditions:*

(1) *K is measurable in all variables;*

(2) *K is an ergodic stationary process in s for fixed t, x;*

(3) *K satisfies Lipschitz condition* (3.3)*;*

(4) *$\big|K(t_1,x,s,\omega) - K(t_2,x,s,\omega)\big| \leq \lambda\big(|t_2 - t_1|, c\big)$ if $t_1 \leq c$, $t_2 \leq c$, $|x| \leq c$, uniformly in ω and s, where $\lambda(\cdot,\cdot) : (R_+)^2 \to R_+$ and $\lambda(0+, c) = 0$ for all $c > 0$;*

(5) *$E|K(t,x,s,\omega)| < \infty$, and both $E|K(t,x,s,\omega)|$ and $|K(t,x,s,\omega)|$ are uniformly integrable in t for fixed x and s.*

Then for any positive constants t_0 and δ, the solutions of (3.4) and (3.6) satisfy

$$\lim_{\varepsilon\to 0} P\left\{\sup_{t\le t_0}\left|x_\varepsilon(t,\omega)-\bar{x}(t)\right|>\delta\right\}=0. \tag{3.7}$$

Proof We have

$$\begin{aligned}\left|x_\varepsilon(t,\omega)-\bar{x}(t)\right| &= \left|\int_0^t \left(K\left(t,x_\varepsilon(s,\omega),\frac{s}{\varepsilon},\omega\right)-\bar{K}(t,\bar{x}(s))\right)ds\right| \\ &\le \int_0^t \left|K\left(t,x_\varepsilon(s,\omega),\frac{s}{\varepsilon},\omega\right)-K\left(t,\bar{x}(s),\frac{s}{\varepsilon},\omega\right)\right| ds \\ &+\left|\int_0^t \left(K\left(t,\bar{x}(s),\frac{s}{\varepsilon},\omega\right)-\bar{K}\left(t,\bar{x}(s)\right)\right)ds\right| \\ &\le l\int_0^t \left|x_\varepsilon(s,\omega)-\bar{x}(s)\right| ds+\left|\Psi_\varepsilon(t,\omega)\right|,\end{aligned}$$

where

$$\Psi_\varepsilon(t,\omega)=\int_0^t\left(K\left(t,\bar{x}(s),\frac{s}{\varepsilon},\omega\right)-\bar{K}(t,\bar{x}(s))\right)ds.$$

The inequality

$$\left|x_\varepsilon(t,\omega)-\bar{x}(t)\right|\le l\int_0^t\left|x_\varepsilon(s,\omega)-\bar{x}(s)\right|ds+\left|\Psi_\varepsilon(t,\omega)\right|$$

implies that

$$\sup_{t\le t_0}\left|x_\varepsilon(t,\omega)-\bar{x}(t)\right|\le\sup_{t\le t_0}\left|\Psi_\varepsilon(t,\omega)\right|\exp\{lt_0\}.$$

Therefore, the proof is a consequence of the following lemma.

□

Lemma 1 *Let Ψ_ε be as defined in the preceding proof. Then*

$$\sup_{t\le t_0}\left|\Psi_\varepsilon(t,\omega)\right|\to 0$$

in probability as $\varepsilon\to 0$.

Proof The function $\bar{x}(t)$ is continuous on $[0,\, t_0]$, and therefore it is bounded on this interval. Define the constant

$$c_1=\sup_{s,t\le t_0}\left|\bar{K}(t,\bar{x}(s))\right| \tag{3.8}$$

and the function

$$K_r(t,x,s,\omega)=\begin{cases} K(t,x,s,\omega) & \text{if } \ |K(t,x,s,\omega)|\le r,\\ rK(t,x,s,\omega)/|K(t,x,s,\omega)| & \text{if } \ |K(t,x,s,\omega)|>r.\end{cases}$$

Let $\bar{K}_r(t,x)=EK_r(t,x,s,\omega)$. Then

$$\Psi_\varepsilon(t,\omega)=\Psi_\varepsilon^r(t,\omega)+\tilde{\Psi}_\varepsilon^r(t,\omega),$$

where

$$\Psi_\varepsilon^r(t,\omega) = \int_0^t \left(K_r\left(t, \bar{x}(s), \frac{s}{\varepsilon}, \omega\right) - \bar{K}_r\left(t, \bar{x}(s)\right)\right) ds,$$

$$\tilde{\Psi}_\varepsilon^r(t,\omega) = \int_0^t \left(K\left(t, \bar{x}(s), \frac{s}{\varepsilon}, \omega\right) - K_r\left(t, \bar{x}(s), \frac{s}{\varepsilon}, \omega\right)\right) ds + \hat{\Psi}^r(t),$$

$$\hat{\Psi}^r(t) = \int_0^t \left(\bar{K}_r\left(t, \bar{x}(s)\right) - \bar{K}\left(t, \bar{x}(s)\right)\right) ds.$$

Note that for $t_1 \le t_0$, $t_2 \le t_0$,

$$\left|\Psi_\varepsilon^r(t_1,\omega) - \Psi_\varepsilon^r(t_2,\omega)\right| \le |t_1 - t_2| \cdot 2r + \lambda\big(|t_2 - t_1), c\big) \cdot t_0,$$

where $c \ge t_0$, $c \ge \sup_{s \le t_0} |\bar{x}(s)|$. Therefore, for any integer $m > 0$,

$$\begin{aligned}
&\sup_{t \le t_0} |\Psi_\varepsilon^r(t,\omega)| \le \sup_{k < mt_0} \left|\Psi_\varepsilon^r\left(\frac{k}{m}\right)\right| + \frac{2r}{m} + t_0\lambda\left(\frac{1}{m}, c\right) \\
&= \frac{2r}{m} + t_0\lambda\left(\frac{1}{m}, c\right) \\
&+ \sum_{k < mt_0} \left|\int_{\frac{k}{m}}^{\frac{k+1}{m}} \left(K_r\left(\frac{k}{m}, \bar{x}(s), \frac{s}{\varepsilon}, \omega\right) - \bar{K}_r\left(\frac{k}{m}, \bar{x}(s)\right)\right) ds\right| \\
&= \frac{2r}{m} + t_0\lambda\left(\frac{1}{m}, c\right) \\
&+ \sum_{k < mt_0} \left|\int_{\frac{k}{m}}^{\frac{k+1}{m}} \left(K_r\left(\frac{k}{m}, \bar{x}(s), \frac{s}{\varepsilon}, \omega\right) - \bar{K}_r\left(\frac{k}{m}, \bar{x}\left(\frac{k}{m}\right)\right)\right) ds\right| \\
&+ \sum_{k < mt_0} 2l \int_{\frac{k}{m}}^{\frac{k+1}{m}} \left|\bar{x}\left(\frac{k}{m}\right) - \bar{x}(s)\right| ds.
\end{aligned}$$

It follows from the ergodic theorem (Theorem 1′ of Chapter 1) that with probability 1 for all k, m,

$$\lim_{\varepsilon \to 0} \int_{\frac{k}{m}}^{\frac{k+1}{m}} K_r\left(\frac{k}{m}, \bar{x}(s), \frac{s}{\varepsilon}, \omega\right) ds = \lim_{\varepsilon \to 0} \varepsilon \int_{\frac{k}{\varepsilon m}}^{\frac{k+1}{\varepsilon m}} K_r\left(\frac{k}{m}, \bar{x}\left(\frac{k}{m}\right), s, \omega\right) ds,$$

$$\frac{1}{m} EK_r\left(\frac{k}{m}, \bar{x}\left(\frac{k}{m}\right), s, \omega\right) = \bar{K}_r\left(\frac{k}{m}, \bar{x}\left(\frac{k}{m}\right)\right).$$

Moreover, for fixed r, c, t_0,

$$\lim_{m \to \infty} \left(\frac{2r}{m} + t_0\lambda\left(\frac{1}{m}, c\right) + \sum_{k < mt_0} 2l \int_{\frac{k}{m}}^{\frac{k+1}{m}} \left|\bar{x}\left(\frac{k}{m}\right) - \bar{x}(s)\right| ds\right) = 0.$$

This implies that

$$\lim_{\varepsilon \to 0} P\left\{\sup_{t \le t_0} |\Psi_\varepsilon^r(t,\omega)| > \delta\right\} = 0$$

for all $\delta > 0$.
Using equality (3.8) we can show that

$$\lim_{\varepsilon \to 0} \sup_{t \le t_0} \left| \int_0^t (\bar{K}_r(t, \bar{x}(s)) - \bar{K}(t, \bar{x}(s)) \, ds \right| = 0.$$

We denote by $[\alpha]$, $\alpha \in R$, the integer k for which $k \le \alpha < k + 1$. Then for $t \le t_0$,

$$\int_0^t \left| K\left(\frac{[mt]}{m}, \bar{x}\left(\frac{[ms]}{m}\right), \frac{s}{\varepsilon}, \omega\right) - K\left(t, \bar{x}(s), \frac{s}{\varepsilon}, \omega\right)\right) \right| ds$$

$$\le t_0 \lambda\left(\frac{1}{m}, c\right) + l \int_0^{t_0} \left| \bar{x}\left(\frac{[ms]}{m}\right) - \bar{x}(s) \right| ds.$$

The same inequality holds for K_r, so

$$\begin{aligned}
\sup_{t \le t_0} |\tilde{\Psi}_\varepsilon^r(t, \omega)| &\le 2t_0 \lambda\left(\frac{1}{m}, c\right) + 2l \int_0^{t_0} \left| \bar{x}\left(\frac{[ms]}{m}\right) - \bar{x}(s) \right| ds \\
&+ \sup_{k \le m t_0} \int_0^{t_0} \left| K\left(\frac{k}{m}, \bar{x}\left(\frac{[ms]}{m}\right), \frac{s}{\varepsilon}, \omega\right) - K_r\left(\frac{k}{m}, \bar{x}\left(\frac{[sm]}{m}\right), \frac{s}{\varepsilon}, \omega\right) \right| ds \\
&\le 2t_0 \lambda\left(\frac{1}{m}, c\right) + 2l \int_0^{t_0} \left| \bar{x}\left(\frac{[ms]}{m}\right) - \bar{x}(s) \right| ds \\
&+ \sum_{k=1}^{m} \int_0^{t_0} \left| K\left(\frac{k}{m}, \bar{x}\left(\frac{[ms]}{m}\right), \frac{s}{\varepsilon}, \omega\right) - K_r\left(\frac{k}{m}, \bar{x}\left(\frac{[sm]}{m}\right), \frac{s}{\varepsilon}, \omega\right) \right| ds.
\end{aligned}$$

Since

$$\lim_{r \to \infty} E \left| K\left(\frac{k}{m}, \bar{x}\left(\frac{i}{m}\right), \frac{s}{\varepsilon}, \omega\right) - K_r\left(\frac{k}{m}, \bar{x}\left(\frac{i}{m}\right), \frac{s}{\varepsilon}, \omega\right) \right| = 0$$

uniformly in s (it does not depend on s!), we have that

$$\lim_{r \to \infty} E \sup_{t \le t_0} \left| \tilde{\Psi}_\varepsilon^r(t, \omega) \right| = 0$$

uniformly in $\varepsilon > 0$.

□

3.1.1 Linear Volterra Integral Equations

Let $K(t, x, s, \omega) = K(t, s, \omega)x$, where $K(t, s, \omega)$ is a measurable function from $R_+ \times R_+$ into $L(R^d)$. (Recall that this is the space of linear operators on R^d.) In this case the Lipschitz condition (3.3) is equivalent to the inequality $\|K(t, s, \omega)\| \le l$. Therefore, Theorem 1 implies the following theorem.

Theorem 2 *Let $x_\varepsilon(t)$ satisfy the equation*

$$x_\varepsilon(t) = \phi(t) + \int_0^t K\left(t, \frac{s}{\varepsilon}, \omega\right) x_\varepsilon(s)\, ds, \tag{3.9}$$

where $\phi(t)$ is a continuous function $R_+ \to R^d$ and $K(t,s,\omega)$ satisfies the conditions:

(1) *K is measurable in all variables;*

(2) *K is an ergodic stationary process in s for fixed t;*

(3) *$\|K(t,s,\omega)\| \le l$, l a constant;*

(4) *$\|K(t_1,s,\omega) - K(t_2,s,\omega)\| \le \lambda\big(|t_2-t_1|,\, |t_1| \vee |t_2|\big)$, where $\lambda : (R_+)^2 \to R$, $\lambda(0+,\, c) = 0$ for all $c > 0$.*

Suppose that the average $\bar{K}(t) = EK(t,s,\omega)$ is a continuous $L(R^d)$-valued function. Let $\bar{x}(t)$ be the solution of the averaged equation

$$\bar{x}(t) = \phi(t) + \int_0^t \bar{K}(t)\bar{x}(s)\, ds. \tag{3.10}$$

Then the relation (3.7) is valid.

We consider condition (3) of the theorem to be a very strong restriction. The following result allows us to avoid it. For this purpose we impose a mixing condition on the stationary process $K(t,s,\omega)$.

Condition M. For any $t_0 > 0$ there exist $c_{t_0} > 0$ and a bounded decreasing function $\rho_{t_0}(u) > 0$ with $\lim_{u\to\infty} \rho_{t_0}(u) = 0$ such that for any $n \ge 2$ and any times t_k if $t_k \le t_0$ and $s_1 > s_2 > \cdots s_n > 0$, we have

$$E\|K(t_1,s_1,\omega)K(t_2,s_2,\omega)\cdots K(t_n,s_n,\omega) - \bar{K}(t_1)\cdots\bar{K}(t_n)\|$$

$$\le c_{t_0}^{n-1} \sum_{k=1}^{n} \rho_{t_0}(s_k - s_{k-1}).$$

Here $\|\cdot\|$ is the norm in $L(R^d)$.

Theorem 3 *Let $x_\varepsilon(t)$ be the solution of equation (3.9) and suppose that the conditions of Theorem 2 are satisfied except for conditions 3) and 4). In addition, we assume Condition M and the following condition:*
(4′) $K(t,s,\omega)$ is continuous in t uniformly in s and ω, and $\bar{K}(t)$ is a continuous function.
Then for any t_0,

$$\lim_{\varepsilon\to 0} \sup_{t \le t_0} E\big|x_\varepsilon(t,\omega) - \bar{x}(t)\big| = 0.$$

Proof The solutions of the linear problems (3.9) and (3.10) can be written explicitly using the resolvent series, respectively, as

$$x_\varepsilon(t) = \phi(t) + \sum_{n=1}^{\infty} \int \cdots \int_{0<s_n<\cdots<s_1\le t} K\left(t, \frac{s_1}{\varepsilon}, \omega\right) \cdots K\left(s_{n-1}, \frac{s_n}{\varepsilon}, \omega\right) \times \phi(s_n)\, ds_1 \cdots ds_n,$$

$$\bar{x}(t) = \phi(t) + \sum_{n=1}^{\infty} \int \cdots \int_{0<s_n<\cdots<s_1\le t} \bar{K}(t)\bar{K}(s_1)\cdots\bar{K}(s_{n-1})\phi(s_n)\, ds_1 \cdots ds_n.$$

So

$$E\big|x_\varepsilon(t) - \bar{x}(t)\big| \le E\left|\int_0^t \left(K\left(t, \frac{s_1}{\varepsilon}, \omega\right) - K(t)\right)\phi(s_1)\, ds_1\right|$$

$$+ \sum_{n=2}^{\infty} \iint_{0<s_n\cdots<s_1\le t} E\left\|K\left(t, \frac{s_1}{\varepsilon}, \omega\right) - \bar{K}(t)\cdots\bar{K}(s_{n-1})\right\| |\phi(s_n)|\, ds_1 \cdots ds_n.$$

It follows from Condition M that for $t \le t_0$,

$$\int\int_{0<s_n<\cdots<s_1\le t} E\left\|K\left(t, \frac{s_1}{\varepsilon}, \omega\right) \cdots K\left(t, \frac{s_n}{\varepsilon}, \omega\right) - \bar{K}(t)\cdots\bar{K}(s_{n-1})\right\|$$
$$\times\, ds_1 \cdots ds_n \le c_{t_0}^{n-1} \sum_{k=1}^{n-1} \int\int_{0<s_n<\cdots<s_1\le t} \rho\left(\frac{s_k - s_{k+1}}{\varepsilon}\right) ds_1 \cdots ds_n$$
$$\le c_{t_0}^{n-1} t_0^n \Psi_{t_0}\left(\frac{t_0}{\varepsilon}\right) \frac{(n-1)}{n!},$$

where $\Psi_{t_0}(T) = \frac{1}{T}\int_0^T \rho_{t_0}(u)du \to 0$, as $T \to \infty$. Let

$$K_r(t,\omega) = \begin{cases} K(t,s,\omega) & \text{if } \|K(t,s,\omega)\| \le r, \\ rK(t,s,\omega)/\|K(t,s,\omega)\| & \text{if } \|K(t,s,\omega)\| > r, \end{cases}$$

and $\bar{K}_r(t) = EK_r(t,s,\omega)$.
Then using the inequality

$$\sup_{t\le t_0} E\left|\int_0^t \left(K\left(t, \frac{s_1}{\varepsilon}, \omega\right) - \bar{K}(t)\right)\phi(s_1)\, ds_1\right|$$
$$\le \sup_{t\le t_0} E\left|\int_0^t \left(K_r\left(t, \frac{s_1}{\varepsilon}, \omega\right) - \bar{K}_r(t)\right)\phi(s_1)\, ds_1\right|$$
$$+ \sup_{t\le t_0} \int_0^t E\left\|K\left(t, \frac{s_1}{\varepsilon}, \omega\right) - K_r\left(t, \frac{s_1}{\varepsilon}, \omega\right)\right\| |\phi(s_1)|\, ds_1$$
$$+ \sup_{t\le t_0} \int_0^t \|\bar{K}(t) - \bar{K}_r(t)\| \cdot |\phi(s)|\, ds$$

and the ergodic theorem, we can prove that

$$\lim_{\varepsilon\to 0}\sup_{t\le t_0} E\left|\int_0^t \left(K\left(t,\frac{s_1}{\varepsilon},\omega\right)-\bar{K}(t)\right)\phi(s_1)\,ds_1\right| = 0.$$

□

The next theorem gives more general conditions under which relation (3.7) is true. Introduce the stochastic processes

$$\mu(s,\omega)=\sup_{t>s}\|K(t,s,\omega)\|$$

and

$$\lambda_h(s,\omega)=\sup_{|t_1-t_2|\le h}\|K(t_1,s,\omega)-K(t_2,s,\omega)\|,$$

where $s\in R_+$ and h is a positive parameter.

Theorem 4 *Let $x_\varepsilon(t)$ be the solution of equation (3.9) with $\phi(t)$ a continuous function and the function $K(t,s,\omega)$ satisfying the conditions (2), (3) of Theorem 2 and the following conditions:*

(5) $E\big|K(t,s,\omega)\big|<\infty$;

(6) *$\mu(s,\omega)$ is an ergodic stationary process with $E\mu(s,\omega)<\infty$;*

(7) *for any $h>0$, $\lambda_h(s,\omega)$ is an ergodic stationary process with*

$$\lim_{h\to 0}E\lambda_h(s,\omega)=0.$$

Then relation (3.7) holds.

Proof Condition (7) implies that $\bar{K}(t)$ is continuous, so $\bar{x}(t)$ is also continuous. We have

$$\begin{aligned}\big|x_\varepsilon(t)-\bar{x}(t)\big| &\le \int_0^t \mu\left(\frac{s}{\varepsilon},\omega\right)\big|x_\varepsilon(s)-\bar{x}(s)\big|\,ds\\ &\quad+\left|\int_0^t\left(K\left(t,\frac{s}{\varepsilon},\omega\right)-\bar{K}(t)\right)x(s)\,ds\right|.\end{aligned}$$

Let

$$\eta_\varepsilon(t_0)=\sup_{t\le t_0}\left|\int_0^t\left(K\left(t,\frac{s}{\varepsilon},\omega\right)-\bar{K}(t)\right)\bar{x}(s)\,ds\right|.$$

Then

$$\begin{aligned}\sup_{t\le t_0}\big|x_\varepsilon(t)-\bar{x}(t)\big| &\le \eta_\varepsilon(t_0)\exp\left\{\int_0^{t_0}\mu\left(\frac{s}{\varepsilon},\omega\right)ds\right\}\\ &\le \eta_\varepsilon(t_0)\exp\left\{\varepsilon\int_0^{t_0/\varepsilon}\mu\Big(s,\omega\Big)\,ds\right\}.\end{aligned}$$

It suffices to prove that $\eta_\varepsilon(t_0) \to 0$ in probability, as $\varepsilon \to 0$. For any positive integer m, we have the following inequality:

$$\eta_\varepsilon(t_0) \le \sup_{k\le m}\left\{\left|\int_0^{\frac{kt_0}{m}}\left(K\left(\frac{kt_0}{m},\frac{s}{\varepsilon},\omega\right)-\bar K\left(\frac{kt_0}{m}\right)\right)\bar x(s)\,ds\right| + \right.$$
$$\sup_{0<h\le\frac{t_0}{m}}\left[\left|\int_0^{\frac{kt_0}{m}+h} K\left(\frac{kt_0}{m}+h,\frac{s}{\varepsilon},\omega\right)\bar x(s)\,ds-\int_0^{\frac{kt_0}{m}} K\left(\frac{kt_0}{m},\frac{s}{\varepsilon},\omega\right)\bar x(s)\,ds\right|\right.$$
$$\left.\left.+\left|\int_0^{\frac{kt_0}{m}+h}\bar K\left(\frac{kt_0}{m}+h\right)\bar x(s)\,ds-\int_0^{\frac{kt_0}{m}}\bar K\left(\frac{kt_0}{m}\right)\bar x(s)\,ds\right|\right]\right\}.$$

Let

$$c=\sup_{s\le t_0}\left\{\|\bar x(s)\|\vee\|\bar K(s)\|\right\},$$

and

$$\delta_m=\sup\left\{\left\|\bar K(t_1)-\bar K(t_2)\right\| : |t_1-t_2|\le\frac{t_0}{m},\ t_1\le t_0,\ t_2\le t_0\right\}.$$

Then

$$\eta_\varepsilon(t_0)\le\sup_{k\le m}\left|\int_0^{\frac{k}{m}t_0}\left(K\left(\frac{k}{m}t_0,\frac{s}{\varepsilon},\omega\right)-\bar K\left(\frac{k}{m}t_0\right)\right)\bar x(s)\,ds\right|$$
$$+c\left[\int_0^{t_0}\lambda_{t_0/m}\left(\frac{s}{\varepsilon},\omega\right)ds+\sup_{k<m}\int_{\frac{k}{m}t_0}^{\frac{k+1}{m}t_0}\mu\left(\frac{s}{\varepsilon},\omega\right)ds+c\frac{t_0}{m}+\delta_m t_0\right].$$

It follows from the ergodic theorem that

$$\int_0^{\frac{k}{m}t_0}K\left(\frac{k}{m}t_0,\frac{s}{\varepsilon},\omega\right)\bar x(s)\,ds\to\int_0^{\frac{k}{m}t_0}\bar K\left(\frac{k}{m}t_0\right)\bar x(s)\,ds,$$
$$\int_0^{t_0}\lambda_{t_0/m}\left(\frac{s}{\varepsilon},\omega\right)ds\to t_0E\lambda_{t_0/m}(s,\omega),$$
$$\int_{\frac{k}{m}t_0}^{\frac{k+1}{m}t_0}\mu\left(\frac{s}{\varepsilon},\omega\right)ds\to\frac{t_0}{m}E\mu(s,\omega)$$

as $\varepsilon\to0$ with probability 1.
Therefore,

$$\limsup_{\varepsilon\to0}P\{\eta_\varepsilon(t_0)>\delta\}\le P\left\{ct_0E\lambda_{t_0/m}(s,\omega)+c\frac{t_0}{m}E\mu(s,\omega)+\frac{c^2t_0}{m}+\delta_m t_0\ge\delta\right\}.$$

This probability equals 0 if m is large enough.

□

3.1.2 Some Nonlinear Equations

Let $y(s)$ be a Y-valued measurable process, where $(Y, \mathcal{C})$ is a measurable space. We consider an integral equation of the form

$$x_\varepsilon(t) = \phi(t) + \int_0^t K\left(t, s, x_\varepsilon(s), y\left(\frac{s}{\varepsilon}\right)\right) ds, \tag{3.11}$$

where $K : R_+ \times R_+ \times R^d \times Y \to R^d$ is a measurable function. In this case we can consider separately the conditions for the function K and the process $y(s)$. We call $y(\cdot)$ the perturbing process and suppose that it satisfies one of the following conditions.

The perturbing process is assumed to be one of two types:

PPa $y(s)$ is a stationary ergodic process.

PPb $y(s)$ is a homogeneous uniformly ergodic Markov process.

We denote by ρ the ergodic distribution for $y(s)$.

Our additional conditions on K in this case are the following:

(LC) (Lipschitz's Condition) For any $c > 0$ there exists $l_c > 0$ for which

$$\big|K(t, s, x_1, y) - K(t, s, x_2, y)\big| \leq l_c |x_1 - x_2|$$

for $y \in Y$, $0 \leq s \leq t \leq c$, $x_1, x_2 \in R^d$.

(TC) (Time Continuity) Let

$$\lambda(h, c) = \sup\big\{\big|K(t_1, s, x, y) - K(t_2, s, x, y)\big| : \ t_1 \leq c, \, t_2 \leq c, \\ |t_1 - t_2| \leq h, \, s \leq c, \, |x| \leq c, \, y \in Y\big\}.$$

Then $\lambda(0+, c) = 0$ for all $c \in R_+$.
Let us introduce the function

$$k_c(y) = \sup_{0 \leq s \leq t \leq c} \big|K(t, s, 0, y)\big|.$$

Then we assume the following condition:

(UI) (Uniform Integrability) $\int k_c(y)\, \rho(dy) < \infty$.

Let

$$\bar{K}(t, s, x) = \int K(t, s, x, y)\, \rho(dy) < \infty.$$

In the same way as we proved Theorem 1 we can prove the following statement.

Theorem 1′ *Let K(t,s,x,y) satisfy conditions* (LC), (TC), *and* (UI).
(1) *If $y(s)$ satisfies condition* (PPa), *then relation* (3.7) *holds for all $t_0 \in R_+$.*

(2) *If $y(s)$ satisfies condition* (PPb) *and*

$$\sup\left\{\int k_c(y')P(s,y,dy'),\ s\in R_+,\ y\in Y\right\}<\infty,$$

where $P(s,y,dy')$ is the transition probability for $y(s)$, then for all $t_0\in R_+$ and $\delta>0$,

$$\lim_{\varepsilon\to 0}\sup_{y\in Y}P_y\left\{\sup_{t\le t_0}\left|x_\varepsilon(t)-\bar{x}(t)\right|>\delta\right\}=0,$$

where P_y is probability under the condition that $y(0)=y$, and $\bar{x}(t)$ is the solution of the averaged integral equation

$$\bar{x}(t)=\phi(t)+\int_0^t\bar{K}(t,s,\bar{x}(s))\,ds.$$

Consider now a linear equation with $K(t,s,x,y) = K(t,s,y)x$, where $K(t,s,y)$ is a measurable function $R_+\times R_+\times Y\to L(R^d)$. Let

$$\mu_c(y)=\sup\{\|K(t,s,y)\|,\ 0\le s\le t\}$$

and

$$\lambda_c(h,y)=\sup\{\|K(t_1,s,y)-K(t_2,s,y)\|,\ t_1\le c,\ t_2\le c,\ |t_1-t_2|\le h,\ s\le c\}.$$

Theorem 4′ *Let $x_\varepsilon(t)$ be the solution of the equation*

$$x_\varepsilon(t)=\phi(t)+\int K\left(t,s,y\left(\frac{s}{\varepsilon}\right)\right)x_\varepsilon(s)\,ds,$$

where $K(t,s,y)$ satisfies the conditions
(1) $\int\mu_c(y)\,\rho(dy)<\infty$ *for* $c\in R_+$,
and
(2) $\lim_{h\to 0}\int\lambda_c(h,y)\,\rho(dy)=0$.
Denote by $\bar{x}(t)$ the solution of the averaged equation

$$\bar{x}(t)=\phi(t)+\int_0^t\bar{K}(t,s)\bar{x}(s)\,ds,$$

where $\bar{K}(t,s)=\int K(t,s,y)\,\rho(dy)$.

I. *If $y(s)$ satisfies the condition* (PPa), *then relation* (3.7) *is valid for all $t_0\in R_+$.*

II. *If $y(s)$ satisfies the condition* (PPb), *and*

$$\sup\left\{\int\mu_c(y')P(s,y,dy'):\ y\in Y,\ s\in R_+\right\}<\infty,$$

and

$$\lim_{h\to 0}\sup\left\{\int\lambda_c(h,y')P(s,y,dy'):\ y\in Y,\ s\in R_+\right\}=0,$$

then relation (3.7) *is valid.*
We will apply these results in various settings later.

3.2 Differential Equations

We consider a system of differential equations of the form

$$\dot{x}_\varepsilon(t) = a\left(x_\varepsilon(t), \frac{t}{\varepsilon}\right), \quad t > 0, \ x_\varepsilon(0) = x_0, \tag{3.12}$$

where $a(x, s)$ is an R^d-valued random field, and we investigate the solution $x_\varepsilon(t)$ which is an R^d-valued stochastic process. We can rewrite (3.12) as an equivalent integral equation:

$$x_\varepsilon(t) = x_0 + \int_0^t a\left(x_\varepsilon(s), \frac{s}{\varepsilon}\right) ds.$$

The following statement is a consequence of Theorem 1.

Statement *Let* $a(x, s)$ *satisfy the following conditions:*

(A) $a(x, s)$ *is an ergodic stationary process for any* $x \in R^d$*;*

(B) $E\left|a(x, s)\right| < \infty$;

(C) $\left|a(x, s) - a(\tilde{x}, s)\right| \le l|x - \tilde{x}|$, *where* $x, \tilde{x} \in R^d$, $s \in R_+$, *and* l *is a positive constant.*

Let $\bar{a}(x) = E(a(x, s))$ (note that this expression does not depend on s because of condition (A)), and let $\bar{x}(t)$ be the solution of the equation

$$\dot{\bar{x}}(t) = \bar{a}\big(\bar{x}(t)\big), \quad \bar{x}(0) = x_0. \tag{3.13}$$

Then

$$\lim_{\varepsilon \to 0} P\left\{\sup_{t \le t_0}\left|x_\varepsilon(t) - \bar{x}(t)\right| > \delta\right\} = 0, \quad \delta > 0, \ t_0 > 0.$$

We will obtain more general results than this in other cases. For example,

Theorem 5 *Let* $a(x, s)$ *satisfy Conditions* (A), (B) *above and the following condition:*
(C') *There exists an ergodic stationary process* $\lambda(s, \omega)$ *for which*

$$\left|a(x, s) - a(\tilde{x}, s)\right| \le \lambda(s, \omega)|x - \tilde{x}|$$

with $E\lambda(s, \omega) < \infty$. *Then for any* $t_0 > 0$,

$$P\left\{\lim_{\varepsilon \to 0} \sup_{s \le t_0}\left|x_\varepsilon(s) - \bar{x}(s)\right| = 0\right\} = 1. \tag{3.14}$$

Proof We have

$$\left|x_\varepsilon(t) - \bar{x}(t)\right| \le \int_0^t \lambda\left(\frac{s}{\varepsilon}, \omega\right) \left|x_\varepsilon(s) - \bar{x}(s)\right| ds + \left|\int_0^t \left(a\left(\bar{x}(s), \frac{s}{\varepsilon}\right) - \bar{a}\left(\bar{x}(s)\right)\right) ds\right|.$$

Therefore, Gronwall's inequality gives

$$\sup_{t\le t_0}\left|x_\varepsilon(t) - \bar{x}(t)\right| \le \exp\left\{\int_0^{t_0} \lambda\left(\frac{s}{\varepsilon}, \omega\right) ds\right\} \times \sup_{t \le t_0}\left|\int_0^t \left(a\left(\bar{x}(s), \frac{s}{\varepsilon}\right) - \bar{a}\left(\bar{x}(s)\right)\right) ds\right|.$$

Since

$$\lim_{\varepsilon\to 0} \int_0^{t_0} \lambda\left(\frac{s}{\varepsilon}, \omega\right) ds = t_0 E\lambda(s, \omega)$$

with probability 1, it suffices to show that with probability 1,

$$\limsup_{\varepsilon\to 0} \sup_{t\le t_0}\left|\int_0^t \left(a\left(\bar{x}(s), \frac{s}{\varepsilon}\right) - \bar{a}\left(\bar{x}(s)\right)\right) ds\right| = 0. \tag{3.15}$$

Set

$$\bar{x}_n(s) = \sum_{k=0}^{\infty} \bar{x}\left(\frac{k}{n}\right) 1_{\{\frac{k}{n} \le t < \frac{k+1}{n}\}}.$$

Then

$$\begin{aligned}\sup_{t\le t_0}&\left|\int_0^t \left(a\left(\bar{x}(s), \frac{s}{\varepsilon}\right) - \bar{a}\left(\bar{x}(s)\right)\right) ds\right| \\ &\le \sup_{t\le t_0}\left|\int_0^t \left(a\left(\bar{x}_n(s), \frac{s}{\varepsilon}\right) - \bar{a}\left(\bar{x}_n(s)\right)\right) ds\right| \\ &\quad + \sup_{t\le t_0}\left|\bar{x}_n(t) - \bar{x}(t)\right| \int_0^{t_0} \left[\lambda\left(\frac{s}{\varepsilon}, \omega\right) + \lambda\right] ds,\end{aligned}$$

where $\lambda = E\lambda(s, \omega)$.
Note that for fixed k and n,

$$\sup_{t\le t_0}\left|\int_{\frac{k}{n}}^{\frac{k+1}{n}} \left(a\left(\bar{x}\left(\frac{k}{n}\right), \frac{s}{\varepsilon}\right) - \bar{a}\left(\bar{x}\left(\frac{k}{n}\right)\right)\right) ds\right| 1_{\{\frac{k}{n} \le t_0 < \frac{k+1}{n}\}}$$

$$\le 2 \sup_{t \le \frac{k+1}{n}} \varepsilon \left|\int_0^{\frac{t}{\varepsilon}} \left(a\left(\bar{x}\left(\frac{k}{n}\right), s\right) - \bar{a}\left(\bar{x}\left(\frac{k}{n}\right)\right)\right) ds\right|,$$

and so the right-hand side tends to zero as $\varepsilon \to 0$ with probability 1. So

$$P\left\{\lim_{\varepsilon\to 0}\sup_{t\le t_0}\left|\int_0^t \left(a\left(\bar{x}_n(s), \frac{s}{\varepsilon}\right) - \bar{a}\big(\bar{x}_n(s)\big)\right) ds\right| = 0\right\} = 1$$

and

$$P\left\{\limsup_{\varepsilon\to 0}\sup_{t\le t_0}\left|\int_0^t \left(a\left(\bar{x}(s), \frac{s}{\varepsilon}\right) - \bar{a}\big(\bar{x}(s)\big)\right) ds\right| ds\right.$$

$$\left.\le 2\sup_{t\le t_0}\big|\bar{x}_n(t) - \bar{x}(t)\big| t_0\lambda\right\} = 1.$$

This implies (3.15) because $\lim_{n\to\infty}\sup_{t\le t_0}\big|\bar{x}_n(t) - \bar{x}(t)\big| = 0$.

□

Now we consider the differential equation of the form

$$\dot{x}_\varepsilon(t) = a\left(t, x_\varepsilon(t), y\left(\frac{t}{\varepsilon}\right)\right), \quad x_\varepsilon(0) = 0 \tag{3.16}$$

where $(Y, \mathcal{C})$ is a measurable space, $a : R_+ \times R^d \times Y \to R^d$ is a measurable function, and the Y-valued process $y(s)$ satisfies the condition (PP) of Section 3.1.

Theorem 6 *Assume that $a(t,x,y)$ satisfies the following conditions:*

I. $\int \big|a(t,x,y)\big|\rho(dy) < \infty$.

II. *There exists a measurable in y function $l(c,y) : R_+ \times Y \to R_+$ for which $\big|a(t,x,y) - a(t,\tilde{x},y)\big| \le l(c,y)|x - \tilde{x}|$ if $t \le c$, $|x| \le c$ and $\int l(c,y)\,\rho(dy) < \infty$.*

III. *There exists a measurable in y function $\lambda(h,c,y) : R_+ \times R_+ \times Y \to R_+$ for which $\big|a(t,x,y) - a(\tilde{t},x,y)\big| \le \lambda(t - \tilde{t}, c, y)$ if $t \le c$, $|x| \le c$ and $\lim_{h\to 0}\int \lambda(h,c,y)\,\rho(dy) = 0$ for all $c \in R_+$.*

Let $\bar{a}(t,x) = \int a(t,x,y)\,\rho(dy)$ and let $\bar{x}(t)$ be the solution of the averaged problem

$$\dot{\bar{x}}(t) = \bar{a}\big(t, \bar{x}(t)\big), \quad t > 0, \ \bar{x}(0) = x_0. \tag{3.17}$$

Then relation (3.14) holds for all $t_0 > 0$.

The proof is the same as the one for Theorem 5.

3.2.1 Linear Differential Equations

We consider the equation

$$\dot{x}_\varepsilon(t) = A\left(t, \frac{t}{\varepsilon}\right)x_\varepsilon(t), \quad x_\varepsilon(0) = x_0 \tag{3.18}$$

where $A(t,s) = A(t,s,\omega) : R_+ \times R_+ \times \Omega \to L(R^d)$ is a measurable function. In addition, it is either
(a) an ergodic stationary $L(R^d)$-valued stochastic process for any $t \in R_+$;
or
(b) $A(t,s,\omega) = A\big(t, y(s,\omega)\big)$, where $y(s,\omega) = y(s)$ is a Y-valued ergodic homogeneous stochastic process with ergodic distribution $\rho(dy)$.
We assume that $E\big\|A(t,s)\big\| < \infty$ in case (a) and $\int \big\|A(t,y)\big\| \rho(dy) < \infty$ in case (b).
Set $\bar{A}(t) = EA(t,s,\omega)$ in case (a) and $\bar{A}(t) = \int A(t,y)\,\rho(dy)$ in case (b).
Theorem 6 implies the following statement.

Theorem 7 *Assume that in case* (a) $A(t,s,\omega)$ *satisfies the condition*

$$\sup_{|t_1 - t_2| \le h} \|A(t_1,s,\omega) - A(t_2,s,\omega)\| \le \lambda(h,c,s,\omega) \quad \textit{if} \quad t_1 \le c,$$

where $\lambda(h,c,s,\omega)$ *is an ergodic stationary process in* s *and*

$$E\lambda(h,c,s,\omega) < \infty;$$

and in case (b) $A(t,y)$ *satisfies the condition*

$$\|A(t_1,y) - A(t_2,y)\| \le \lambda(h,c,y) \quad \textit{if} \quad t_1 \le c, \ |t_1 - t_2| \le h$$

and $\lim_{h\to\infty} \int \lambda(h,c,y)\,\rho(dy) = 0$.
Let $\bar{x}(t)$ *be the solution of the averaged problem*

$$\dot{\bar{x}}(t) = \bar{A}(t)\bar{x}(t), \quad \bar{x}(0) = x_0.$$

Then relation (3.14) *is true for all* $t_0 > 0$.

3.3 Difference Equations

Let $\varphi_n(x,\omega) : \mathbf{Z}_+ \times R^d \times \Omega \to R^d$ satisfy the following condition:

$\{\varphi_n(x,\omega),\, n = 0,1,2,\dots\}$ is an ergodic stationary sequence for any $x \in R^d$, and $\varphi_n(x,\omega)$ is $\mathcal{B}(R^d) \otimes \mathcal{F}$ measurable for all $n \in \mathbf{Z}_+$.

We consider the sequence of random vectors $\big\{x_n^\varepsilon,\, n = 0,1,\dots\big\}$ that satisfies the relations

$$x_0^\varepsilon = x_0, \quad x_{n+1}^\varepsilon = x_n^\varepsilon + \varepsilon\varphi_n(x_n^\varepsilon,\omega). \tag{3.19}$$

Suppose that $E\big|\varphi_n(x,\omega)\big| < \infty$ and let $\bar{\varphi}(x) = E\varphi_n(x,\omega)$.
Let $\bar{x}_n^\varepsilon$ satisfy the relations:

$$\bar{x}_0^\varepsilon = x_0, \quad \bar{x}_{n+1}^\varepsilon = \bar{x}_n^\varepsilon + \varepsilon\bar{\varphi}(\bar{x}_n^\varepsilon). \tag{3.20}$$

Lemma 2 *[73] Suppose that* $\bar{\varphi}(x)$ *satisfies a Lipschitz condition with constant* l.

Then for any $t_0 > 0$,

$$\lim_{\varepsilon \to 0} \sup_{k \le t_0/\varepsilon} \left| \bar{x}_k^\varepsilon - \bar{x}(k\varepsilon) \right| = 0,$$

where $\bar{x}(t)$ *is the solution of the equation*

$$\dot{\bar{x}}(t) = \bar{\varphi}\big(\bar{x}(t)\big), \quad \bar{x}(0) = x_0. \tag{3.21}$$

Proof Note that

$$\bar{x}\big((k+1)\varepsilon\big) - \bar{x}(k\varepsilon) = \int_{k\varepsilon}^{(k+1)\varepsilon} \bar{\varphi}\big(\bar{x}(s)\big)\, ds.$$

Therefore, if $k\varepsilon \le t_0$, then

$$\bar{x}\big((k+1)\varepsilon\big) - \bar{x}_{k+1}^\varepsilon = \bar{x}(k\varepsilon) - \bar{x}_k^\varepsilon + \int_{k\varepsilon}^{(k+1)\varepsilon} \big(\bar{\varphi}\big(\bar{x}(s)\big) - \bar{\varphi}(\bar{x}_k^\varepsilon)\big)\, ds$$

and

$$\left|\bar{x}\big((k+1)\varepsilon\big) - \bar{x}_{k+1}^\varepsilon\right| \le \left|\bar{x}(k\varepsilon) - \bar{x}_k^\varepsilon\right| + \varepsilon l \left|\bar{x}(k\varepsilon)\right| + \varepsilon l \alpha_\varepsilon, \tag{3.22}$$

where $\alpha_\varepsilon = \sup\big\{\left|\bar{x}(s) - \bar{x}(t)\right| :\ t \le t_0,\ |s-t| \le \varepsilon\big\}$.
Then (3.22) implies that

$$\sup_{k\varepsilon \le t_0} \left|\bar{x}(k\varepsilon) - \bar{x}_k^\varepsilon\right| \le l t_0 \alpha_\varepsilon \exp\{l t_0\}.$$

□

Theorem 8 *Let* $\varphi_n(x,\omega)$ *be as in the previous theorem and satisfy in addition the following condition: There exists a stationary ergodic sequence* $l_n(\omega)$ *for which* $El_n(\omega) < \infty$ *and*

$$\left|\varphi_n(x,\omega) - \varphi_n(\tilde{x},\omega)\right| \le l_n(\omega)|x - \tilde{x}|. \tag{3.23}$$

Then for any $t_0 \in R_+$,

$$P\left\{\lim_{\varepsilon \to 0} \sup_{\varepsilon k \le t_0} \left|x_k^\varepsilon - \bar{x}(\varepsilon k)\right| = 0\right\} = 1, \tag{3.24}$$

where $\bar{x}(t)$ *is the solution of equation* (3.21).

Proof Note that $\left|\bar{\varphi}(x) - \bar{\varphi}(\tilde{x})\right| \le El_n(\omega)|x - \tilde{x}|$, so the condition of Lemma 2 is fulfilled. Therefore, it remains to prove that

$$P\left\{\lim_{\varepsilon \to 0} \sup_{\varepsilon k \le t_0} \left|x_k^\varepsilon - \bar{x}_k^\varepsilon\right| = 0\right\} = 1. \tag{3.25}$$

It follows from relations (3.19) and (3.20) that

$$\begin{aligned} x_{n+1}^\varepsilon - \bar{x}_{n+1}^\varepsilon &= \varepsilon \sum_{k=1}^n \left[\varphi_k(x_k^\varepsilon, \omega) - \bar{\varphi}(\bar{x}_k^\varepsilon)\right] \\ &= \varepsilon \sum_{k=1}^n \varphi_k(x_k^\varepsilon, \omega) + \varepsilon \sum_{k=1}^n \left[\varphi_k(\bar{x}_k^\varepsilon, \omega) - \bar{\varphi}(\bar{x}_k^\varepsilon)\right], \end{aligned}$$

so

$$\left|x_{n+1}^{\varepsilon}-\bar{x}_{n+1}^{\varepsilon}\right| \le \varepsilon\sum_{k=1}^{n} l_k(\omega)|x_k^{\varepsilon}-\bar{x}_k^{\varepsilon}| + \varepsilon\left|\sum_{k=1}^{n}\left[\varphi_k(\bar{x}_k^{\varepsilon},\omega)-\bar{\varphi}(\bar{x}_k^{\varepsilon})\right]\right|,$$

and

$$\left|x_{n+1}^{\varepsilon}-\bar{x}_{n+1}^{\varepsilon}\right| \le \sup_{m\le n}\left|\sum_{k=1}^{m}\varphi_k(\bar{x}_k^{\varepsilon},\omega)\right| \exp\left\{\varepsilon\sum_{k=1}^{n} l_k(\omega)\right\}.$$

Since

$$P\left\{\lim_{\varepsilon\to 0}\sup_{\varepsilon n\le t_0}\exp\left\{\varepsilon\sum_{k=1}^{n} l_k(\omega)\right\} = \exp\{lt_0\}\right\} = 1,$$

where $l = El_n(\omega)$, (3.25) is a consequence of the relation

$$P\left\{\limsup_{\varepsilon\to 0}\sup_{m\varepsilon\le t_0}\left|\sum_{k=1}^{m}\varphi_k(\bar{x}_k^{\varepsilon},\omega)\right| = 0\right\} = 1. \tag{3.26}$$

Note that

$$\left|\sum_{k=1}^{m}\left[\varphi_k(\bar{x}_k^{\varepsilon},\omega)-\bar{\varphi}(\bar{x}_k^{\varepsilon})\right] - \sum_{k=1}^{m}\left[\varphi(\bar{x}(k\varepsilon),\omega)-\bar{\varphi}(\bar{x}(k\varepsilon))\right]\right|$$

$$\le \sum_{k=1}^{m}\left(l_k(\omega)+l\right)\sup_{i\le m}\left|\bar{x}_i^{\varepsilon}-\bar{x}(i\varepsilon)\right|,$$

so (3.26) is equivalent to the relation

$$P\left\{\limsup_{\varepsilon\to 0}\sup_{m\varepsilon\le t_0}\left|\sum_{k=1}^{m}\left[\varphi_k(\bar{x}(k\varepsilon,\omega)-\bar{\varphi}(\bar{x}(k\varepsilon))\right]\right| = 0\right\} = 1, \tag{3.27}$$

and this can be proved in the same way as (3.15) in Theorem 5.

□

Remark *Assume that* $\varphi_k(x,\omega)=\varphi(x,\xi_k)$*, where* $\{\xi_k,\ k=0,1,\dots\}$ *is a sequence of* Y*-valued random variables* ($(Y,\mathcal{C})$ *is a measurable space*) *that satisfies one of the following conditions:*
(A) φ_k *is an ergodic homogeneous Markov chain,*
or
(B) φ_k *is a stationary ergodic sequence.*
Let the ergodic distribution of ξ_k *be denoted by the measure* $\rho(dy)$ *on* $\mathcal{C}$*. Suppose that the function*

$$\varphi(x,y):\ R^d\times Y\to R^d$$

is measurable and $\int|\varphi(x,y)|\rho(dy)<\infty$*. We let*

$$\bar{\varphi}(x) = \int \varphi(x,y)\,\rho(dy).$$

Suppose that $|\varphi(x,y) - \varphi(\tilde{x},y)| \le l(y)|x - \tilde{x}|$ *and* $\int l(y)\,\rho(dy) < \infty$. *Then the statement of Theorem 8 is true.*

3.3.1 Linear Difference Equations

Assume $\varphi_k(x,\omega) = \phi_k(\omega)x$, where $\{\phi_k(\omega),\ k = 0,1,2,\dots\}$ is a discrete–time (R^d)-valued ergodic stationary process. Suppose that $E\|\phi_k(\omega)\| = l < \infty$. Then the process $l_k(\omega) = \|\phi_k(\omega)\|$ satisfies the conditions of Theorem 8. Let $\bar{\varphi}(x) = \bar{\phi}x$, where $\bar{\phi} = E\phi_k(\omega)$, $x_k^\varepsilon = (I + \varepsilon\bar{\phi})^k x_0$, and $\bar{x}(t) = \exp\{t\bar{\phi}\}x_0$.
The statement of Theorem 8 remains true in this case.
We can also study in the same way the problem when $\varphi_k(x,\omega)$ is of the form $\phi(\xi_k)x$, where $\phi(y) : Y \to L(R^d)$ and ξ_k satisfies one of the conditions (A) or (B).

3.4 Large Deviation for Differential Equations

We consider in this section the differential equation

$$\dot{x}_\epsilon(t) = a(x_\epsilon(t), y(t/\epsilon)) \tag{3.28}$$

in R^d with the initial condition

$$x_\epsilon(0) = x_0 \in R^d,$$

where $y(t)$ is a homogeneous Markov process in a compact space Y with transition probability $P(t,y,B)$ satisfying Feller's condition (see (FC) Section 2.4), the process is ergodic with ergodic distribution ρ and satisfies the conditions of Lemma 4 in Chapter 2. Assume that the function $a(\cdot,\cdot) : R^d \times Y \to R^d$ satisfies the following conditions:

(a1)

$$\sup\{|a(x,y)| + ||a_x(x,y)|| : x \in R^d,\ \ y \in Y\} < \infty.$$

(a2) For any $x \in R^d$ the function $a(x,y)$ is continuous in y for almost all y with respect to the measure $\rho(dy)$.

(a3) Denote by $\mathbf{M}$ the set of all probability measures on $\mathcal{B}(Y)$, and by $\mathcal{M}_\rho$ the set of measures $m \in \mathbf{M}$ that are absolutely continuous with respect to the measure ρ. Set

$$A(x) = \left\{ z \in R^d : z = \int a(x,y)m(dy), m \in \mathcal{M}_\rho \right\}.$$

Then for all $x \in R^d$, $A(x)$ is a closed bounded convex set in R^d with nonempty interior.
Note that the only condition in (a3) is that $A(x)$ contain an interior point for each x .

Here we consider large deviations of the solutions to equation (3.26). We will use the notation and results of Section 2.4. In particular, the function $I(m)$ is given by formula (2.10).

3.4.1 Some Auxiliary Results

Denote by $(\mathcal{AC})(x_0, T)$ the set of continuous functions $x(\cdot) : [0, T] \to R^d$ that have the derivatives $x'(t)$ satisfying the relation $x'(t) \in A(x(t))$ for $t < T$ and $x(0) = x_0$. Denote by $(\mathcal{AD})(x_0, T)$ the set of functions of the form

$$x(t) = x_0 + \int_0^t \sum_k 1_{\{t_k \le s < t_{k+1}\}} v_k \, ds, \tag{3.29}$$

where $t_0 = 0 < t_1 < \cdots < t_n = T$ and vectors $v_0, \ldots, v_{n-1}$ from R^d satisfy the relations $v_k \in A(x(t))$ for $t \in [t_k, t_{k+1})$.
We consider $\mathbf{M}$ to be a metric space with a distance d, convergence in which is equivalent to the weak convergence of measures.
To construct a function from $(\mathcal{AD})(x_0, T)$ let us introduce the function

$$u(x, v) = \sup\{t : v \in A(x + sv)\}, s \le t\}. \tag{3.30}$$

Then the function $x(t)$ given by formula (3.29) is a function from $(\mathcal{AD})(x_0, T)$ if and only if $v_k \in A(x(t_k))$ and $t_{k+1} - t_k \le u(x(t_k, v_k)$ for $k < n$. Using condition (a3) we can prove the next statement.

Lemma 1 *Let $x(\cdot) \in (\mathcal{AC})(x_0, T)$. Then for any $h > 0$ a function $x^*(\cdot) \in (\mathcal{AD})(x_0, T)$ exists for which*

$$\sup_{s \in [0,T]} |x(s) - x^*(s)| \le h.$$

Now we consider a method for constructing functions from $(\mathcal{AC})(x_0, T)$. Denote by $\mathcal{C}(T, b, \mathcal{M})$ the set of continuous functions $m : [0, T] \to \mathbf{M}$ satisfying the condition

$$\sup_{t \in [0,T]} I(m(t)) \le b.$$

Let $m(\cdot) \in \mathcal{C}(T, b, \mathcal{M})$, and let

$$\bar{a}_{m(\cdot)}(t, x) = \int a(x, y) m(t, dy).$$

We consider the solution to the differential equation

$$\dot{x}_{m(\cdot)}(t) = \bar{a}_{m(\cdot)}(t, x_{m(\cdot)}(t))$$

with the initial condition $x_{m(\cdot)}(0) = x_0$. It is easy to see that

$$x_{m(\cdot)}(\cdot) \in (\mathcal{AC})(x_0, T).$$

For $v \in A(x)$ set

$$M(x,v) = \left\{ m : \int a(x,y)m(dy) = v \right\},$$
$$S(x,v) = \inf \left\{ I(m) : m \in M(x,v) \right\},$$

and

$$M_r^*(x,v) = M(x,v) \cap C_{S(x,v)+r}(I).$$

For $h \in (0,1)$ denote by H_h the set in $R^d \times R^d$ that is determined by the relation

$$H_h = \{(x;v), x \in R^d, v \in R^d : (1-h)^{-1}(v - h\bar{a}(x)) \in A(x)\}.$$

Remark Note that $\bar{a}(x)$ is an interior point of the set $A(x)$. If $(x;v) \in H_h$ and $h > 0$, then v is an interior point of the set $A(x)$, and for any interior point v of the set $A(x)$ the relation $(x;v) \in H_h$ is fulfilled for some $h > 0$.

Lemma 2 *The function $S(x,v)$ is continuous on the set H_h for all $h > 0$, and the set-valued functions $M(x,v), M^*(x,v)$ are continuous on the same set in the metric*

$$d(A,B) = \sup_{m \in A} \inf_{m' \in B} d(m,m'), \quad A \in \mathcal{B}(\mathcal{M}), \quad B \in \mathcal{B}(\mathcal{M}),$$

where $\mathcal{B}(\mathcal{M})$ is the Borel σ-algebra in $\mathbf{M}$.

The proof of the lemma rests on convexity of the sets $M(x,v), M_r^*(x,v)$ and of the function $I(m)$.
Using the last lemma and Remark 1 we can prove the next statement.

Lemma 3 *Assume that $x(\cdot) \in (\mathcal{AC})(x_0,T)$ satisfies the condition*

$$(x(t), \dot{x}(t)) \in H_h, t \in [0,T]$$

for some $h > 0$. Then
(1) For any $r > 0$ there exists a number $b > 0$ and a function

$$m(\cdot) \in \mathcal{C}(T, b, \mathcal{M})$$

satisfying the conditions

$$m(t) \in M_r^*(x(t), x'(t)), t \in [0,T]$$

and

$$x'(t) = \int a(x(t),y)m(t,dy), \quad t \in [0,T].$$

(2) If $\dot{x}(t)$ is a continuous function, then for any $\delta > 0$ there exists a function

$$x^*(t) \in \mathcal{AD}(x_0,T)$$

for which $\dot{x}^*(t)$ *is constant on the intervals* $(k\delta, (k+1)\delta \wedge T), k = 0, 1, \ldots,$ *and*

$$\sup_{t \le T} |x(t) - x^*(t)| \le c\delta^2$$

for some $c > 0$.

Introduce a random variable

$$\nu_t(C) = \frac{1}{t} \int_0^t 1_{\{y(s) \in C\}}, \quad C \in \mathcal{B}(Y),$$

in the space $\mathbf{M}$ and let Q_y^t be the distribution of ν_t under the condition $y(0) = y$:

$$Q_y^t(B) = P\{\nu_t \in B / y(0) = y\}, \quad B \in \mathcal{B}(M).$$

We need some additional statements concerning the asymptotic behavior of ν_t as a function of t. Denote by $B_r(m)$ the closed ball in $\mathbf{M}$ that is centered at m and has radius r.

Lemma 4 *For all* $m \in \mathbf{M}$ *the relation*

$$\limsup_{r \to 0} \limsup_{t \to \infty} \sup_y \left| \frac{1}{t} \log Q_y^t(B_r(m)) + I(m) \right| = 0 \tag{3.31}$$

is fulfilled.

The proof follows from Theorems 2.7 and 2.8.

Corollary 1 *For given* $\delta > 0$ *there exist numbers* $T > 0, \rho > 0$ *for which the inequalities*

$$- \delta t_k - \sum_{j \le k} (t_j - t_{j-1}) I(m_j)$$
$$\le \log P_y \left(\bigcap_{j \le k} \left\{ \frac{t_j \nu_{t_j} - t_{j-1} \nu_{t_{j-1}}}{t_j - t_{j-1}} \in B_r(m_j) \right\} \right) \le \delta t_k - \sum_{j \le k} (t_j - t_{j-1}) I(m_j) \tag{3.32}$$

are valid for all natural numbers k, $y \in Y$, $0 < t_1 < \cdots < t_k$ *satisfying the conditions* $t_j - t_{j-1} > T$, $j \le k$, *and* $r < \rho$.

Introduce the family of probability measures

$$\nu_t^\Delta(s, B) = \frac{1}{t\Delta} \int_{st}^{t(s+\Delta)} 1_B(y(u)) du, \quad t > 0, s \in [0, 1], \Delta \in (0, 1]. \tag{3.33}$$

Theorem 10 *Let* $m(\cdot) : [0, 1] \to \mathbf{M}$ *be a continuous function for which*

$$\sup_{s \in [0,1]} I(m(s)) < \infty.$$

Then the relation

$$\lim_{\Delta\to 0}\limsup_{r\to 0}\limsup_{t\to\infty}\sup_{y}\left|\frac{1}{t}\log P_y\left\{\sup_{s\in[0,1]} d(\nu_t^\Delta(s), m(s))\le r\right\}\right. \\ \left. + \int_0^1 I(m(s))\,ds\right| = 0 \tag{3.34}$$

is fulfilled.

Proof Let

$$q_y^n(t,r) = P_y\left\{\max_{k\le n} d\left(\nu_t^\Delta\left(\frac{k}{n}\right), m\left(\frac{k}{n}\right)\right)\right\}\le r. \tag{3.35}$$

Note that

$$\operatorname{var}\left(\nu_t^\Delta(s_1) - \nu_t^\Delta(s_2)\right)\le 2\frac{|s_1-s_2|}{\Delta}.$$

Here var$(\cdot)$ denotes the variation of a signed measure. Since the convergence of measures in variation implies their weak convergence, we can assume that $\operatorname{var}(m_1-m_2)\ge d(m_1,m_2)$. So

$$d\left(\nu_t^\Delta(s_1), \nu_t^\Delta(s_2)\right)\le 2\frac{|s_1-s_2|}{\Delta}.$$

Set

$$\alpha_n = \sup_{|s_1-s_2|\le n^{-1}} d(m(s_1), m(s_2)).$$

Then

$$\sup_{s\in[0,1]} d(\nu_t^\Delta(s), m(s))\le \max_{k\le n} d\left(\nu_t^\Delta\left(\frac{k}{n}\right), m\left(\frac{k}{n}\right)\right) + \frac{2}{n\Delta} + \alpha_n. \tag{3.36}$$

For n large enough we can write the inequality

$$q_y^n(t,r)\ge P_y\left\{\sup_{s\in[0,1]} d(\nu_t^\Delta(s), m(s))\le r\right\}\ge q_y^n\left(t, r-\frac{2}{n\Delta}-\alpha_n\right). \tag{3.37}$$

It follows from Corollary 1 that

$$\lim_{r\to 0}\limsup_{t\to\infty}\sup_{y}\left|\frac{1}{t}\log q_y^n(t,r) + n^{-1}\sum_{k\le n} I\left(m\left(\frac{k}{n}\right)\right)\right| = 0. \tag{3.38}$$

This relation and inequalities (3.37) imply the proof of the theorem.

□

3.4.2 Main Theorem.

Theorem 11 *(i) For any function $m(\cdot) \in \mathcal{C}(T, b, \mathcal{M})$, the relation*

$$\lim_{r\to 0}\lim_{\varepsilon\to 0}\left|\varepsilon \log P\{\sup_{t\in[0,T]}|x_\varepsilon(t) - x_{m(\cdot)}(t)| \le r\} + \int_0^T I(m(s))\,ds\right| = 0 \quad (3.39)$$

is fulfilled.
(ii) For any function $x(\cdot) \in \bigcup_{h>0} H_h$ having a continuous derivative $\dot{x}(\cdot)$, the relation

$$\lim_{r\to 0}\lim_{\varepsilon\to 0}\left|\varepsilon \log P\{\sup_{t\in[0,T]}|x(t) - x_\varepsilon(t)| \le r\} + \int_0^T S(x(t), x'(t)\,dt\right| = 0 \quad (3.40)$$

holds.

Proof We use notation

$$\nu(\varepsilon, h, s, \cdot) = \nu^h_{\varepsilon^{-1}}(s, \cdot)$$

(see formula (3.33)). Let $T = nh$. Since

$$\int_{kh}^{(k+1)h} g\left(y\left(\frac{t}{\varepsilon}\right)\right) dt = h\int g(y)\nu(\varepsilon, h, kh, dy)$$

for all bounded measurable functions $g : Y \to R$, we can write the relations

$$x_\varepsilon((k+1)h) - x_\varepsilon(kh) = h\int a(x_\varepsilon(kh), y)\nu(\varepsilon, h, kh, dy) + O(h^2), \quad (3.41)$$

and

$$x_{m(\cdot)}((k+1)h) - x_{m(\cdot)}(kh) = h\int a(x_{m(\cdot)}(kh), y)m(kh, dy) + O(h^2). \quad (3.42)$$

These relations imply the existence of a constant $c_1 > 0$ for which the inequality

$$\max_{k\le n}\left|x_\varepsilon(kh) - x_{m(\cdot)}(kh)\right|$$
$$\le c_1\left(h + h\sum_{k<n}\sup_{|x|\le|x_0|+\bar{c}T}\left|\int a(x,y)\nu(\varepsilon, h, kh, dy) - \int a(x,y)m(kh, dy)\right|\right) \quad (3.43)$$

is fulfilled. Let

$$\theta^\varepsilon_h = \sup_{s\in[0,T]} d(\nu(\varepsilon, h, s), m(s)).$$

Then for any $r > 0$ we can find $\delta > 0$ and $h_0 > 0$ for which

$$\max_{s\in[0,T]}|x_\varepsilon(s) - x_{m(\cdot)}(s)| \le r$$

if $\theta_h^\varepsilon < \delta, \ h < h_0$. So Theorem 1 implies the inequality

$$\lim_{r\to 0} \liminf_{\varepsilon\to 0} \left(\varepsilon \log P \left\{ \sup_{t\in[0,T]} |x_\varepsilon(t) - x_{m(\cdot)}(t)| \le r \right\} \right) \ge - \int_0^T I(m(s))\, ds. \tag{3.44}$$

Let $x(\cdot)$ satisfy (ii) and let

$$x^*(\cdot) \in (\mathcal{AD})(x_0, T), \quad \sup_{t\in[0,T]} |x^*(t) - x(t)| \le r,$$

and

$$x^*(t) = x_k^* + (t - t_k)v_k, \quad t \in [t_k, t_{k+1}], \quad t_k = k\delta, \quad k = 0, 1, \ldots, n, \quad \delta = \frac{T}{n}.$$

Set

$$F_k = \left\{ m : \left| \int a(x_k^*, y) m(dy) - v_k \right| \le \hat{\delta}_k \right\},$$

where the $\hat{\delta}_k$ are chosen in such a way that the relations

$$\nu(\varepsilon, t_{k+1} - t_k, t_k) \in F_k, k = 1, \ldots, n,$$

imply the inequalities

$$|x_\varepsilon(t_k) - x^*(t_k)| \le 2r, \quad k = 1, \ldots, n.$$

This implies the relation

$$P\left\{ \sup_{t\in[0,T]} |x_\varepsilon(t) - x(t)| \le r \right\} \le P\left\{ \bigcap_{k<n} \{\nu(\varepsilon, t_{k+1} - t_k, t_k, \cdot) \in F_k\} \right\}. \tag{3.45}$$

It follows from Corollary 1 that

$$\limsup_{\varepsilon\to 0} \log P\left\{ \bigcup_{k<n} \{\nu(\varepsilon, t_{k+1} - t_k, t_k, \cdot) \in F_k\} \right\} \le - \sum_{k<n} \inf_{m\in F_k} I(m)(t_{k+1} - t_k). \tag{3.46}$$

It can be proved that

$$\lim_{r\to 0} \limsup_{n} \sum_{k<n} (t_{k+1} - t_k) | \inf_{m\in F_k} I(m) - S(x_k, v_k)| = 0. \tag{3.47}$$

So we obtain the inequality

$$\lim_{r\to 0} \limsup_{\varepsilon\to 0} \varepsilon \log P\left\{ \sup_{t\in[0,T]} |x(t) - x_\varepsilon(t)| \le r \right\} \le - \int_0^T S(x(t), x'(t))\, dt. \tag{3.48}$$

The proof of the theorem follows from inequalities (3.44) and (3.46), and the relation

$$I(m(t)) = S\left(x_{m(\cdot)}(t), \dot{x}_{m(\cdot)}\right),$$

which is true for all

$$m(\cdot) \in \bigcup_{b>0} \mathcal{C}(T, b, \mathcal{M}).$$

□

Corollary 2 *(i) For all $x \in R^d$ the relation*

$$\lim_{r\to 0} \limsup_{\varepsilon\to 0} \Big| \varepsilon \log P\{|x_\varepsilon(T) - x| < r\} + \inf\left\{\int_0^T S(x(t), x'(t))dt : x(\cdot) \in (\mathcal{AC})(x_0, T), x(T) = x\right\}\Big| = 0$$

is fulfilled.
(ii) For any closed set $F \subset R^d$ we have the relation

$$\lim_{\varepsilon\to 0} (\epsilon \log P\{\inf_{t\in[0,T]} \rho(x_\epsilon(t), F) = 0\}) = -\inf\left\{\int_0^t S(x(s), x'(s))\, ds : x(\cdot) \in (\mathcal{AC}), x(t) \in F, t \in [0, T]\right\},$$

where

$$\rho(x, F) = \inf\{|x - x'| : x' \in F\}.$$

3.4.3 Systems with Additive Perturbations

Assume that

$$a(x, y) = a(x) + B(x)b(y),$$

where $a(\cdot) : R^d \to R^d$ and $B(\cdot) : R^d \to L(R^d)$ are continuous functions with continuous bounded derivatives $a_x(x), B_x(x)$; $B(x)$ is an invertible matrix for all $x \in R^d$; and $b(\cdot) : Y \to R^d$ is a continuous function for which $\int b(y)\, \rho(dy) = 0$. Let

$$V = \left\{v \in R^d : v = \int b(y)\, m(dy), m \in \mathcal{M}_\rho\right\}.$$

Set

$$I^*(v) = \inf\left\{I(m) : \int b(y)\, m(dy) = v, \quad v \in V\right\}.$$

The main properties of this function are presented in the following statement.

Lemma 5 *$I^*(v)$ is a nonnegative convex function, and $I^*(0) = 0$.*

Proof We prove the convexity of this function. Let $v_k \in V, k = 1, 2,$ and $s \in [0, 1]$. For given $h > 0$, the measures $m_k \in \mathcal{M}_\rho, k = 1, 2,$ satisfy the relations

$$|I^*(v_k) - I(m_k)| < h, \quad \int b(y) m_k(dy),\ k = 1, 2.$$

Since we have the relation

$$\int b(y)(s\, m_1(dy) + (1-s)\, m_2(dy)) = sb_1 + (1-s)b_2,$$

we can write the inequality

$$\begin{aligned} I^*(sv_1 + (1-s)v_2) &\leq I(sm_1 + (1_s)m_2) \leq sI(m_1) + (1-s)I(m_2) \\ &\leq sI^*(v_1) + (1-s)I^*(v_2) + h, \end{aligned}$$

which completes the proof.

□

Remark The function $S(x, v)$ is determined by the relations

$$S(x, v) = I^*(B^{-1}(x)(v - a(x)))$$

if $B^{-1}(x)(v - a(x)) \in V$, and $S(x, v) = +\infty$ otherwise. Introduce for $T > 0$ the function from $C_{[0,T]}(R^d)$ into $[0, +\infty]$ by the relations

$$H_T(f) = \int_0^T S(f(t), \dot{f}(t))\, dt$$

if $f(\cdot)$ is absolutely continuous, and $H_T(f) = +\infty$ otherwise. Then the relations

$$\lim_{r \to 0} \limsup_{\varepsilon \to 0} \left| \varepsilon \log P\{ \sup_{t \in [0,T]} |x_\varepsilon(t) - f(t)| \leq r\} + H_T(f) \right| = 0 \tag{3.49}$$

and

$$\lim_{r \to 0} \limsup_{\varepsilon \to 0} \left| \varepsilon \log \{|x_\varepsilon(T) - x| \leq r\} + \inf\{H_T(f) : f(0) = x_0, f(T) = x\} \right| = 0 \tag{3.50}$$

hold.

4
Normal Deviations

In Chapter 3 we derived the leading–order approximation to the function x_ε by solving an associated averaged problem for $\bar{x}$ and estimating the difference between them. In this chapter we consider the remainder $x_\varepsilon(t,\omega) - \bar{x}(t,\omega)$ in greater detail. We refer to this remainder as the first–order deviation, or simply the *deviation*. The deviation tends to zero in some sense if an averaging theorem is true. When that is the case, we can investigate the asymptotic behavior of $\left(x_\varepsilon(t,\omega) - \bar{x}(t,\omega)\right)/b_\varepsilon$ as $\varepsilon \to 0$, where b_ε is a small scaling factor that is to be determined. If the scaled deviation is (asymptotically) a Gaussian random variable, which is plausible from the central limit theorem, we will say that the deviation is normal. As a rule, we can choose $b_\varepsilon = \sqrt{\varepsilon}$.

4.1 Volterra Integral Equations

We consider first the Volterra integral equation in (3.4) and the corresponding averaged equation (3.6). We define the scaled deviation to be

$$\tilde{x}_\varepsilon(t) = \frac{x_\varepsilon(t,\omega) - \bar{x}(t)}{\sqrt{\varepsilon}}. \tag{4.1}$$

We suppose that the conditions of Theorem 1 of Section 3.1 are satisfied, and we investigate additional conditions under which $\tilde{x}_\varepsilon(t)$ converges weakly to a Gaussian stochastic process.

First we will obtain an equation for $\tilde{x}_\varepsilon(t)$, and then transform it to a suitable form. It is easy to see from formulas (3.4) and (3.6) that $\tilde{x}_\varepsilon(t)$ satisfies the relation

$$\begin{aligned}\tilde{x}_\varepsilon(t) &= \int_0^t \varepsilon^{-1/2}\left(K\left(t, \bar{x}(s) + \varepsilon^{1/2}\tilde{x}_\varepsilon(s), \frac{s}{\varepsilon}, \omega\right) - \bar{K}\left(t, \bar{x}(s)\right)\right) ds \\ &\equiv \int_0^t L_\varepsilon\left(t, s, \tilde{x}_\varepsilon, \frac{s}{\varepsilon}, \omega\right) ds + z_\varepsilon(t),\end{aligned} \tag{4.2}$$

where

$$z_\varepsilon(t) = \varepsilon^{-1/2}\int_0^t \left(K\left(t, \bar{x}(s), \frac{s}{\varepsilon}, \omega\right) - \bar{K}\left(t, \bar{x}(s)\right)\right) ds \tag{4.3}$$

and

$$L_\varepsilon(t, s, x, u, \omega) = \varepsilon^{-1/2}\left(K(t, \bar{x}(s) + \varepsilon^{1/2}x, u, \omega) - K(t, \bar{x}(s), u, \omega)\right).$$

Note that $L_\varepsilon(t, s, x, u, \omega)$ satisfies the Lipschitz condition

$$|L_\varepsilon(t, s, x, u, \omega) - L_\varepsilon(t, s, \tilde{x}, u, \omega)| \le l|x - \tilde{x}|$$

if K satisfies relation (3.3), and the constant l is the same as in (3.3), so it does not depend on ε. Suppose that the derivative $K_x(t, x, s, \omega)$ exists as an $L(R^d)$-valued random variable. Then, as $\varepsilon \to 0$,

$$L_\varepsilon(t, s, \bar{x}, u, \omega) \to K_x(t, \bar{x}(s), u, \omega)x.$$

Furthermore, the process $z_\varepsilon(t)$ converges to a Gaussian process $\hat{z}(t)$ under some general conditions.

We first consider the linear equation

$$x_\varepsilon^*(t) = \int_0^t L\left(t, s, \frac{s}{\varepsilon}, \omega\right) x_\varepsilon^*(s)\, ds + z_\varepsilon(t)$$

where $L\left(t, s, \frac{s}{\varepsilon}, \omega\right) = K_x\left(t, \bar{x}(s), \frac{s}{\varepsilon}, \omega\right)$. This equation is obtained by substituting L for L_ε in equation (4.2). Set $\hat{L}(t, s) = EL\left(t, s, \frac{s}{\varepsilon}, \omega\right)$; let $\hat{x}_\varepsilon(t)$ be the solution of the equation

$$\hat{x}_\varepsilon(t) = \int_0^t \hat{L}(t, s)\hat{x}_\varepsilon(s)\, ds + z_\varepsilon(t); \tag{4.4}$$

and consider the equation

$$\hat{x}(t) = \int_0^t \hat{L}(t, s)\hat{x}(s)\, ds + \hat{z}(t). \tag{4.5}$$

It is easy to find conditions under which $\tilde{x}_\varepsilon(t) - x_\varepsilon^*(t)$ and $x_\varepsilon^*(t) - \hat{x}_\varepsilon$ converge to zero in probability. Once that is done, we prove that $\tilde{x}_\varepsilon(t)$ converges weakly to $\hat{x}(t)$. Since $\hat{x}(t)$ is the solution of a linear equation, it can be expressed as a linear transformation of a Gaussian stochastic process $\hat{z}(t)$, so $\tilde{x}_\varepsilon(t)$ is asymptotically a Gaussian stochastic process.

The conditions in the next theorem ensure that this plan works.

Theorem 1 *Let $K(t, x, s, \omega)$ satisfy the conditions:*

(1) *K is measurable in all variables; it is a stationary ergodic process in s for fixed t and x, with $E|K(t, x, s, \omega)| < \infty$; and it satisfies conditions (1) and (2) of Section 2.3.4.*

(2) *K satisfies a Lipschitz condition in x: For some positive real constant l,*

$$|K(t, x_1, s, \omega) - K(t, x_2, s, \omega)| \le l|x_1 - x_2|.$$

(3) *Let $\bar{K}(t, x) = EK(t, x, s, \omega)$ and $\tilde{K}(t, x, s, \omega) = K(t, x, s, \omega) - \bar{K}(t, x)$. Then there exists an $L(R^d)$-valued continuous function*

$$B(t_1, x_1, t_2, x_2, v) : R_+ \times R^d \times R_+ \times R^d \times R \to L(R^d)$$

for which

$$E\left(\tilde{K}(t_1, x_1, s, \omega), z_1\right)\left(\tilde{K}(t_2, x_2, s + v, \omega), z_2\right) = (B(t_1, x_1, t_2, x_2, v)z_1, z_2) \tag{4.6}$$

and

$$\int_{-\infty}^{\infty} \|B(t_1, x_1, t_2, x_2, v)\| dv \tag{4.7}$$

converges uniformly with respect to $t_1 \le c$, $|x_1| \le c$, $t_2 \le c$, $|x_2| \le c$ for every $c > 0$.

(4) *There exists the derivative $K_x(t, x, s, \omega)$, and for $z \in R^d$,*

$$\left|K_x(t, x, s, \omega)z - \varepsilon^{-1}\left(K(t, x + \varepsilon z, s, \omega) - K(t, x, s, \omega)\right)\right| \le \alpha(\varepsilon)(1+|z|), \tag{4.8}$$

where $\alpha(\varepsilon) \to 0$ as $\varepsilon \to 0$,

(5) *$K_x(t, x, s, \omega)$ is an ergodic stationary $L(R^d)$-valued stochastic process for fixed t and x, and*

$$\|K_x(t_1, x_1, s, \omega) - K_x(t_2, x_2, s, \omega)\| \le \lambda\left(|t_1 - t_2|, c\right) + \lambda\left(|x_1 - x_2|, c\right) \tag{4.9}$$

for $t_1 \le c$, $t_2 \le c$, $x_1 \le c$, $x_2 \le c$, where the function $\lambda(h, c) : R_+^2 \to R$ satisfies the relation $\lambda(0+, c) = 0$.

Denote by $\hat{z}(t)$ a Gaussian R^d-valued process for which $E\hat{z}(t) = 0$ and

$$E\left(\hat{z}(t_1), z_1\right)\left(\hat{z}(t_2), z_2\right) = \left(B(t_1, t_2)z_1, z_2\right), \quad z_1, z_1 \in R^d,$$

where

$$B(t_1, t_2) = \int_0^{t_1 \wedge t_2} ds \int_{-\infty}^{\infty} B(t_1, \bar{x}(s), t_2, \bar{x}(s), v) dv. \tag{4.10}$$

Then the stochastic process $\tilde{x}_\varepsilon(t)$ converges weakly to the stochastic process $\hat{x}(t)$, the solution to integral equation (4.5).

Proof For the proof, we follow the plan outlined above and use the notation introduced at the beginning of this section. Note that

$$\begin{aligned}|\tilde{x}_\varepsilon(t) - x^*_\varepsilon(t)| \leq &\int_0^t \left| L_\varepsilon\left(t, s, \tilde{x}_\varepsilon(s), \frac{s}{\varepsilon}, \omega\right) - L_\varepsilon\left(t, s, x^*_\varepsilon(s), \frac{s}{\varepsilon}, \omega\right)\right| ds \\ &+ \int_0^t \left| L_\varepsilon\left(t, s, x^*_\varepsilon(s), \frac{s}{\varepsilon}, \omega\right) x^*_\varepsilon(s)\right| ds \\ \leq\, & l \int_0^t |\tilde{x}_\varepsilon(s) - x^*_\varepsilon(s)|\, ds + \alpha(\varepsilon) \int_0^t (1 + |\bar{x}(s)|)\, |x^*_\varepsilon(s)|\, ds.\end{aligned}$$

This inequality implies that

$$\int_0^{t_0} |\tilde{x}_\varepsilon(s) - x^*_\varepsilon(s)|^2\, ds \leq A(t_0)\alpha(\varepsilon)\left(1 + \int_0^{t_0} |x^*_\varepsilon(s)|^2\, ds\right), \tag{4.11}$$

where $A(t_0)$ is a constant. Since $\left\|L\left(t, s, \frac{s}{\varepsilon}, \omega\right)\right\| \leq l$, there exists a constant $B(t_0)$ for which

$$\int_0^{t_0} |x^*_\varepsilon(s)|^2\, ds \leq B(t_0) \int_0^{t_0} |z_\varepsilon(s)|^2\, ds. \tag{4.12}$$

Now we consider the difference $x^*_\varepsilon(t) - \hat{x}_\varepsilon(t)$. We can write

$$\begin{aligned}|x^*_\varepsilon(t) - \hat{x}_\varepsilon(t)| \leq\, & l \int_0^t |x^*_\varepsilon(s) - \hat{x}_\varepsilon(s)|\, ds \\ &+ \left|\int_0^t \left(L\left(t, s, \frac{s}{\varepsilon}, \omega\right) - \hat{L}(t, s)\hat{x}_\varepsilon(s)\right) ds\right|,\end{aligned}$$

which implies that for some constant $c(t_0)$,

$$\int_0^{t_0} |x^*_\varepsilon(s) - \hat{x}_\varepsilon(s)|^2 \leq c(t_0) \int_0^{t_0} \left(\int_0^{t_0} \left(L\left(t, \frac{s}{\varepsilon}, \omega\right) - \hat{L}(t, s)\right) \hat{x}_\varepsilon(s)\, ds\right)^2 dt. \tag{4.13}$$

Let $\varphi(t)$ be the solution of the Volterra equation

$$\varphi(t) = \int_0^t \hat{L}(t, s)\varphi(s)\, ds + \psi(t),$$

where $\psi(t) \in L_2(0, t_0)$. The solution of this equation can be written as

$$\varphi(t) = \psi(t) + \int_0^t \Gamma(t, s)\psi(s)\, ds,$$

where Γ is the resolvent kernel

$$\begin{aligned}\Gamma(t, s) =& \hat{L}(t, s) + \int_{s_1}^t \hat{L}(t, s_1)\hat{L}(s_1, s)\, ds_1 + \cdots \\ &+ \iint_{s < x_1 < \cdots < s_n < t} \hat{L}(t, s_n) \cdots \hat{L}(s_1, s)\, ds_1 \cdots ds_n + \cdots,\end{aligned}$$

and $\Gamma(t, s)$ is a bounded continuous function.

It follows from (4.4) that

$$\hat{x}_\varepsilon(t) = z_\varepsilon(t) + \int_0^t \Gamma(t,s) z_\varepsilon(s)\, ds. \tag{4.14}$$

We next investigate the asymptotic properties of the process $z_\varepsilon(t)$. For this purpose we prove the following lemma.

Lemma 1 *The stochastic process $z_\varepsilon(t)$ converges weakly to the stochastic process $\hat{z}(t)$.*

Proof It follows from formula (4.3) that if $B_\varepsilon(t_1, t_2)$ is an $L(R^d)$-valued function for which $E\left(z_\varepsilon(t_1), z_1\right)\left(z_\varepsilon(t_2), z_2\right) = \left(B_\varepsilon(t_1,t_2) z_1, z_2\right)$, then for $t_1 < t_2$,

$$B_\varepsilon(t_1, t_2) = \int_0^{t_1} ds \int_{-\frac{s}{\varepsilon}}^{\frac{t_2 - s}{\varepsilon}} B\left(t_1, \bar{x}(s), t_2, \bar{x}(s + \varepsilon v), v\right) dv,$$

where B is given by equation (4.6). It is easy to see that $B_\varepsilon(t_1, t_2) \to B(t_1, t_2)$ as $\varepsilon \to 0$ uniformly for $t_1, t_2 \le c < \infty$.
The proof of the lemma is a consequence of Theorem 6, Section 4.2.

□

Now, to prove the theorem it suffices to prove that

$$\int_0^{t_0} \left(\int_0^t \left| L\left(t, s, \frac{s}{\varepsilon}, \omega\right) - \hat{L}(t,s) \right| \hat{x}_\varepsilon(s)\, ds \right)^2 dt \to 0 \tag{4.15}$$

in probability as $\varepsilon \to 0$. Set $\tilde{L} = L(t, s, \frac{s}{\varepsilon}, \omega) - \hat{L}(t,s)$ and

$$\tilde{L}_n(t, s, u, \omega) = \tilde{L}\left(\frac{[nt]}{n}, \frac{[ns]}{n}, u, \omega \right),$$

where $[n\alpha]$ is the integer part of $n\alpha$ ($= k$ if $k \le n\alpha < k+1$).
It follows from condition (5) that for $s \le t_0$ and $t \le t_0$,

$$\left\| \tilde{L}_n(t, s, u, \omega) - \tilde{L}(t, s, u, \omega) \right\| \le \beta_n,$$

where $\beta_n \to 0$, since $\sup_{s \le t_0} |\bar{x}(s)| < \infty$.
So, we have to prove that

$$\int_0^{t_0} |\hat{x}_\varepsilon(s)|^2\, ds \tag{4.16}$$

is bounded in probability and that

$$\int_0^{t_0} \left| \int_0^t \tilde{L}_n\left(t, s, \frac{s}{\varepsilon}, \omega\right) \hat{x}_\varepsilon(s)\, ds \right|^2 dt \to 0$$

in probability. Let $r_\varepsilon(t_1, t_2) = E(\hat{x}_\varepsilon(t_1), \hat{x}_\varepsilon(t_2))$. Then it follows from (4.14) that

$$r_\varepsilon(t_1,t_2) = \mathrm{Tr}B_\varepsilon(t_1,t_2) + \int_0^{t_1} \mathrm{Tr}B_\varepsilon(t_2,s)\Gamma(t_1,s)\,ds + \int_0^{t_2} \mathrm{Tr}B_\varepsilon(t_2,s)\Gamma(t_2,s)\,ds + \int_0^{t_1}\int_0^{t_2} \mathrm{Tr}\Gamma(t_1,s_1)\Gamma(t_2,s_2)B_\varepsilon(s_1,s_2)\,ds_1\,ds_2 \tag{4.17}$$

(here Tr B is the trace of the matrix B, i.e., the sum of the elements on the main diagonal of B). So

$$E\int_0^{t_0} |\hat{x}_\varepsilon(s)|^2 = \int_0^{t_0}\int_0^{t_0} r_\varepsilon(s_1,s_2)\,ds_1\,ds_2$$

is bounded. It is easy to see that $r_\varepsilon(t_1,t_2) \to r(t_1,t_2)$ uniformly if $t_1 \le t_0$, $t_2 \le t_0$, where $r(t_1,t_2)$ is defined by (4.17) when we replace B_ε by B. Note that

$$\begin{aligned}
&E\left|\int_0^t \tilde{L}_n\left(t,s,\frac{s}{\varepsilon},\omega\right)\left(\hat{x}_\varepsilon(s) - \hat{x}_\varepsilon\left(\frac{[ns]}{n}\right)\right) ds\right|^2 \\
&\le 4l^2 \int_0^t E\left|\hat{x}_\varepsilon(s) - \hat{x}_\varepsilon\left(\frac{[ns]}{n}\right)\right|^2 ds \\
&= 4l^2 \int_0^t \left[r_\varepsilon(s,s) + r_\varepsilon\left(\frac{[ns]}{n},\frac{[ns]}{n}\right) - 2r_\varepsilon\left(s,\frac{[ns]}{n}\right)\right] ds,
\end{aligned}$$

since $\|L(\cdots)\| \le l$ is a consequence of condition (2).
Therefore,

$$\lim_{n\to\infty} E\int_0^{t_0}\left(\int_0^t \tilde{L}_n\left(t,s,\frac{s}{\varepsilon},\omega\right)\left(\hat{x}_\varepsilon(s) - \hat{x}_\varepsilon\left(\frac{[ns]}{n}\right)\right) ds\right)^2 ds = 0.$$

The proof of the theorem is now a consequence of the fact that for each k,

$$\lim_{\varepsilon\to 0}\int_{\frac{k}{n}}^{\frac{k+1}{n}} \tilde{L}_n\left(t,s,\frac{s}{\varepsilon},\omega\right) ds = 0$$

with probability 1.

□

Remark If we could prove that $z_\varepsilon(t)$ converges weakly to $\hat{z}(t)$ in $C_{[0,t_0]}$, then we could conclude that $\tilde{x}_\varepsilon(t)$ converges weakly in $C_{[0,t_0]}$ to $\hat{x}(t)$. But we have only the weak convergence of $\tilde{x}_\varepsilon(t)$ to $\hat{x}(t)$, that is, the convergence of finite–dimensional distributions.

4.2 Differential Equations

Next, we consider a system of differential equations of the form

$$\dot{x}_\varepsilon(t) = a\left(x_\varepsilon(t), \frac{t}{\varepsilon}, \omega\right), \quad t > 0,\ x_\varepsilon(0) = x_0, \tag{4.18}$$

where $a(x, s, \omega)$ is an R^d-valued random field and the solution $x_\varepsilon(t)$ is an R^d-valued stochastic process. As was shown in Section 3.2, we can transform equation (4.18) into a Volterra integral equation. After this, we can use Theorem 1 from Section 4.1.

Theorem 2 *Let $a(x, t, \omega)$ satisfy the following conditions:*

(1) *a is measurable in all variables; it is a stationary ergodic process in t for each fixed x, $E\,|a(x, t, \omega)| < \infty$; and it satisfies conditions (1) and (2) of Section 2.2.*

(2) *Let $\bar{a}(x) = Ea(x, t, \omega)$ and $\tilde{a}(x, t, \omega) = a(x, t, \omega) - \bar{a}(x)$. Then there exists an $L(R^d)$-valued function $B(x_1, x_2, t)$ for which*

$$E\left(\tilde{a}(x_1, t_1, \omega), z_1\right)\left(\tilde{a}(x_2, t_2, \omega), z_2\right) = \left(B(x_1, x_2, t_2 - t_1, \omega)z_1, z_2\right)$$

and $\int_{-\infty}^{\infty} \|B(x_1, x_2, t)\|\, dt$ converges uniformly for $|x_1|, |x_2| \le c$ for any $c > 0$.

(3) *The derivative $a_x(x, t, \omega)$ exists as a stationary ergodic process in t for fixed x; there exists a function $\lambda(\theta, c) : R_+^2 \to R_+$ for which $\lambda(0+, c) = 0$; and for any $c > 0$ there is an ergodic process $\rho_c(t)$ with $E\rho_c(t) < \infty$ such that for any small real number θ,*

$$\left|\theta^{-1}\left(a(x + \theta z, t) - a(x, t)\right) - a_x(x, t)z\right| \le \rho_c(t)\lambda(\theta, c)|z|$$

and

$$\|a_x(x_1, t) - a_x(x_2, t)\| \le \rho_c(t)\lambda(|x_1 - x_2|, c) \text{ if } |x_1| \le c,\ |x_2| \le c.$$

Then the process $\tilde{x}_\varepsilon(t) = \varepsilon^{-1/2}\left(x_\varepsilon(t) - \bar{x}(t)\right)$, where $\bar{x}(t)$ is the solution of the problem

$$\dot{\bar{x}}(t) = \bar{a}\left(\bar{x}(t)\right), \quad x(0) = x_0, \tag{4.19}$$

and $\bar{a}(x) = Ea(x, t)$, converges weakly in $C_{[0,t_0]}$ to the Gaussian process $\hat{x}(t)$, which is the solution of the linear stochastic differential equation

$$d\hat{x}(t) = a_x\left(\bar{x}(t)\right)\hat{x}(t)\, dt + d\hat{z}(t), \quad \hat{x}(0) = 0, \tag{4.20}$$

where $\hat{z}(t)$ is an R^d-valued Gaussian process with independent increments for which $E\hat{z}(t) = 0$ and

$$E\left(\hat{z}(t), z\right)^2 = \int_0^t \int_{-\infty}^{\infty} \left(B\left(\bar{x}(s), \bar{x}(s), v\right) z, z\right) dv\, ds. \tag{4.21}$$

Proof We have

$$\begin{aligned}\tilde{x}_\varepsilon(t) = &\int_0^t a_x\left(\bar{x}(s), \frac{s}{\varepsilon}, \omega\right)\tilde{x}_\varepsilon(s)\,ds + \int_0^t \frac{1}{\sqrt{\varepsilon}}\tilde{a}\left(\bar{x}(s), \frac{s}{\varepsilon}, \omega\right)\,ds \\ &+ \int_0^t \left(\frac{1}{\sqrt{\varepsilon}}\left[a\left(\bar{x}(s)+\sqrt{\varepsilon}\tilde{x}_\varepsilon(s), \frac{s}{\varepsilon}, \omega\right) - a\left(\bar{x}(s), \frac{s}{\varepsilon}, \omega\right)\right]\right. \\ &\left. - a_x\left(\bar{x}(s), \frac{s}{\varepsilon}, \omega\right)\right)\tilde{x}_\varepsilon(s)\,ds.\end{aligned} \tag{4.22}$$

Let

$$z_\varepsilon(t) = \frac{1}{\sqrt{\varepsilon}}\int_0^t \tilde{a}\left(\bar{x}(s), \frac{s}{\varepsilon}, \omega\right)\,ds. \tag{4.23}$$

It follows from Theorem 6 of Chapter 2 that $z_\varepsilon(t)$ converges weakly in $C_{[0,t_0]}$ to the process $\hat{z}(t)$. Equation (4.22) implies that for $t \le t_0$,

$$|\tilde{x}_\varepsilon(t)| \le \int_0^t \left(\left\|a_x\left(\bar{x}(s), \frac{s}{\varepsilon}, \omega\right)\right\| + \rho_c\left(\frac{s}{\varepsilon}\right)\lambda(\sqrt{\varepsilon}, c)\right)|\tilde{x}_\varepsilon(s)|\,ds + |z_\varepsilon(t)|,$$

where $c > \sup_{t\le t_0}|\bar{x}(t)|$. So, $u = \sup_{t\le t_0}|\tilde{x}_\varepsilon(t)|$, then

$$\begin{aligned}u \le &\sup_{t\le t_0}|z_\varepsilon(t)|\exp\left\{\int_0^{t_0}\left(\left\|a_x\left(\bar{x}(s), \frac{s}{\varepsilon}, \omega\right)\right\| + \rho_c\left(\frac{s}{\varepsilon}\right)\lambda(\sqrt{\varepsilon}, c)\right)\,ds\right\} \\ \equiv &\,\alpha_\varepsilon(t).\end{aligned} \tag{4.24}$$

Let

$$\begin{aligned}u_\varepsilon(t) = \int_0^t &\left(\frac{1}{\sqrt{\varepsilon}}\left[a\left(\bar{x}(s)+\sqrt{\varepsilon}\tilde{x}_\varepsilon(s), \frac{s}{\varepsilon}, \omega\right) - a\left(\bar{x}(s), \frac{s}{\varepsilon}, \omega\right)\right]\right. \\ &\left. - a_x\left(\bar{x}(s), \frac{s}{\varepsilon}, \omega\right)\right)\,ds.\end{aligned}$$

It follows from condition (3) that

$$\sup_{t\le t_0}|u_\varepsilon(t)| \le \lambda(\sqrt{\varepsilon}, c)\alpha_\varepsilon(t_0)\int_0^{t_0}\rho_c\left(\frac{s}{\varepsilon}\right)\,ds. \tag{4.25}$$

Let $\hat{x}_\varepsilon(t)$ be the solution of the equation

$$\hat{x}_\varepsilon(t) = \int_0^t a_x\left(\bar{x}(s), \frac{s}{\varepsilon}, \omega\right)\hat{x}_\varepsilon(s)\,ds + z_\varepsilon(t). \tag{4.26}$$

Then

$$|\hat{x}_\varepsilon(t) - \tilde{x}_\varepsilon(t)| \le \int_0^t \left\|a_x\left(\bar{x}(s), \frac{s}{\varepsilon}, \omega\right)\right\|\,|\hat{z}_\varepsilon(s) - \tilde{x}_\varepsilon(s)|\,ds + |u_\varepsilon(t)|.$$

Therefore,

$$\sup_{t\le t_0}|\hat{x}_\varepsilon(t) - \tilde{x}_\varepsilon(t)| \le \exp\left\{\int_0^{t_0}\left\|a_x\left(\bar{x}(s), \frac{s}{\varepsilon}, \omega\right)\right\|\,ds\right\}\sup_{t\le t_0}|u_\varepsilon(t)|. \tag{4.27}$$

Now denote by $x_\varepsilon^*(t)$ the solution of the equation

$$x_\varepsilon^*(t) = \int_0^t \bar{a}_x(\bar{x}(s))\, x_\varepsilon^*(s)\, ds + z_\varepsilon(t). \tag{4.28}$$

It is easy to see that $x_\varepsilon^*(t)$ converges weakly to $\hat{x}(t)$ in $C_{[0,t_0]}$, since we have the same result for $z_\varepsilon(t)$. In addition,

$$\begin{aligned} \sup_{t\le t_0} |x_\varepsilon^*(t) - \hat{x}_\varepsilon(t)| \le \sup_{t\le t_0} &\left| \int_0^t \left(a_x\left(\bar{x}(s), \frac{s}{\varepsilon}, \omega\right) - \bar{a}_x(\bar{x}(s))\right) ds \right| \\ &\times \exp\left\{ \int_0^{t_0} \left\| a_x\left(\bar{x}(s), \frac{s}{\varepsilon}, \omega\right)\right\| ds \right\}. \end{aligned} \tag{4.29}$$

So to prove the theorem, it suffices to prove the following two lemmas.

Lemma 2

$$\lim_{\varepsilon\to 0} \int_0^{t_0} \left\| a_x\left(\bar{x}(s), \frac{s}{\varepsilon}, \omega\right)\right\| ds = \int_0^{t_0} \left\|\bar{a}_x(\bar{x}(s))\right\| ds.$$

Lemma 3

$$\sup_{t\le t_0} \left| \int_0^t a_x\left(\bar{x}(s), \frac{s}{\varepsilon}, \omega\right) ds - \int_0^{t_0} \bar{a}_x(\bar{x}(s))\, ds \right| \to 0$$

in probability as $\varepsilon \to 0$.

Suppose that we have proved these lemmas. Then $\hat{x}_\varepsilon(t)$ converges weakly to $\hat{x}(t)$ in $C_{[0,t_0]}$ because of the inequality (4.29). Since $\sup_{t\le t_0} |z_\varepsilon(t)|$ is bounded in probability, $\sup_{t\le t_0} |u_\varepsilon(t)| \to 0$ as $\varepsilon \to 0$ because of (4.25). This implies that $\tilde{x}_\varepsilon(t)$ converges weakly to $\hat{x}(t)$ in $C_{[0,t_0]}$ because of inequality (4.27).
This completes the proof of the theorem.

□

Proof of Lemma 2
Note that it follows from condition 3) that

$$\left\| a_x\left(\bar{x}_n(t), \frac{t}{\varepsilon}, \omega\right) - \bar{a}_x\left(\bar{x}(t), \frac{t}{\varepsilon}, \omega\right)\right\| \le \rho_c\left(\frac{t}{\varepsilon}\right) \lambda\left(|\bar{x}_n(t) - \bar{x}(t)|, c\right)$$

if $\bar{x}_n(t)$ is an R^d-valued function for which $|\bar{x}_n(t)| \le c$ and $|\bar{x}(t)| \le c$. Set

$$\bar{x}_n(t) = \bar{x}\left(\frac{k}{n}\right)$$

if $t \in [\frac{k}{n}, \frac{k+1}{n})$. Then

$$\begin{aligned} &\left| \int_0^{t_0} \left\| a_x\left(\bar{x}(s), \frac{s}{\varepsilon}, \omega\right)\right\| ds - \int_0^{t_0} \left\| a_x\left(\bar{x}_n(s), \frac{s}{\varepsilon}, \omega\right)\right\| ds \right| \\ &\qquad \le \lambda\left(\sup_{t\le t_0} |\bar{x}_n(t) - \bar{x}(t)|, c\right) \int_0^{t_0} \rho_c\left(\frac{t}{\varepsilon}\right) dt. \end{aligned}$$

It follows from the ergodic theorem that

$$\lim_{\varepsilon\to 0}\int_0^{t_0}\left\|a_x\left(\bar{x}_n(s),\frac{s}{\varepsilon},\omega\right)\right\|\,ds=\int_0^{t_0}\|\bar{a}_x\left(x_n(s)\right)\|\,ds.$$

The proof of the lemma is a consequence of the relations

$$\lim_{n\to\infty}\sup_{t\leq t_0}|\bar{x}_n(t)-\bar{x}(t)|=0$$

and

$$\lim_{n\to\infty}\int_0^{t_0}\|a_x\left(x_n(s)\right)\|\,ds=\int_0^{t_0}\|a_x\left(x(s)\right)\|\,ds.$$

□

Proof of Lemma 3
We note that

$$\sup_{t\leq t_0}\int_0^t\left|a_x\left(\bar{x}(s),\frac{s}{\varepsilon},\omega\right)\,ds-a_x\left(\bar{x}_n(s),\frac{s}{\varepsilon},\omega\right)\right|ds$$
$$\leq\lambda\left(\sup_{t\leq t_0}|\bar{x}_n(t)-\bar{x}(t)|\,,c\right)\int_0^{t_0}\rho_c\left(\frac{t}{\varepsilon}\right)dt,$$

and for any n,

$$\lim_{\varepsilon\to 0}\sup_{t\leq t_0}\left|\int_0^t a_x\left(\bar{x}_n(s),\frac{s}{\varepsilon},\omega\right)\,ds-\int_0^t\bar{a}_x\left(x_n(s)\right)\,ds\right|=0,$$

which follows from the ergodic theorem.

□

Next, we apply these results to a linear problem.

Theorem 3 *Consider a linear initial value problem of the form*

$$\dot{x}_\varepsilon(t)=A\left(\frac{t}{\varepsilon},\omega\right)x_\varepsilon(t),\quad x_\varepsilon(0)=x_0,\tag{4.30}$$

where $A(t)$ is an ergodic, $L(R^d)$–valued stationary process satisfying conditions (1) and (2) of Section 2.2.
Let $\bar{A}=EA(t)$, $\bar{x}(t)=\exp\{t\bar{A}\}x_0$, and $B(t)=EA^*(t,\omega)A(0,\omega)$.
Assume that

$$\int_0^\infty\|B(t)\|\,dt<\infty.$$

Then the process $\tilde{x}_\varepsilon(t)=\left(x_\varepsilon(t)-\bar{x}(t)\right)/\sqrt{\varepsilon}$ converges weakly in $C_{[0,t_0]}$ to the Gaussian process $\hat{x}(t)$ defined by the formula

$$\hat{x}(t)=\int_0^t\exp\left\{(t-s)\bar{A}\right\}(d\hat{Z}(s))\bar{x}(s),\tag{4.31}$$

where $\hat{Z}(t)$ is an $L(R^d)$–valued Gaussian process with independent increments for which

$$E\hat{Z}(t) = 0, \quad E\hat{Z}^*(t)\hat{Z}(t) = t\int_{-\infty}^{\infty} B(v)\,dv. \tag{4.32}$$

Proof The proof of this statement is a consequence of Theorem 2 when we set

$$a(x,t,\omega) = A(t,\omega)x.$$

The conditions on $A(t,\omega)$ imply that all of the conditions of Theorem 2 are satisfied. Denote by $\hat{Z}_\varepsilon(t)$ the $L(R^d)$-valued stochastic process

$$\hat{Z}_\varepsilon(t) = \frac{1}{\sqrt{\varepsilon}}\int_0^t \left(A\left(\frac{s}{\varepsilon}\right) - \bar{A}\right) ds.$$

This process converges weakly in $C_{[0,t_0]}$ to the process $\hat{Z}(t)$, and $\tilde{x}_\varepsilon(t)$ converges weakly in $C_{[0,t_0]}$ to the process $\hat{x}(t)$, which is the solution of the equation

$$d\hat{x}(t) = \bar{A}\,\hat{x}(t)\,dt + d\hat{Z}(t)\,\bar{x}(t), \quad \hat{x}(0) = 0. \tag{4.33}$$

The solution $\hat{x}(t)$ can be rewritten in the form shown in equation (4.31). □

Remark 2 *Let*

$$\hat{x}_\varepsilon(t) = \bar{x}(t) + \sqrt{\varepsilon}\hat{x}(t). \tag{4.34}$$

It follows from Theorem 3 that the distributions of the processes

$$\frac{x_\varepsilon(t) - \bar{x}(t)}{\sqrt{\varepsilon}} \text{ and } \frac{\hat{x}_\varepsilon(t) - \bar{x}(t)}{\sqrt{\varepsilon}}$$

coincide asymptotically as $\varepsilon \to 0$. So the distributions of $x_\varepsilon(t)$ and $\hat{x}_\varepsilon(t)$ are also close in some sense.
There are several useful consequences of Theorem 3.

I. *Consider a function $\phi(x(\cdot)) : C_{[0,t_0]} \to R$, and suppose it satisfies the Lipschitz condition*

$$|\phi(x(\cdot) + y(\cdot)) - \phi(x(\cdot))| \le l_\phi \|y(\cdot)\|_{C_{[0,t_0]}}$$

for all $x(\cdot)$ and $y(\cdot)$ in $C_{[0,t_0]}$, where l_ϕ is a constant and where

$$\|y(\cdot)\|_{C_{[0,t_0]}} = \sup_{t\in[0,t_0]} |y(t)|.$$

Then

$$|E\phi(x_\varepsilon(\cdot)) - E\phi(\hat{x}_\varepsilon(\cdot))| = o(\sqrt{\varepsilon}), \tag{4.35}$$

since

$$\lim_{\varepsilon\to 0}\left(E\frac{\phi\left(x_\varepsilon(\cdot)\right)-\phi(\bar{x}(\cdot))}{\sqrt{\varepsilon}}-E\frac{\phi\left(\hat{x}_\varepsilon(\cdot)\right)-\phi\left(\bar{x}(\cdot)\right)}{\sqrt{\varepsilon}}\right)=0. \tag{4.36}$$

The last relation follows from the facts that the family of functions

$$\left\{G_\varepsilon\left(y(\cdot)\right)=\frac{1}{\sqrt{\varepsilon}}\left(\phi\left(\bar{x}(\cdot)+\sqrt{\varepsilon}y(\cdot)\right)-\phi(\bar{x}(\cdot))\right),\, \varepsilon>0\right\}$$

is uniformly locally bounded in $C_{[0,t_0]}$ *and that all these functions satisfy the Lipschitz condition with the same constant* l_ϕ.

II. *Let* $\phi\left(x(\cdot)\right)$ *satisfy the condition of statement* I *and suppose there exists the derivative*

$$\phi'\left(x(\cdot),y(\cdot)\right)=\lim_{h\to 0}\frac{1}{h}\left(\phi\left(x(\cdot)+hy(\cdot)\right)-\phi\left(x(\cdot)\right)\right). \tag{4.37}$$

Suppose further that the functions

$$\left\{H_\varepsilon\left(y(\cdot)\right)=\frac{1}{\varepsilon}\left(\phi\left(\bar{x}(\cdot)+\sqrt{\varepsilon}\,y(\cdot)\right)-\phi(\bar{x})-\sqrt{\varepsilon}\,\phi'\left(\bar{x}(\cdot),y(\cdot)\right)\right),\,\varepsilon>0\right\} \tag{4.38}$$

satisfy a Lipschitz condition with the same constant (this is true, for example, if ϕ'' is bounded). *Then*

$$\left|E\phi\left(x_\varepsilon(\cdot)\right)-E\phi\left(\hat{x}_\varepsilon(\cdot)\right)\right|=o(\varepsilon). \tag{4.39}$$

Remark 3 *The process* $\hat{x}_\varepsilon(t)$ *satisfies the stochastic differential equation*

$$d\hat{x}_\varepsilon(t)=\bar{A}\,\hat{x}_\varepsilon(t)\,dt+\sqrt{\varepsilon}\,d\hat{Z}(t)\,\hat{x}_\varepsilon(t)-\varepsilon\,d\hat{Z}(t)\,\hat{x}(t).$$

We introduce the process $\hat{\hat{x}}_\varepsilon(t)$, *which is the solution of the stochastic differential equation*

$$d\hat{\hat{x}}_\varepsilon(t)=\bar{A}\,\hat{\hat{x}}_\varepsilon(t)\,dt+\sqrt{\varepsilon}\,d\hat{Z}(t)\,\hat{\hat{x}}_\varepsilon(t).$$

If $\hat{x}_\varepsilon(0)=\hat{\hat{x}}_\varepsilon(0)=\bar{x}(0)$, *then*

$$\sup_{t\le t_0}\left|\hat{x}_\varepsilon(t)-\hat{\hat{x}}_\varepsilon(t)\right|=O(\varepsilon).$$

These remarks show that we can approximate the process $x_\varepsilon(t)$ using a solution of a linear stochastic differential equation of diffusion type.

We consider next construction of the solution of the nonlinear equation (4.18) guided by our work on linear problems. First, we transform the process $\hat{Z}(t)$ that was introduced in Theorem 2. Let $B(x)=\int B(x,x,v)dv$, so $B(x)$ is an $L_+(R^d)$-valued function on R^d, and $B(x)$ is a nonnegative symmetric operator. We assume that $\det B(x)>0$, so we may write $B(x)=$

$Q^2(x)$, where $Q(x)$ is also a nonnegative symmetric operator. Next, we introduce an R^d-valued process

$$w(t) = \int_0^t Q^{-1}(\bar{x}(s))\, d\hat{Z}(s). \tag{4.40}$$

Then $w(t)$ is a Gaussian process with independent increments, and $Ew(t) = 0$. Moreover,

$$\begin{aligned} E(w(t), z)^2 &= E\left(\int_0^t Q^{-1}(\bar{x}(s))\, d\hat{Z}(s), z\right)^2 \\ &= \int_0^t \left(B(\bar{x}(s))\, Q^{-1}(\bar{x}(s))\, z, Q^{-1}(\bar{x}(s))\, z\right) ds \equiv t(z,z), \end{aligned}$$

for any vector z. So $w(t)$ is a Wiener process in R^d, and $\hat{Z}(t)$ can be expressed as

$$\hat{Z}(t) = \int_0^t Q(\bar{x}(s))\, dw(s). \tag{4.41}$$

With these preliminaries, we state further important results in the following theorem.

Theorem 4 *Let $\hat{x}_\varepsilon(t)$ be the solution of the stochastic differential equation*

$$d\hat{x}_\varepsilon(t) = \bar{a}\left(\hat{x}_\varepsilon(t)\right) dt + \sqrt{\varepsilon}\, Q\left(\hat{x}_\varepsilon(t)\right) dw(t) \tag{4.42}$$

with the initial condition $\hat{x}_\varepsilon(0) = x_0$, and suppose that the function $Q(x)$ satisfies the Lipschitz condition $\|Q(x_1) - Q(x_2)\| \le q|x_1 - x_2|$. We assume that the conditions of Theorem 2 are fulfilled and that $x_\varepsilon(t)$ is the solution of equation (4.18).
Then:
I. *If $\phi(x(\cdot)) : C_{[0,t_0]} \to R$ satisfies the Lipschitz condition, then (4.35) holds.*
II. *If ϕ has a continuous bounded derivative $\phi'(x,y)$ for which the function $H_\varepsilon(y(\cdot))$, which is represented by formula (4.38), also satisfies the Lipschitz condition, then (4.35) holds.*

The proof of these statements can be obtained in the same way as statements I and II in Remark 2. We have only to take into account the fact that

$$\sup_{t \le t_0} \left|\bar{x}(t) + \sqrt{\varepsilon}\, \hat{x}_\varepsilon(t) - \hat{x}_\varepsilon(t)\right| = O(\varepsilon), \tag{4.43}$$

which relation can be proved as was done in Remark 3 for the linear equation.

4.2.1 Markov Perturbations

Let $(Y, \mathcal{C})$ be a measurable space, and let $y(t)$ be a homogeneous Markov process in this space having transition probability

$$P(t, y, C) = P\{y(t) \in C / y(0) = y\}, \quad C \in \mathcal{C}. \tag{4.44}$$

We assume further that $y(t)$ is an ergodic process with an ergodic distribution $\rho(dy)$ satisfying the mixing condition SMC II from Section 2.3 and

$$R(y, C) = \int_0^\infty [P(t, y, C) - \rho(C)]\, dt. \tag{4.45}$$

We consider the initial value problem

$$\dot{x}_\varepsilon(t) = a\left(t, x_\varepsilon(t), y\left(\frac{t}{\varepsilon}\right)\right), \quad x_\varepsilon(0) = x_0, \tag{4.46}$$

where $a(t, x, y) : R_+ \times R^d \times Y \to R^d$ is a measurable function that satisfies additionally the following conditions:
(a1) There exists a function $\lambda(h, c) : R_+^2 \to R_+$, increasing in h, for which $\lambda(0+, c) = 0$ and

$$|a(t_1, x, y) - a(t_2, x, y)| \le \lambda\left(|t_1 - t_2|, t_1 \vee t_2 \vee |x|\right).$$

(a2) There exists a function $l(c, y) : R_+ \times Y \to R_+$, increasing in c, for which

$$|a(t, x_1, y) - a(t, x_2, y)| \le l(c, y)|x_1 - x_2| \text{ for } |x_1| \le c,\ |x_2| \le c \tag{4.47}$$

and

$$\int l(c, y)\rho(dy) < \infty, \qquad \int |a(t, x, y)|\, \rho(dy) < \infty.$$

(a3) There exists the derivative $a_x(t, x, y)$, with $\int \|a_x(t, x, y)\|\, \rho(dy) < \infty$, and $a_x(t, x, y)$ is locally uniformly continuous in t and x uniformly with respect to $y \in Y$.

Let

$$\bar{a}(t, x) = \int a(t, x, y)\rho(dy). \tag{4.48}$$

Then

$$\bar{a}_x(t, x) = \int a_x(t, x, y)\, \rho(dy).$$

Let $\bar{x}(t)$ be the solution of the initial value problem

$$\dot{\bar{x}}(t) = \bar{a}\left(t, \bar{x}(t)\right), \quad \bar{x}(0) = x_0,$$

and define

$$\tilde{x}_\varepsilon(t) = \frac{x_\varepsilon(t) - \bar{x}(t)}{\sqrt{\varepsilon}}.$$

Denote by $B(t, x_1, x_2)$ the operator in $L(R^d)$ defined by

$$(B(t,x_1,x_2)z_1,z_2) = \iint (a(t,x_1,y),z_1)\,(a(t,x_2,y'),z_2) \\ \times \{R(y',dy)\,\rho(dy') + R(y,dy')\,\rho(dy)\}\,.$$

Suppose that $\det B(t,x,x) > 0$, and let $Q(t,x) = B^{1/2}(t,x,x)$, where $Q(t,x)$ is a positive matrix. Let $\hat{Z}(t)$ be defined by the formula

$$\hat{Z}(t) = \int_0^t Q\,(s, \bar{x}(s))\; dw(s),$$

where $w(t)$ is the Wiener process in R^d, and denote by $\hat{x}(t)$ the solution of the integral equation

$$\hat{x}(t) = \int_0^t a_x\,(s,\bar{x}(s))\,\hat{x}(s)\,ds + \hat{Z}(t).$$

Let $\hat{x}(t)$ be the solution of the stochastic differential equation

$$d\hat{x}(t) = \bar{a}\,(t,\hat{x}(t))\;dt + Q\,(t,\hat{x}(t))\,dw(t). \tag{4.49}$$

With these preliminaries, we state the following theorem.

Theorem 5 (1) *The process $\tilde{x}_\varepsilon(t)$ converges to the process $\hat{x}(t)$ weakly in $C_{[0,t_0]}$ for all t_0.*
(2) *The statements* I *and* II *of Theorem* 4 *hold.*

The proof is the same as the proofs of Theorems 2 and 4, and it is not presented here.

4.3 Difference Equations

We consider a difference equation of the form (3.19), where the sequence $\{\varphi_n(x,\omega),\ n = 1,2,\dots\}$ satisfies the following conditions:

(φ**1**) For $n = 1,2,\dots$, the function $\varphi_n(x,\omega) : R^d \times \Omega \to R^d$ is $\mathcal{B}_{R^d} \otimes \mathcal{F}$-measurable and $E\,|\varphi_n(x,\omega)|^2 < \infty$.

(φ**2**) For fixed $x \in R^d$, $\{\varphi_n(x,\omega),\ n = 1,2,\dots\}$ is an ergodic stationary sequence, and it satisfies the conditions of Theorem 5 of Chapter 2.

(φ**3**) There exists an ergodic stationary sequence $\{l_n(\omega), n = 1,2,\dots\}$ in R_+ for which

$$E\,|\varphi_n(x,\omega) - \varphi_n(x',\omega)| \le l_n(\omega)|x - x'|.$$

(φ**4**) There exists the derivative

$$\frac{\partial \varphi_n}{\partial x}(x,\omega),$$

and for fixed x the sequence

$$\left\{\frac{\partial \varphi_n}{\partial x}(x,\omega),\ n=1,2,\dots\right\}$$

is an $L(R^d)$-valued stationary sequence. Moreover, there exist a function $\lambda(\theta,c):R_+^2\to R_+$, increasing in θ, and a constant c for which $\lambda(0+,c)=0$ and

$$\left|\theta^{-1}\left(\varphi_n(x+\theta z,\omega)-\varphi_n(x,\omega)\right)-\frac{\partial}{\partial x}\varphi_n(x,\omega)z\right|\le l_n(\omega)\lambda(\theta,|z|).$$

(φ**5**) There exists a sequence of $L(R^d)$-valued functions

$$\{B_k(x,x'),\ k=0,\pm1,\pm2,\dots\}$$

for which

$$(B_{n-m}(x,x')z,z')=E\left(\tilde\varphi_n(x,\omega),z\right)\left(\tilde\varphi_m(x',\omega),z'\right), \tag{4.50}$$

where $\tilde\varphi_n(x,\omega)=\varphi_n(x,\omega)-\bar\varphi(x)$, $\bar\varphi(x)=E\varphi_n(x,\omega)$, and

$$\sum_k\|B_k(x,x')\|<\infty.$$

If condition (φ**5**) is satisfied, we set

$$B(x,x')=\sum_k B_k(x,x'). \tag{4.51}$$

Denote by $\bar x_n^\varepsilon$ the sequence that is determined by relation (3.20), and let $\tilde x_n^\varepsilon=(x_n^\varepsilon-\bar x_n^\varepsilon)/\sqrt{\varepsilon}$.
Let

$$\tilde x^\varepsilon(t)=\sum_{\varepsilon k\le t}(\tilde x_k^\varepsilon-\tilde x_{k-1}^\varepsilon), \tag{4.52}$$

and let $\bar x(t)$ be the solution of equation (3.21).

Theorem 6 *Let conditions (φ**1**)–(φ**5**) be satisfied. Then the process $\tilde x^\varepsilon(t)$ converges weakly in $C_{[0,t_0]}$ to the process $\hat x(t)$, which is the solution of the equation*

$$\hat x(t)=\int_0^t\frac{\partial}{\partial x}\bar\varphi\left(\bar x(s)\right)\hat x(s)\,ds+\hat Z(t), \tag{4.53}$$

where $\hat Z(t)$ is a Gaussian process with independent increments in R^d for which $E\hat Z(t)=0$ and

$$E\left(\hat Z(t),x\right)^2=\int_0^t B\left(\bar x(s),\bar x(s)\right)\,ds. \tag{4.54}$$

Proof Note that $\tilde{x}_n^\varepsilon$ satisfies the relation

$$\tilde{x}_n^\varepsilon = \sum_{k<n} \sqrt{\varepsilon}\,(\varphi_k(x_k^\varepsilon,\omega) - \bar{\varphi}(x_k^\varepsilon)) = \sum_{k<n} \sqrt{\varepsilon}\,(\varphi_k(\bar{x}_k^\varepsilon + \sqrt{\varepsilon}\tilde{x}_k^\varepsilon,\omega) - \varphi_k(\bar{x}_k^\varepsilon)) + \sum_{k<n} \sqrt{\varepsilon}\,(\varphi_k(\bar{x}_k^\varepsilon,\omega) - \bar{\varphi}(\bar{x}_k^\varepsilon)) . \tag{4.55}$$

Let

$$z^\varepsilon(t) = \sum_{\varepsilon k \le t} \sqrt{\varepsilon}\,(\varphi_k(\bar{x}_k^\varepsilon,\omega) - \bar{\varphi}_k^\varepsilon) .$$

It follows from Theorem 5 of Chapter 2 that $\hat{Z}^\varepsilon(t)$ converges weakly in $C_{[0,t_0]}$ to the process $\hat{Z}(t)$.
Since

$$\left|\varphi_k(x+\sqrt{\varepsilon}\,y,\omega) - \varphi_k(x,\omega)\right| \le \sqrt{\varepsilon}\,l_k(\omega)|y|,$$

we have that

$$|\tilde{x}_n^\varepsilon| \le \varepsilon \sum_{k<n} l_k(\omega)|\tilde{x}_k^\varepsilon| + \sup_{t\le\varepsilon n} |z^\varepsilon(t)| ,$$

and so

$$\sup_{\varepsilon n\le t_0} |\tilde{x}_n^\varepsilon| \le \exp\left\{\varepsilon \sum_{\varepsilon k\le t_0} l_k(\omega)\right\} \sup_{t\le\varepsilon n} |z^\varepsilon(t)| . \tag{4.56}$$

It follows from condition (φ**4**) that

$$\left|\sum_{k<n} \sqrt{\varepsilon}\,(\varphi_k(\bar{x}_k^\varepsilon + \sqrt{\varepsilon}\tilde{x}_k^\varepsilon,\omega) - \varphi_k(\bar{x}_k^\varepsilon)) - \frac{\partial \varphi_k}{\partial x}(\bar{x}_k^\varepsilon,\omega)\tilde{x}_k^\varepsilon\right| \le \varepsilon \sum_{k<n} l_k(\omega)\lambda\left(\sqrt{\varepsilon}, |\tilde{x}_k^\varepsilon|\right) \tag{4.57}$$

and

$$\sup_{\varepsilon n\le t_0} \left|\varepsilon \sum_{k<n} l_k(\omega)\lambda\left(\sqrt{\varepsilon}, |\tilde{x}_k^\varepsilon|\right)\right| \le \lambda\left(\sqrt{\varepsilon},\ \sup_{\varepsilon n\le t_0} |\tilde{x}_n^\varepsilon|\right) . \tag{4.58}$$

Since

$$\limsup_{\varepsilon\to 0} \varepsilon \sum_{\varepsilon k\le t_0} l_k(\omega) = t_0\, El_1(\omega), \tag{4.59}$$

and

$$\limsup_{\varepsilon\to 0} \sum_{\varepsilon k\le t_0} |\gamma_k^\varepsilon| = 0, \tag{4.60}$$

where

$$\gamma_k^\varepsilon = \sqrt{\varepsilon}\,\left(\varphi_k(\bar{x}_k^\varepsilon + \sqrt{\varepsilon}\,\tilde{x}_k^\varepsilon, \omega) - \varphi_k(\bar{x}_k^\varepsilon)\right) - \frac{\partial \varphi_k}{\partial x}(\bar{x}_k^\varepsilon, \omega)\,\tilde{x}_k^\varepsilon, \tag{4.61}$$

it follows from (4.55) that

$$\tilde{x}_n^\varepsilon = \sum_{k<n} \varepsilon \frac{\partial \varphi_k}{\partial x}(\bar{x}_k^\varepsilon, \omega)\tilde{x}_k^\varepsilon + \sum_{k<n} \gamma_k^\varepsilon + \hat{Z}^\varepsilon(\varepsilon n). \tag{4.62}$$

Let

$$\beta_k^\varepsilon = \left(\frac{\partial \varphi_k}{\partial x}(\bar{x}_k^\varepsilon, \omega) - \frac{\partial \bar{\varphi}}{\partial x}(\bar{x}_k^\varepsilon)\right)\tilde{x}_k^\varepsilon.$$

Then

$$\tilde{x}_n^\varepsilon = \sum_{k<n} \varepsilon \frac{\partial \bar{\varphi}}{\partial x}(\bar{x}_k^\varepsilon, \omega)\tilde{x}_k^\varepsilon + \sum_{k<n} (\gamma_k^\varepsilon + \beta_k^\varepsilon) + \hat{Z}^\varepsilon(\varepsilon n). \tag{4.63}$$

Using the ergodicity of $\frac{\partial \varphi_k}{\partial x}(x, \omega)$, the boundedness of $\sup_{\varepsilon k \le t_0} |\tilde{x}_k^\varepsilon|$, and the proofs of Lemmas 2 and 3 in Section 4.2, it is easy to show that

$$\sup_{n\varepsilon < t_0} \left|\sum_{k<n} \beta_k^\varepsilon\right| \to 0$$

in probability as $\varepsilon \to 0$. If $\hat{x}_n^\varepsilon$ is the solution of the equation

$$\hat{x}_n^\varepsilon = \varepsilon \sum_{k<n} \frac{\partial \bar{\varphi}}{\partial x}(\bar{x}_k^\varepsilon)\hat{x}_k^\varepsilon + \hat{Z}^\varepsilon(\varepsilon n), \tag{4.64}$$

then it follows from (4.63) and (4.64) that

$$\sup_{\varepsilon n \le t_0} |\hat{x}_n^\varepsilon - \tilde{x}_n^\varepsilon| \le \exp\{ct_0\} \sup_{\varepsilon n < t_0} \left|\sum_{k<n} (\gamma_k^\varepsilon + \beta_k^\varepsilon)\right|,$$

where

$$c = \sup_{\varepsilon k < t_0} \left\|\frac{\partial \bar{\varphi}}{\partial x}(\bar{x}_k^\varepsilon)\right\|.$$

With this, the proof of the theorem follows from the next lemma. □

Lemma 4 *Set*

$$\hat{x}^\varepsilon(t) = \sum_n \hat{x}_n^\varepsilon\, 1_{\{\varepsilon n \le t < \varepsilon(n+1)\}}.$$

Then the process $\hat{x}^\varepsilon(t)$ *converges weakly in* $C_{[0,t_0]}$ *to the process* $\hat{x}(t)$.

Proof Let $\bar{x}^\varepsilon(t) = \sum_n \bar{x}_n^\varepsilon\, 1_{\{\varepsilon n \le t < \varepsilon(n+1)\}}$. Let $\hat{\hat{x}}^\varepsilon(t)$ be the solution of the equation

$$\hat{\hat{x}}^\varepsilon(t) = \int_0^t \frac{\partial \bar{\varphi}}{\partial x}\,(\bar{x}^\varepsilon(s))\,\hat{\hat{x}}^\varepsilon(s)\,ds + \hat{Z}^\varepsilon(t),$$

and let $x_\varepsilon^*(t)$ be the solution of the equation

$$x_\varepsilon^*(t) = \int_0^t \frac{\partial \bar{\varphi}}{\partial x}(\bar{x}(s))\, x_\varepsilon^*(s)\, ds + \hat{Z}^\varepsilon(t). \tag{4.65}$$

Then

$$\sup_{t \le t_0} \left| \hat{x}^\varepsilon(t) - \hat{\hat{x}}^\varepsilon(t) \right| + \sup_{t \le t_0} \left| \hat{\hat{x}}^\varepsilon(t) - x_\varepsilon^*(t) \right| \to 0$$

in probability. Since $x_\varepsilon^*(t)$ converges weakly in $C_{[0,t_0]}$ to $\hat{x}(t)$, we have the same for $\hat{x}^\varepsilon(t)$.

□

Remark 4 *We can extend Theorem 4 to the case of difference equations without any additional assumptions.*

5
Diffusion Approximation

In this chapter we consider randomly perturbed systems of differential and difference equations whose averaged systems are static. So, the solution of the perturbed equation is expected to converge as $\varepsilon \to 0$ to a constant determined by the initial position of the system. It follows from the theorem on normal deviations that the deviations of such solutions from the initial position are asymptotically Gaussian random variables whose variance is of order εt. Here we consider the problem farther out in time. We prove, under some reasonable assumptions, that the stochastic process $x_\varepsilon(t/\varepsilon)$, where $x_\varepsilon(t)$ is the solution of the perturbed system at time t, is asymptotically a diffusion process as $\varepsilon \to 0$. We derive here a detailed description of these processes. Moreover, if the averaged equation has a first integral (i.e., a function φ exists that is constant on any trajectory of the averaged equation), then for any fixed time, say t_1, the function $\varphi(x_\varepsilon(t_1))$ is asymptotically constant, but the function $\varphi(x_\varepsilon(t/\varepsilon))$ is approximately a diffusion process.

5.1 Differential Equations

We study differential equations involving various kinds of random perturbations that are either Markov or stationary processes.

5.1.1 Markov Jump Perturbations

Consider the solution $x_\varepsilon(t)$ in R^d of the initial value problem

$$\dot{x}_\varepsilon(t) = a\left(x_\varepsilon(t), y\left(\frac{t}{\varepsilon}\right)\right), \quad x_\varepsilon(0) = x_0, \tag{5.1}$$

where $y(t)$ is a homogeneous Markov process in a measurable space $(Y, \mathcal{C})$ and $a(x,y) : R^d \times Y \to R^d$ is a measurable function. Furthermore, suppose that $y(t)$ is a jump process having a transition probability function $P(t,y,C)$, for $t > 0$, $y \in Y$, and $C \in \mathcal{C}$, that satisfies the relation

$$\lim_{t\to 0} \frac{1}{t}\left(P(t,y,C) - 1_C(y)\right) = \pi(y,C), \tag{5.2}$$

where the limit π is of bounded variation, i.e.,

$$\sup_y \text{Variation}\ \pi(y,\cdot) < \infty.$$

In addition, suppose that $y(t)$ is an ergodic process with ergodic distribution ρ satisfying SMC II of Section 2.3. As before, we define

$$R(y,C) = \int_0^\infty \left[P(t,y,C) - \rho(C)\right] dt. \tag{5.3}$$

The following lemma is a direct consequence of formula (5.3)

Lemma 1 *Let $g : Y \to R$ be a bounded measurable function for which $\int g(y)\rho(dy) = 0$. Then*

$$\lim_{t\to 0} \frac{1}{t}\left[\int P(t,\tilde{y},dy) \int R(y,dy')g(y') - \int R(\tilde{y},dy')g(y')\right] = -g(\tilde{y}). \tag{5.4}$$

The following corollary results from this lemma.

Corollary For any $y \in Y$ and any $C \in \mathcal{C}$,

$$\int \pi(y,dy')\, R(y',C) = -1_C(y). \tag{5.5}$$

Next, we formulate our assumptions about the function $a(x,y)$:

(a1) For all $x, \tilde{x} \in R^d$, $\int |a(x,y)|\,\rho(dy) < \infty$, and there exists a measurable function $l : Y \to R_+$ for which $\int l(y)\,\rho(dy) < \infty$ and

$$\left|a(x,y) - a(\tilde{x},y)\right| \le l(y)|x - \tilde{x}|$$

for each $y \in Y$;

(a2) $\int a(x,y)\,\rho(dy) = 0$.

Condition (a1) implies the existence and uniqueness of a solution to (5.1).

We are interested in the case where the noise in the system acts on a faster time scale than that on which the system responds. So, we consider random perturbations of the form $y_\varepsilon(t) = y\left(t/\varepsilon\right)$.

Lemma 2 *The process* $\big(x_\varepsilon(t), y_\varepsilon(t)\big)$ *is a homogeneous Markov process in the phase space* $R^d \times Y$. *Its generator is determined by the relation*

$$\begin{aligned}\lim_{t\to 0}\frac{1}{t}\,&\big(E_{x,y}g\big(x_\varepsilon(t), y_\varepsilon(t)\big) - g(x,y)\big)\\ &= (g_x(x,y), a(x,y)) + \frac{1}{\varepsilon}\int g(x,\tilde{y})\pi(y, d\tilde{y}),\end{aligned} \tag{5.6}$$

which is defined on the functions $g : R^d \times Y \to R$ *for which* $g(x,y)$ *and* $g_x(x,y)$ *are measurable and bounded;* $E_{x,y}$ *is the expectation under the conditions that* $x_\varepsilon(0) = x$, $y_\varepsilon(0) = y$.

Proof Denote by $\nu_\varepsilon(t)$ the number of jumps of the process $y_\varepsilon(t)$ on the interval $[0,t]$. Let P_y, E_y be the probability and the expectation, respectively, related to the Markov process $y_\varepsilon(t)$ under the condition $y_\varepsilon(0) = y$. It follows from the theory of jump Markov processes (see [64, Chapter 7]) that

$$P_y\{\nu_\varepsilon(t) > 1\} = o(t),\ P_y\{\nu_\varepsilon(t) = 1\} = \frac{1}{\varepsilon}\,\pi(y, Y\setminus\{y\}) + o(t)$$

and

$$E_y g(y_\varepsilon(t)) - g(y)\,1_{\{\nu_\varepsilon(t)=1\}} = t\,\frac{1}{\varepsilon}\int (g(y') - g(y))\,\pi(y, dy') + o(t).$$

So

$$\begin{aligned}E_{x,y}\left\{g(x_\varepsilon(t), y_\varepsilon(t)) - g(x,y)\right\} &= E_{x,y}\int_0^t (g_x(x_\varepsilon(s), y), a(x_\varepsilon(s), y))\,ds\\ \times 1_{\{\nu_\varepsilon(t)=0\}} + \frac{t}{\varepsilon}\int (g(x,y') - g(x,y))\,\pi(y, dy') + o(t).\end{aligned}$$

This completes the proof of the lemma.

□

Denote by $\mathcal{F}_t^\varepsilon$ the σ-algebra generated by $\big(x_\varepsilon(s), y_\varepsilon(s)\big)$, $s \le t$. It follows from (5.6) that for functions g that are bounded and measurable and that have the same g_x, the following relation is valid: If $t_1 < t_2$, then

$$\begin{aligned}&E\Big(g\big(x_\varepsilon(t_2), y_\varepsilon(t_2)\big) - g\big(x_\varepsilon(t_1), y_\varepsilon(t_1)\big)/\mathcal{F}_{t_1}^\varepsilon\Big)\\ &= E\bigg(\int_{t_1}^{t_2}\Big[\Big(g_x\big(x_\varepsilon(s), y_\varepsilon(s)\big), a\big(x_\varepsilon(s), y_\varepsilon(s)\big)\Big)\\ &\quad + \frac{1}{\varepsilon}\int g\big(x_\varepsilon(s), \tilde{y}\big)\pi\big(y_\varepsilon(s), \tilde{y}\big)\Big]\,ds/\mathcal{F}_{t_1}^\varepsilon\bigg).\end{aligned} \tag{5.7}$$

Note that the averaging theorem (see Chapter 3, Theorem 5) is valid for $x_\varepsilon(t)$. It follows from condition (a2) that $\bar{a}(x) = 0$. Therefore, as $\varepsilon \to 0$, $x_\varepsilon(t) \to x_0$ uniformly with probability 1 over any finite interval of t.

We will investigate $x_\varepsilon(t)$ for "large" t, which means here for t to be of order $1/\varepsilon$. It will be shown that the process $x_\varepsilon(t/\varepsilon)$ converges weakly in C to a diffusion process under some additional assumptions. This diffusion process then is referred to as a diffusion approximation to $x_\varepsilon(t)$.

Theorem 1 *Let $y(t)$ satisfy condition SMC* II *of Section 2.3, and let $a(x,y)$ satisfy conditions* (a1) *and* (a2) *and the following additional condition:*
(a3) *There exist the first derivative $a_x(x,y)$,*

$$\sup_{x,y} \big[|a(x,y)| + \|a_x(x,y)\|\big] < \infty,$$

and

$$\lim_{\delta\to 0} \sup\{\|a_x(x_1,y) - a_x(x_2,y)\| : |x_1 - x_2| \le \delta, y \in Y\} = 0.$$

We define the second–order differential operator L acting on functions

$$f \in C^{(2)}(R)$$

by the relation

$$Lf(x) = \iint \left(\frac{\partial}{\partial x} (f'(x), a(x,\tilde{y}))\, R(y, d\tilde{y}), a(x,y) \right) \rho(dy). \tag{5.8}$$

We assume that L is the generator of a diffusion process $\hat{x}(t)$ in R^d. Finally, define the process $\hat{x}_\varepsilon(t) = x_\varepsilon(t/\varepsilon)$, where $x_\varepsilon(t)$ is the solution to equation (5.1) with $x_\varepsilon(0) = x_0$. Then $\hat{x}_\varepsilon(t)$ converges weakly in C as $\varepsilon \to 0$ to the diffusion process $\hat{x}(t)$ with initial value $\hat{x}(0) = x_0$.

Proof Let $0 \le t_1 < t_2 < \infty$. Define

$$\Delta(t_1,t_2) = E\big(f(x_\varepsilon(t_2)) - f(x_\varepsilon(t_1))/\mathcal{F}^\varepsilon_{t_1}\big),$$

where $f \in C^{(3)}(R^d)$. Using the relation

$$\frac{d}{dt} f(x_\varepsilon(t)) = \big(f'(x_\varepsilon(t)), a(x_\varepsilon(t), y_\varepsilon(t))\big)$$

we can write

$$\Delta(t_1,t_2) = E\left(\int_{t_1}^{t_2} \big(f'(x_\varepsilon(s)), a(x_\varepsilon(s), y_\varepsilon(s))\big)\, ds/\mathcal{F}^\varepsilon_{t_1} \right). \tag{5.9}$$

The function $A(x,y) = \int a(x,\tilde{y}) R(y,d\tilde{y})$ satisfies the equation

$$\int \pi(y,dy) A(x,\tilde{y}) = -a(x,\tilde{y}). \tag{5.10}$$

It follows from formula (5.7) that

$$E\left((f'(x_\varepsilon(t_2)), A(x_\varepsilon(t_2), y_\varepsilon(t_2))) - (f'(x_\varepsilon(t_1)), A(x_\varepsilon(t_1), y_\varepsilon(t_1)))/\mathcal{F}^\varepsilon_{t_1}\right)$$
$$= E\left(\int_{t_1}^{t_2}\left[\left(\frac{\partial}{\partial x}(f'(x_\varepsilon(s)), A(x_\varepsilon(s), y_\varepsilon(s))), a(x_\varepsilon(s), y_\varepsilon(s))\right)\right.\right.$$
$$\left.\left.-\frac{1}{\varepsilon}(f'(x_\varepsilon(s), a(x_\varepsilon(s), y_\varepsilon(s)))\right] ds/\mathcal{F}^\varepsilon_{t_1}\right). \tag{5.11}$$

(Here relation (5.10) was used.) It follows from (5.11) and (5.9) that

$$\Delta(t_1, t_2) = \varepsilon E\Bigg(F(x_\varepsilon(t_2), y_\varepsilon(t_2)) - F(x_\varepsilon(t_1), y_\varepsilon(t_1))$$
$$+ \int_{t_1}^{t_2}(F_x(x_\varepsilon(s), y_\varepsilon(s)), a(x_\varepsilon(s), y_\varepsilon(s)))\, ds/\mathcal{F}^\varepsilon_{t_1}\Bigg),$$

where

$$F(x, y) = (f'(x), A(x, y)).$$

Set

$$c(x, y) = (F_x(x, y), a(x, y)) - Lf(x).$$

Then

$$E\left(f\left(x_\varepsilon\left(\frac{t_2}{\varepsilon}\right)\right) - f\left(x_\varepsilon\left(\frac{t_1}{\varepsilon}\right)\right) - \int_{t_1}^{t_2} Lf\left(x_\varepsilon\left(\frac{s}{\varepsilon}\right)\right) ds/\mathcal{F}^\varepsilon_{t_1/\varepsilon}\right)$$

$$= E\left(\Delta\left(\frac{t_1}{\varepsilon}, \frac{t_2}{\varepsilon}\right) - \varepsilon\int_{t_1/\varepsilon}^{t_2/\varepsilon} Lf(x_\varepsilon(s))\, ds/\mathcal{F}^\varepsilon_{t_1/\varepsilon}\right)$$

$$= O(\varepsilon) + E\left(\varepsilon\int_{t_1/\varepsilon}^{t_2/\varepsilon} c\big(x_\varepsilon(s), y_\varepsilon(s)\big)\, ds/\mathcal{F}^\varepsilon_{t_1/\varepsilon}\right). \tag{5.12}$$

Note that $\int c(x, y)\rho(dy) = 0$. Define the function

$$C(\varepsilon, T) = \sup\left\{\left|E_{xy}\int_0^t c(x_\varepsilon(s), y_\varepsilon(s))\, ds\right| : x \in R^d, y \in Y, \varepsilon t \le T\right\},$$

for $\varepsilon > 0, T > 0$. Set

$$\lambda(\delta) = \sup\{|c(x_1, y) - c(x_2, y)| : |x_1 - x_2| \le \delta, y \in Y\}.$$

It follows from condition (a3) that $\lambda(\delta) \to 0$ as $\delta \to 0$. We estimate $C(\varepsilon, T)$ and show that $\varepsilon\, C(\varepsilon, T) \to 0$ as $\varepsilon \to 0$. This implies the relation

$$\lim_{\varepsilon\to 0} E\left|E\left(f(\hat{x}_\varepsilon(t_2)) - f(\hat{x}_\varepsilon(t_1)) - \int_{t_1}^{t_2} Lf(\hat{x}_\varepsilon(s))\, ds/\hat{\mathcal{F}}^\varepsilon_{t_1}\right)\right| = 0, \tag{5.13}$$

where $\hat{\mathcal{F}}_t^\varepsilon = \mathcal{F}_{t/\varepsilon}^\varepsilon$. Using the relation $\big|x_\varepsilon(t_2) - x_\varepsilon(t_1)\big| \le \alpha|t_2 - t_1|$, where $\alpha = \sup_{x,y}|a(x,y)|$, we have

$$\begin{aligned}
&\left|E_{x,y}\int_0^t c\big(x_\varepsilon(s), y_\varepsilon(s)\big)\,ds\right| \\
&= \left|E_{x,y}\int_0^t \big[c\big(x_\varepsilon(s), y_\varepsilon(s)\big) - c(x, y_\varepsilon(s))\big]\,ds \right.\\
&\qquad \left. + \int_0^t c(x,y')\big(P(\frac{s}{\varepsilon}, y, dy') - \rho(dy')\big)\,ds\right| \\
&\le \lambda(\alpha t)\,t + \varepsilon\left|\int_0^{t/\varepsilon} c(x,y')\big[P(s,y,dy') - \rho(dy')\big]\,ds\right|.
\end{aligned}$$

It follows from SMC II that $\int_0^t c(x,y')\big[P(s,y,dy') - \rho(dy')\big]\,ds = O(1)$ uniformly in t, x, and y. So

$$\left|E_{x,y}\int_0^t c\big(x_\varepsilon(s), y_\varepsilon(s)\big)\,ds\right| \le \lambda(\alpha t)\,t + c^*\,\varepsilon,$$

where c^* is a constant. Therefore,

$$\left|E_{x,y}\left(\int_{t_1}^{t_2} c\big(x_\varepsilon(s), y_\varepsilon(s)\big)\,ds/\mathcal{F}_{t_1}^\varepsilon\right)\right| \le \lambda\big(\alpha(t_2 - t_1)\big)\,(t_2 - t_1) + c^*\,\varepsilon$$

with probability 1. Set $\tau_k = kh$, $h = t/n\varepsilon$. Then

$$\left|E_{x,y}\left(\int_{\tau_k}^{\tau_{k+1}} c\big(x_\varepsilon(s), y_\varepsilon(s)\big)\,ds/\mathcal{F}_{\tau_k}^\varepsilon\right)\right| \le \lambda(\alpha h)\,h + c^*\,\varepsilon,$$

$$\begin{aligned}
\varepsilon\left|E_{x,y}\int_0^{t/\varepsilon} c\big(x_\varepsilon(s), y_\varepsilon(s)\big)\,ds\right| &\le \varepsilon n\big[\lambda(\alpha h)\,h + c^*\,\varepsilon\big] \\
&= c^*\,\varepsilon^2 n + t\,\lambda\left(\alpha\frac{t}{n\varepsilon}\right).
\end{aligned}$$

Let $\varepsilon^2 n \to 0$ and $n\varepsilon \to \infty$. Then $\varepsilon^2 n + t\,\lambda\big(\alpha t/n\varepsilon\big) \to 0$. So relation (5.13) is proved.

It follows from Theorem 2 of Chapter 2 that the process $\hat{x}_\varepsilon(t)$ converges weakly to the process $\hat{x}(t)$. To prove the theorem, we will prove that the family of stochastic processes $\{\hat{x}_\varepsilon(t), \varepsilon \in (0,1)\}$ is weakly compact in C. We need some auxiliary results to do this.

Lemma 3 *Let $\phi(x) = 1 - \exp\{-|x|^2\}$. Then a constant $C > 0$ exists for which*

$$E_y\phi\left(x_\varepsilon\left(\frac{t}{\varepsilon}\right) - x_0\right) \le Ct \tag{5.14}$$

and

$$E_y\phi^2\left(x_\varepsilon\left(\frac{t}{\varepsilon}\right) - x_0\right) \le C(\varepsilon^2 + t^2) \tag{5.15}$$

Proof Applying the relations (5.9) and (5.11) to the function $f(x) = \phi(x - x_0)$, with $t_1 = 0$ and $t_2 = t/\varepsilon$, we obtain the relations (here we write t_ε for t/ε)

$$\begin{aligned} E_y\phi\left(x_\varepsilon\left(t_\varepsilon\right) - x_0\right) &= -\varepsilon E_y\left(\phi'\left(x_\varepsilon\left(t_\varepsilon\right) - x_0\right), A\left(x_\varepsilon\left(t_\varepsilon\right), y_\varepsilon\left(t_\varepsilon\right)\right)\right) \\ &\quad + E_y\int_0^{t_\varepsilon}\left(\frac{\partial}{\partial x}\left(\phi'(x_\varepsilon(s) - x_0), A(x_\varepsilon(s), y_\varepsilon(s))\right), a(x_\varepsilon(s), y_\varepsilon(s))\right) ds \\ &= O(t) - \varepsilon E_y\left[\left(\phi'\left(x_\varepsilon\left(t_\varepsilon\right) - x_0\right), A\left(x_\varepsilon\left(t_\varepsilon\right), y_\varepsilon\left(t_\varepsilon\right)\right) - A\left(x_0, y_\varepsilon\left(t_\varepsilon\right)\right)\right)\right. \\ &\quad \left. + \left(\phi'\left(x_\varepsilon\left(t_\varepsilon\right) - x_0\right), A\left(x_0, y\left(t_\varepsilon\right)\right)\right)\right] \\ &\le O\left(t + \varepsilon E_y\phi\left(x_\varepsilon\left(t_\varepsilon\right) - x_0\right)\right). \end{aligned}$$

Here we have used the inequalities: $|x_\varepsilon(t/\varepsilon) - x_0| \le K\,t$ and

$$|(\phi'(x - x_0), A(x, y) - A(x_0, y))| \le K\,\phi(x - x_0),$$

which are true for some $K > 0$. This completes the proof of (5.14).
Define $\Psi(x) = \phi^2(x)$. Then in the same way as for ϕ we can obtain the estimate

$$\begin{aligned} E_y\Psi\left(x_\varepsilon\left(t_\varepsilon\right) - x_0\right) &\le \varepsilon E_y\left(\Psi'\left(x_\varepsilon\left(t_\varepsilon\right) - x_0\right), A\left(x_0, y\left(t_\varepsilon\right)\right)\right) \\ &\quad + c_1 E_y\left(\varepsilon\Psi\left(x_\varepsilon\left(t_\varepsilon\right) - x_0\right) + \varepsilon\int_0^{t_\varepsilon}\phi(x_\varepsilon(s))\,ds\right) \\ &\le \varepsilon c_2 t + c_1\varepsilon E_y\Psi\left(x_\varepsilon\left(t_\varepsilon\right) - x_0\right) + c_1\varepsilon\int_0^{t_\varepsilon} c_\varepsilon s\,ds, \end{aligned}$$

since $|\Psi'(x)| \le K_1\Phi(x)$ for some K_1. This completes the proof of (5.15). □

It follows from Lemma 3 that

$$\begin{aligned} E_y\phi^2(\hat{x}_\varepsilon(t_2) - x_\varepsilon(t_1)) &= E_y E(\phi^2(x_\varepsilon(t_2) - x_\varepsilon(t_1))/\hat{\mathcal{F}}^\varepsilon_{t_1}) \\ &\le c(\varepsilon^2 + (t_2 - t_1)^2). \end{aligned}$$

Since

$$\sup_{|t_2 - t_1| \le \delta} |\hat{x}_\varepsilon(t_2) - \hat{x}_\varepsilon(t_1)| \le \hat{K}\,\delta/\varepsilon,$$

where $\hat{K} = \sup_{x,y}|a(x, y)|$, the family of stochastic processes $\{\hat{x}_\varepsilon(t), \varepsilon \in (0,1)\}$ is compact in C, which follows from Theorem 1 of Chapter 2. □

Remark 1 *The differential operator L defined in (5.8) is of the form*

$$Lf(x) = \big(f'(x), \hat{a}(x)\big) + \frac{1}{2}\mathrm{Tr}\,f''(x)\hat{B}(x),$$

where

$$\hat{a}(x) = \iint a'_x(x, \tilde{y})a(x, y)R(y, d\tilde{y})\,\rho(dy)$$

and

$$\big(\hat{B}(x)z, z\big) = 2 \iint \big(a(x,y), z\big)\big(a(x,\tilde{y}), z\big) R(y, d\tilde{y})\, \rho(dy), \quad z \in R^d.$$

The function $\hat{a}(x)$ *is bounded and continuous, and* $\hat{B}(x)$ *has a bounded and continuous derivative. Under these conditions the generator L determines the transition probabilities of the process (and there exists a diffusion process with this generator) if the following additional condition is satisfied:*

(a4) $\inf_{|z|=1}\big(\hat{B}(x)z, z\big) > 0$ *for all* $x \in R^d$.

5.1.2 Some Generalizations

We consider a family of Markov processes $\{(x_\varepsilon(t), y_\varepsilon(t)), \varepsilon > 0\}$ in the space $R^d \times Y$ satisfying the following conditions:

(i) Let $P^\varepsilon(s, x, y, t, B, C)$, for $0 \le s < t$, $x \in R^d$, $y \in Y$, $B \in \mathcal{B}(R^d)$, and for $C \in \mathcal{C}$ be the transition probability function for the Markov process $(x_\varepsilon(t), y_\varepsilon(t))$, i.e.,

$$P^\varepsilon(s, x, y, t, B, C) = P\{x_\varepsilon(t) \in B, y_\varepsilon(t) \in C / x_\varepsilon(s) = x, y_\varepsilon(s) = y\}.$$

Then the following relation holds for any function $g(x,y)$ for which $g(x,y)$ and $g_x(x,y)$ are bounded and continuous in x and are $\mathcal{C}$-measurable in y:

$$\lim_{h \to 0} \frac{1}{h}\bigg(\int P^\varepsilon(t, x, y, t+h, dx', dy') g(x', y') - g(x,y) \bigg)$$

$$= (a(t,x,y), g_x(x,y)) + \frac{1}{\varepsilon} \int g(x, y') \pi^x(y, dy'),$$

where $a(t,x,y)$ is a bounded function from $R_+ \times R^d \times Y$ into R^d, and $\pi^x(y, C)$ is a signed measure of bounded variation for all $y \in Y$ and $x \in R^d$.

(ii) The function $a(t,x,y)$ is continuous in t and x and is $\mathcal{C}$-measurable in y, $\sup_{x,y} \operatorname{var} \pi^x(y, C) < \infty$, and there exists a constant l for which

$$|a(t, x_1, y) - a(t, x_2, y)| \le l\, |x_1 - x_2|.$$

Remark 2 *Under condition* (ii) *the transition probability* P^ε *satisfies the backward Kolmogorov equation: For any function* g *that satisfies the conditions listed in* (i), *define the function*

$$G^\varepsilon(t, x, y) = \iint g(x', y') P^\varepsilon(t, x, y, T, dx', dy').$$

This function satisfies the equation

$$-\frac{\partial G^\varepsilon}{\partial t}(t,x,y) = (a(t,x,y), G^\varepsilon_x(t,x,y)) + \frac{1}{\varepsilon}\int G^\varepsilon(t,x,y')\pi^x(y,dy')$$

if $t < T$ *and* $\lim_{t\to T} G^\varepsilon(t,x,y) = g(x,y)$.
This equation can be used to evaluate the transition probability G^ε. *In particular, the processes* $(x_\varepsilon(t), y_\varepsilon(t))$ *satisfy the relation*

$$\dot{x}_\varepsilon(t) = a(t, x_\varepsilon(t), y_\varepsilon(t)).$$

(iii) Introduce the operator

$$\Pi^x g(y) = \int g(y')\pi^x(y,dy') \text{ for } g \in \mathbf{B}(Y)$$

where $\mathbf{B}(Y)$ is the space of bounded $\mathcal{C}$–measurable functions $g : Y \to R$. This is the generator of some homogeneous Markov jump process in Y for any $x \in X$. We assume that the Markov process $y^x(t)$ with the generator Π^x is ergodic and satisfies condition SMC II of Section 2.3. We denote by $P^x(t,y,C)$ the transition probability function for the Markov process $y^x(t)$, and by $\rho^x(c)$ its ergodic distribution. Set

$$R^x g(y) = \int_0^\infty g(y')(P^x(t,y,dy') - \rho^x(dy'))\,dt \quad \text{for } g \in \mathbf{B}.$$

Remark 3 *If* $g(x,y)$ *is a function for which* $g(x,\cdot) \in \mathbf{B}(Y)$ *for any fixed* $x \in R^d$, *then Lemma* 1 *implies that*

$$\Pi^x R^x g(x,y) = -g(x,y).$$

Note that in the case $a(t,x,y) = a(x,y)$ *and* $\pi^x(y,C) = \pi(y,C)$, *the stochastic processes* $x_\varepsilon(t)$ *and* $y_\varepsilon(t)$ *are the same as in Section 5.1.1. The following theorem is an extension of Theorem* 1 *to the more general case considered here.*

Theorem 2 *Let conditions* (i)–(iii) *be satisfied. In addition, assume that the following conditions hold:*

(iv) $\int a(t,x,y)\,\rho^x(dy) = 0$ *for all* t, x.

(v) *There exist bounded and continuous derivatives in* x *and* t,

$$\frac{\partial a}{\partial x}(t,x,y),\ \frac{\partial^2 a}{\partial x^2}(t,x,y),\ \frac{\partial a}{\partial t}(t,x,y),\ \frac{\partial^2 a}{\partial t^2}(t,x,y),\ \frac{\partial R^x}{\partial x}g(y),$$

for any bounded measurable function $g : Y \to R$.

(vi) *For $f \in C^{(2)}(R^d)$ there exists the limit*

$$\begin{aligned}
&\tilde{L}f(x) \\
&= \lim_{T\to\infty,\, S\to\infty} \frac{1}{T}\int_T^{T+S} \left(\frac{\partial R^x}{\partial x}(f'(x), a(t,x,y)), a(t,x,y)\right)\rho^x(dy)\,dt \\
&= (\tilde{a}(x), f'(x)) + \frac{1}{2}\mathrm{Tr}\tilde{B}(x)f''(x).
\end{aligned} \tag{5.16}$$

(vii) *The differential operator $\tilde{L}$ is the generator of a unique diffusion process in R^d.*

Then the process $\hat{x}_\varepsilon(t) = x_\varepsilon(t/\varepsilon)$ converges weakly in C as $\varepsilon \to 0$ to the diffusion process $\hat{x}(t)$ having generator L and initial value x_0.

Proof Denote by $\mathcal{F}_t^\varepsilon$ the σ-algebra generated by $\{(x_\varepsilon(s), y_\varepsilon(s)) : s \le t\}$. We will establish that for $0 < t_1 < t_2$,

$$\lim_{\varepsilon\to 0} E\left|E\left(f(\tilde{x}_\varepsilon(t_2)) - f(\tilde{x}_\varepsilon(t_1)) - \int_{t_1}^{t_2} \tilde{L}f(\tilde{x}_\varepsilon(s))\,ds/\mathcal{F}^\varepsilon_{t_1/\varepsilon}\right)\right| = 0. \tag{5.17}$$

For this purpose we need some auxiliary statements that we establish in four steps:

(1) Let $g : R_+ \times R^d \times Y \to R$ be a measurable bounded function that has bounded derivatives $g_t(t,x,y)$, $g_x(t,x,y)$ that are continuous in t, x. Then for $0 \le t_1 < t_2$ the following relation holds:

$$\begin{aligned}
&E\big(g(t_2, x_\varepsilon(t_2), y_\varepsilon(t_2)) - g(t_1, x_\varepsilon(t_1), y_\varepsilon(t_1))/\mathcal{F}^\varepsilon_{t_1}\big) \\
&= E\left[\int_{t_1}^{t_2} \big(g_t(t, x_\varepsilon(t), y_\varepsilon(t)) + H_\varepsilon^t g(t, x_\varepsilon(t), y_\varepsilon(t))\big)\,dt/\mathcal{F}^\varepsilon_{t_1}\right],
\end{aligned} \tag{5.18}$$

where

$$H_\varepsilon^t g(x,y) = \big(g_x(x,y), a(t,x,y)\big) + \frac{1}{\varepsilon}\Pi^x g(x,y).$$

To prove this we observe that it follows from (ii) that the derivatives of the right-hand side and the left–hand side of (5.18) with respect to t_2 coincide.

(2) Let $\Psi : R_+ \times R^d \times Y \to R$ be a bounded measurable function having the derivatives $\Psi_t(t,x,y)$, $\Psi_x(t,x,y)$ that are bounded and continuous in t and x. Assume that $\int \Psi(t,x,y)\rho^x(dy) = 0$ for all $t \ge 0$ and $x \in R^d$. Then for any $t \ge 0$ and $h > 0$,

$$\left|E_{t,x,y}\int_t^{t+h} \Psi(s, x_\varepsilon(s), y_\varepsilon(s))\,ds\right| = O(\varepsilon + \varepsilon h), \tag{5.19}$$

where $E_{t,x,y}$ is the conditional expectation under the conditions that $x_\varepsilon(t) = x$, $y_\varepsilon(t) = y$. To prove this statement we introduce the function

$$Z(t,x,y) = R^x\Psi(t,x,y).$$

Then $Z_t(t,x,y)$ and $Z_x(t,x,y)$ are bounded and continuous in t and x. Therefore, using (5.18) we write

$$E_{t,x,y}\big(Z(t+h,x_\varepsilon(t+h),y_\varepsilon(t+h)) - Z(t,x,y)\big)$$
$$= E_{t,x,y}\int_t^{t+h}\bigg[Z_s(s,x_\varepsilon(s),y_\varepsilon(s)) + \big(Z_x(s,x_\varepsilon(s),y_\varepsilon(s)),a(s,x_\varepsilon(s),y_\varepsilon(s)\big)$$
$$+\ \frac{1}{\varepsilon}\Pi^x Z(s,x_\varepsilon(s),y_\varepsilon(s))\bigg]\,ds.$$

Since $\Pi^x\, Z(s,x,y) = -\Psi(s,x,y)$,

$$E_{t,x,y}\int_t^{t+h}\Psi\big(s,x_\varepsilon(s),y_\varepsilon(s)\big)\,ds = \varepsilon E_{t,x,y}\big(Z(t,x,y) - Z(t+h,x_\varepsilon(t+h)\big)$$

$$+\varepsilon E_{t,x,y}\int_t^{t+h}\big[Z_s(s,x_\varepsilon(s),y_\varepsilon(s))+(Z_x(s,x_\varepsilon(s),y_\varepsilon(s)),a(s,x_\varepsilon(s),y_\varepsilon(s)))\big]\,ds.$$

Relation (5.19) is a consequence of this relation.
(3) Now we apply relation (5.18) to the function $g(x,y) = f(x)$:

$$\begin{aligned}&E\big(f(x_\varepsilon(t_2)) - f(x_\varepsilon(t_1))/\mathcal{F}^\varepsilon_{t_1}\big)\\ &= E\bigg(\int_{t_1}^{t_2}\big(f'(x_\varepsilon(s)),a(s,x_\varepsilon(s),y_\varepsilon(s))\big)\,ds/\mathcal{F}^\varepsilon_{t_1}\bigg).\end{aligned} \tag{5.20}$$

Set $F(t,x,y) = R^x\big(f'(x),a(t,x,y)\big)$. Then

$$\Pi^x F(t,x,y) = -\big(f'(x),a(t,x,y)\big).$$

Applying relation (5.18) to the function $F(t,x,y)$, we have

$$E\big(F(t_2,x_\varepsilon(t_2),y_\varepsilon(t_2)) - F(t_1,x_\varepsilon(t_1),y_\varepsilon(t_1))/\mathcal{F}^\varepsilon_{t_1}\big)$$
$$= E\bigg(\int_{t_1}^{t_2}\big[F_t(s,x_\varepsilon(s),y_\varepsilon(s)) + \big(F_x(s,x_\varepsilon(s),y_\varepsilon(s)),a(s,x_\varepsilon(s),y_\varepsilon(s))\big)$$
$$-\ \frac{1}{\varepsilon}(f'(x_\varepsilon(s)),a(s,x_\varepsilon(s),y_\varepsilon(s)))\big]\,ds/\mathcal{F}^\varepsilon_{t_1}\bigg),$$

so

$$E\bigg(\int_{t_1}^{t_2}\big(f'(x_\varepsilon(s)),a(s,x_\varepsilon(s),y_\varepsilon(s))\big)\,ds/\mathcal{F}^\varepsilon_{t_1}\bigg)$$
$$= \varepsilon E\bigg[F(t_1,x_\varepsilon(t_1),y_\varepsilon(t_1)) - F(t_2,x_\varepsilon(t_1),y_\varepsilon(t_1))$$
$$+\int_{t_1}^{t_2}\big(F_t(s,x_\varepsilon(s),y_\varepsilon(s)) + \big(F_x(s,x_\varepsilon(s),y_\varepsilon(s)),a(s,x_\varepsilon(s),y_\varepsilon(s))\big)\big)\,ds/\mathcal{F}^\varepsilon_{t_1}\bigg]. \tag{5.21}$$

Set

$$\Phi(s,x,y) = \big(F_x(s,x,y), a(s,x,y)\big),$$

$$\bar{\Phi}(s,x) = \int \Phi(s,x,y)\,\rho^x(dy), \quad \tilde{\Phi}(s,x,y) = \Phi(s,x,y) - \bar{\Phi}(s,x).$$

$$\Psi(t,x,y) = F_t(t,x,y) + \tilde{\Phi}(s,x,y).$$

It follows from relations (5.20) and (5.21) that

$$E\big(f(x_\varepsilon(t_2)) - f(x_\varepsilon(t_1))/\mathcal{F}^\varepsilon_{t_1}\big)$$
$$= E\left(\varepsilon \int_{t_1}^{t_2} \bar{\Phi}(s, x_\varepsilon(s))\,ds + \varepsilon \int_{t_1}^{t_2} \Psi(s, x_\varepsilon(s), y_\varepsilon(s))\,ds/\mathcal{F}^\varepsilon_{t_1}\right) + O(\varepsilon). \tag{5.22}$$

Note that $O(\varepsilon) \le \sup\big|F(t,x,y)\big|\varepsilon$.
Since $\Psi(s,x,y)$ satisfies the conditions of statement (2), relation (5.19) is satisfied, and

$$\left|E\bigg(\int_{t_1}^{t_2} \Psi(s, x_\varepsilon(s), y_\varepsilon(s))\,ds/\mathcal{F}^\varepsilon_{t_1}\bigg)\right| = O\big(\varepsilon(t_2 - t_1) + \varepsilon\big). \tag{5.23}$$

We substitute t_1/ε and t_2/ε for t_1 and t_2 in (5.22) and use relation (5.23). In this way we get that

$$E\big(f(\tilde{x}_\varepsilon(t_2)) - f(\tilde{x}_\varepsilon(t_1))/\mathcal{F}^\varepsilon_{t_1/\varepsilon}\big) = O(\varepsilon(t_2 - t_1) + \varepsilon^2)$$

$$+\varepsilon\, E\bigg(\int_{t_1/\varepsilon}^{t_2/\varepsilon} \bar{\Phi}(s, x_\varepsilon(s))\,ds/\mathcal{F}^\varepsilon_{t_1/\varepsilon}\bigg),$$

and after changing s to s/ε in the integral on the right-hand side we obtain

$$E\big(f(\tilde{x}_\varepsilon(t_2)) - f(\tilde{x}_\varepsilon(t_1)) - \int_{t_1}^{t_2} \bar{\Phi}(\varepsilon s, \tilde{x}_\varepsilon(s))\,ds/\mathcal{F}^\varepsilon_{t_1/\varepsilon}\big) = O(\varepsilon). \tag{5.24}$$

(4) To prove (5.17) it is enough to show that

$$\lim_{\varepsilon\to 0} E\left|E\bigg(\int_{t_1}^{t_2} \big(\bar{\Phi}(\varepsilon s, x_\varepsilon(s)) - Lf(x_\varepsilon(s))\big)\,ds|\mathcal{F}^\varepsilon_{t_1/\varepsilon}\bigg)\right| = 0. \tag{5.25}$$

For this we show that

$$\lim_{\varepsilon\to 0} \frac{1}{h_\varepsilon} \sup_{t,x,y} E_{\frac{t}{\varepsilon},x,y} \int_t^{t+h_\varepsilon} \big(\bar{\Phi}(\varepsilon s, x_\varepsilon(s)) - Lf(x_\varepsilon(s))\big)\,ds = 0 \tag{5.26}$$

for any $h_\varepsilon > 0$ for which $h_\varepsilon \to 0$ as $\varepsilon \to 0$.
Note that it follows from (5.24) that

$$E_{\frac{t}{\varepsilon},x,y}\big(f(x_\varepsilon(t+h)) - f(x_\varepsilon(t))\big) = O(\varepsilon + h). \tag{5.27}$$

Therefore, for some constant c_1 the following relation is true:

$$P_{\frac{t}{\varepsilon},x,y}\big\{\big|x_\varepsilon(t+h) - x_\varepsilon(t)\big| > \delta\big\} \le \frac{c_1}{\delta}(\varepsilon + h). \tag{5.28}$$

To prove this we apply relation (5.27) to the function

$$f(x) = 1 - \exp\{-|x - \bar{x}|^2\}$$

under the condition that $x_\varepsilon(t/\varepsilon) = \bar{x}$. The function $\phi(\varepsilon s, x) - Lf(x)$ has a bounded derivative in x, so

$$\begin{aligned} \Big| E_{\frac{t}{\varepsilon},x,y} \int_t^{t+h} & \left[\left(\bar{\Phi}(\varepsilon s, x_\varepsilon(s)) - Lf(x_\varepsilon(s)) \right) \right. \\ & \left. - \left(\bar{\Phi}(\varepsilon s, x) - Lf(x) \right) \right] ds \Big| = O\left(h\delta + \frac{1}{\delta^2}(\varepsilon + h) \right). \end{aligned} \tag{5.29}$$

Now,

$$\frac{1}{h_\varepsilon} \int_t^{t+h_\varepsilon} \left(\bar{\Phi}(\varepsilon s, x) - \tilde{L}f(x) \right) ds = \frac{\varepsilon}{h_\varepsilon} \int_{t/\varepsilon}^{t/\varepsilon + h_\varepsilon/\varepsilon} \Phi(s, x)\, ds - \tilde{L}f(x) \to 0$$

uniformly in t, x, and y if $h_\varepsilon/\varepsilon \to \infty$, due to condition (vi). The function $\phi(\varepsilon s, x) - Lf(x)$ has a bounded derivative in x, so

$$\begin{aligned} & \left| E_{t/\varepsilon,x,y} \int_t^{t+h} \left[\left(\bar{\Phi}(\varepsilon s, x_\varepsilon(s)) - Lf(x_\varepsilon(s)) \right) - \left(\bar{\Phi}(\varepsilon s, x) - Lf(x) \right) \right] ds \right| \\ & \qquad = O\left(h\delta + \frac{1}{\delta^2}(\varepsilon + h) \right). \end{aligned} \tag{5.30}$$

Set $h_\varepsilon = \sqrt{\varepsilon}$, $\delta = \delta_\varepsilon = \varepsilon^{1/6}$. Then

$$E_{\frac{t}{\varepsilon},x,y} \frac{1}{h_\varepsilon} \int_t^{t+h_\varepsilon} \left(\bar{\Phi}(\varepsilon s, \hat{x}_\varepsilon(s)) - Lf(\hat{x}_\varepsilon(s)) \right) ds = O(\varepsilon^{1/6}) + o(1),$$

where $o(1) \to 0$ as $\varepsilon \to 0$ uniformly in t, x, and y. So relation (5.17) is proved, and the process $\hat{x}_\varepsilon(t)$ converges weakly to the process $\hat{x}(t)$ because of Theorem 2 of Chapter 2. To prove the theorem we have to prove the compactness in C of the family of stochastic processes $\{\hat{x}_\varepsilon(t), \varepsilon \in (0,1)\}$. This can be done in the same way as in Theorem 1 using the relation

$$E_{\frac{t}{\varepsilon},x,y} \phi(\hat{x}_\varepsilon(t+h) - \hat{x}_\varepsilon(t)) = O(\varepsilon + h), \tag{5.31}$$

where $\phi(x) = 1 - \exp\{-x^2\}$, which follows from equation (5.24).

□

5.1.3 General Markov Perturbations

Let Y now be a finite–dimensional Euclidean space. We consider a Markov process $\left(x_\varepsilon(t), y_\varepsilon(t)\right)$ in $R^d \times Y$ that satisfies the following conditions:

(A1) Denote by $P^\varepsilon(s, x, y, t, dx', dy')$ the transition probability of the process. Then for $g \in C^{(2)}(R^d \times Y)$,

$$\lim_{h \to 0} \frac{1}{h} \left(\int P^\varepsilon(t, x, y, t+h, dx', dy') g(x', y') - g(x, y) \right)$$

$$= \left(g_x(x,y), a(t,x,y)\right) + \frac{1}{\varepsilon} A^x g(x,y), \tag{5.32}$$

where $a(t,x,y)$ satisfies condition (ii) from Section 5.1.2 and A^x, for fixed x, is the generator of a homogeneous Markov process in Y, which is defined for $g \in C^{(2)}(Y)$. Note that $A^x g(x,y)$ is the result of the action of A^x on the function $g(x,y)$ as a function of y.

(A2) Denote by $y^x(t)$ the Markov process in Y having generator A^x, and let $Q^x(t,y,dy')$ denote its transition probability. We assume that $y^x(t)$ is an ergodic process and that its ergodic distribution is $\rho^x(dy)$. Finally, we assume that $y^x(t)$ satisfies condition SMC II of Section 2.3. If

$$R^x(y,C) = \int_0^\infty \left(Q^x(t,y,C) - \rho^x(C)\right) dt, \tag{5.33}$$

then $\int g(x,y')R^x(y,dy') \in C^{(2)}(R^d \times Y)$, $\int g(x,y)\,\rho^x(dy) \in C^{(2)}(R^d)$ for $g \in C^{(2)}(X \times Y)$, and

$$A^x \int g(x,y')R^x(y,C) = \int g(x,y)\,\rho^x(dy) - g(x,y). \tag{5.34}$$

(A3) We assume that $a(t,x,y)$ satisfies conditions (iv), (v), (vi), and (vii) in Section 5.1.2.

Theorem 3 *Under conditions* (A1), (A2), (A3) *the statement of Theorem 2 is valid.*

The proof is the same as in Theorem 2, and it is not presented here.

5.1.4 Stationary Perturbations

Let $a(t,x,\omega)$ be an R^d-valued random function on $R \times R^d$. We assume that $\frac{\partial a}{\partial x}(t,x,\omega)$ is a random function that is locally bounded in t and x. Let $x_\varepsilon(t)$ be the solution of the initial value problem

$$\dot{x}_\varepsilon(t) = a(t/\varepsilon, x_\varepsilon(t), \omega), \quad x_\varepsilon(0) = x_0. \tag{5.35}$$

Assume that the random function $a(t,x,\omega)$ satisfies the following conditions:

(F1) a is a stationary function in t; in particular, for any fixed h the random functions $a(t,x,\omega)$ and $a(t+h,x,\omega)$ have the same finite–dimensional distributions. It follows that $\frac{\partial a}{\partial x}(t,x,\omega)$ is also a stationary random function. We assume that

$$P\left\{\sup_{t,x} \left\|\frac{\partial a}{\partial x}(t,x,\omega)\right\| < \infty\right\} = 1.$$

(F2) Denote by $\mathcal{F}_t$ the σ-algebra generated by $\{a(s,x,\omega),\ s \le t\}$ and by $\mathcal{F}^u$ the σ-algebra generated by $\{a(s,x,\omega),\ s \ge u\}$. Let $L_2^a(P)$ be the space of $\mathcal{F}_\infty$-measurable random variables ξ for which $E\xi^2 < \infty$. We introduce the linear operator θ_s in $L_2^a(P)$ for which $\theta_s\big(a(t,x,\omega),z\big) = \big(a(t+s,x,\omega),z\big)$, $t \in R$, $x \in R^d$, $z \in R^d$. We assume that the following weak mixing condition is satisfied: If ξ_1 is $\mathcal{F}_0$-measurable, ξ_2 is $\mathcal{F}^0$-measurable, and ξ_1, $\xi_2 \in L_2^a(P)$, then

$$\lim_{t\to\infty} (E\xi_1\theta_t\xi_2 - E\xi_1 E\xi_2) = 0. \tag{5.36}$$

Under condition (F2) the random operator θ_s satisfies the ergodic property for all $\mathcal{F}$-measurable ξ for which $E|\xi| < \infty$, namely,

$$\lim_{T\to\infty} \frac{1}{T}\int_0^T \theta_s\xi\,ds = E\xi. \tag{5.37}$$

Denote by $\bar{a}(x) = Ea(x,\omega)$. It follows from Theorem 5 of Chapter 3 that $x_\varepsilon(t)$ converges uniformly as $\varepsilon \to 0$ to the solution $\bar{x}(t)$ of the averaged problem

$$\dot{\bar{x}}(t) = \bar{a}\big(\bar{x}(t)\big), \quad \bar{x}(0) = x_0 \tag{5.38}$$

on any finite interval $[0,t_0]$. We will assume that $\bar{a}(x) = Ea(t,x,\omega) = 0$. The behavior of $x_\varepsilon(t/\varepsilon)$ will be investigated under this condition and some additional conditions that are formulated below.

(F3) We assume that for any $\mathcal{F}_0$-measurable random variable ξ_0 with $E|\xi_0| < \infty$ and $E\xi_0 = 0$ there exists $\int_0^\infty E(\theta_s\xi_0/\mathcal{F}_0)\,ds = \Pi_0\xi_0$. Define $\Pi_u\xi_0 = \theta_u\Pi_0\xi_0$. Then $\Pi_u\xi_0$ is a stationary process. Set

$$\Pi_0^T\xi_0 = \int_0^T E(\theta_s\xi_0/\mathcal{F}_0)\,ds, \quad T > 0.$$

We assume additionally that $\lim_{T\to\infty} E|\Pi_0^T\xi_0 - \Pi_0\xi_0| = 0$.

(F4) Introduce the functions

$$\hat{a}(x) = E\big(\Pi_0 a_x(0,x,\omega)\big)a(0,x,\omega) \tag{5.39}$$

(this function is R^d-valued because a_x is an $L(R^d)$-valued function) and

$$\hat{B}(x) = 2E\big(\Pi_0 a(0,x,\omega)\big) * a(0,x,\omega), \tag{5.40}$$

where $z_1 * z_2$ is the operator from $L(R^d)$ for which

$$(z_1 * z_2)x = (z_1,x)z_2,\, z_1,\, z_2 \in R^d.$$

Then $\Pi_0 a_x(0,x)$ and $\Pi_0 a(0,x)$ are defined, because $Ea(0,x,\omega) = 0$ and $Ea_x(0,x,\omega) = 0$. We assume that the functions $\hat{a}$ and $\hat{B}$ are bounded and continuous and that the differential operator

$$\hat{L}f(x) = \big(f'(x),\hat{a}(x)\big) + \frac{1}{2}\mathrm{Tr}\, f''(x)\hat{B}(x) \tag{5.41}$$

is the generator of a diffusion process; so, the transition probabilities of this process are determined by $\hat{a}$ and $\hat{B}$.

(F5) We assume that

$$\sup_x \big(|a(0,x,\omega)| + \|a_x(0,x,\omega)\|\big) < \infty$$

and

$$\lim_{T\to\infty} E \sup_x \big(|\Pi_0^T a(0,x,\omega) - \Pi_0 a(0,x,\omega)| + \|(\Pi_0^T - \Pi_0)\, a_x(0,x,\omega)\|\big) = 0.$$

Theorem 4 *Let conditions* (F1)–(F5) *be fulfilled. Then the process* $\hat{x}_\varepsilon(t) = x_\varepsilon(t/\varepsilon)$ *converges weakly as* $\varepsilon \to 0$ *to the diffusion process* $\hat{x}(t)$ *having initial value* x_0 *and generator* $\hat{L}$ *given by (5.40).*

Proof We consider

$$\begin{aligned} &E\big(f(\hat{x}_\varepsilon(t_2)) - f(\hat{x}_\varepsilon(t_1))/\mathcal{F}_{t_1/\varepsilon^2}\big) \\ &\quad = E\bigg(\int_{t_1/\varepsilon}^{t_2/\varepsilon} \big(f'(x_\varepsilon(s)), a(s/\varepsilon, x_\varepsilon(s), \omega)\big)\, ds/\mathcal{F}_{t_1/\varepsilon^2}\bigg). \end{aligned} \tag{5.42}$$

We transform the right-hand side of (5.42) using the following relations: For $0 \le \alpha < \beta$,

$$\begin{aligned} \int_\alpha^\beta \big(f'(x_\varepsilon(s)), a(s/\varepsilon, x_\varepsilon(s), \omega)\big)\, ds &= \int_\alpha^\beta \big(f'(x_\varepsilon(\alpha)), a(s/\varepsilon, x_\varepsilon(\alpha), \omega)\big)\, ds \\ + \int_\alpha^\beta \int_\alpha^s &\big(f''(x_\varepsilon(u)) a(u/\varepsilon, x_\varepsilon(u), \omega), a(s/\varepsilon, x_\varepsilon(u), \omega)\big)\, du\, ds \\ + \int_\alpha^\beta \int_\alpha^s &\big(f'(x_\varepsilon(u)), a_x(s/\varepsilon, x_\varepsilon(u), \omega) a(u/\varepsilon, x_\varepsilon(u), \omega)\big)\, du\, ds. \end{aligned}$$

Note that

$$\begin{aligned} &E\bigg(\int_\alpha^\beta \big(f'(x_\varepsilon(\alpha)), a(s/\varepsilon, x_\varepsilon(\alpha), \omega)\big)\, ds/\mathcal{F}_{\alpha/\varepsilon}\bigg) \\ &\quad = \varepsilon\big(f'(x), \theta_{\alpha/\varepsilon} \Pi_0^{\varepsilon^{-1}(\beta-\alpha)} a(0,x,\omega)\big)_{x = x_\varepsilon(\alpha)}, \\ &E\bigg(\int_u^\beta \big(f''(x_\varepsilon(u)) a(u/\varepsilon, x_\varepsilon(u), \omega), a(s/\varepsilon, x_\varepsilon(u), \omega)\big)\, ds/\mathcal{F}_{\frac{u}{\varepsilon}}\bigg) \\ &\quad = \varepsilon\big(f''(x) a(u/\varepsilon, x, \omega), \theta_{\frac{u}{\varepsilon}} \Pi_0^{\varepsilon^{-1}(\beta-u)} a(0,x,\omega)\big)_{x = x_\varepsilon(u)}, \\ &E\bigg(\int_u^\beta \big(f'(x_\varepsilon(u)), a_x(s/\varepsilon, x_\varepsilon(u), \omega) a(u/\varepsilon, x_\alpha(u), \omega)\big)\, ds/\mathcal{F}_{\frac{u}{\varepsilon}}\bigg) \\ &\quad = \varepsilon\big(f'(x), \theta_{\frac{u}{\varepsilon}} \Pi_0^{\varepsilon^{-1}(\beta-u)} a_x(0,x,\omega) a(u/\varepsilon, x, \omega)\big)_{x = x_\varepsilon(u)}. \end{aligned}$$

Therefore,

$$E\big(f(\hat{x}_\varepsilon(t_2)) - f(\hat{x}_\varepsilon(t_1))/\mathcal{F}_{\frac{t_1}{\varepsilon^2}}\big) = \varepsilon\big(f'(x), \theta_{\frac{t_1}{t_2}} \Pi_0^{\varepsilon^{-2}(t_2-t_1)} a(0,x,\omega)\big)_{x=x_\varepsilon(t_1/\varepsilon)}$$
$$+ \varepsilon E\left(\int_{\frac{t_1}{\varepsilon}}^{\frac{t_2}{\varepsilon}} \left(f''(x) a\left(\frac{u}{\varepsilon}, x, \omega\right), \theta_{\frac{u}{\varepsilon}} \Pi_0^{\varepsilon^{-2}(t_2-\varepsilon u)} a(0,x,\omega)\right)_{x=x_\varepsilon(u)} du/\mathcal{F}_{t_1/\varepsilon^2}\right)$$
$$+ \varepsilon E\bigg(\int_{t_1/\varepsilon}^{t_2/\varepsilon} \big(f'(x), \big[\theta_{\frac{u}{\varepsilon}} \Pi_0^{\varepsilon^{-2}(t_2-\varepsilon u)} a_x(0,x,\omega)\big]\big)_{x=x_\varepsilon(u)} du/\mathcal{F}_{t_1/\varepsilon^2}\bigg).$$

We introduce the following random functions: For $f \in C^{(2)}(R^d)$ let

$$Z_0 f(x) = \big(f'(x), \big[\Pi_0 a_x(0,x,\omega)\big] a(0,x,\omega)\big) + \big(f''(x) a(0,x,\omega), \Pi_0 a(0,x,\omega)\big) \tag{5.43}$$

and

$$Z_0^T f(x) = \big(f'(x), \big[\Pi_0^T a_x(0,x,\omega)\big] a(0,x,\omega)\big) + \big(f''(x) a(0,x,\omega), \Pi_0^T a(0,x,\omega)\big). \tag{5.44}$$

It follows from condition (F5) that

$$\sup_x \big|Z_0^T f(x) - Z_0 f(x)\big| \le \lambda_1(T),$$

where $\lambda_1(T)$ is a $\mathcal{F}_0$-measurable random variable for which $\lim_{T\to\infty} E\lambda_1(T) = 0$. Then

$$\varepsilon E \int_{t_1/\varepsilon}^{t_2/\varepsilon} \left| \big[\theta_{u/\varepsilon}\big(Z_0^{\varepsilon^{-2}(t_2-\varepsilon u)} f(x) - Z_0 f(x)\big)\big]_{x=x_\varepsilon(u)} \right| du$$

$$\le \varepsilon \int_{t_1/\varepsilon}^{t_2/\varepsilon} E\lambda_1\big(\frac{t_2}{\varepsilon} - \frac{u}{\varepsilon}\big)\, du = \int_0^{t_2-t_1} E\lambda_1\big(\frac{t_2-t_1-v}{\varepsilon^2}\big)\, dv. \tag{5.45}$$

Using (5.45) and condition (F5) we can obtain from relation (5.42) the following statement:

$$E\left|E\bigg(f(\hat{x}_\varepsilon(t_2)) - f(\hat{x}_\varepsilon(t_1)) - \varepsilon \int_{t_1/\varepsilon}^{t_2/\varepsilon} \big[\theta_{u/\varepsilon} Z_0 f(x)\big]_{x=x_\varepsilon(u)} du/\mathcal{F}_{t_1/\varepsilon}\bigg)\right|$$

$$\le O\bigg(\varepsilon + \int_0^{t_2-t_1} E\lambda_1\big(\frac{t_2-t_1-v}{\varepsilon^2}\big)\, dv\bigg).$$

So

$$\lim_{\varepsilon\to 0} E\left|E\bigg(f(\hat{x}_\varepsilon(t_2)) - f(\hat{x}_\varepsilon(t_1)) - \int_{t_1}^{t_2} [\theta_{u/\varepsilon^2} Z_0 f(x)]_{x=\hat{x}_\varepsilon(u)}\, du/\mathcal{F}_{t_1/\varepsilon^2}\bigg)\right| = 0. \tag{5.46}$$

Now we prove that

$$\lim_{\varepsilon\to 0} E\left|E\bigg(\int_{t_1}^{t_2} \big([\theta_{u/\varepsilon^2} Z_0 f(x)]_{x=\hat{x}_\varepsilon(u)} - \hat{L} f(\hat{x}_\varepsilon(u))\big)\, du\Big/\mathcal{F}_{t_1/\varepsilon^2}\bigg)\right| = 0. \tag{5.47}$$

For any $\delta > 0$ set

$$\lambda_2(\delta, \omega) = \sup\{|Z_0 f(x) - Z_0 f(x')| : |x - x_1| \leq \delta\}.$$

Then $\lambda_2(\delta, \omega)$ is an $\mathcal{F}_0$- measurable random variable for which $E\lambda_2(\delta, \omega) \to 0$ as $\delta \to 0$. Note that $|\hat{x}_\varepsilon(u) - \hat{x}_\varepsilon(v)| \leq (c/\varepsilon)|u - v|$, where c is the constant from condition (F1). This implies the inequality

$$\left| [\theta_{u/\varepsilon^2} Z_0 f(x)]_{x=\hat{x}_\varepsilon(u)} - [\theta_{u/\varepsilon^2} Z_0 f(x)]_{x=\hat{x}_\varepsilon(v)} \right| \leq \theta_{u/\varepsilon^2} \lambda_2\left(\frac{\delta c}{\varepsilon}, \omega\right),$$

if $|u - v| \leq \delta$ from which it follows that for $\delta > 0$,

$$E\left| \int_{t_1}^{t_2} \frac{1}{\delta} \int_v^{v+\delta} ([\theta_{u/\varepsilon^2} Z_0 f(x)]_{x=\hat{x}_\varepsilon(u)} - [\theta_{u/\varepsilon^2} Z_0 f(x)]_{x=\hat{x}_\varepsilon(v)})\, du\, dv \right|$$

$$\leq (t_2 - t_1) E\lambda_2\left(\frac{c}{\varepsilon}\delta, \omega\right). \tag{5.48}$$

It follows from the ergodic theorem that for all x,

$$\lim_{\varepsilon \to 0} \frac{1}{\delta} \int_v^{v+\delta} \theta_{u/\varepsilon^2} Z_0 f(x)\, du = \hat{L} f(x) \quad \text{if } \delta = \delta_\varepsilon$$

satisfies the condition $\delta_{\varepsilon/\varepsilon^2} \to \infty$.
Since

$$E\left| \sup_{|x - x'| \leq \rho} \frac{1}{\delta} \int_v^{v+\delta} (\theta_{u/\varepsilon^2} Z_0 f(x) - \theta_{u/\varepsilon^2} Z_0 f(x))\, du \right| \leq E\lambda_2(\rho, \omega),$$

we conclude that

$$\lim_{\varepsilon \to 0} E \sup_{|x| \leq r} \left| \frac{1}{\delta_\varepsilon} \int_0^{\delta_\varepsilon} \theta_{u/\varepsilon^2} Z_0 f(x)\, du - \hat{L} f(x) \right| = 0 \tag{5.49}$$

for any $r > 0$ if $\delta_{\varepsilon/\varepsilon^2} \to \infty$ as $\varepsilon \to 0$. Then (5.49) and (5.48) imply (5.47), and the last relation and (5.46) give us the relation

$$\lim_{\varepsilon \to 0} E \left| E\left(f(\hat{x}_\varepsilon(t_2)) - f(\hat{x}_\varepsilon(t_1) - \int_{t_1}^{t_2} \hat{L} f(\hat{x}_\varepsilon(s))\, ds \Big/ \mathcal{F}_{t_1/\varepsilon^2} \right) \right| = 0. \tag{5.50}$$

With this the statement of the theorem follows as a consequence of Theorem 2 of Chapter 2.

□

5.1.5 *Diffusion Approximations to First Integrals*

Consider the solution of equation (5.1) with the same assumptions on the Markov process $y(t)$ that were stated in Section 5.1.1. Assume that the function $a(x, y)$ satisfies condition (a1) of that section, and we write $\bar{a}(x) =$

$\int a(x,y)\,\rho(dy)$. (We do not assume here that $\bar{a}(x) = 0$.) We consider the case where the averaged equation

$$\dot{\bar{x}}(t) = \bar{a}\big(\bar{x}(t)\big) \tag{5.51}$$

has a first integral, say denoted by $\varphi(x)$. That is, the function $\varphi : R^d \to R$ is continuous and differentiable, and it satisfies the differential equation

$$\big(\varphi'(x), \bar{a}(x)\big) = 0. \tag{5.52}$$

Then $\varphi\big(\bar{x}(t)\big)$ = constant, and

$$\int \big(\varphi'(x), a(x,y)\big)\rho(dy) = 0.$$

We will investigate in this section the stochastic process $\varphi(x_\varepsilon(t))$. Note that condition (a1) ensures that Theorem 5 of Chapter 3 can be applied to study equation (5.1). Therefore, $\varphi(x_\varepsilon(t))$ converges to $\varphi(x_0)$ uniformly on any finite interval as $\varepsilon \to 0$ with probability 1. In this section we derive a diffusion approximation to $\varphi(x_\varepsilon(t))$.

Some preliminary work is needed to formulate the results. First, assume that the function $a(x,y)$ satisfies, in addition, condition (a2) of Section 5.1.1 and that the derivatives φ_x, φ_{xx} exist as continuous and bounded functions. Then, for any twice continuously differentiable function $\theta :\ R \to R$, we have

$$\begin{aligned} &E\big(\theta\big(\varphi(x_\varepsilon(t_2))\big) - \theta\big(\varphi(x_\varepsilon(t_1))\big)/\mathcal{F}_{t_1/\varepsilon}\big) \\ &= \int_{t_1}^{t_2} \theta'\left(\varphi(x_\varepsilon(s))\right)\left(\varphi'(x_\varepsilon(s)), a(x_\varepsilon(s), y_\varepsilon(s))\right) ds \\ &= \varepsilon\Bigg[G(x_\varepsilon(t_1), y_\varepsilon(t_1)) - G(x_\varepsilon(t_2), y_\varepsilon(t_2)) \\ &\quad + \int_{t_1}^{t_2} \left(G_x(x_\varepsilon(s), y_\varepsilon(s)), a(x_\varepsilon(s), y_\varepsilon(s))\right) ds/\mathcal{F}_{t_1/\varepsilon}\Bigg], \end{aligned} \tag{5.53}$$

where

$$G(x,y) = \theta'(\varphi(x)) \int R(y, dy')\big(\varphi'(x), a(x,y')\big).$$

(We used formula (5.7) here.) So,

$$\begin{aligned} &E\big(\theta\big(\varphi(x_\varepsilon(t_2/\varepsilon))\big) - \theta\big(\varphi(x_\varepsilon(t_1/\varepsilon))\big)/\mathcal{F}_{t_1/\varepsilon}\big) \\ &= \varepsilon E\bigg(\int_{t_1/\varepsilon}^{t_2/\varepsilon} H\left(\varphi(x_\varepsilon(s)), x_\varepsilon(s), y_\varepsilon(s)\right) ds/\mathcal{F}_{t_1/\varepsilon}\bigg) + O(\varepsilon), \end{aligned} \tag{5.54}$$

where

$$H(\varphi, x, y) = \theta''(\varphi) \int (R(y, dy')(\varphi'(x), a(x, y')) \cdot (\varphi'(x), a(x, y))$$
$$+ \theta'(\varphi) \int (R(y, dy')(\varphi''(x)a(x, y'), a(x, y))$$
$$+ \theta'(\varphi) \int (R(y, dy')(\varphi'(x), a_x(x, y')a(x, y)).$$

Set $\hat{H}(\varphi, x) = \int H(\varphi, x, y)\rho(dy)$. Since $\int [H(\varphi, x, y) - \hat{H}(\varphi, x)]\rho(dy) = 0$, it follows from the proof of Theorem 1 that

$$\lim_{\varepsilon \to 0} \varepsilon \left| \int_{t_1/\varepsilon}^{t_2/\varepsilon} \left(H\big(\varphi(x_\varepsilon(s)), x_\varepsilon(s), y_\varepsilon(s)\big) - \hat{H}\big(\varphi(x_\varepsilon(s)), x_\varepsilon(s)\big)\right) ds \right| = 0. \tag{5.55}$$

Define

$$\beta(x) = \iint R(y, dy')\rho(dy)\big(\varphi'(x), a(x, y')\big)\big(\varphi'(x), a(x, y)\big)$$

and

$$\frac{\alpha(x)}{2} = \iint R(y, dy')\rho(dy)\big[\big(\varphi''(x)a(x, y'), a(x, y)\big) + \big(\varphi'(x), a_x(x, y')a(x, y)\big)\big].$$

Then, it follows from relations (5.54) and (5.55) that

$$0 = \lim_{\varepsilon \to 0} E\bigg| E\bigg(\theta\big(\varphi(x_\varepsilon(t_2/\varepsilon))\big) - \theta\big(\varphi(x_\varepsilon(t_1))\big)$$
$$- \varepsilon \int_{t_1/\varepsilon}^{t_2/\varepsilon} \left[\alpha(x_\varepsilon(s))\theta'\big(\varphi(x_\varepsilon(s))\big) + \frac{1}{2}\beta(x_\varepsilon(s))\theta''\big(\varphi(x_\varepsilon(s))\big)\right] ds/\mathcal{F}_{t_1/\varepsilon^2}\bigg)\bigg|. \tag{5.56}$$

We need an assumption concerning the ergodic behavior of the solution of the averaged equation:
EAE1 Denote by $\bar{x}(t, x_0)$ the solution of equation (5.51) for which

$$\bar{x}(0, x_0) = x_0.$$

For any constant c for which the set $\Phi_c = \{x : \varphi(x) = c\}$ is not empty, this set is compact, and there exists a probability measure $m_c(dx)$ on R^d for which

$$\lim_{T \to \infty} \frac{1}{T} \int_0^T g\big(\bar{x}(t, x_0)\big)\, dt = \int g(x)\, m_{\varphi(x_0)}(dx)$$

uniformly in x_0, $|x_0| \le r$ for any $r > 0$ and every $g \in C(R^d)$.
Set

$$\hat{\alpha}(c) = \int \alpha(x)\, m_c(dx)$$

and

$$\hat{\beta}(c) = \int \beta(x)\, m_c(dx),$$

and define $\hat{c}_- = \inf_x \varphi(x)$, $\hat{c}_+ = \sup \varphi(x)$. The functions $\hat{\alpha}(z)$ and $\hat{\beta}(z)$ are defined for $z \in (\hat{c}_-, \hat{c}_+)$. The conditions on $a(x, y)$ and $\varphi(x)$ ensure that these functions are bounded, and condition (**EAE1**) implies that $\hat{\alpha}(z)$ and $\hat{\beta}(z)$ are continuous. Note that $\hat{\beta}(z) \geq 0$. Assume that $\hat{\beta}(z) > 0$ for $z \in (\hat{c}_-, \hat{c}_+)$. Then the differential operator L^φ, which is defined by the relation

$$L^\varphi \theta(z) = \hat{\alpha}(z)\theta'(z) + \frac{1}{2}\hat{\beta}(z)\theta''(z), \quad z \in R, \tag{5.57}$$

for $\theta \in C^{(2)}(\hat{c}_-, \hat{c}_+)$, is the generator of a diffusion Markov process in the interval $(\hat{c}_-, \hat{c}_+)$ with absorbing boundary.

Theorem 5 *Assume that the following conditions are fulfilled:*

(i) *The function* $a(x, y)$ *satisfies the conditions* (a1) *and* (a2) *of Section 5.1.1.*

(ii) *The Markov process* $y(t)$ *satisfies condition SMC* II *of Section 2.3.*

(iii) *Condition* (EAE1) *is satisfied by a function* $\varphi : R^d \to R$ *that has continuous bounded derivatives* φ_x, φ_{xx} *and satisfies equation (5.52).*

(iv) *The differential operator* L^φ *given by the formula (5.57) has positive coefficient* $\beta(z)$ *for* $z \in (\hat{c}_-, \hat{c}_+)$, *where* $\hat{c}_- = \inf_x \varphi(x)$, $\hat{c}_+ = \sup \varphi(x)$, *so* L^φ *is the generator of the diffusion process on the interval* $(\hat{c}_-, \hat{c}_+)$ *with absorbing boundary. Denote by* $\tau_\varepsilon = \max\{t : x_\varepsilon(s) \in (\hat{c}_-, \hat{c}_+), s < t\}$.

Then the stochastic process $z_\varepsilon(t) = \varphi(x_\varepsilon(t \wedge \tau_\varepsilon/\varepsilon))$ *converges weakly in* C *to the diffusion process with generator* L^φ *and absorbing boundary satisfying the initial condition* $z(0) = \varphi(x_0)$, *where*

$$x_0 = x_\varepsilon(0) \quad \textit{for all} \quad \varepsilon > 0.$$

Proof First we prove the relation

$$\lim_{\varepsilon \to 0} E \left| \left(\theta(z_\varepsilon(t_2)) - \theta(z_\varepsilon(t_1)) - \int_{t_1}^{t_2} L^\varphi \theta(z_\varepsilon(s))\, ds \Big/ \mathcal{F}_{t_1/\varepsilon^2} \right) \right| = 0 \tag{5.58}$$

for any function $\theta(z) \in C^{(2)}(R)$ for which such a $z_1 < z_2$ exists for $\hat{c}_- < z_1 < z_2 < \hat{c}_+$ that $\theta(z) = 0$ if $z \leq z_1$, $z \geq z_2$. We derive next some auxiliary statements.

Lemma 4

$$P\{|z_\varepsilon(t) - z_\varepsilon(0)| > c\} \leq O\left(\frac{\varepsilon + t\lambda}{1 - e^{-\lambda c^2}} \right) \tag{5.59}$$

for every $\lambda > 0$ *and* $c > 0$.

Proof The proof follows from formula (5.54) if we set $t_1 = 0$, $t_2 = t$, and $\theta(z) = \exp\{-\lambda z^2\}$ there and take into account that $H = O(\lambda)$.

□

Lemma 5 *Let $\Psi \in C(R^d)$ and $\hat{\Psi}(z) = \int \Psi(x)\, m_z(dx)$. Then*

$$\lim_{\varepsilon \to 0} E\left| \varepsilon \int_{t_1/\varepsilon}^{t_2/\varepsilon} \Psi(x_\varepsilon(s))\, ds - \int_{t_1}^{t_2} \hat{\Psi}(z_\varepsilon(s))\, ds \right| = 0.$$

Proof It follows from Theorem 5 of Chapter 3 that for any $T > 0$,

$$\lim_{\varepsilon \to 0} E\left| \frac{1}{T} \int_0^T [\Psi(\bar{x}(x_\varepsilon(s), u) - \Psi(x_\varepsilon(s+u))]\, du \right| = 0$$

uniformly in s. So

$$\lim_{\varepsilon \to 0} E\left| \varepsilon \int_{t_1/\varepsilon}^{t_2/\varepsilon} \Psi(x_\varepsilon(s))\, ds - \varepsilon \int_{t_1/\varepsilon}^{t_2/\varepsilon} \frac{1}{T} \int_0^T \Psi(\bar{x}(x_\varepsilon(s), u))\, du\, ds \right| = 0.$$

It follows from condition (**EAE1**) that for any $c > 0$,

$$\lim_{T \to \infty} \sup_{s < (t/\varepsilon)} E\left| \frac{1}{T} \int_0^T \Psi(\bar{x}(x_\varepsilon(s), u)\, du - \hat{\Psi}(z_\varepsilon(s)) \right| 1_{\{|z_\varepsilon(s)| \le c\}} = 0.$$

This relation and Lemma 3 imply the relation

$$\limsup_{\varepsilon \to 0, T \to \infty} E\left| \varepsilon \int_{t_1/\varepsilon}^{t_2/\varepsilon} \frac{1}{T} \int_0^T \Psi(\bar{x}(x_\varepsilon(s), u))\, du - \int_{t_1}^{t_2} \hat{\Psi}(z_\varepsilon(s))\, ds \right| = 0.$$

□

We return now to the proof of the theorem. Relation (5.58) is a consequence of relation (5.56) and Lemma 5. Relation (5.58) and Remark 8 of Chapter 2 imply the weak convergence of the processes $z_\varepsilon(t)$ to the process $z(t)$ as $\varepsilon \to 0$. To prove that the family of stochastic processes $\{z_\varepsilon(t), \varepsilon \in (0,1)\}$ is weakly compact in C, we use the same method as in Theorem 1 using Lemma 3.

□

Remark 4 *The behavior of the diffusion process $z(t)$ in the case $(\hat{c}_-, \hat{c}_+) = (0, \infty)$ (i.e., $\inf_x \varphi(x) = 0$ and $\varphi(x) \to +\infty$ as $|x| \to \infty$) is determined by the coefficients of the differential operator L^φ: Set*

$$\mathcal{U}(z) = \exp\left\{ - \int_{z_0}^{z} 2\hat{\alpha}(u) \hat{\beta}^{-1}(u)\, du \right\}, \quad z_0 > 0,$$

$$I_1 = \int_0^{z_0} \mathcal{U}(z)\, dz, \quad I_2 = \iint_{0<u<z<z_0} \frac{\mathcal{U}(z)}{\mathcal{U}(u)\hat{\beta}(u)}\, du\, dz.$$

(i) *If* $I_1 = +\infty$, $I_2 = +\infty$, *then* 0 *is a natural boundary for* $z(t)$ (*this means that* $P\{\tau_0 = +\infty\} = 1$, $\tau_0 = \inf\{t : z(t) = 0\}$, *and* $P\{\lim_{t\to\infty} z(t) = 0\} = 0$). *The point* $z = 0$ *is inaccessible.*

(ii) *If* $I_1 < \infty$, $I_2 < \infty$, *then* 0 *is a regular boundary,* $P\{\tau_0 < \infty\} > 0$, *and* $P\{\tau_0 < \infty/z(0)\} \to 1$ *as* $z(0) \to 0$.

(iii) *If* $I_1 < \infty$, $I_2 = +\infty$, *then* 0 *is an attracting boundary:*

$$P\{\tau_0 < \infty\} = 0, \qquad \lim_{z(0)\to 0} P\left(\lim_{t\to\infty} z(t) = 0/z(0)\right) = 1.$$

(iv) *If* $I_1 = +\infty$, $I_2 < \infty$, *then* 0 *is a repelling boundary; let* $\tau(0, z)$ *be the exit time from the interval* $(0, z)$. *Then for* $u \in (0, z)$, $P(z(\tau(0, z) = z/z(0) = 0) = 1$, *and* $E(\tau(0, z)/z(0) = u)$ *is uniformly bounded in* $u \in (0, z)$. *The point* 0 *is inaccessible.*

The proof of these remarks is available in [40].

Now we consider the case when equation (5.51) has m first integrals, say $\varphi_1(x), \ldots, \varphi_m(x)$ satisfying equation (5.51). We assume that the mapping $x \to (\varphi_1(x), \ldots, \varphi_m(x))$ of R^d into R^m is continuous. Denote by ϕ the image of R^d under this mapping: If $(z_i, \ldots, z_m) \in \phi$, then $x \in R^d$ exists for which $\varphi_k(x) = z_k$, $k = 1, \ldots, m$.

We introduce an extension of **EAE1** to this case:

EAE2 Denote by $\phi_{z_1,\ldots,z_m}$ the set of those x for which $\varphi_k(x) = z_{k_1}$, $k = 1, \ldots, m$, $(z_1, \ldots, z_m) \in \phi$. Assume that $\phi_{z_1,\ldots,z_m}$ is a compact set and that a measure m_z exists, $z = (z_1, \ldots, z_m) \in R^m$ for which

$$\lim_{T\to\infty} \frac{1}{T} \int_0^T g(\bar{x}(t, x_0))\, dt = \int g(x) m_z(dx) \tag{5.60}$$

if $(\varphi_1(x_0), \ldots, \varphi_m(x_0)) = z$ for $g \in C^{(2)}(R^m)$ uniformly in z, $|z| \le C$, for all $C > 0$.

Define a second–order differential operator $\tilde{L}$ by the relation

$$\begin{aligned} \tilde{L}\Psi(z) = \iiint &\left(\frac{\partial}{\partial x}([\Psi(\varphi_1(x), \ldots, \varphi_m(x))]_x, a(x, y')), a(x, y)\right) \\ &\times R(y, dy')\, \rho(dy)\, m_z(dx) \end{aligned} \tag{5.61}$$

for functions $\Psi \in C^{(2)}(R^m)$ for which $\operatorname{supp}\Psi \subset \operatorname{int}\phi$, provided that continuous bounded derivatives $\frac{\partial \varphi_k}{\partial x}(x)$, $\frac{\partial^2 \varphi_k}{\partial x^2}(x)$, $k = 1, \ldots, m$, exist. Here $\operatorname{supp}\Psi$ is the closure of the set $\{z : |\Psi(z) > 0\}$ and $\operatorname{int}\phi$ is the set of interior points of ϕ.

Theorem 6 *Assume that the following conditions are fulfilled:*

(i-ii) (i) *and* (ii) *of Theorem* 1 *hold.*

(iii) *Condition* (EAE2), *and* $\frac{\partial}{\partial x}\varphi_k(x)$, $\frac{\partial^2}{\partial x}\varphi_k(x)$ *are bounded and continuous for* $k = 1, \ldots, m$.

(iv) *The differential operator $\tilde{L}$ is the generator of a diffusion process $z(t)$ in the set* $\mathrm{int}\,\phi$ *with absorbing boundary.*

Set

$$z_\varepsilon(t) = \left(\varphi_1\left(x\left(\frac{t}{\varepsilon}\right)\right), \dots, \varphi_m\left(x\left(\frac{t}{\varepsilon}\right)\right)\right), \tau_\varepsilon = \max\{t : z_\varepsilon(s) \in \mathrm{int}\phi, s < t\}.$$

Then the stochastic process $z_\varepsilon(t \wedge \tau_\varepsilon)$ converges weakly in C as $\varepsilon \to 0$ to the stochastic process $z(t)$ with generator L^φ, absorbing boundary, and with initial value $z(0) = (\varphi_1(x_0), \dots, \varphi_m(x_0))$.

The proof is the same as the proof of Theorem 5 and is not presented here.

5.2 Difference Equations

In this section we consider discrete–time systems described by difference equations that are also perturbed by random noise.

5.2.1 *Markov Perturbations*

Let $\{y_n, n \in \mathbf{Z}_+\}$ be a homogeneous Markov process in a measurable space $(Y, \mathcal{C})$. Denote by $P_k(y, C)$ its transition probability for k steps. We consider the difference equation

$$x^\varepsilon_{n+1} - x^\varepsilon_n = \varepsilon\varphi(x^\varepsilon_n, y_n), \quad n \ge 0, \quad x^\varepsilon_0 = x_0, \tag{5.62}$$

where $\varphi : R^d \times Y \to R^d$ is a measurable function.
Assume that $\{y_n\}$ is an ergodic process with ergodic distribution $\rho(dy)$, that the function φ satisfies the condition $\int |\varphi(x, y)|\, \rho(dy) < \infty$, and that

$$\int \varphi(x, y)\, \rho(dy) = 0.$$

We define a continuous–time stochastic process in R^d by

$$\tilde{x}^\varepsilon(t) = \sum_n 1_{\{\varepsilon^2 n \le t < \varepsilon^2(n+1)\}} \left[(t\,\varepsilon^{-2} - n)x^\varepsilon_{n+1} + (n + 1 - t\,\varepsilon^{-2})x^\varepsilon_n\right], \tag{5.63}$$

so $\tilde{x}^\varepsilon(t)$ is a continuous function of t and

$$\tilde{x}^\varepsilon(n\varepsilon^2) = x^\varepsilon_n.$$

We will prove that under general conditions the process $\tilde{x}^\varepsilon(t)$ converges weakly in C to a diffusion process in R^d.

Theorem 7 *Assume that the Markov process $\{y_n, n \in \mathbf{Z}_+\}$ satisfies condition SMC* II *of Section 2.3 and that φ is a measurable function that satisfies the following conditions:*

(a) $\varphi(x, y)$ *is continuous in* x *and has a continuous derivative* $\varphi_x(x, y)$ *with*

$$\sup_{x,y}(|\varphi(x, y)| + \|\varphi_x(x, y)\|) < \infty;$$

(b) $\lim_{\delta \to 0} \sup\{\|\varphi_x(x, y) - \varphi_x(x', y)\| : |x - x'| < \delta, y \in Y\} = 0$;
(c) $\int \varphi(x, y)\, \rho(dy) = 0$.
Let an operator L *be defined for functions* $f \in C^{(2)}(R^d)$ *by the formula*

$$\begin{aligned} Lf(x) = &\int \left(\frac{\partial}{\partial x} \int (f_x(x), \varphi(x, \tilde{y})) R(y, d\tilde{y}), \varphi(x, y) \right) \rho(dy) \\ &+ \frac{1}{2} \int (f_{xx}(x) \varphi(x, y), \varphi(x, y)) \rho(dy), \end{aligned} \tag{5.64}$$

where

$$R(y, C) = 1_{\{y \in C\}} + \sum_{k=1}^{\infty} (P_k(y, C) - \rho(C)), \quad c \in C.$$

Assume that L *is the generator of a diffusion process* $\tilde{x}(t)$*. Then as* $\varepsilon \to 0$ *the stochastic process* $\tilde{x}^{\varepsilon}(t)$ *converges weakly in* C *to the diffusion process* $\tilde{x}(t)$ *having initial condition* $\tilde{x}(0) = x_0$.

Before proving this theorem, we establish several supporting facts.

Lemma 6 *Let* $g : Y \to R$ *be a measurable function for which* $\int g(y)\, \rho(dy) = 0$*. Then the operator* $Rg(y) = \int R(y, dy') g(y')$ *satisfies the relation*

$$\int P(y, dy') Rg(y') = Rg(y) - g(y).$$

Proof

$$Rg(y) = \sum_{k=1}^{\infty} \int P_k(y, dy') g(y') + g(y),$$

since $\int g(y)\, \rho(dy) = 0$. So,

$$\int P(y, dy') Rg(y') = \sum_{k=1}^{\infty} \int P_k(y, dy') g(y').$$

This completes the proof of Lemma 6.

□

Lemma 7 *Let* $\psi(x, y) : R^d \times Y \to R$ *satisfy the following conditions:*

(α) $\psi(x, y)$ *and* $\psi_x(x, y)$ *are bounded measurable functions continuous in* x.

(β) *If* $\lambda(\delta)$ *is defined by the relation*

$$\lambda(\delta) = \sup\{|\Psi_x(x, y) - \Psi_x(x', y)| : |x - x'| < \delta, y \in Y\},$$

then $\lim_{\delta \to 0} \lambda(\delta) = 0$.

(γ) $\int \psi(x,y)\,\rho(dy) = 0.$

Set

$$\Psi(x,y) = \int \psi(x,\tilde{y})R(y,d\tilde{y}). \tag{5.65}$$

Denote by $\mathcal{F}_n$ *the* σ*-algebra generated by* $\{y_0,\dots,y_n\}$*. Then for any* $n_1 < n_2$,

$$\begin{aligned} E\biggl(\sum_{k=n_1}^{n_2} \psi(x_k^\varepsilon, y_k)\Big/\mathcal{F}_{n_1}\biggr) =& E\Bigl(\Psi(x_{n_1}^\varepsilon, y_{n_1}) - \Psi(x_{n_2+1}^\varepsilon, y_{n_2+1})\Big/\mathcal{F}_{n_1}\Bigr) \\ &+ \varepsilon E\biggl(\sum_{k=n_1}^{n_2}\bigl(\Psi_x(x_k^\varepsilon, y_{k+1}), \varphi(x_k^\varepsilon, y_{k+1})\bigr)\Big/\mathcal{F}_{n_1}\biggr) \\ &+ O\bigl(\varepsilon\lambda(\varepsilon)(n_2-n_1)\bigr). \end{aligned}$$

Proof It follows from Lemma 6 that

$$\begin{aligned} &E\bigl(\Psi(x_{k+1}^\varepsilon, y_{k+1}) - \Psi(x_k^\varepsilon, y_k)/\mathcal{F}_k\bigr) \\ &= E\bigl(\Psi(x_k^\varepsilon, y_{k+1}) - \Psi(x_k^\varepsilon, y_k)/\mathcal{F}_k\bigr) + E\bigl(\Psi(x_{k+1}^\varepsilon, y_{k+1}) - \Psi(x_k^\varepsilon, y_{k+1})/\mathcal{F}_k\bigr) \\ &= -\psi(x_k^\varepsilon, y_k) + \varepsilon E\bigl(\bigl(\Psi_x(x_k^\varepsilon, y_{k+1}), \varphi(x_k^\varepsilon, y_k)\bigr)/\mathcal{F}_k\bigr) \\ &\quad + E\Bigl(\Psi(x_k^\varepsilon + \varepsilon\varphi(x_k^\varepsilon, y_{k+1}), y_k) - \Psi(x_k^\varepsilon, y_{k+1}) \\ &\quad - \varepsilon\bigl(\Psi_x(x_k^\varepsilon, y_{k+1}), \varphi(x_k^\varepsilon, y_k)\bigr)\Big/\mathcal{F}_k\Bigr). \end{aligned}$$

It follows from condition (β) that

$$\bigl|\Psi(x+\varepsilon\varphi(x,y),y) - \Psi(x,y) - \varepsilon(\Psi_x(x,y),\varphi(x,y))\bigr| = O\bigl(\varepsilon\lambda(\varepsilon)\bigr).$$

So

$$\begin{aligned} &E\Bigl(\Psi(x_{n_2+1}^\varepsilon, y_{n_2+1}) - \Psi(x_{n_1}^\varepsilon, y_{n_1})/\mathcal{F}_{n_1}\Bigr) \\ &= E\biggl(\sum_{k=n_1}^{n_2}\bigl[\varepsilon(\Psi_x(x_k^\varepsilon, y_{k+1}), \varphi(x_k^\varepsilon, y_k)) - \psi(x_k^\varepsilon, y_k)\bigr]\Big/\mathcal{F}_{n_1}\biggr) \\ &\quad + O\bigl((n_2-n_1)\varepsilon\lambda(\varepsilon)\bigr). \end{aligned}$$

This completes the proof of Lemma 7.

□

Lemma 8 *Assume that a function* $\theta(x,y) : R^d \times Y \to R$ *satisfies the following conditions:*

(1) $\theta(x,y)$ *is a measurable bounded real-valued function.*

(2) *Let* $\rho(\delta) = \sup\{|\theta(x,y) - \theta(x',y)| : |x-x'| \le \delta, y \in Y\}$*. Then* $\lim_{\delta\to 0}\rho(\delta) = 0$.

Set $\bar\theta(x) = \int \theta(x,y)\rho(dy)$*. Then for any integers* $n_1 < n_2$*,*

$$E\Big(\sum_{k=n_1}^{n_2} \theta(x_k^\varepsilon, y_k)\Big/\mathcal{F}_{n_1}\Big) = E\Big(\sum_{k=n_1}^{n_2} \bar\theta(x_k^\varepsilon)\Big/\mathcal{F}_{n_1}\Big) + o(n_2-n_1) + (n_2-n_1)\beta(\varepsilon),$$

where $\lim_{\varepsilon\to 0}\beta(\varepsilon) = 0$.

Proof Denote by E_y the conditional expectation operator under the condition $y_0 = y$. It follows from condition SMC II that

$$\limsup_{x\in R^d, y\in Y} E_y\left|\frac{1}{n}\sum_{k=0}^{n-1}\theta(x,y_k) - \bar\theta(x)\right| = 0.$$

Let $m < n_2 - n_1$. Then

$$\begin{aligned}
&\sum_{k=n_1}^{n_2}\theta(x_k^\varepsilon, y_k)\\
&= \sum_{k=n_1}^{n_2}\Big(\frac{1}{m}\sum_{i=0}^{m-1}\theta(x_k^\varepsilon, y_{k+i}) + \frac{1}{m}\sum_{i=0}^{m-1}\big(\theta(x_{k+1}^\varepsilon, y_{k+i}) - \theta(x_k^\varepsilon, y_{k+1})\big)\Big)\\
&\quad + O(m).
\end{aligned}$$

If $|\varphi(x,y)| \le c$, then $|\theta(x_{k+i}^\varepsilon, y) - \theta(x_k^\varepsilon, y)| \le \rho(cm\varepsilon)$ for $0 \le i \le m$ due to condition (2). So

$$\begin{aligned}
&E\Big(\sum_{k=n_1}^{n_2}\big(\theta(x_{k_1}^\varepsilon y_k) - \bar\theta(x_k^\varepsilon)\big)\Big/\mathcal{F}_{n_1}\Big)\\
&\le E\Big(\sum_{k=n_1}^{n_2} E\Big(\Big|\frac{1}{m}\sum_{k=0}^{m-1}\theta(x_k^\varepsilon, y_{k+i}\Big)\Big) - \bar\theta(x_k^\varepsilon)\Big|\Big/\mathcal{F}_k\Big)\Big/\mathcal{F}_{n_1}\Big)\\
&\quad + O(m + (n_2-n_1)\rho(cm\varepsilon)) = O(m + (n_2-n_1)(\rho(cm\varepsilon) + \alpha_m)),
\end{aligned}$$

where

$$\alpha_m = \sup\left\{E_y\left|\frac{1}{m}\sum_{k=0}^{m-1}\theta(x,y_k) - \bar\theta(x)\right| : x\in R^d, y\in Y\right\}.$$

As a result, $\alpha_m \to 0$ as $m\to\infty$. This completes the proof of Lemma 8. □

We now return to prove the theorem.

Proof

Let $f \in C^{(3)}(R^d)$. Then

$$\begin{aligned}
f(x_{k+1}^\varepsilon) - f(x_k^\varepsilon) =& \varepsilon\big(f_x(x_k^\varepsilon), \varphi(x_k^\varepsilon, y_{k+1})\big)\\
&+ \frac{1}{2}\varepsilon^2\big(f_{xx}(x_k^\varepsilon)\varphi(x_k^\varepsilon, y_{k+1}), \varphi(x_k^\varepsilon, y_{k+1})\big) + O(\varepsilon^3).
\end{aligned} \tag{5.66}$$

Set $\psi(x,y) = \int (f_x(x), \varphi(x,\tilde{y}))P(y,d\tilde{y})$.
It follows from conditions (a) and (b) that the function $\psi(x,y)$ satisfies the conditions of Lemma 7 with $\lambda(\varepsilon) = c_1\varepsilon$, $c_1 > 0$ a constant. So for $n_1 < n_2$,

$$E\Big(\sum_{k=n_1}^{n_2} \psi(x_k^\varepsilon, y_k) \Big/ \mathcal{F}_{n_1}\Big) = O(1) + O\big(\varepsilon^2(n_2 - n_1)\big)$$
$$+ \varepsilon E\Big(\sum_{k=n_1}^{n_2} \big(\Psi_x(x_k^\varepsilon, y_{k+1}), \varphi(x_k^\varepsilon, y_{k+1})\big) \Big/ \mathcal{F}_{n_1}\Big),$$

where $\Psi(x,y)$ is determined by formula (5.65).
So for $0 < n_1 < n_2$ we have

$$\begin{aligned} E\big(f(x_{n_2}^\varepsilon) - f(x_{n_1}^\varepsilon)/\mathcal{F}_{n_1}\big) &= O(\varepsilon) + O\big(\varepsilon^3(n_2 - n_1)\big) \\ &\quad + \varepsilon^2 E\Big(\sum_{k=n_1}^{n_2-1} \big[\tfrac{1}{2}\big(f_{xx}(x_k^\varepsilon)\varphi(x_k^\varepsilon, y_{k+1}), \varphi(x_k^\varepsilon, y_{k+1})\big) \\ &\quad + \big(\Psi_x(x_k^\varepsilon, y_{k+1}), \varphi(x_k^\varepsilon, y_{k+1})\big)\big] \Big/ \mathcal{F}_{n_1}\Big). \end{aligned} \tag{5.67}$$

Let

$$\theta(x,y) = \int \Big[\frac{1}{2}\big(f_{xx}\varphi(x,\tilde{y}), \varphi(x,\tilde{y})\big) + \big(\Psi_x(x,\tilde{y}), \varphi(x,\tilde{y})\big)\Big] P(y,d\tilde{y}).$$

Then the function $\theta(x,y)$ satisfies the conditions of Lemma 8. We apply Lemma 8 to the last sum in equality (5.67). This gives us the relation

$$E\big(f(x_{n_2}^\varepsilon) - f(x_{n_1}^\varepsilon)/\mathcal{F}_{n_1}\big) = \varepsilon^2 E\Big(\sum_{k=n_1}^{n_2} \bar{\theta}(x_k^\varepsilon) \Big/ \mathcal{F}_{n_1}\Big) + O\big(\varepsilon^3(n_2 - n_1)\big) + O(\varepsilon), \tag{5.68}$$

where

$$\begin{aligned} \bar{\theta}(x) &= \int \theta(x,y)\rho(dy) \\ &= \int \Big[\frac{1}{2}\big(f_{xx}(x)\varphi(x,y), \varphi(x,y)\big) + \big(\Psi_x(x,y), \varphi(x,y)\big)\Big] \rho(dy) \\ &= Lf(x). \end{aligned}$$

Let $\mathcal{F}_t^\varepsilon = \bigvee_{k \le \varepsilon^2 n} \mathcal{F}_k$ and assume that $t_i \in [n_i\varepsilon^2, (n_i+1)\varepsilon^2)$, $i = 1,2$. Then it follows from (5.68) that

$$\begin{aligned} &E\big(f(\tilde{x}^\varepsilon(t_2)) - f(\tilde{x}^\varepsilon(t_1))/\mathcal{F}_{t_1}^\varepsilon\big) \\ &= E\Big(\int_{t_1}^{t_2} Lf\big(x_\varepsilon(t)\big)\,dt \Big/ \mathcal{F}_{t_1}^\varepsilon\Big) + O(\varepsilon) + O\big(\varepsilon(t_2 - t_1)\big). \end{aligned} \tag{5.69}$$

Theorem 2 of Chapter 2 implies that the stochastic processes $\tilde{x}^\varepsilon(t)$ converge weakly to the process $\tilde{x}(t)$.

Now we prove that the family of stochastic processes $\{\tilde{x}^{\varepsilon}(t), \varepsilon \in (0,1)\}$ is weakly compact in C. Let
$\phi(x) = 1 - \exp\{-|x - x_0|^2\}$. First we prove that

$$E_y(\phi(x_n^{\varepsilon}) - x_0) = O(\varepsilon^2 n).$$

Applying Lemma 7 to the function

$$\psi(x,y) = (\phi_x^{(x)}, \varphi(x,y)),$$

we can write

$$\begin{aligned} E_g\phi(x_n^{\varepsilon}) &= \varepsilon E_y \sum_{k=0}^{n} \Psi(x_k^{\varepsilon}, y_k) + O(\varepsilon^2 n) \\ &= -\varepsilon E_y \Psi(x_n^{\varepsilon}, y_n) + O(\varepsilon^2 n) = O(\varepsilon^2 n), \end{aligned}$$

since $|\Psi(x_n^{\varepsilon}, y)| = O(|x_n^{\varepsilon} - x_0|) = O(\varepsilon n)$. This implies the inequality

$$E_y\phi(\tilde{x}^{\varepsilon}(t)) \leq kt,$$

where k does not depend on $\varepsilon \in (0,1)$. In the same way as in Lemma 3 we can prove that

$$E_y\phi^2(\tilde{x}^{\varepsilon}(t)) \leq k(\varepsilon^2 + t).$$

Now we can apply Theorem 1 of Chapter 2 in the same way as in the proof of compactness in Theorem 1.
This completes the proof of the theorem.

□

5.2.2 *Diffusion Approximations to First Integrals*

We consider the averaged differential equations we obtained for the difference equation (5.62): It is of the form

$$\dot{\bar{x}}(t) = \bar{\varphi}(\bar{x}(t)), \quad \bar{x}(0) = x_0, \tag{5.70}$$

where

$$\bar{\varphi}(x) = \int \varphi(x,y)\,\rho(dy). \tag{5.71}$$

It is assumed that the function $\varphi(x,y)$ satisfies conditions (a) and (b) of Theorem 7, except possibly for the condition $\int \varphi(x,y)\,\rho(dy) = 0$.
Let $\phi(x)$ be a first integral for equation (5.70), so $\phi(x)$ has a first derivative $\phi'(x)$ that satisfies the relation

$$(\phi'(x), \bar{\varphi}(x)) = 0.$$

Denote by $\bar{x}(x_0, t)$ the solution to the initial value problem (5.70) satisfying the initial condition $\bar{x}(x_0, 0) = x_0$. Then $\phi(\bar{x}(x_0,t)) = \phi(x_0)$. It follows from the averaging theorem for difference equations (Theorem 8 of Chapter 3) that $|x_n^\varepsilon - \bar{x}(x_0, \varepsilon n)| \to 0$ with probability 1 as $\varepsilon \to 0$ uniformly in $n \le u\varepsilon^{-1}$ for all $u > 0$. So

$$\sup_{n \le u\varepsilon^{-1}} |\phi(x_0) - \phi(x_n^\varepsilon)| \to 0$$

with probability 1 as $\varepsilon \to 0$.
Set $c_- = \inf\{\phi(x) : x \in R^d\}$, $c_+ = \sup\{\phi(x) : x \in R^d\}$, $z_n^\varepsilon = \phi(x_{n \wedge \nu_\varepsilon}^\varepsilon)$, and $\nu_\varepsilon = \sup\{n : \phi(x_k^\varepsilon) \in (c_-, c_+), k \le n\}$ for $n = 0, 1, \ldots$. We will investigate the asymptotic behavior of the stochastic process

$$z^\varepsilon(t) = \sum_n [(t\varepsilon^{-2} - n) z_{n+1}^\varepsilon + (n + 1 - t\varepsilon^{-2}) z_n^\varepsilon] 1_{\{\varepsilon^2 n \le t \le \varepsilon^2(n+1)\}}. \quad (5.72)$$

In particular, we will show that $z^\varepsilon(t)$ converges weakly in C to a diffusion process on an interval in R.
First, we introduce some restrictions on the function $\phi(x)$.

(ϕ**1**) $\phi(x)$ is a continuous function with the bounded continuous derivatives $\phi'(x)$ and $\phi''(x)$, and $\phi''(x)$ is uniformly continuous in R^d.

(ϕ**2**) Denote by ϕ_c the level set $U_c = \{x : \phi(x) = c\}$. For all $c \in (c_-, c_+)$ the set U_c is compact in R^d, and a probability measure m_c exists in R^d for which

$$\lim_{T \to \infty} \sup_{|x_0| \le r} \left| \frac{1}{T} \int_0^T g(\bar{x}(x_0, t))\, dt - \int g(x')\, m_{\phi(x_0)}(dx') \right| = 0$$

for $g \in C(R^d)$, and $r > 0$. For all $g \in C$, $\int g(x)\, m_u(dx)$ is continuous in u.

We introduce the differential operator $L^z\theta(u)$ in $C^{(2)}(R)$ by the formula

$$\iiint \left(\frac{\partial}{\partial x} [\theta'(\phi(x))(\phi'(x), \varphi(x, y'))], \varphi(x, y) \right) R(y, dy')\, \rho(dy)\, m_u(dx)$$
$$+ \frac{1}{2} \iint \left(\frac{\partial^2}{\partial x^2} \theta(\phi(x)) \varphi(x, y), \varphi(x, y) \right) \rho(dy)\, m_u(dx). \quad (5.73)$$

Theorem 8 *Assume that the stochastic process $\{y_n, n \in \mathbf{Z}_+\}$ satisfies condition* SMC II *of Section 2.3, the function $\varphi(x, y)$ satisfies conditions* (a) *and* (b) *of Theorem 7; the function $\phi(x)$ satisfies conditions* (ϕ**1**), (ϕ**2**), *and the differential operator $L\theta(u)$ is the generator of a diffusion process $z(t)$ in the interval (c_-, c_+) with absorbing boundary. Then the stochastic process $z^\varepsilon(t)$ converges weakly in C as $\varepsilon \to 0$ to the diffusion process $z(t)$ with the initial value $z(0) = \phi(x_0)$, where $x_0 = x_0^\varepsilon$ for all $\varepsilon > 0$.*

Proof Let

$$G(x) = \theta(\phi(x)),$$

where $\theta \in C^{(3)}(R)$ and $\operatorname{supp}\theta \subset (c_-, c_+)$. Then

$$\theta(z_{k+1}^\varepsilon) - \theta(z_k^\varepsilon) = \varepsilon(G'(x_k^\varepsilon), \varphi(x_k^\varepsilon, y_k)) + \frac{1}{2}\varepsilon^2(G''(x_k^\varepsilon)\varphi(x_k^\varepsilon, y_k), \varphi(x_k^\varepsilon, y_k)) + o(\varepsilon^2).$$

Set $\Psi(x, y) = (G'(x), \varphi(x, y))$. Then

$$\int \Psi(x, y)\,\rho(dy) = \theta'(\phi(x))(\phi'(x), \bar{\varphi}(x)) = 0.$$

It follows from Lemma 7 that

$$E\big(\theta(z_{n_2+1}^\varepsilon) - \theta(z_{n_1}^\varepsilon)/\mathcal{F}_{n_1}\big) = E\Big(\varepsilon^2 \sum_{k=n_1}^{n_2} \psi(x_k^\varepsilon, y_k)\Big/\mathcal{F}_{n_1}\Big) + \frac{1}{2}\varepsilon^2 E\Big(\sum_{k=n_1}^{n_2} (G''(x_k^\varepsilon)\varphi(x_k^\varepsilon, y_k)), \varphi(x_k^\varepsilon, y_k)\Big/\mathcal{F}_{n_1}\Big) + O(\varepsilon),$$

where $\Psi(x, y)$ is related to $\psi(x, y)$ by formula (5.66).
Set

$$H(x, y) = \Psi(x, y) + \frac{1}{2}(G''(x)\varphi(x, y), \varphi(x, y))$$

and

$$\overline{H}(x) = \int H(x, y)\,\rho(dy).$$

It follows from Lemma 8 that

$$\left|E\Big(\sum_{k=n_1}^{n_2} \big(H(x_k^\varepsilon, y_k) - \overline{H}(x_k^\varepsilon)\big/\mathcal{F}_{n_1}\big)\Big)\right| = o(n_2 - n_1) + (n_2 - n_1)\beta(\varepsilon),$$

where $\beta(\varepsilon) \to 0$ as $\varepsilon \to 0$. This implies that

$$E\big(\theta(z_{n_2+1}^\varepsilon) - \theta(z_{n_1}^\varepsilon)/\mathcal{F}_{n_1}\big) = \varepsilon^2 E\Big(\sum_{k=n_1}^{n_2} \overline{H}(x_k^\varepsilon)\Big/\mathcal{F}_{n_1}\Big) + O(\varepsilon) + o(n_2 - n_1) + (n_2 - n_1)\beta(\varepsilon). \tag{5.74}$$

At this point, we need another auxiliary result.

Lemma 9 *Let*

$$\hat{H}(u) = \int \overline{H}(x)\,m_u(dx).$$

Then

$$\lim_{\varepsilon \to 0} \sup_{n_2 \le u/\varepsilon^2} E\left|\varepsilon^2 \sum_{k=n_1}^{n_2} \big(\hat{H}(z_k^\varepsilon) - \overline{H}(x_k^\varepsilon)\big)\right| = 0. \tag{5.75}$$

Proof For any N we have the inequality

$$E\left|\sum_{k=n_1}^{n_2}\left(\hat{H}(z_k^\varepsilon)-\sum_{k=n_1}^{n_2}\overline{H}(x_k^\varepsilon)\right)\right|\le E\left|\sum_{k=n_1}^{n_2}\left(\frac{1}{N}\sum_{i=1}^{N}\overline{H}(x_{k+i}^\varepsilon)-\hat{H}(z_k^\varepsilon)\right)\right|$$

$$+O(N)\le E\left|\sum_{k=n_1}^{n_2}\frac{1}{N}\left(\sum_{i=1}^{N}\overline{H}(x_{k+i}^\varepsilon)-\overline{H}(\bar{x}(x_k^\varepsilon,i\varepsilon))\right)\right|$$

$$+E\left|\sum_{k=n_1}^{n_2}\left(\frac{1}{N}\sum_{i=1}^{N}\overline{H}(\bar{x}(x_k^\varepsilon,i\varepsilon))-\hat{H}(z_k^\varepsilon)\right)\right|+O(N).$$

It follows from the averaging theorem for difference equations (Theorem 8 of Chapter 3) that a function $N(\varepsilon):(0,1)\to\mathbf{Z}_+$ exists for which $\varepsilon N(\varepsilon)\to\infty$ and

$$\lim_{\varepsilon\to 0}\sup_{x,y}E_y\left|\frac{1}{N(\varepsilon)}\sum_{i=1}^{N(\varepsilon)}\overline{H}(\tilde{x}_i^\varepsilon)-\overline{H}(\bar{x}(x,i\varepsilon))\right|=0.$$

Here $\tilde{x}_i^\varepsilon$ is the solution to the difference equation (5.63) with $x_0=x$. This implies that

$$\lim_{\varepsilon\to 0}\sup_{n_2\le u/\varepsilon^2}E\left|\sum_{k=n_1}^{n_2}\frac{1}{N_\varepsilon}\sum_{i=1}^{N_\varepsilon}\overline{H}(x_{k+i}^\varepsilon)-\overline{H}(\bar{x}(x_k^\varepsilon,i\varepsilon))\right|=0. \tag{5.76}$$

It follows from condition ($\phi(\mathbf{2})$ that for any $r>0$,

$$\lim_{\varepsilon\to 0}\sup_{|x|\le r}E\sum_{k=n_1}^{n_2}\left(\frac{1}{N_\varepsilon}\sum_{i=1}^{N_\varepsilon}(\overline{H}(\bar{x}(x_k^\varepsilon,i\varepsilon))-\hat{H}(z_k^\varepsilon))\right)1_{\{|x_k^\varepsilon|\le r\}}=0 \tag{5.77}$$

if $\varepsilon N_\varepsilon\to 0$.
Using relation (5.74) we can prove in the same way as in Lemma 4 that

$$\lim_{r\to\infty}\limsup_{\varepsilon\to 0}\sup_{n_2\le u/\varepsilon^2}E\sum_{k=n_1}^{n_2}\varepsilon^2 1_{\{|x_k^\varepsilon|>r\}}=0. \tag{5.78}$$

Relations (5.76), (5.77), and (5.78) imply relation (5.75), and the proof of the lemma, if $\varepsilon N_\varepsilon\to\infty$, $\varepsilon^2 N_\varepsilon\to 0$.

□

We return to the proof of the theorem. It follows from equations (5.74) and (5.75) that for $0\le t_1<t_2$,

$$\lim_{\varepsilon\to 0}E\left|E(\theta(z^\varepsilon(t_2))-\theta(z^\varepsilon(t_1))-\int_{t_1}^{t_2}L^z(\theta(z^\varepsilon(s))\,ds\,\mathcal{F}_{t_1}^\varepsilon)\right|=0, \tag{5.79}$$

where $\mathcal{F}_t^\varepsilon=\vee_{\varepsilon^2k\le t}\mathcal{F}_k$. Here we have used the fact that

$$\hat{H}(u)=L^z\theta(u).$$

This is true for all $\theta \in C^{(2)}$, so Theorem 2 of Chapter 2 implies that the process $z^{\varepsilon}(t)$ converges to the process $z(t)$ weakly. To prove the theorem we have to show that the family of stochastic processes $\{z_1^{\varepsilon}(t), \varepsilon \in (0,1)\}$ is weakly compact in C. This can be proved in the same way as in Theorems 5 and 7. This completes the proof of Theorem 8.

□

Remark 5 *The differential operator $L^z\theta(u)$ can be represented in the form*

$$L^z\theta(u) = \alpha(u)\theta'(u) + \frac{1}{2}\beta(u)\theta''(u), \tag{5.80}$$

where

$$\begin{aligned}\alpha(u) = &\iiint \left(\frac{\partial}{\partial x}(\phi'(x), \varphi(x,y')), \varphi(x,y)\right) R(y,dy')\,\rho(dy)\,m_u(dx)\\ &+ \frac{1}{2}\iint (\phi''(x)\varphi(x,y), \varphi(x,y))\,\rho(dy)\,m_u(dx)\end{aligned} \tag{5.81}$$

and

$$\begin{aligned}\beta(u) = 2&\iiint (\phi'(x), \varphi(x,y'))(\phi'(x), \varphi(x,y))R(y,dy')\,\rho(dy)\,m_u(dx)\\ &+ \iint (\phi'(x), \varphi(x,y))^2\,\rho(dy)\,m_u(dx).\end{aligned} \tag{5.82}$$

It follows from the conditions of Theorem 8 that $\alpha(u)$ and $\beta(u)$ are continuous.

Remark 6 *Assume that $\phi(x) \geq 0$, $\phi(x) \to \infty$ as $|x| \to \infty$, and there is a unique point $\bar{x}$ for which $\phi(\bar{x}) = 0$. Then $m_0(B) = 1_{\{\bar{x}\in B\}}$. We can calculate $\alpha(0)$ and $\beta(0)$; namely,*

$$\begin{aligned}\alpha(0) &= \frac{1}{2}\int (\phi''(\bar{x})\varphi(\bar{x},y), \varphi(\bar{x},y))\,\rho(dy),\\ \beta(0) &= 0.\end{aligned}$$

Assume in addition that $\int |\varphi(\bar{x},y)|^2\,\rho(dy) > 0$ and the Jacobian determinant $\det\phi''(\bar{x}) > 0$. Then the point 0 is either a natural or a repelling boundary for the diffusion process $z(t)$ on the interval $(0,\infty)$. (See Remark 3.) Also, $+\infty$ is the natural boundary for this process. Let τ be the stopping time for which $\tilde{z}(s) \in (0,+\infty)$ and

$$\sup_{s\leq\tau} \tilde{z}(s) \wedge (\tilde{z}(s))^{-1} = 0.$$

Then $P\{\tau = +\infty\} = 1$.

Remark 7 *Assume that $\varphi(x, y_1)$ has a continuous conditional distribution for all $x \in R^d$ and $y_0 \in Y$. Then $P\{\nu_\varepsilon = +\infty\} = 1$.*

5.2.3 Stationary Perturbations

We consider the difference equation

$$x_{k+1}^\varepsilon - x_k^\varepsilon = \varepsilon \varphi_{k+1}(x_k^\varepsilon, \omega), \quad k = 0, 1, \ldots, \quad x_0^\varepsilon = x_0, \tag{5.83}$$

where $x_0 \in R^d$ is a nonrandom initial value, and the sequence of random functions $\{\varphi_k(x, \omega),\ k = 1, 2, \ldots\}$ satisfies the following conditions:

(1) There exists an $\mathcal{F}$-measurable transformation $T : \Omega \to \Omega$ that preserves the measure P (here $(\Omega, \mathcal{F}, P)$ is the probability space on which the sequence $\{\varphi_k(x, \omega)\}$ is defined) for which $\varphi_k(x, \omega) = \varphi_0(x, T^k \omega)$.

(2) There exists a σ-algebra $\mathcal{F}_0$ satisfying the condition that for all $\mathcal{F}_0$-measurable bounded random variables $\xi(\omega)$ the series

$$R\xi(\omega) \equiv \sum_{k=0}^{\infty} \big[E(\xi(T^k(\omega))/\mathcal{F}_0) - E\xi\big] \tag{5.84}$$

is convergent and $\sup_\omega |R\xi(\omega)| < \infty$.

(3) $\varphi_0(x, \omega)$ is $\mathcal{F}_0$-measurable and $E\varphi_0(x, \omega) = 0$.

(4) $\varphi_0(x, \omega)$ is a twice differentiable function in x, and $\varphi_0(x, \omega)$, $\frac{\partial \varphi_0}{\partial x}(x, \omega)$, $\frac{\partial^2 \varphi_0}{\partial x^2}(x, \omega)$ are bounded continuous functions of x.

(5) the transformation T is ergodic.

Theorem 9 *Assume that conditions (1)–(5) are satisfied. Then the stochastic process $\tilde{x}^\varepsilon(t)$ defined by equation (5.63) with $\{x_n^\varepsilon\}$ given in equation (5.83) converges in C, as $\varepsilon \to 0$, to the diffusion process $\tilde{x}(t)$ having generator $\tilde{L}$ that is defined for functions $f \in C^{(2)}(R^d)$ by the formula*

$$\tilde{L}f(x) = E\left[\left(\frac{\partial}{\partial x}(f'(x), R\varphi_0(x, T\omega)), \varphi_0(x, \omega)\right) + \frac{1}{2}(f''(x)\varphi_0(x, \omega), \varphi_0(x, \omega))\right] \tag{5.85}$$

with initial condition $\tilde{x}(0) = x_0$.

Proof Let $n_1 < n_2$. Then for $f \in C^{(3)}(R^d)$ we can write

$$f(x^\varepsilon_{n_2}) - f(x^\varepsilon_{n_1}) = \sum_{k=n_1}^{n_2-1} \left[f(x^\varepsilon_{k+1}) - f(x^\varepsilon_k)\right]$$

$$= \sum_{k=n_1}^{n_2-1} \Big[\varepsilon\big(f'(x^\varepsilon_k), \varphi_0(x^\varepsilon_k, T^{k+1}\omega)\big)$$

$$+ \frac{\varepsilon^2}{2}\big(f''(x^\varepsilon_k)\varphi_0(x^\varepsilon_k, T^{k+1}\omega), \varphi_0(x^\varepsilon_k, T^{k+1}\omega)\big) + O(\varepsilon^3)\Big].$$

Let $\psi(x,\omega) = \big(f'(x), \varphi_0(x,\omega)\big)$. We consider the sum

$$S_1 = \sum_{k=n_1}^{n_2-1} \psi(x^\varepsilon_k, T^{k+1}\omega).$$

Note that for any bounded $\mathcal{F}$-measurable random variable ξ,

$$E\big(R\xi(T\omega) - R\xi(\omega)/\mathcal{F}_0\big) = -\xi(\omega).$$

So

$$\psi(x, T\omega) = E\big(R\psi(x, T\omega) - R\psi(x, T^2\omega)/\mathcal{F}_1\big),$$

and

$$\psi(x^\varepsilon_k, T^{k+1}\omega) = E\big(R\psi(x^\varepsilon_k, T^{k+1}\omega) - R\psi(x^\varepsilon_k, T^{k+2}\omega)/\mathcal{F}_{k+1}\big)$$

$$= E\big(R\psi(x^\varepsilon_k, T^{k+1}\omega) - R\psi(x^\varepsilon_{k+1}, T^{k+2}\omega)$$

$$+ \varepsilon R\big(\frac{\partial\psi}{\partial x}(x^\varepsilon_k, T^{k+2}\omega), \varphi_0(x^\varepsilon_k, T^{k+1}\omega)\big)/\mathcal{F}_{k+1}\big) + O(\varepsilon^2).$$

Using the last equality we obtain the following representation for $E(S_1/\mathcal{F}_{n_1})$:

$$E(S_1/\mathcal{F}_{n_1}) = \varepsilon E\Bigg(\sum_{k=n_1}^{n_2-1} R\bigg(\frac{\partial\psi}{\partial x}(x^\varepsilon_k), T^{k+2}\omega\bigg), \varphi_0(x^\varepsilon_k, T^{k+1}\omega)\bigg)$$

$$+ R\psi(x^\varepsilon_{n_1}, T^{n_1+1}\omega) - R\psi(x^\varepsilon_{n_2}, T^{n_2+1}\omega)\Big/\mathcal{F}_{n_1}\Bigg)$$

$$+ (n_2 - n_1)O(\varepsilon^2).$$

Let $n_k - 1 < t_k\varepsilon^{-2} \le n_k$, $k = 1, 2$. Then using the boundedness of $R\psi(x,\omega)$ we can write

$$E\big(f(\tilde{x}^\varepsilon(t_2)) - f(\tilde{x}^\varepsilon(t_1)/\mathcal{F}^\varepsilon_{t_1}\big)$$
$$= E\bigg(\varepsilon^2 \sum_{k=n_1}^{n_2-1}\Big[R\big(\frac{\partial\psi}{\partial x}(x^\varepsilon_k), T^{k+2}\omega), \varphi_0(x^\varepsilon_k, T^{k+1}\omega)\big)$$
$$+ \frac{1}{2}\big(f''(x^\varepsilon_k)\varphi_0(x^\varepsilon_k, T^{k+1}\omega), \varphi_0(x^\varepsilon_k, T^{k+1}\omega)\big)\Big]\Big/\mathcal{F}_{n_1}\bigg) + O(\varepsilon^2)$$
$$= E\bigg(\int_{t_1}^{t_2} \tilde{L}f(\tilde{x}^\varepsilon(t))\,dt\Big/\mathcal{F}^\varepsilon_{t_1}\bigg) + O(\varepsilon^2) + \Delta(\varepsilon), \tag{5.86}$$

where

$$\Delta_\varepsilon = E\bigg(\varepsilon^2 \sum_{k+n_1}^{n_2-1}\big[g(x^\varepsilon_k, T^{k+1}\omega) - \tilde{L}f(x^\varepsilon_k)\big]\Big/\mathcal{F}_{n_1}\bigg) \tag{5.87}$$

and

$$g(x,\omega) = E\bigg(R\Big(\frac{\partial\psi}{\partial x}(x, T\omega), \varphi_0(x,\omega)\Big)\Big/\mathcal{F}_0\bigg) + \frac{1}{2}\big(f''(x)\varphi_0(x,\omega), \varphi_0(x,\omega)\big). \tag{5.88}$$

Now we can prove the relation

$$\lim_{\varepsilon\to 0} E\bigg|E\big(f(\tilde{x}^\varepsilon(t_2)) - f(\tilde{x}^\varepsilon(t_1)) - \int_{t_1}^{t_2} \tilde{L}f\big(\tilde{x}^\varepsilon(t)\big)\,dt\Big/\mathcal{F}^\varepsilon_{t_1}\big)\bigg| = 0 \tag{5.89}$$

for $f \in C^{(3)}(R^d)$, where $\mathcal{F}^\varepsilon_t = \vee_{k<t\varepsilon^{-2}}\mathcal{F}_k$. Since $Eg(x,\omega) = \tilde{L}f(x)$, to prove this we will prove that

$$\lim_{\varepsilon\to 0} E|\Delta_\varepsilon| = 0.$$

This is a consequence of the following lemma:

Lemma 10 *Assume that a bounded function $h : R^d \times \Omega \to R$ satisfies the following conditions:*

(a) *it is $\mathcal{F}_0$-measurable in ω;*

(b) *there exists a constant l for which*

$$|h(x_1,\omega) - h(x_2,\omega)| \le l|x_1 - x_2|$$

for all x_1, $x_2, \in R^d$ and $\omega \in \Omega$. Let $\hat{h}(x) = Eh(x,\omega)$.

Then for any $u > 0$,

$$\lim_{\varepsilon\to 0} E\varepsilon^2\bigg|\sum_{k\le u/\varepsilon^2}\big(h(x^\varepsilon_k, T^{k+1}\omega) - \hat{h}(x)\big)\bigg| = 0. \tag{5.90}$$

Proof It follows from relations (5.86), (5.87), and (5.88) that

$$E(1 - \exp\{-\lambda|\tilde{x}^\varepsilon(t) - x_0|^2\}) = O(\varepsilon) + O(\lambda t). \tag{5.91}$$

To obtain this we apply these relations to the function

$$f(x) = 1 - \exp\{-\lambda|x - x_0|^2\}$$

and $t_1 = 0$, $t_2 = t$ and take into account that $f'(x) = O(\lambda)$, $f''(x) = O(\lambda)$. So

$$P\{|x_k^\varepsilon - x_0| > r\} = O\biggl(\frac{\varepsilon + \lambda}{1 - e^{-\lambda r^2}}\biggr) \quad \text{if } k \le u\varepsilon^{-2},$$

and

$$P\{|x_k^\varepsilon - x_0| > r\} = O\biggl(\varepsilon + \frac{k\varepsilon^2}{\sqrt{r}}\biggr). \tag{5.92}$$

Condition (5) implies the relation

$$\lim_{\mathcal{N}\to\infty} \frac{1}{\mathcal{N}} \sum_{k=1}^{\mathcal{N}} h(x, T_\omega^k) = \hat{h}(x)$$

with probability 1 for all $x \in R^d$. Since the function

$$\frac{1}{\mathcal{N}} \sum_{k=1}^{\mathcal{N}} h(x, T^k\omega) - \hat{h}(x)$$

is uniformly continuous in x with respect to ω due to condition (b), the function

$$\alpha(N, r) = E \sup_{|x|\le r} \biggl|\sum_{k=1}^{N} h(x, T^k\omega) - \hat{h}(x)\biggr| \tag{5.93}$$

satisfies the relation

$$\lim_{N\to\infty} \alpha(N, r) = 0, \quad r > 0. \tag{5.94}$$

Let N be a fixed integer in $\mathbf{Z}_+$. Then

$$\varepsilon^2 E \sum_{k \le u\varepsilon^2} \bigl(h(x_k^\varepsilon, T_\omega^{k+1}) - \hat{h}(x_k^\varepsilon)\bigr) = Z_{\varepsilon,N}^1 + Z_{\varepsilon,N}^2 + Z_{\varepsilon,N}^3, \tag{5.95}$$

where

$$Z_{\varepsilon,N}^1 = \varepsilon^2 E \sum_{k \le u\varepsilon^{-2}} \biggl(\frac{1}{N}\sum_{i=1}^{N} h(x_k^\varepsilon, T_\omega^{k+i}) - \hat{h}(x_k^\varepsilon)\biggr), \tag{5.96}$$

$$Z_{\varepsilon,N}^2 = \varepsilon^2 E \sum_{k \le u\varepsilon^{-2}} \frac{1}{N}\sum_{i=1}^{N} \bigl(h(x_{k+i-1}^\varepsilon, T_\omega^{k+i}) - h(x_k^\varepsilon, T_\omega^{k+i})\bigr), \tag{5.97}$$

and

$$Z^3_{\varepsilon,N} = \varepsilon^2 E \sum_{k \le u\varepsilon^{-2}} \Big(-\frac{1}{N} \sum_{i=1}^{N} h(x^\varepsilon_{k+i-1}, T^{k+i}_\omega) + h(x^\varepsilon_{k,l} T^{k+1}\omega) \Big). \quad (5.98)$$

We have the next inequalities for $|Z^j_{\varepsilon,N}|$, $j = 1, 2, 3$:

$$|Z^1_{\varepsilon,N}| \le u\Big(\alpha(N,r) + 2c_2 \frac{u}{\sqrt{r - |x_0|}} \Big) + O(\varepsilon), \quad (5.99)$$

$$|Z^2_{\varepsilon,N}| \le ulNc_1\varepsilon, \quad (5.100)$$

and

$$|Z^3_{\varepsilon,N}| \le 2Nc_2\varepsilon^2, \quad (5.101)$$

where $c_1 = \sup_{x,\omega} |\varphi_0(x,\omega)|$ and $c_2 = \sup_{x,\omega} h(x,\omega)$. Inequality (5.99) is a consequence of the relation

$$E\Big| \frac{1}{N} \sum_{i=1}^{N} h\big(x^\varepsilon_k, T^{k+i}(\omega)\big) - \hat{h}(x^\varepsilon_k) \Big| \le \alpha(N,r) + 2c_2 P\{|x^\varepsilon_k| > r\}$$

and (5.92). Inequality (5.100) is a consequence of condition (b) and the inequality

$$|x^\varepsilon_k - x^\varepsilon_l| \le c_1\varepsilon|k - l|.$$

Inequality (5.101) follows from the identity

$$\sum_{k=0}^{n} a_k - \sum_{k=0}^{n} \frac{1}{N} \sum_{i=0}^{N-1} a_{k+i} = \sum_{i=0}^{N-1} \frac{N-1-i}{N} a_i + \sum_{i=0}^{N-1} \frac{N-i}{N} a_{n+i}.$$

Relation (5.94) implies that

$$\lim_{N\to\infty} \limsup_{\varepsilon\to 0} \Big(\sum_{l=1}^{3} |Z^l_{\varepsilon,N}| \Big) = 0.$$

□

We return to the proof of the theorem. Relation (5.90) and Theorem 2 of Chapter 2 imply the weak convergence of the process $\tilde{x}^\varepsilon(t)$ to the process $\tilde{x}(t)$. To prove the theorem we have to prove the weak compactness of the family of stochastic problems $\{\tilde{x}^\varepsilon(t), \varepsilon \in (0,1)\}$. It follows from relation (5.87) that

$$E(1 - \exp\{|\tilde{x}^\varepsilon(t) - x_0|^2\}) = O(\varepsilon + t).$$

This equation implies that

$$E(1 - \exp\{|\tilde{x}^\varepsilon(t) - \tilde{x}^\varepsilon(0)|^2\})^2 = O(\varepsilon^2 + t^2)$$

and therefore

$$E(1 - \exp\{-|\tilde{x}^\varepsilon(t_2) - \tilde{x}^\varepsilon(t_1)|^2\})^2 = O(\varepsilon^2 + |t_1 - t_2|^2)$$

for $0 \le t_1 < t_2$. Note that

$$|\tilde{x}^\varepsilon(t_2) - \tilde{x}^\varepsilon(t_1)| \le \frac{c_1}{\varepsilon}|t_2 - t_1|.$$

So compactness of $\{\tilde{x}^\varepsilon(t), \varepsilon \in (0,1)\}$ follows from Theorem 1 of Chapter 2.

□

Consider difference equation (5.83) with the function $\varphi_k(x,\omega)$ satisfying conditions (1), (2), (4), (5) and the additional condition
(3′) $\varphi_0(x,\omega)$ is $\mathcal{F}_0$-measurable.
Let $\bar{\varphi}(x) = E\varphi_0(x,\omega)$. The conditions listed here are sufficient to use the averaging theorem (Theorem 8 of Chapter 3). We assume that the averaged equation

$$\dot{\bar{x}}(t) = \bar{\varphi}(\bar{x}(t)) \tag{5.102}$$

has a first integral $\phi(x)$ that satisfies conditions (ϕ**1**) and (ϕ**2**) of Section 5.2.2. Set

$$z_k^\varepsilon = \phi(x_{k\wedge\nu_\varepsilon}^\varepsilon),$$

where ν_ε is determined in Section 5.2.2, and define $z^\varepsilon(t)$ by formula (5.73), with

$$c_- = \inf\{\phi(x) : x \in R^d\}, \qquad c_+ = \sup\{\phi(x) : x \in R^d\}.$$

With these preliminaries, we have the following theorem.

Theorem 10 *Assume that the sequence of functions $\{\varphi_k(x,\omega), k = 0,1\}$ satisfies conditions (1), (2), (3'), (4), (5), that the first integral $\phi(x)$ to equation (5.102) satisfies condition (ϕ**1**), (ϕ**2**), and that the differential operator $L^z\theta$, which is defined by formula (5.73), is the generator of a diffusion process $z(t)$ in the interval (c_-, c_+) with absorbing boundary. Then the stochastic process $z^\varepsilon(t)$ converges weakly in C as $\varepsilon \to 0$ to the diffusion process $z(t)$ with the initial value $z(0) = \phi(x_0)$, where $x_0 = x_0^\varepsilon$ for all $\varepsilon > 0$.*

The proof of the theorem is the same as the proof of Theorem 8 if we use Lemma 10 instead of Lemma 8. The details are left to the reader.

6
Stability

In this chapter we consider a variety of stability problems for differential and difference equations when they are perturbed by random noise. This entails an investigation of the behavior of solutions $x_\varepsilon(t)$ as $t \to \infty$ for small values of ε. In addition, we describe the growth of solutions to certain randomly perturbed convolution equations. A number of examples of stability phenomena are presented here, also.

6.1 Stability of Perturbed Differential Equations

To begin the discussion of stability properties, we consider systems of differential equations in R^d of the form

$$\dot{x}_\varepsilon(t) = a\left(y\left(\frac{t}{\varepsilon}\right), x_\varepsilon(t)\right), \tag{6.1}$$

where the perturbing process $y(t)$ is either an ergodic Markov process in a measurable space $(Y, \mathcal{C})$ or a Y-valued ergodic stationary process. We will assume that the corresponding averaged equation has an equilibrium point $\bar{x}$, so $\bar{x}(t) = \bar{x}$ is a solution to the averaged equation, and we investigate the behavior of the solution to equation (6.1) as $t \to \infty$ for $x_\varepsilon(0)$ close to $\bar{x}$ and for ε small. The solution to equation (6.1) is stable in a neighborhood of the stationary point $\bar{x}$ if for any $\delta > 0$ there are numbers $\varepsilon_0 > 0$ and $\varepsilon_1 > 0$ for which

$$P\left\{\lim_{t\to\infty} x_\varepsilon(t) = \bar{x}\right\} > 1 - \delta \quad \text{if } \varepsilon < \varepsilon_0 \text{ and } |x_\varepsilon(0) - \bar{x}| < \varepsilon_1.$$

6.1.1 *Jump Perturbations of Nonlinear Equations*

First, consider a linear initial value problem of the form

$$\dot{x}_\varepsilon(t) = A\left(y\left(\frac{t}{\varepsilon}\right)\right) x_\varepsilon(t), \quad x_\varepsilon(0) = x_0, \tag{6.2}$$

where $A(y)$ is a measurable function from Y to $L(R^d)$. (Note that in this case, $x \equiv 0$ is a static state of the system.)
Our first stability results are summarized in the following theorem.

Theorem 1 *Let $y(t)$ be a Markov process that is uniformly ergodic, and suppose that A is bounded: For all $y \in Y$, $\|A(y)\| \le b$ for some finite constant b. Let $\bar{A} = \int_Y A(y)\,\rho(dy)$. If $x_\varepsilon(t)$ is the solution to equation* (6.2), *then the following statements are equivalent:*

(i) *There exists $t_0 > 0$ for which $\left\|e^{t_0\bar{A}}\right\| < 1$.*

(ii) *There exist $\delta > 0$ and $\varepsilon_0 > 0$ for which*

$$P\left\{\sup_{t>0} \left|x_\varepsilon(t)\right| e^{\delta t} < \infty\right\} = 1 \tag{6.3}$$

for any $\varepsilon \le \varepsilon_0$ and $x_0 \in R^d$.

(iii) *For any $\alpha > 0$, $x_0 \in R^d$, and $c > 0$ there exist $\varepsilon_0 > 0$ and $t_0 > 0$ for which*

$$P\left\{\left|x_\varepsilon(t)\right| > c\right\} < \alpha$$

for $t \ge t_0$, $\varepsilon \le \varepsilon_0$.

Proof Using the relation

$$\frac{d}{dt}\left(x_\varepsilon(t), x_\varepsilon(t)\right) = 2\left(A\left(y\left(\frac{t}{\varepsilon}\right)\right) x_\varepsilon(t), x_\varepsilon(t)\right) \le 2b\left(x_\varepsilon(t), x_\varepsilon(t)\right)$$

we can prove that

$$\left|x_\varepsilon(t)\right| \le |x_0| \exp\{bt\}. \tag{6.4}$$

It follows from the averaging theorem (Theorem 6 of Chapter 3) that

$$\sup_{t \le t_1} \left|x_\varepsilon(t) - e^{t\bar{A}} x_0\right| \to 0$$

in probability as $\varepsilon \to 0$ for any $t_1 > 0$, uniformly in $y = y(0)$, where $\bar{x}(t) = e^{t\bar{A}} x_0$ is the solution of the averaged equation, $\dot{\bar{x}}(t) = \bar{A}\bar{x}(t)$, having initial condition $\bar{x}(0) = x_0$.
Therefore,

$$\lim_{\varepsilon \to 0} E_y \left|x_\varepsilon(t)\right| \le \left\|e^{t\bar{A}}\right\| \cdot |x_0|. \tag{6.5}$$

Here E_y is the expectation under the condition that $y(0) = y$.

(i) $\Rightarrow$ (ii). If (i) is satisfied, then there are numbers $q < 1$ and $\varepsilon_0 > 0$ for which

$$E_y\big|x_\varepsilon(t)\big| \le q|x_0| \quad \text{for} \quad \varepsilon \le \varepsilon_0.$$

Let $U_\varepsilon(t_1, t)$ be an $L(R^d)$-valued stochastic process that is a fundamental solution of the differential equation

$$\frac{d}{dt}U_\varepsilon(t_1,t) = A\left(y\left(\frac{t}{\varepsilon}\right)\right)U_\varepsilon(t_1,t) \quad \text{for} \quad t > t_1 \tag{6.6}$$

with the initial value $U_\varepsilon(t_1, t_1) = I$. Then

$$E\big(\big|U_\varepsilon(t_1, t_1 + t_0)x_0\big|/\mathcal{F}_{t_1/\varepsilon}\big) \le q|x_0|,$$

so

$$\begin{aligned} E_y\big|x_\varepsilon(kt_0)\big| &= E\big|E_yU_\varepsilon\big((k-1)t_0, kt_0\big)x_\varepsilon\big((k-1)t_0\big)\big| \\ &\le qE_y\big|x_\varepsilon\big((k-1)t_0\big)\big| \le q^k|x_0| \end{aligned} \tag{6.7}$$

if $\varepsilon \le \varepsilon_0$. Let $t \in [kt_0, (k+1)t_0]$. Then

$$E_y\big|x_\varepsilon(t)\big| \le q^kE_y\big|x_\varepsilon(t-kt_0)\big| \le q^ke^{bt_0}|x_0| \le Ce^{-\alpha t}|x_0|, \tag{6.8}$$

where $C = e^{bt_0}/q$ and $\alpha = -\log q/t_0$, since $q^{k+1} \le q^{(t/t_0)} = \exp\{t\log q/t_0\}$. Let $0 < \delta < \alpha$. Then

$$\frac{d}{dt}(e^{\delta t}x_\varepsilon(t)) = \delta e^{\delta t}x_\varepsilon(t) + e^{\delta t}A\left(y\left(\frac{t}{\varepsilon}\right)\right)x_\varepsilon(t).$$

Integrating gives

$$e^{\delta t_2}x_\varepsilon(t_2) - e^{\delta t_1}x_\varepsilon(t_1) = \int_{t_1}^{t_2}\left(\delta x_\varepsilon(t) + A\left(y\left(\frac{t}{\varepsilon}\right)\right)x_\varepsilon(t)\right)e^{\delta t}\,dt,$$

and so

$$e^{\delta t_1}\big|x_\varepsilon(t_1)\big| \le \int_{t_1}^{t_2}(\delta\big|x_\varepsilon(t)\big| + C\big|x_\varepsilon(t)\big|)e^{\delta t}\,dt + e^{\delta t_2}\big|x_\varepsilon(t_2)\big|.$$

So using (6.8) we can write

$$e^{\delta|t_1|}\big|x_\varepsilon(t_1)\big| \le (C+\delta)\int_{t_1}^{\infty}\big|x_\varepsilon(t)\big|e^{\delta t}\,dt, \tag{6.9}$$

so

$$\sup_{t_1>0} e^{\delta|t_1|}\big|x_\varepsilon(t_1)\big| \le (C+\delta)\int_0^{\infty}\big|x_\varepsilon(t)\big|e^{\delta t}\,dt; \tag{6.10}$$

and relation (6.3) is proved.
(ii) $\Rightarrow$ (iii). This is obvious, and the proof is left to the reader.
(iii) $\Rightarrow$ (i). Let $\{e_1, \ldots, e_d\}$ be a basis in R^d, and let $x_\varepsilon^k(t)$ be the solution of equation (6.2) with $x_0 = e_k$ for $k = 1, \ldots, d$. If (iii) is satisfied, then

numbers $\varepsilon_0 > 0$ and t_0 exist such that

$$P\left\{|x_\varepsilon^k(t)| > \frac{1}{2d}\right\} \le \frac{1}{2d},$$

for $\varepsilon \le \varepsilon_0$ and $t \ge t_0$, where d is the dimension of R^d. Let

$$x_0 = \sum_{k=1}^{d} \alpha_k e_k.$$

Then

$$P\left\{\sup_{\alpha_1,\ldots,\alpha_d} |x_\varepsilon(t_0)| > \frac{1}{2}|x_0|\right\} \le \sum_{k=1}^{d} P\left\{|x_\varepsilon^k(t)| > \frac{1}{2d}\right\} \le \frac{1}{2}.$$

Therefore,

$$\|e^{t_0\bar{A}}\| = \lim_{\varepsilon\to 0} \sup\{|x_\varepsilon(t_0)| : \alpha_1^2 + \cdots + \alpha_d^2 \le 1\} \le \frac{1}{2}.$$

This means that (i) is satisfied, and the proof of Theorem 1 is complete. □

Remark 1 *If the solution $x_\varepsilon(t)$ satisfies condition* (ii), *it is said to be exponentially stable, and if it satisfies condition* (iii), *it is said to be stable in probability. Condition* (i) *means that the solution to the averaged equation is asymptotically stable. The theorem shows that stability in probability, exponential stability of the perturbed system, and asymptotic stability of the averaged system are equivalent for this linear system.*

Next we consider a nonlinear initial value problem of the form (6.1):

$$\dot{x}_\varepsilon(t) = a\left(y\left(\frac{t}{\varepsilon}\right), x_\varepsilon(t)\right) \quad x_\varepsilon(0) = x_0, \tag{6.11}$$

with the same assumptions on $y(t)$. We assume also that the function $a : Y \times R^d \to R$ satisfies the following conditions:

(1) There exists a continuous (in x) derivative $a_x(y,x)$ and $\|a_x(y,x)\| \le b$, where b is a constant.

(2) $\sup_{y\in Y} |a(y,0)| \le b_1$, for some constant $b_1 > 0$.

Set $\bar{a}(x) = \int a(y,x)\,\rho(dy)$, and let $\bar{x}(t)$ be the solution of the averaged initial value problem

$$\dot{\bar{x}}(t) = \bar{a}(\bar{x}(t)), \quad \bar{x}(0) = x_0. \tag{6.12}$$

Conditions (1) and (2), the assumptions on $y(t)$, and Theorem 6 of Chapter 3 imply the uniform convergence as $\varepsilon \to 0$ of $x_\varepsilon(t)$ to $\bar{x}(t)$ with probability 1 on any finite interval.

(3) Assume that $\bar{x}$ is a stationary point for equation (6.12), i.e., $\bar{a}(\bar{x}) = 0$, and that $\bar{A} = \bar{a}_x(\bar{x})$ satisfies the condition

$$\lim_{t\to\infty} \left\| e^{t\bar{A}} \right\| = 0. \tag{6.13}$$

It is known in the theory of stability for ordinary differential equations (see [134, p.321]) that in this case there exists $\delta > 0$ for which

$$\lim_{t\to\infty} \left| \bar{x}(t) - \bar{x} \right| = 0 \quad \text{if} \quad \left| \bar{x}(0) - \bar{x} \right| < \delta. \tag{6.14}$$

We will formulate some additional conditions on $a(y, x)$ under which $x_\varepsilon(t)$ is stable at the point $\bar{x}$ in a probabilistic sense. Consider the following condition:

(4) $a(y, \bar{x}) = 0$ for all $y \in Y$ and

$$\sup_{y\in Y} \left| a(y, x) - A(y)(x - \bar{x}) \right| = o(|x - \bar{x}|), \tag{6.15}$$

where $A(y) = a_x(y, \bar{x})$ and $\int A(y)\rho(dy) = \bar{A}$.

With these preliminaries, we have the following theorem:

Theorem 2 *Assume that the Markov process $y(t)$ satisfies condition* SMC II *of Section 2.3 and that $a(y, x)$ satisfies conditions* (1) *through* (4). *Then there exists $\varepsilon_0 > 0$ for which*

$$\lim_{x_0\to\bar{x}} P\left\{ \lim_{t\to\infty} \sup |x_\varepsilon(t) - \bar{x}| = 0 \right\} = 1 \tag{6.16}$$

for all $\varepsilon \le \varepsilon_0$.

Proof We divide the proof into several parts.
(1) There exists a symmetric positive matrix $C \in L(R^d)$ for which $\bar{B} = C\bar{A}^* + \bar{A}C$ is a strictly negative matrix (for example, we can set $C = -(\bar{A} + \bar{A}^* + \mu I)^{-1}$, where $\mu > 0$ is small enough).
Set $\tilde{A}(y) = A(y) - \bar{A}$, $\tilde{B}(y) = C\tilde{A}^*(y) + \tilde{A}(y)C$.
We introduce the function

$$\Psi_\varepsilon(x, y) = (C\tilde{x}, \tilde{x}) + \varepsilon\left(\int \tilde{B}(y')R(y, dy')\tilde{x}, \tilde{x} \right) = (C\tilde{x}, \tilde{x}) + \varepsilon\big(\tilde{D}(y)\tilde{x}, \tilde{x}\big),$$

where $\tilde{x} = x - \bar{x}$ and the function $R(y, dy')$ was introduced in Section 2.2. Under the assumptions of the theorem there exists a constant $b_1 > 0$ for which $\left|(\tilde{D}(y)\tilde{x}, \tilde{x})\right| \le b_1(\tilde{x}, \tilde{x})$ for all $\tilde{x} \in R^d$, $y \in Y$.
(2) Recall that $\big(x_\varepsilon(t), y(t/\varepsilon)\big)$ is a homogeneous Markov process in $X \times Y$ with generator

$$G_\varepsilon f(x, y) = \big(a(y, x), f_x(x, y)\big) + \frac{1}{\varepsilon} \int \big[f(x, y') - f(x, y)\big]\pi(y, dy'), \tag{6.17}$$

which is defined for functions f that are bounded, measurable, and continuous (in x) and whose derivative $f_x(x, y)$ has the same properties as stated in Lemma 2, Chapter 5.

It follows from Lemmas 1 and 2 in Chapter 5 that

$$\begin{aligned}
G_\varepsilon \Psi_\varepsilon(x,y) &= \big(Ca(y,x),\tilde{x}\big) + \big(C\tilde{x}, a(y,x)\big) + \varepsilon\big(\tilde{D}(y)A(y)\tilde{x},\tilde{x}\big) \\
&\quad + \varepsilon\big(\tilde{D}(y)\tilde{x}, A(y)\tilde{x}\big) - \big(\tilde{B}(y)\tilde{x},\tilde{x}\big) \\
&= \big(CA(y)\tilde{x},\tilde{x}\big) + \big(A^*(y)C\tilde{x},\tilde{x}\big) - \big(\tilde{B}(y)\tilde{x},\tilde{x}\big) + o(|\tilde{x}|^2) + O(\varepsilon|\tilde{x}|^2) \\
&= \big(\overline{B}\tilde{x},\tilde{x}\big) + o(|\tilde{x}|^2) + O(\varepsilon|\tilde{x}|^2).
\end{aligned}$$

This relation implies that $\delta > 0$, $\varepsilon_0 > 0$, $b_2 > 0$, and $b_3 > 0$ exist such that for $\varepsilon \le \varepsilon_0$ and $|x - \bar{x}| \le \delta$ we have

$$\Psi_\varepsilon(x,y) \ge b_2(\tilde{x},\tilde{x}) \quad \text{and} \quad G_\varepsilon \Psi_\varepsilon(x,y) \le -b_3(\tilde{x},\tilde{x}).$$

(3) Let $\tilde{x}_\varepsilon(t) = x_\varepsilon(t) - \bar{x}$, and let τ_ε be the stopping time with respect to the filtration $(\mathcal{F}_t^\varepsilon, t \ge 0)$ generated by the stochastic process $y(t/\varepsilon)$ defined by the relation $\tau_\varepsilon = \inf\{t : |\tilde{x}_\varepsilon(t)| > \delta\}$. Then for $t \le \tau_\varepsilon$,

$$\begin{aligned}
|\tilde{x}_\varepsilon(t)|^2 &\le \frac{1}{b_2}\Psi_\varepsilon\left(x_\varepsilon(t), y\left(\frac{t}{\varepsilon}\right)\right) \\
&= \frac{1}{b_2}\left[\int_0^t G_\varepsilon \Psi_\varepsilon\left(x_\varepsilon(s), y\left(\frac{s}{\varepsilon}\right)\right) ds + \Psi_\varepsilon(x_0,y_0) + \zeta_\varepsilon(t)\right],
\end{aligned}$$

where

$$\zeta_\varepsilon(t) = \Psi_\varepsilon\left(x_\varepsilon(t), y\left(\frac{t}{\varepsilon}\right)\right) - \int_0^t G_\varepsilon \Psi_\varepsilon\left(x_\varepsilon(s), y\left(\frac{s}{\varepsilon}\right)\right) ds - \Psi_\varepsilon(x_0,y_0)$$

is a locally bounded martingale with respect to the filtration and $E\zeta_\varepsilon(t) = 0$.

As a result of this,

$$\begin{aligned}
|\tilde{x}_\varepsilon(t\wedge\tau_\varepsilon)|^2 &\le \frac{1}{b_2}\left(\int_0^{t\wedge\tau_\varepsilon} G_\varepsilon \Psi_\varepsilon\left(x_\varepsilon(s), y\left(\frac{s}{\varepsilon}\right)\right) ds + \Psi_\varepsilon(x_0,y_0) + \zeta_\varepsilon(t)\right) \\
&\le \frac{1}{b_2}\left(\Psi_\varepsilon(x_0,y_0) - \int_0^{t\wedge\tau_\varepsilon} b_3\left|\tilde{x}_\varepsilon(s)\right|^2 ds\right) + \zeta_\varepsilon(t\wedge\tau_\varepsilon).
\end{aligned} \tag{6.18}$$

Therefore,

$$P\{\tau_\varepsilon < \infty\} \le \frac{1}{\delta^2 b_2} b_4 |x_0 - \bar{x}|^2, \tag{6.19}$$

where $b_4 = \|C\| + \varepsilon_0 b_1$ and

$$\lim_{x_0 \to \bar{x}} P\{\tau_\varepsilon < \infty\} = 0. \tag{6.20}$$

(4) The function $\Psi_\varepsilon\big(x_\varepsilon(t), y(t/\varepsilon)\big)$ is a bounded supermartingale (see, for example, [90] for definitions) on the interval $[0,\tau_\varepsilon)$; therefore, there exists

$$\lim_{t\to\tau_\varepsilon} \Psi_\varepsilon\left(x_\varepsilon(t), y\left(\frac{t}{\varepsilon}\right)\right).$$

Equation (6.18) implies the inequality

$$E\int_0^{t\wedge\tau_\varepsilon} |\tilde{x}_\varepsilon(s)|^2\, ds < \infty,$$

and $b_2(\tilde{x},\tilde{x}) \le \Psi_\varepsilon(x,y) \le b_4(\tilde{x},\tilde{x})$, so

$$E\int_0^{t\wedge\tau_\varepsilon} \Psi_\varepsilon\left(x_\varepsilon(s), y\left(\frac{s}{\varepsilon}\right)\right) ds < \infty.$$

Therefore,

$$P\left\{\lim_{t\to\infty} |\tilde{x}_\varepsilon(t)| = 0/\tau_\varepsilon = +\infty\right\} = 1. \tag{6.21}$$

Equations (6.20) and (6.21) complete the proof of the theorem.

□

Remark 2 *Let $\lambda < b_3/b_4$, and set $\eta_\varepsilon(t) = e^{\lambda t}\Psi_\varepsilon\left(x_\varepsilon(t), y\left(t/\varepsilon\right)\right)$. The function $\eta_\varepsilon(t)$ is a supermartingale on the interval $[0,\tau_\varepsilon)$, since*

$$\eta_\varepsilon(t) = \eta_\varepsilon(0) + \int_0^t e^{\lambda s} d\zeta_\varepsilon(s)$$
$$+ \int_0^t e^{\lambda s}\left[G_\varepsilon \Psi_\varepsilon\left(x_\varepsilon(s), y\left(\frac{s}{\varepsilon}\right)\right) + \lambda\Psi_\varepsilon\left(x_\varepsilon(s), y\left(\frac{s}{\varepsilon}\right)\right)\right] ds,$$

The first integral is a martingale, and the second integral is a decreasing function for $t < \tau_\varepsilon$, since for $|\tilde{x}| < \delta$ we have

$$G_\varepsilon\Psi_\varepsilon(\tilde{x},y) + \lambda\Psi_\varepsilon(\tilde{x},y) \le -b_3(\tilde{x},\tilde{x}) + \lambda b_4(\tilde{x},\tilde{x}) \le (\lambda b_4 - b_3)(\tilde{x},\tilde{x}),$$

and $\lambda b_4 - b_3 < 0$. In the same way as in part (4) of the proof we can prove that

$$\lim_{x_0\to\bar{x}} P\left\{\lim_{t\to\infty} |\tilde{x}_\varepsilon(t)|^2 e^{t\lambda} = 0\right\} = 1. \tag{6.22}$$

We consider next a generalization of Theorem 2.

Theorem 3 *Assume that the Markov process $y(t)$ satisfies condition* SMC II *of Section 2.3 and that $a(y,x)$ satisfies the following conditions:*

(A) *$a(y,x)$ and $a_x(y,x)$ are measurable and locally bounded in x.*

(B) *There exists a function $\Psi_\varepsilon(x,y)$ that is continuously differentiable in x and measurable in y with the properties that*

$$\lim_{x\to\bar{x}} \sup_y \Psi_\varepsilon(x,y) = 0,$$

$$\Psi_\varepsilon(x,y) \ge g(x), \quad \textit{and} \quad G_\varepsilon\Psi_\varepsilon(x,y) \le -g(x)$$

for $y \in Y$, $|x-\bar{x}| \le \delta$, where $g(x)$ is a continuous function, $g(x) > 0$, for $|x-\bar{x}| > 0$, $g(\bar{x}) = 0$, and G_ε is as in (6.17), and δ is sufficiently small.

Then

$$\lim_{x_0 \to \bar{x}} P\left\{ \lim_{t\to\infty} g\left(x_\varepsilon(t)\right) e^{\lambda t} = 0 \right\} = 1 \tag{6.23}$$

for any $\varepsilon > 0$ and $\lambda < 1$.

Proof We consider the stopping time τ_ε introduced in the proof of Theorem 2. Then $\Psi_\varepsilon\left(x_\varepsilon(t), y\left(t/\varepsilon\right)\right)$ is a supermartingale on the interval $[0, \tau_\varepsilon)$. Set $\alpha = \inf\{g(x) : |x-\bar{x}| = \delta\} > 0$ and $\varphi(x) = \sup_y \Psi_\varepsilon(x,y)$. Since

$$\alpha P\{\tau_\varepsilon < \infty\} \le \varphi(x_0),$$

we have $\lim_{x_0 \to \bar{x}} P\{\tau_\varepsilon = \infty\} = 1$.
For any $\lambda < 1$, we can prove in the same way as in Remark 2 that $\Psi_\varepsilon\big(x_\varepsilon(t), y(t/\varepsilon)\big)e^{\lambda t}$ is a supermartingale on the interval $[0, \tau_\varepsilon)$. So

$$P\left\{ \sup_{t<\tau_\varepsilon} \left[\Psi_\varepsilon\left(x_\varepsilon(t), y\left(\frac{t}{\varepsilon}\right)\right) e^{\lambda t} \right] < \infty \right\} = 1.$$

Therefore,

$$P\left\{ \sup_{t<\tau_\varepsilon} \left[g\left(x_\varepsilon(t)\right) e^{\lambda t} \right] < \infty \right\} = 1,$$

and

$$P\left\{ \sup_t g\left(x_\varepsilon(t)\right) e^{\lambda t} < \infty \right\} > 1 - P\{\tau_\varepsilon < \infty\}.$$

This completes the proof of the theorem.

□

Suppose that there exists a Liapunov function for the averaged equation

$$\dot{\bar{x}}(t) = \bar{a}\big(\bar{x}(t)\big)$$

at the point $\bar{x}$, i.e., there exists a function $f_0(x)$ that satisfies the following conditions:

(α) $f_0(x)$ is continuous, $f_0(\bar{x}) = 0$, and $f_0(x) > 0$ for $x \ne \bar{x}$;

(β) $f_0(x)$ has a continuous derivative for $x \ne \bar{x}$ and

$$(\bar{a}(x), \nabla f_0(x)) < 0 \quad \text{for} \quad x \ne \bar{x}.$$

We consider the function $\Psi_\varepsilon(x,y) = f_0(x) + \varepsilon f_1(x,y)$, where

$$f_1(x,y) = \int R(y, dy')\left(a(x,y'), \nabla f_0(x)\right).$$

Then

$$G_\varepsilon \Psi_\varepsilon(x,y) = (a(x,y), \nabla f_0(x)) - \Pi f_1(x,y) + \varepsilon\,(a(x,y), \nabla f_1(x,y))$$
$$= \left(\bar a(x), \frac{\partial f_0}{\partial x}(x)\right) + \varepsilon \int R(y,dy')\Bigg[\left(a_x(x,y')a(x,y), \frac{\partial f_0}{\partial x}(x)\right) + \left(\frac{\partial^2 f_0}{\partial x^2}(x)a(x,y), a(x,y')\right)\Bigg].$$

Remark 3 *Assume that there exist constants c_1, c_2, c_3 for which*

$$\big|(a(x,y), \nabla f_0(x))\big| \le c_1 f_0(x), \tag{6.24}$$

$$\big|(a_x(x,y')a(x,y), \nabla f_0(x))\big| \le c_2\big|(\bar a(x), \nabla f_0(x))\big|, \tag{6.25}$$

and

$$\left|\left(\frac{\partial^2 f_0}{\partial x^2}(x)a(x,y), a(x,y')\right)\right| \le c_3\left|\left(\bar a(x), \frac{\partial f_0}{\partial x}(x)\right)\right| \tag{6.26}$$

for $0 < |x-\bar x| \le \delta$ and y, $y' \in Y$. Then $\Psi_\varepsilon(x,y)$ satisfies the conditions of Theorem 3 with the function

$$g(x) = [(1-c_1\varepsilon)f_0(x)] \wedge \left[-(1-(c_1+c_3)\varepsilon)\left(\bar a(x), \frac{\partial f_0}{\partial x}(x)\right)\right]$$

if $\varepsilon < (1/c_1) \wedge (1/(c_2+c_3))$.

Example 1 Liapunov Function for a Noisy Planar System
Consider a system in the plane R^2, where we write $x = (x_1, x_2)$, $a = (a_1, a_2)$. In particular, we take

$$a_1(x,y) = \alpha(y)x_1 + \beta(y)x_2^3, \quad a_2(x,y) = \gamma(y)x_2^3 + \delta(y)x_1,$$

where $\alpha(y), \beta(y), \gamma(y), \delta(y)$ are bounded measurable functions. Suppose that

$$\bar\alpha = \int \alpha(y)\rho(dy) < 0,\ \bar\gamma = \int \gamma(y)\rho(dy) < 0,\ \bar\beta = \bar\delta = 0.$$

The averaged system is

$$\dot{\bar x}_1 = \bar\alpha \bar x_1, \quad \dot{\bar x}_2 = \bar\gamma \bar x_2^3,$$

for which the function $f_0(x) = x_1^2/2 + x_2^4/4$ is a Liapunov function at $(0,0)$. Then

$$\left(a(x,y), \frac{\partial f_0}{\partial x}(x)\right)$$
$$= \alpha(y)(x_1)^2 + \gamma(y)(x_2)^6 + (\beta(y)+\gamma(y))\,x_1(x_2)^3 = O\left(f_0(x)\right),$$
$$a_x(x,y) = \begin{pmatrix} a(x,y) & 3\beta(y)(x_2)^2 \\ \delta(y) & 3\gamma(y)(x_2)^2 \end{pmatrix}, \quad \frac{\partial^2 f_0}{\partial x^2} = \begin{pmatrix} 1 & 0 \\ 0 & 3(x_2)^2 \end{pmatrix},$$

where $\left(\bar{a}(x), \frac{\partial f_0}{\partial x}(x)\right) = \bar{\alpha}(x_1)^2 + \bar{\gamma}(x_2)^6$ and $|a(x,y)| = O\left((x_1)^2 + (x_2)^6\right)$. So the inequalities (6.24)–(6.26) are satisfied, and the statement of Theorem 3 holds with $g(x) = \theta\big[x_1^2 + x_2^6\big]$ if θ is small enough.

Example 2 Randomly Perturbed Gradient Systems in R^d
Consider a gradient system of the form

$$\dot{x}(t) = -\nabla F\left(x(t)\right), \tag{6.27}$$

where $F : R^d \to R$ is a sufficiently smooth function and where

$$\nabla F = \left(\frac{\partial F}{\partial x_1}, \dots, \frac{\partial F}{\partial x_d}\right)^T$$

is the gradient of F. If $\bar{x}$ is a local minimum of the function F, then $F(x) - F(\bar{x})$ is a Liapunov function for equation (6.27), since

$$\big(\nabla F, -\nabla F(x)\big) = -\big|\nabla F(x)\big|^2. \tag{6.28}$$

Let $x_\varepsilon(t)$ be the solution of the equation

$$\dot{x}_\varepsilon(t) = -\nabla F\left(y\left(\frac{t}{\varepsilon}\right), x_\varepsilon(t)\right), \tag{6.29}$$

where ∇ contains derivatives only with respect to the components of x, the function $F : Y \times R^d \to R$ is measurable in y, and it satisfies the following additional conditions:

(i) $F(y, \bar{x}) = 0$ for all $y \in Y$, the first and second derivatives of F with respect to the components of x exist, the functions $F(y,x)$, $\nabla F(y,x)$, $\nabla\nabla^T F(y,x)$ are continuous in x and locally bounded for $|x - \bar{x}| < \delta$.

(ii) $\nabla F(y, \bar{x}) = 0$ for all $y \in Y$.

(iii) If $\bar{F}(x) = \int F(y,x)\,\rho(dy)$, then for $0 < |x - \bar{x}| < \delta$, $\big|\nabla \bar{F}(x)\big| > 0$, $\bar{F}(x) > 0$,

$$|\nabla F(y,x)| = O\left(\big|\nabla \bar{F}(x)\big|\right),$$

and

$$\big|\nabla \bar{F}(x)\big|^2 = O\left(\bar{F}(x)\right).$$

Set $\Psi_\varepsilon(x,y) = f_0(x) + \varepsilon f_1(x,y)$, where $f_0(x) = \bar{F}(x)$ and

$$f_1(x,y) = \int R(y, dy')\left(-\nabla F(y', x), \nabla \bar{F}(x)\right).$$

Then

$$|f_1(x,y)| = O\left(\big|\nabla \bar{F}(x)\big|^2\right) = O\left(f_0(x)\right)$$

and

$$|(\nabla F(y,x), \nabla f_1(x,y))| = O\left(\big|\nabla \bar{F}(x)\big|^2\right) = O\left(\big|(\nabla \bar{F}(x), \nabla f_0(x))\big|\right).$$

It follows from these assumptions that the statement of Theorem 3 holds for this example.

6.1.2 Stationary Perturbations of Differential Equations

First, we consider linear differential equations of the form

$$\dot{x}_\varepsilon(t) = A\left(\frac{t}{\varepsilon}\right) x_\varepsilon(t), \tag{6.30}$$

where $A(t)$ is an $L(R^d)$-valued stationary process. Set $\bar{A} = EA(t)$. We prove the following theorem.

Theorem 4 *Suppose that*

(1) $\lim_{t\to\infty} \|e^{t\bar{A}}\| \to 0,$
and

(2) *$A(t)$ is an ergodic stationary process satisfying the condition $E\exp\{\lambda\|A(t)\|\} < \infty$ for all $\lambda > 0$.*

Denote by $U_\varepsilon(t)$ the $L(R^d)$-valued stochastic process that solves the differential equation

$$\dot{U}_\varepsilon(t) = A\left(\frac{t}{\varepsilon}\right) U_\varepsilon(t) \tag{6.31}$$

with the initial condition

$$U_\varepsilon(0) = I. \tag{6.32}$$

Then $\varepsilon_0 > 0$ exists for which

$$P\left\{\lim_{t\to\infty} \|U_\varepsilon(t)\| = 0\right\} = 1 \tag{6.33}$$

for all $\varepsilon \le \varepsilon_0$.

Proof Note that the solution of equation (6.30) satisfying the initial condition $x_\varepsilon(0) = x_0$ can be written in the form

$$x_\varepsilon(t) = U_\varepsilon(t)x_0. \tag{6.34}$$

Condition (2) of the theorem implies that the statement of Theorem 5 in Chapter 3 is satisfied, so $\left\|U_\varepsilon(t) - \exp\{t\bar{A}\}\right\| \to 0$ uniformly in t on any finite interval with probability 1. Using the relations

$$U_\varepsilon(t) = I + \int_0^t A\left(\frac{s}{\varepsilon}\right)U_\varepsilon(s)\, ds$$

and

$$\|U_\varepsilon(t)\| \le 1 + \int_0^t \left\|A\left(\frac{s}{\varepsilon}\right)\right\| \|U_\varepsilon(s)\|\, ds$$

and Gronwall's inequality we can prove that

$$\|U_\varepsilon(t)\| \le \exp\left\{\int_0^t \left\|A\left(\frac{s}{\varepsilon}\right)\right\| ds\right\}. \tag{6.35}$$

It follows from condition (2) and Jensen's inequality that

$$\begin{aligned} E\|U_\varepsilon(t)\|^\alpha &\le E\exp\left\{\alpha t \cdot \frac{1}{t}\int_0^t \left\|A\left(\frac{s}{\varepsilon}\right)\right\| ds\right\} \\ &\le E\frac{1}{t}\int_0^t \exp\left\{\alpha t\left\|A\left(\frac{s}{\varepsilon}\right)\right\|\right\} ds < \infty. \end{aligned}$$

So $\lim_{\varepsilon\to 0} E\|U_\varepsilon(t)\| = \exp\{t\bar{A}\}$ for any $t > 0$; hence, there exist ε_0, $q < 1$ and $\bar{t}$ for which $E\|U_\varepsilon(\bar{t})\| \le q$ for $\varepsilon \le \varepsilon_0$.

Let $U_\varepsilon(t_0, t)$ for $t > t_0$ be the solution of the differential equation

$$\dot{U}_\varepsilon(t_0, t) = A\left(\frac{t}{\varepsilon}\right) U_\varepsilon(t_0, t),$$

with the initial condition $U_\varepsilon(t_0, t_0) = I$. It is easy to see using the semigroup property of U_ε that for $0 < t_1 < \cdots < t_k$ the following relation holds:

$$U_\varepsilon(t_k) = U_\varepsilon(t_{k-1}, t_k)\cdots U_\varepsilon(0, t_1).$$

Denote by $\mathcal{A}$ the σ-algebra generated by $\{A(t),\, t \in R\}$. Let $R(\mathcal{A})$ be the linear space of $\mathcal{A}$-measurable random variables. We introduce the family of mappings $\{\theta_h,\, h > 0\}$ acting in $R(\mathcal{A})$ that is defined by the relation

$$\theta_h\big(A(t)x_1, x_2\big) = \big(A(t+h)x_1, x_2\big), \quad x_1, x_2 \in R^d,\ t \in R.$$

Then $U_\varepsilon(t_0, t) = \theta_{t_0/\varepsilon} U_\varepsilon(0, t - t_0)$ (θ_h acts on all elements of the matrix $U_\varepsilon(0, t - t_0)$).

The sequence of $L(R^d)$-valued random variables

$$V_\varepsilon(k) = U_\varepsilon\big((k-1)\bar{t}, k\bar{t}\big)$$

is stationary and ergodic. In the same way as in the derivation of inequality (6.35), we obtain the following inequality:

$$\|U_\varepsilon(t_0, t)\| \le \exp\left\{\int_{t_0}^t \left\|A\left(\frac{s}{\varepsilon}\right)\right\| ds\right\}.$$

If $t \in [n\bar{t}, (n+1)\bar{t}]$, then

$$\|U_\varepsilon(t)\| \le \exp\left\{\sum_{k=1}^n \log\|V_\varepsilon(k)\| + \int_{n\bar{t}}^{(n+1)\bar{t}} \left\|A\left(\frac{s}{\varepsilon}\right)\right\| ds\right\}. \tag{6.36}$$

Note that

$$E\log\|V_\varepsilon(k)\| = E\log\|U_\varepsilon(\bar{t})\| \le \log q < 0$$

and

$$\frac{1}{n}\left(\int_{n\bar{t}}^{(n+1)\bar{t}}\left\|A\left(\frac{s}{\varepsilon}\right)\right\|ds\right)$$
$$=\frac{1}{n}\int_{0}^{(n+1)\bar{t}}\left\|A\left(\frac{s}{\varepsilon}\right)\right\|ds-\frac{1}{n}\int_{0}^{n\bar{t}}\left\|A\left(\frac{s}{\varepsilon}\right)\right\|ds\to 0$$

with probability 1 as $n\to\infty$ due to the ergodic theorem. So

$$\frac{1}{n}\left(\sum_{k=1}^{n}\log V_\varepsilon(k)+\int_{n\bar{t}}^{(n+1)\bar{t}}\left\|A\left(\frac{s}{\varepsilon}\right)\right\|ds\right)\to E\log\|U_\varepsilon(\bar{t})\|<0,$$

as $n\to\infty$. This completes the proof of the theorem.

□

Next, we consider a nonlinear differential equation of the form

$$\dot{x}_\varepsilon(t)=a\left(\frac{t}{\varepsilon},x_\varepsilon(t)\right), \tag{6.37}$$

where $a(t,x)$ is an R^d-valued random function that is defined on $R_+\times R^d$. We assume that this random function satisfies the following conditions:

(a1) For fixed x, $a(t,x)$ is an ergodic stationary process in t; there exists the derivative $a_x(t,x)$; and for fixed x, a_x is an $L(R^d)$-valued ergodic stationary process.

(a2) There exists a stationary process $\lambda(t)$ for which $|a(t,x)|+\|a_x(t,x)\|\le\lambda(t)$ and $E\lambda^2(t)<\infty$.

Under these conditions Theorem 5 of Chapter 3 may be used; therefore, if $\bar{a}(x)=Ea(t,x)$ and $\bar{x}(t)$ is the solution of the equation $d\dot{\bar{x}}=\bar{a}(\bar{x}(t))$ with the initial condition $x(0)=x_0$, then with probability 1,

$$\sup_{t\le T}|x_\varepsilon(t)-\bar{x}(t)|\to 0$$

as $\varepsilon\to 0$, for all $T>0$, where $x_\varepsilon(t)$ is the solution of equation (6.37) with the initial condition $x_\varepsilon(0)=x_0$.
Suppose that $\bar{x}$ is a stable stationary point for the averaged equation; i.e., there exists $\delta>0$ for which $\lim_{t\to\infty}\bar{x}(t)=\bar{x}$ if $|x_0-\bar{x}|<\delta$.
We investigate next additional conditions under which $\bar{x}$ is a stable stationary point for equation (6.37). In particular, we require some additional assumption about the random field $a(t,x)$.

(a3) For all n and $x_1,x_2,\dots,x_n\in R^d$ the stochastic process

$$(a(t,x_1),a(t,x_2),\dots,a(t,x_n))$$

in $(R^d)^n$ is stationary and ergodic.

Let $\mathcal{F}_s$ be the σ-algebra generated by the history of a up to time s, $\{a(t,x),\, t \le s,\, x \in R^d\}$. We will make use of the translation operator θ_h that was introduced in the proof of Theorem 4.
Denote by D_2^0 the set of $\mathcal{F}$-measurable random variables ξ for which $E\xi = 0$, $E\xi^2 < \infty$, and $\lim_{h\to 0} E(\theta_h\xi - \xi)^2 = 0$. Suppose there exists an $\mathcal{F}$-measurable random variable $R(\xi)$ for which

$$\lim_{T\to\infty} E\left(\int_0^T E(\theta_s\xi/\mathcal{F}_0)\,ds - R(\xi)\right)^2 = 0. \tag{6.38}$$

We write this as

$$R(\xi) = \int_0^\infty E(\theta_s\xi/\mathcal{F}_0)\,ds, \tag{6.39}$$

where the integral on the right-hand side of (6.39) converges in the mean square sense.

Lemma 1 *Let $\xi \in D_2^0$. Set*

$$\mu_t(\xi) = \theta_t R(\xi) - R(\xi) + \int_0^t \theta_s\xi\,ds. \tag{6.40}$$

Then $\mu_t(\xi)$ is a square integrable martingale, and its square characteristic $\langle\mu(\xi)\rangle_t$ satisfies the relation

$$E\langle\mu(\xi)\rangle_t = 2tE(\xi R(\xi)). \tag{6.41}$$

Proof $\mu_t(\xi)$ is square integrable because $\xi \in D_2^0$, so

$$E(\theta_t R(\xi))^2 = E\big(R(\xi)\big)^2 < \infty, \quad E\left(\int_0^t E(\theta_s\xi/\mathcal{F}_0)\,ds\right)^2 < \infty.$$

For $t \ge 0$, $h > 0$ we have

$$\begin{aligned}
&E\big(\theta_{t+h}R(\xi) - \theta_t R(\xi)/\mathcal{F}_t\big) = \theta_t E\big(\theta_h R(\xi) - R(\xi)/\mathcal{F}_0\big)\\
&= \theta_t E\left(\int_h^\infty E(\theta_s\xi/\mathcal{F}_h)\,ds - \int_0^\infty E(\theta_s\xi/\mathcal{F}_0\,ds/\mathcal{F}_0\right)\\
&= -\theta_t E\left(\int_0^h E(\theta_s\xi/\mathcal{F}_0)\,ds\right) = -E\left(\int_t^{t+h} \theta_s\xi\,ds/\mathcal{F}_0\right).
\end{aligned}$$

So for $0 \le t_1 < t_2$,

$$E\left(\theta_{t_2}R(\xi) - \theta_{t_1}R(\xi) + \int_{t_1}^{t_2} \theta_s\xi\,ds/\mathcal{F}_{t_1}\right) = 0. \tag{6.42}$$

This relation means that $\mu_t(\xi)$ is a martingale.
Next, we calculate $E\big(\theta_h R(\xi) - R(\xi)\big)^2 \equiv \Delta_h$.

We have

$$\begin{aligned}\Delta_h &= E\big(\theta_h R(\xi)\big)^2 + E\big(R(\xi)\big)^2 - 2ER(\xi)\theta_h R(\xi)\\ &= 2E\big(R(\xi)\big)^2 - 2ER(\xi)E\big(\theta_h R(\xi)/\mathcal{F}_0\big) = 2ER(\xi)E\big(R(\xi) - \theta_h R(\xi)/\mathcal{F}_0\big)\\ &= 2ER(\xi)\int_0^h \theta_s\xi\, ds\end{aligned}$$

(here we used formula (6.42)). Note that

$$E\Big(\int_0^h \theta_s\xi\, ds\Big)^2 = \int_0^h\int_0^h E\theta_s\xi\theta_u\xi\, ds\, du \le E\xi^2 h^2.$$

Therefore,

$$E\big(\mu_{t+h}(\xi) - \mu_t(\xi)\big)^2 = E(\mu_h(\xi))^2 = 2ER(\xi)\int_0^h \theta_s\xi\, ds + O(h^2).$$

For all n,

$$E\mu_t^2(\xi) = nE\mu_{\frac{t}{n}}^2(\xi) = 2nER(\xi)\int_0^{t/n} \theta_s\xi\, ds + O\Big(n\frac{t^2}{n^2}\Big).$$

This relation implies formula (6.41).

□

Lemma 2 *Let $\xi \in D_2^0$ and $g \in C^{(1)}(R^d)$. Set*

$$G(t,x) = \theta_t R(\xi) g(x).$$

Assume that conditions (1)–(3) are fulfilled. Then for $0 \le t_1 < t_2$ and $\varepsilon > 0$,

$$\begin{aligned}&E\left(G\left(\frac{t_2}{\varepsilon}, x_\varepsilon(t_2)\right) - G\left(\frac{t_1}{\varepsilon}, x_\varepsilon(t_1)\right)/\mathcal{F}_{t_1/\varepsilon}\right)\\ &= E\bigg(\int_{t_1}^{t_2}\bigg[-\frac{\theta_{\frac{s}{\varepsilon}}}{\varepsilon}\xi\cdot g(x_\varepsilon(s)) + \big[\theta_{\frac{s}{\varepsilon}} R(\xi)\big]\\ &\quad\times\left(g_x(x_\varepsilon(s)), a\left(\frac{s}{\varepsilon}, x_\varepsilon(s)\right)\right)\bigg] ds/\mathcal{F}_{t_1/\varepsilon}\bigg).\end{aligned} \tag{6.43}$$

Proof For $t \ge 0$, $h > 0$ we have

$$\begin{aligned}&G\left(\frac{t+h}{\varepsilon}, x_\varepsilon(t+h)\right) - G\left(\frac{t}{\varepsilon}, x_\varepsilon(t)\right) = \theta_{\frac{t}{\varepsilon}}\left[G\left(\frac{h}{\varepsilon}, x_\varepsilon(h)\right) - G\left(0, x_\varepsilon(0)\right)\right]\\ &= \theta_{\frac{t}{\varepsilon}}\left[\theta_{\frac{h}{\varepsilon}} R(\xi) - R(\xi)\right]\left[g(x_\varepsilon(h)) - g(x_\varepsilon(0))\right] +\\ &\theta_{\frac{t}{\varepsilon}}\left[\theta_{\frac{h}{\varepsilon}} R(\xi) - R(\xi)\right] g(x_\varepsilon(0)) + \theta_{\frac{t}{\varepsilon}} R(\xi)\int_0^h \left(g_x(x_\varepsilon(s)), a\left(\frac{s}{\varepsilon}, x_\varepsilon(s)\right)\right) ds.\end{aligned}$$

Denote the function that is integrated on the right-hand side of equation (6.43) by $H_\varepsilon(s,\xi,g)$. Then

$$E\left(G\left(\frac{t+h}{\varepsilon}, x_\varepsilon(t+h)\right) - G\left(\frac{t}{\varepsilon}, x_\varepsilon(t)\right) - \int_t^{t+h} H_\varepsilon(s,\xi,g)\,ds/\mathcal{F}_{t/\varepsilon}\right)$$
$$= E\Bigg(\Bigg\{\theta_{\frac{t}{\varepsilon}}\left[\theta_{\frac{h}{\varepsilon}} R(\xi) - R(\xi)\right][g(x_\varepsilon(h)) - g(x_\varepsilon(0))]$$
$$- \int_t^{t+h} \left(g_x(x_\varepsilon(s)), a\left(\frac{s}{\varepsilon}, x_\varepsilon(s)\right)\right)\left(\theta_{\frac{s}{\varepsilon}} R(\xi) - \theta\frac{t}{\varepsilon} R(\xi)\right) ds$$
$$+ \frac{1}{\varepsilon}\int_t^{t+h} \theta_{\frac{s}{\varepsilon}}\xi\left(g(x_\varepsilon(s)) - g(x_\varepsilon(t))\right) ds\Bigg\}/\mathcal{F}_{t/\varepsilon}\Bigg).$$

Using the inequalities

$$E\big[\theta_h R(\xi) - R(\xi)\big]^2 \le c_1 h$$

and

$$E\big[g(x_\varepsilon(h)) - g(x_\varepsilon(0))\big]^2 \le c_2 h^2$$

we can prove that

$$E\left(G\left(\frac{t+h}{\varepsilon}, x_\varepsilon(t+h)\right) - G\left(\frac{t}{\varepsilon}, x_\varepsilon(t)\right) - \int_t^{t+h} H_\varepsilon(s,\xi,g)\,ds/\mathcal{F}_{t/\varepsilon}\right) = o(h)$$

uniformly in t. This implies relation (6.43).

□

Remark 4 *Let $g(t,x) = \theta_t \xi g(x)$. Then $g(t,x)$ is a stationary process for all x and $g(0,x) \in D_2^0$.*
Set

$$G(t,x) = \theta_t R\big(g(0,x)\big). \tag{6.44}$$

The statement of Lemma 2 is equivalent to the following one: The stochastic process

$$\eta_\varepsilon(t) = G\left(\frac{t}{\varepsilon}, x_\varepsilon(t)\right)$$
$$+ \int_0^t \left[\frac{1}{\varepsilon} g\left(\frac{s}{\varepsilon}, x_\varepsilon(s)\right) - \left(\frac{\partial}{\partial x} G\left(\frac{s}{\varepsilon}, x_\varepsilon(s)\right), a\left(\frac{s}{\varepsilon}, x_\varepsilon(s)\right)\right)\right] ds \tag{6.45}$$

is a martingale.

We show next that relation (6.44) implies that the stochastic process $\eta_\varepsilon(t)$ given by formula (6.45) is a martingale under more general conditions.

Lemma 3 *Assume that $g(t,x) = \theta_t g(0,x)$, where $g(0,x) \in D_2^0$ for all x, and that $G(t,x)$ is defined by formula (6.44). We assume, in addition, that $g(t,x)$ and $G(t,x)$ satisfy the following conditions:*

I. *$\left|g(t,x_1) - g(t,x_2)\right| \le \gamma(t)|x_1 - x_2|$, where $\gamma(t)$ is a stationary stochastic process, $E\gamma^2(t) < \infty$.*

II. *There exists $\frac{\partial G}{\partial x}(t,x)$as a continuous function of x, and $\left|\frac{\partial G}{\partial x}(t,x)\right| \le \gamma(t)$, where $\gamma(t)$ is the same as in condition I. Moreover,*

$$E\left(\left|\frac{\partial G}{\partial x}(t+h,x) - \frac{\partial G}{\partial x}(t,x)\right| / \mathcal{F}_t\right) \le \alpha(h)\gamma(t),$$

where $\alpha(h)$ is a nonrandom increasing function from R_+ to R and $\alpha(0+) = 0$.

Then the process $\eta_\varepsilon(t)$ determined by (6.45) is a martingale.

Proof Let $0 \le t < t+h$ and $\varepsilon > 0$. Then

$$\begin{aligned}
&G\left(\frac{t+h}{\varepsilon}, x_\varepsilon(t+h)\right) - G\left(\frac{t+h}{\varepsilon}, x_\varepsilon(t)\right)\\
&= \int_t^{t+h} \left(G_x\left(\frac{t+h}{\varepsilon}, x_\varepsilon(s)\right), a\left(\frac{s}{\varepsilon}, x_\varepsilon(s)\right)\right) ds\\
&= \int_t^{t+h} \left(G_x\left(\frac{s}{\varepsilon}, x_\varepsilon(s)\right), a\left(\frac{s}{\varepsilon}, x_\varepsilon(s)\right)\right) ds\\
&\quad + \int_t^{t+h} \left(G_x\left(\frac{t+h}{\varepsilon}, x_\varepsilon(s)\right) - G_x\left(\frac{s}{\varepsilon}, x_\varepsilon(s)\right), a\left(\frac{s}{\varepsilon}, x_\varepsilon(s)\right)\right) ds
\end{aligned}$$

and

$$\begin{aligned}
&E\left|\int_t^{t+h} \left(G_x\left(\frac{t+h}{\varepsilon}, x_\varepsilon(s)\right) - G_x\left(\frac{s}{\varepsilon}, x_\varepsilon(s)\right), a\left(\frac{s}{\varepsilon}, x_\varepsilon(s)\right)\right) ds\right|\\
&\le \alpha\left(\frac{h}{\varepsilon}\right) \int_t^{t+h} E\gamma\left(\frac{s}{\varepsilon}\right)\lambda\left(\frac{s}{\varepsilon}\right) ds \le \alpha\left(\frac{h}{\varepsilon}\right) ch,
\end{aligned}$$

where $c > 0$ is a constant. Further,

$$\begin{aligned}
&E\left(G\left(\frac{t+h}{\varepsilon}, x_\varepsilon(t)\right) - G\left(\frac{t}{\varepsilon}, x_\varepsilon(t)\right) / \mathcal{F}_{t/\varepsilon}\right)\\
&= E\left(-\frac{1}{\varepsilon}\int_t^{t+h} g\left(\frac{s}{\varepsilon}, x_\varepsilon(t)\right) ds / \mathcal{F}_{t/\varepsilon}\right)\\
&= E\left(-\frac{1}{\varepsilon}\int_t^{t+h} g\left(\frac{s}{\varepsilon} x_\varepsilon(s)\right) ds\right.\\
&\quad \left. + \frac{1}{\varepsilon}\int_t^{t+h} \left(g\left(\frac{s}{\varepsilon}, x_\varepsilon(s)\right) - g\left(\frac{s}{\varepsilon}, x_\varepsilon(t)\right)\right) ds / \mathcal{F}_{t/\varepsilon}\right)
\end{aligned}$$

and

$$
\begin{aligned}
&E\left|\int_t^{t+h}\left(g\Big(\frac{s}{\varepsilon},x_\varepsilon(s)\Big)-g\Big(\frac{s}{\varepsilon},x_\varepsilon(t)\Big)\right)ds\right| \\
&\le E\int_t^{t+h}\gamma\Big(\frac{s}{\varepsilon}\Big)\big|x_\varepsilon(s)-x_\varepsilon(t)\big|\,ds \\
&\le E\int_t^{t+h}\gamma\Big(\frac{s}{\varepsilon}\Big)\int_t^s\left|a\Big(\frac{u}{\varepsilon},x_\varepsilon(u)\Big)\right|du\,ds\le c_1h^2.
\end{aligned}
$$

So

$$
\begin{aligned}
&E\big|E\big(\eta_\varepsilon(t+h)-\eta_\varepsilon(t)/\mathcal{F}_{t/\varepsilon}\big)\big| \\
&=E\bigg|E\bigg(\left[G\left(\frac{t+h}{\varepsilon},x_\varepsilon(t+h)\right)-G\left(\frac{t}{\varepsilon},x_\varepsilon(t)\right)\right] \\
&\quad+\left[G\left(\frac{t+h}{\varepsilon},x_\varepsilon(t)\right)-G\left(\frac{t}{\varepsilon},x_\varepsilon(t)\right)\right]+\frac{1}{\varepsilon}\int_t^{t+h}g\left(\frac{s}{\varepsilon},x_\varepsilon(s)\right)ds \\
&\quad-\int_t^{t+h}\left(G_x\left(\frac{s}{\varepsilon},x_\varepsilon(s)\right),a\left(\frac{s}{\varepsilon},x_\varepsilon(s)\right)\right)ds\bigg)\bigg|\le h\left(c\alpha\left(\frac{h}{\varepsilon}\right)+\frac{1}{\varepsilon}c_1h\right).
\end{aligned}
$$

This implies the statement of the lemma.

□

Next we make some additional assumptions about the random function $a(t,x)$.

(a4) Assume that $a(t,0)=0$ for all t,

$$
a(0,x)-\bar a(x)\in D_2^0,\ \frac{\partial a}{\partial x}(0,x)-\frac{\partial\bar a}{\partial x}(x)\in D_2^0\quad\text{for all}\quad x\in R^d;
$$

a constant $c>0$ exists for which $\|\frac{\partial a}{\partial x}(t,0)\|\le c$,

$$
\left\|\frac{\partial R}{\partial x}(a(0,x)-\bar a(x))\right\|\le c,\quad |R(a(0,x)-\bar a(x))|\le c|x|,
$$

and

$$
|a(t,x)-a_x(t,0)x|\le c|x|^2\quad\text{if } |x|\le 1.
$$

Theorem 5 *Let conditions* (a1)–(a4) *be fulfilled. Set* $\bar A=EA(t)$. *Assume that there exists a positive matrix* $C\in L(R^d)$ *such that the matrix* $\bar AC+C\bar A^*$ *is strictly negative.*
Then the point $\bar x=0$ *is stable for* $x_\varepsilon(t)$ *with probability* 1: *That is,*

$$
\lim_{x_0\to 0}P\left\{\limsup_{t\to\infty}|x_\varepsilon(t)|=0\right\}=1.
$$

Proof Define $g\in C^{(2)}(R^d)$ by $g(x)=(Cx,x)$ for $|x|\le 1$.
Let us introduce the random fields

$$
g(0,x)=\big(g_x(x),a(0,x)\big),\quad G(0,x)=R\big(g(0,x)\big),\quad G(t,x)=\theta_tG(0,x).
$$

It follows from Lemma 3 that the stochastic process

$$\eta_\varepsilon(t) = G\Big(\frac{t}{\varepsilon}, x_\varepsilon(t)\Big) + \frac{1}{\varepsilon}\int_0^t \Big(\frac{\partial g}{\partial x}(x_\varepsilon(s)), a\Big(\frac{s}{\varepsilon}, x_\varepsilon(s)\Big) - \bar{a}(x_\varepsilon(s))\Big)\, ds$$
$$- \int_0^t \Big(\frac{\partial G}{\partial x}\Big(\frac{s}{\varepsilon}, x_\varepsilon(s)\Big), a\Big(\frac{s}{\varepsilon}, x_\varepsilon(s)\Big)\Big)\, ds$$

is a martingale. Using the relation

$$\int_0^t \Big(\frac{\partial g}{\partial x}(x_\varepsilon(s)), a\Big(\frac{s}{\varepsilon}, x_\varepsilon(s)\Big)\Big)\, ds = g\big(x_\varepsilon(t)\big) - g(x_0)$$

we can write

$$g\big(x_\varepsilon(t)\big) = g(x_0) + \int_0^t \Big(\bar{a}(x_\varepsilon(s)), \frac{\partial g}{\partial x}(x_\varepsilon(s))\Big)\, ds + \varepsilon\eta_\varepsilon$$
$$+ \varepsilon G\Big(\frac{t}{\varepsilon}, x_\varepsilon(t)\Big) + \varepsilon\int_0^t \Big(\frac{\partial G}{\partial x}\Big(\frac{s}{\varepsilon}, x_\varepsilon(s)\Big), a\Big(\frac{s}{\varepsilon}, x_\varepsilon(s)\Big)\Big)\, ds. \tag{6.46}$$

Assume that $|x_0| < \delta < 1$ and set

$$\tau_\varepsilon = \inf\{t : |x_\varepsilon(t)| = \delta\}.$$

For $t \le \tau_\varepsilon$ we have

$$\big(Cx_\varepsilon(t), x_\varepsilon(t)\big) = (Cx_0, x_0) + \int_0^t \big([\bar{A}C + C\bar{A}^*]x_\varepsilon(s), x_\varepsilon(s)\big)\, ds$$
$$+ \int_0^t \big[\big(C\bar{\alpha}(x_\varepsilon(s)), x_\varepsilon(s)\big) + \big(\bar{\alpha}(x_\varepsilon(s)), Cx_\varepsilon(s)\big)\big]\, ds$$
$$+ \varepsilon\int_0^t \Big(\frac{\partial G}{\partial x}\Big(\frac{s}{\varepsilon}, x_\varepsilon(s)\Big), a\Big(\frac{s}{\varepsilon}, x_\varepsilon(s)\Big)\, ds + \varepsilon G\Big(\frac{t}{\varepsilon}, x_\varepsilon(t)\Big) + \varepsilon\eta_\varepsilon(t), \tag{6.47}$$

where $\bar{\alpha}(x) = \bar{a}(x) - \bar{A}x$, $|\bar{\alpha}(x)| \le c|x|^2$.
Using condition (a4) we obtain the inequality

$$\big(Cx_\varepsilon(t), x_\varepsilon(t)\big) \le (Cx_0, x_0) + \int_0^t \big([\bar{A}C + C\bar{A}^*]x_\varepsilon(s), x_\varepsilon(s)\big)\, ds$$

$$+2\|C\|c\delta\int_0^t |x_\varepsilon(s)|^2\, ds + \varepsilon c_1\int_0^t |x_\varepsilon(s)|^2 + \varepsilon c_1\|x_\varepsilon(t)\|^2 + \varepsilon\eta_\varepsilon(t)$$

if $t < \tau_\varepsilon$. There exist $\lambda > 0$ for which $\big((\bar{A}C + C\bar{A}^*)x, x\big) \le -\lambda(x,x)$ and $\mu > 0$ for which $(Cx, x) \ge \mu(x,x)$. So

$$(\mu - \varepsilon c_1)\big(x_\varepsilon(t), x_\varepsilon(t)\big) \le (Cx_0, x_0) - (\lambda - \varepsilon c_1 - 2\|C\|c\delta)\int_0^t |x_\varepsilon(s)|^2\, ds + \varepsilon\eta_\varepsilon(t)$$

for $t \le \tau_\varepsilon$.

Let $\varepsilon_0 > 0$ and $\delta > 0$ satisfy the inequalities

$$c_1\varepsilon_0 \le \frac{1}{2}\mu, \quad c_1\varepsilon_0 + 2\|C\|c\delta < \frac{1}{2}\lambda.$$

Then for $\varepsilon \le \varepsilon_0$ and $t \le \tau_\varepsilon$,

$$\frac{\mu}{2}|x_\varepsilon(t)|^2 \le (Cx_0, x_0) - \frac{\lambda}{2}\int_0^t |x_\varepsilon(s)|^2\, ds + \varepsilon\eta_\varepsilon(t). \tag{6.48}$$

The proof of Theorem 5 follows from this relation in the same way as the proof of Theorem 2 followed from relation (6.18).

□

Let conditions (a1), (a2), (a3), (a4) be fulfilled and let the averaged equation have a Liapunov function g at the point $x = 0$ (see Remark 3). We will show that under some additional assumptions the solution $x_\varepsilon(t)$ of the perturbed equation is stable at the point $x = 0$ with probability 1; i.e., the statement of Theorem 5 is true.

We use the functions $g(0, x)$, $G(0, x)$, and $G(t, x)$ that were introduced in the proof of the theorem, but now using for g the Liapunov function for the averaged equation. Then relation (6.46) is true, where $\eta_\varepsilon(t)$ is a martingale.

Remark 5 *Assume that there exist $\delta > 0$ and a constant $c > 0$ for which*

$$|G(0,x)| \le cg(x), \qquad \left|\left(\frac{\partial G}{\partial x}(0,x), a(0,x)\right)\right| \le c\,|(g_x(x), \bar{a}(x))|$$

for $|x| \le \delta$. Then the statement of Theorem 5 is valid.

To prove this, we note that for $t \le \tau_\varepsilon$, where τ_ε was introduced in the proof of Theorem 5, we have

$$g\big(x_\varepsilon(t)\big) \le g(x_0) + \int_0^t h\big(x_\varepsilon(s)\big)\, ds + \varepsilon c g\big(x_\varepsilon(t)\big) + \varepsilon\int_0^t c\big|h\big(x_\varepsilon(s)\big)\big|\, ds + \varepsilon\eta_\varepsilon(t),$$

where $h(x) = -\big(\bar{a}(x), g(x)\big) > 0$.

So

$$(1 - c\varepsilon)g\big(x_\varepsilon(t)\big) \le g(x_0) - (1 - c\varepsilon)\int_0^t h\big(x_\varepsilon(s)\big)\, ds + \varepsilon\eta_\varepsilon(t).$$

Let $c\varepsilon < 1$. Then

$$Eg\big(x_\varepsilon(\tau_\varepsilon)\big) \le \frac{g(x_0)}{1 - c\varepsilon}.$$

Let $\inf\{g(x) : |x| = \delta\} = m > 0$. Then

$$mP\{\tau_\varepsilon < \infty\} \le \frac{g(x_0)}{1 - c\varepsilon}, \qquad \lim_{x_0 \to 0} P\{\tau_\varepsilon < \infty\} = 0. \tag{6.49}$$

Set

$$f(x) = (1 - c\varepsilon)g(x) \wedge h(x).$$

Since the function $f(x)$ is strictly positive for $0 < |x| \le \delta$,

$$f\big(x_\varepsilon(t\wedge\tau_\varepsilon)\big) \le g(x_0) - \int_0^{t\wedge\tau_\varepsilon} f\big(x_\varepsilon(s)\big)\,ds + \varepsilon\eta_\varepsilon(t\wedge\tau_\varepsilon)$$

and

$$E\int_0^{\tau_\varepsilon} f\big(x_\varepsilon(s)\big)\,ds \le g(x_0). \tag{6.50}$$

The relations (6.49) and (6.50) imply the statement of Remark 5.

Example 3 Stationary Gradient System
We consider a gradient system that is perturbed by stationary noise: Say,

$$\dot{x}_\varepsilon(t) = -\nabla F\big(t, x_\varepsilon(t)\big), \tag{6.51}$$

where $F(t,x)$ is a stationary random field in t that has first and second derivatives with respect to the components of x. Let $a(t,x) = -\nabla F(t,x)$, and suppose that this function satisfies conditions (a1), (a2), (a3), (a4). Set $\bar{F}(x) = EF(t,x)$. Then the averaged equation

$$\dot{\bar{x}}(t) = -\nabla\bar{F}\big(\bar{x}(t)\big)$$

has the Liapunov function $\bar{F}(x) - \bar{F}(0)$ at the point 0, if 0 is a local minimum for $\bar{F}(x)$. In this case

$$h(x) = -\Big(\nabla\bar{F}(x), -\nabla\bar{F}(x)\Big) = \big|\nabla\bar{F}(x)\big|^2.$$

We have assumed that the second derivatives $\nabla\nabla^T F(0,x)$ are defined. So,

$$\begin{aligned} G(0,x) &= -\big(\nabla\bar{F}, R\big(\nabla F(0,x) - \nabla\bar{F}(x)\big)\big), \\ \big(\nabla G(0,x), a(0,x)\big) &= \big(\nabla\nabla^T\bar{F}(x)\nabla F(0,x), R\big(\nabla F(0,x) - \nabla\bar{F}(x)\big)\big) \\ &\quad + \big(R\big(\nabla\nabla^T F(0,x) - \nabla\nabla^T\bar{F}(x)\big)\nabla\bar{F}(x), \nabla F(0,x)\big). \end{aligned}$$

Suppose that the following conditions are fulfilled for $|x| < \delta$:

$$\big|\nabla\bar{F}(x)\big|^2 \le c\big|\bar{F}(x) - F(0)\big)\big|^2$$

and

$$\begin{aligned} &|\nabla F(0,x)|^2 + \big|R(\nabla F(0,x) - \nabla\bar{F}(x))\big| + \big\|R(\nabla\nabla^T F(0,x) - \nabla\nabla^T\bar{F}(x))\big\| \\ &\quad \le c|\nabla\bar{F}(x)|^2, \end{aligned}$$

where $\delta > 0$ and $c > 0$ are some constants.
Under these assumptions all the conditions of Remark 4 are fulfilled, so

$$\lim_{x_0\to\bar{x}} P\left\{\limsup_{t\to\infty} |x_\varepsilon(t)| = 0\right\} = 1. \tag{6.52}$$

6.2 Stochastic Resonance for Gradient Systems

Let Y be a compact space, and let $y(t)$, for $t \in R_+$, be a homogeneous Markov process in Y satisfying all conditions of Section 2.4. In particular, ρ is the ergodic distribution of the process, and the function $I(m)$ is defined by formula (2.10). We consider the stochastic process $x_\varepsilon(t)$ in R^d satisfying the differential equation

$$\dot{x}_\varepsilon(t) = \nabla F(x_\varepsilon(t), y(t/\varepsilon)), \tag{6.53}$$

with initial condition $x_\varepsilon(0) = x_0$, where ε is a positive number and $F : R^d \times Y \to R$ is a continuous function with continuous derivatives in x of the first and the second order. We use in this section the notations of vector field theory; in particular, by ∇F we denote the gradient of the function F with respect to the components of x; i.e., $\nabla F = F_x$. We will investigate the asymptotic behavior of the function $x_\varepsilon(t)$ as $\varepsilon \to 0$ and $t \to \infty$.
Set

$$\bar{F}(x) = \int F(x,y)\,\rho(dy).$$

The equation

$$\dot{\bar{x}}(t) = \nabla \bar{F}(\bar{x}(t)) \tag{6.54}$$

is the averaged equation for equation (6.53). We assume that the function $\bar{F}$ satisfies the conditions

(1) $\bar{F}(x) \to \infty$ as $|x| \to \infty$, and $(x, \nabla \bar{F}(x)) > 0$ if $|x|$ is sufficiently large,

(2) the set $\{x : \nabla \bar{F}(x) = 0\}$ is finite.

Note that a point $\bar{x} \in R^d$ is a stable static state for the averaged system if and only if it is a local minimum of the function $\bar{F}$.

6.2.1 *Large Deviations near a Stable Static State*

Assume that $\bar{x}$ is a local minimum of the function $\bar{F}$, and

$$\iint (\nabla F(\bar{x}, y), \nabla F(\bar{x}, y') R(y, dy') \rho(dy) > 0,$$

where the function

$$R(y, B) \qquad \text{for} \quad y \in Y, \quad B \in \mathcal{B}(Y)$$

is defined in Section 2.3. It follows from Theorem 2 of Chapter 4 (equation (4.19)) that with the initial condition $x_0 = \bar{x}$ the stochastic process

$$\frac{x_\varepsilon(t) - \bar{x}}{\sqrt{\varepsilon}}$$

converges weakly in $C_{[0,T]}$ to a Gaussian stochastic process with independent increments, $z(t)$, in R^d for all $t > 0$. The distribution of the random variable $z(t)$ is determined by the formula

$$E(z(t), v)^2 = 2t \iint (v, \nabla F(\bar{x}, y))(v, \nabla F(\bar{x}, y') R(y, dy')\, \rho(dy). \tag{6.55}$$

So $\bar{x}$ cannot be a static point for equation (6.53) if ε is small enough.
We will consider gradient systems with additive noise; that is, the function $F(x, y)$ in equation (6.53) has the form

(GSAN)

$$F(x, y) = \bar{F}(x) + (a(x), b(y)),$$

where the functions $\bar{F} : R^d \to R$ and $a : R^d \to R^d$ are continuous and have continuous bounded derivatives of the first and the second order, and the Jacobian matrix $B(x) = a_x(x)$ is invertible for all $x \in R^d$. The function $b : Y \to R^d$ is continuous and satisfies the relation $\int b(y)\rho(dy) = 0$. As before, the notation (a, b) is the usual dot product of the two vectors.
We will use the notation of Section 3.4.3 for the set $V \subset R^d$ and the functions $I^*(v)$, $S(x, v)$, and $H_T(f)$. Using Theorem 10 of Chapter 3 we can prove the following lemma.

Lemma 11 *Let $x \neq \bar{x}$. Set*

$$\tau(x_0, \bar{x}, \varepsilon, \delta) = \inf\{t : |x_\varepsilon(t) - x| \le \delta\},$$

where $x_0 = x_\varepsilon(0)$, $\delta > 0$. We set $\inf\{\emptyset\} = +\infty$. Let

$$G_T(x_0, x, \delta) \\ = \inf\{H_u(f) : f \in C_{[0,T]}(R^d), f(0) = x_0, |f(u) - x| \le \delta, u \in [0, T]\}.$$

Then we have that

$$\lim_{\varepsilon \to 0} \varepsilon \log P\{\tau(x_0, x, \varepsilon, \delta) \le T\} = -G_T(x_0, x, \delta).$$

Introduce the functions

$$G_T(x_0, x) = \inf_{f,u}\{H_u(f) : f \in C_{[0,T]}(R^d), f(0) = x(0), f(u) = x, u \in [0, T]\}$$

and

$$G(x_0, x) = \lim_{T \to \infty} G_T(x_0, x).$$

The existence of the last limit follows from the inequality $G_T(x_0, x) \ge G_{T'}(x_0, x)$ for $T' > T$. It is easy to see that

$$\lim_{\delta \to 0} G_T(x_0, x, \delta) = G_T(x_0, x),$$

and that the function $G(x, x')$ satisfies the inequality

$$G(x_1, x_2) \le G(x_1, x) + G(x, x_2), \quad x_1, x_2, x \in R^d. \tag{6.56}$$

Set

$$\mathcal{RA}(\bar{x}) = \left\{x_0 \in R^d : \lim_{t\to\infty} \bar{x}(t) = \bar{x}\right\},$$

where $\bar{x}(t)$ is the solution to equation (6.54) satisfying the initial condition $\bar{x}(0) = x_0$. We need an additional assumption about the function $I^*(\cdot)$:

(PI) The relation $I^*(v) = 0$ holds only for $v = 0$.

Remark Let condition **(PI)** be satisfied. Set

$$\theta(s) = \inf\{I^*(v) : |v| = s\}, \quad s \in [0,b], \quad b = \sup\{|v| : v \in V\}.$$

Then $\theta(s)$ is a continuous increasing function, and it is positive for $s > 0$. Introduce the function $\theta^*(s)$ by the formula

$$\inf\{\alpha\theta(u_1) + (1-\alpha)\theta(u_2) : u_1, u_2 \in [0,b], \alpha \in [0,1], \alpha u_1 + (1-\alpha)u_2 = s\},$$

for $s \in [0,b]$. Then $\theta^*(s)$ is a convex positive function for $s > 0$ satisfying the inequality $\theta^*(s) \le \theta(s)$, for $s \in [0,b]$.

Lemma 12 *(i) For all $x_0 \in \mathcal{RA}(\bar{x})$ the equation $G(x_0, x) = G(\bar{x}, x)$ holds. (ii) $G(\bar{x}, x) > 0$ if $x \ne \bar{x}$.*

Proof Let $\bar{x}(t)$ be the solution of equation (6.54) satisfying the initial condition $\bar{x}(0) = x_0$. Then

$$G_{T+t}(x_0, x) \le G_T(\bar{x}(t), x),$$

so

$$G(x_0, x) \le G(\bar{x}(t), x).$$

It is easy to see that the function $G(x, x')$ is continuous, so

$$G(x_0, x) \le \lim_{t\to\infty} G(\bar{x}(t), x) = G(x_0, \bar{x}).$$

To prove statement (ii) we use the formula

$$H_T(f) = \int_0^T I^*(B^{-1}(f(t))(f'(t) + \nabla\bar{F}(f(t))))\,dt, \tag{6.57}$$

which follows from condition **(GSAN)**, equation (6.53), and Remark 2 of Chapter 3. We estimate $H_u(f)$ for a function $f \in C_{[0,u]}(R^d)$ satisfying the relations

$$f(0) = \bar{x},\ f(u) = x,\ |x - \bar{x}| = \delta,\ |f(t) - \bar{x}| \le \delta,\ t \in [0,u],$$

where $\delta > 0$ is small. The functions $B^{-1}(x')$ and $\nabla\bar{F}(x')$ are continuous and have continuous derivatives, so

$$B^{-1}(x') = \bar{B} + O(\delta),\ \nabla\bar{F}(x') = \bar{A}x'(1 + O(\delta))$$

for $|x' - \bar{x}| \le \delta$, where $\bar{B} = B^{-1}(\bar{x})$ is an invertible matrix, and $\bar{A} = \bar{F}_{xx}(\bar{x})$ is a positive matrix, since $\bar{x}$ is a minimum of the function $\bar{F}$. We can write

$$H_u(f) = \int_0^u I^*(\bar{B}(f'(t) + \bar{A}f(t)))\, dt\, (1 + O(\delta)).$$

The remark preceding the statement of the lemma implies the inequality

$$\begin{aligned}\int_0^u I^*(\bar{B}(f'(t) + \bar{A}f(t)))\, dt &\ge \int_0^u \theta^*\left(c|f'(t) + \bar{A}f(t)|\right)\, dt \\ &\ge u\theta^*\left(\frac{c}{u}\int_0^u |f'(t) + \bar{A}f(t)|\, dt\right) \\ &\ge u\theta^*\left(\frac{c_1}{u}\int_0^u |f'(t) + \bar{A}f(t)|^2\right)\, dt,\end{aligned}$$

since the function θ^* is convex. Here c_1 is another constant. Using the methods of the calculus of variations, we can prove that there exists a positive constant $c(\delta, \bar{A})$ depending only on δ and on a positive matrix $\bar{A}$ for which

$$\int_0^1 |f'(t) + \bar{A}f(t)|^2\, dt \ge c(\delta, \bar{A}).$$

As a result, we have the inequality

$$H_u(f) \ge \theta^*(c(\delta, \bar{A})),$$

which completes the proof.

□

Now we investigate the distribution of the stopping times $\tau(\bar{x}, x, \varepsilon, \delta)$ for small δ as $\varepsilon \to 0$ and $T \to \infty$.

Theorem 11 *Set*

$$U_T(\bar{x}, x, \varepsilon, \delta) = P\{\tau(\bar{x}, x, \varepsilon, \delta) < T\}.$$

Let $T(\cdot) : R_+ \to R_+$ be a decreasing function for which

$$\lim_{\varepsilon \to 0} T(\varepsilon) = +\infty.$$

Then
(i) the relation

$$\lim_{\varepsilon \to 0} U_{T(\varepsilon)}(\bar{x}, x, \varepsilon, \delta) = 0$$

holds if

$$\limsup_{\varepsilon \to 0} (T(\varepsilon) U_T(\bar{x}, x, \varepsilon, \delta)) = 0;$$

(ii) the relation

$$\lim_{\varepsilon \to 0} U_{T(\varepsilon)}(\bar{x}, x, \varepsilon, \delta) = 1$$

holds if

$$\liminf_{\varepsilon \to 0}(T(\varepsilon)U_T(\bar{x}, x, \varepsilon, \delta)) = +\infty.$$

The proof of this theorem rests on the following lemma.

Lemma 13 *Let $\{\mathcal{F}_k, k = 1, 2, \dots\}$ be an increasing sequence of σ-algebras, and let $A_k \in \mathcal{F}_k$ be a sequence of subsets satisfying the relations*

$$c_1\, p_k < P\{A_k/\mathcal{F}_{k-1}\} < c_2\, p_k,\ k = 2, \dots,$$

where $0 < c_1 < c_2$ are constants. Then

$$P\left\{\bigcup_k A_k\right\} > 1 - \frac{c_2 \sum_k p_k}{c_1^2(\sum_k p_k - 1)^2}$$

if $\sum_k p_k$ is large enough, and

$$P\left\{\bigcup_k A_k\right\} < c_2 \sum_k p_k.$$

Proof The second statement of the lemma is evident. To prove the first we introduce the random variables $\xi_k = 1_{A_k}$, $k = 1, 2, \dots$. Then we have the relation

$$P\left\{\bigcup_k A_k\right\} = P\left\{\sum_k \xi_k \geq 1\right\}.$$

Using Chebyshev's inequality we can write

$$\begin{aligned} P\left\{\sum_k \xi_k < 1\right\} &= P\left\{\sum_k (\xi_k - E(\xi_k/\mathcal{F}_{k-1})) < -\sum_k (E\xi_k/\mathcal{F}_{k-1}) + 1\right\} \\ &\leq P\left\{|\sum_k (\xi_k - E(\xi_k/\mathcal{F}_k| > c_1 \sum_k p_k - 1\right\} \\ &\leq \frac{c_2 \sum_k p_k}{c_1^2(\sum_k p_k - 1)^2}. \end{aligned}$$

The lemma is proved.

□

For the proof of the theorem we consider the sequence of stopping times τ_k^ε, where $\tau_0^\varepsilon = 0$, and

$$\tau_k^\varepsilon = \inf\{t \geq \tau_{k-1}^\varepsilon + T : |x_\varepsilon(t) - \bar{x}| \leq \delta\},\ k > 0,$$

where $T > 0$ is sufficiently large and $\delta > 0$ is sufficiently small. We set

$$\mathcal{F}_k = \mathcal{F}^\varepsilon_{\tau_k^\varepsilon},$$

where $\mathcal{F}_t^\varepsilon$ is the σ-algebra generated by $y(s)$, $s \leq t/\varepsilon$, and

$$A_k = \{\inf\{|x_\varepsilon(t) - \bar{x}| \leq \delta : t \in [\tau_{k-1}^\varepsilon, \tau_k^\varepsilon]\}\}.$$

Remark Let the function $T(\varepsilon)$ be such that $\lim(\varepsilon \log T(\varepsilon)) = \lambda$. Then the relation

$$\lim_{\delta\to 0}\lim_{\varepsilon\to 0} P\{\tau(\bar{x}, x, \varepsilon, \delta) \le T(\varepsilon)\} = 1_{\{\lambda \ge G(\bar{x},x)\}}$$

holds.

6.2.2 Transitions Between Stable Static States

Denote by $\{\bar{x}_k, k = 1, \dots, m\}$ the set of all stable static states of the unperturbed system. It can be proved that

$$\lim_{\varepsilon\to 0}\limsup_{T\to\infty} E\frac{1}{T}\int_0^T 1_{\{\min_{k\le m}|x_\varepsilon(t)-\bar{x}_k|\ge\delta\}}\,dt = 0.$$

This means that the perturbed systems is almost all of the time in a neighborhood of the set of stable static states, but the system can change from near one stable static state to a neighborhood of another. To describe these changes we need some preliminaries.
Assume that for all $k \le N$ we have

$$G(\bar{x}_k, \bar{x}_i) \ne G(\bar{x}_k, \bar{x}_j)$$

if $i \ne j$. Introduce the function $\iota : \{1, \dots, N\} \to \{1, \dots, N\}$ for which $\iota(k) = i$ if

$$\inf_{j\le N} G(\bar{x}_k, \bar{x}_j) = G(\bar{x}_k, \bar{x}_i).$$

We use the notation

$$\iota^1(\cdot) = \iota(\cdot),\ \ \iota^n(\cdot) = \iota(\iota^{n-1}(\cdot)),\ \ n > 1$$

for iterates of ι.
The following theorem shows that if $\iota(i) = k$, then the solution probably passes from near x_i to near x_k before any other minimum, and it describes the (approximate) distribution of passage times.

Theorem 12 *Let $i \le N$, $\iota(i) = k$, and let $\delta > 0$ satisfy the relation*

$$B_\delta(\bar{x}_l) \in \mathcal{RA}(\bar{x}_l),\ l \le N,$$

where $B_r(\cdot)$ is the ball of the radius r. Then
(i) for all $x \in \mathcal{RA}(\bar{x}_i)$, and all $j \le N$, $j \ne i$, $j \ne k$, we have the relation

$$\lim_{\varepsilon\to 0} P\{\tau(x, \bar{x}_k, \varepsilon, \delta) < \tau(x, \bar{x}_j, \varepsilon, \delta)\} = 1;$$

(ii) for some function $\Lambda_i(\cdot) : R_+ \to R_+$ we have the relation

$$\lim_{\varepsilon\to 0} P\{\tau(x, \bar{x}_k, \varepsilon, \delta) > s\Lambda(\varepsilon)\} = e^{-s},\ s > 0.$$

Proof Statement (i) follows from the inequality $G(\bar{x}_i, \bar{x}_k) < G(\bar{x}_i, \bar{x}_j)$ and the above remark, which implies the relations

$$\lim_{\varepsilon \to 0} P\left\{\tau(x, \bar{x}_k, \varepsilon, \delta) < \exp\left\{\frac{g}{\varepsilon}\right\}\right\} = 1$$

and

$$\lim_{\varepsilon \to 0} P\left\{\tau(x, \bar{x}_j, \varepsilon, \delta) > \exp\left\{\frac{g}{\varepsilon}\right\}\right\} = 1$$

for $g \in (G(\bar{x}_i, \bar{x}_k), G(\bar{x}_i, \bar{x}_j))$.
To prove statement (ii) we note that the random variable $\tau(\bar{x}_i, \bar{x}_k, \varepsilon, \delta)$ has a continuous distribution, so for some positive value $\Lambda_i(\varepsilon)$ the relation

$$P\{\tau(\bar{x}_i, \bar{x}_k, \varepsilon, \delta) > \Lambda_i(\varepsilon)\} = e^{-1}$$

holds. Now the proof follows from the relations

$$\begin{aligned} &P\{\tau(\bar{x}_i, \bar{x}_k, \varepsilon, \delta) > (s+t)\Lambda_i(\varepsilon)\} \\ &= E1_{\{\tau(\bar{x}_i, \bar{x}_k > t\Lambda_i(\varepsilon)\}} P\{\tau(\bar{x}_i, \bar{x}_k, \varepsilon, \delta) > (s+t)\Lambda_i(\varepsilon)/\mathcal{F}^{\varepsilon}_{t\Lambda_i(\varepsilon)}\} \\ &\approx P\{\tau(\bar{x}_i, \bar{x}_k, \varepsilon, \delta) > t\Lambda_i(\varepsilon)\} P\{\tau(\bar{x}_i, \bar{x}_k, \varepsilon, \delta) > \Lambda_i(\varepsilon)\}. \end{aligned}$$

This completes the proof. □

6.2.3 Stochastic Resonance

Now consider a gradient system that depends on some parameter, say α:

$$\dot{x}(t) = \nabla F(\alpha, x(t)),$$

where the function $F(\alpha, x)$ for fixed $\alpha \in R$ satisfies the conditions introduced in the preceding section, and it is a continuous, periodic function of α with period 1. We assume that α is slowly varying in time, say $\alpha(t) = t/T(\varepsilon)$, and we consider this system perturbed by additive noise, i.e., satisfying the additional condition **(GSAN)**, which has the form

$$\dot{x}_\varepsilon(t) = \nabla F(t/T(\varepsilon), x_\varepsilon(t)) + B(x_\varepsilon(t))v(y(t/\varepsilon)).$$

Theorem 13 *Let the function $F(\alpha, x)$ for all α have two minima, $\bar{x}_k(\alpha)$, $k = 1, 2$, and let the functions*

$$g_1(\alpha) = G(\bar{x}_1(\alpha), \bar{x}_2(\alpha)), \quad g_2(\alpha) = G(\bar{x}_2(\alpha), \bar{x}_1(\alpha))$$

satisfy the relations

$$g_1(0) > g_1(\alpha) > g_1\left(\frac{1}{2}\right) < g_1\left(\frac{1}{2} + \alpha\right) < g_1(1)$$

and

$$g_2(0) < g_2(\alpha) < g_2\left(\frac{1}{2}\right) > g_2\left(\frac{1}{2} + \alpha\right) > g_2(1)$$

for $\alpha \in \left(0, \frac{1}{2}\right)$, *and* $g_2(0) = g_1\left(\frac{1}{2}\right) = g$. *Let the function* $T(\varepsilon)$ *satisfy the relation*

$$\lim_{\varepsilon \to 0} \varepsilon \log T(\varepsilon) = g.$$

Assume that the initial conditions satisfy the relation $x_\varepsilon(0) \in \mathcal{RA}(x_1(0))$. *Then, for all* $t > 0$ *we have that*

$$\lim_{\varepsilon \to 0} E \int_0^t |x_\varepsilon(sT(\varepsilon)) - \chi(s)|\, ds = 0,$$

where

$$\chi(\alpha) = \bar{x}_{k(\alpha)}(\alpha), \qquad k(\alpha) = \sum_n (1 + 1_{\{\alpha \in [n-1/2, n)\}}).$$

Functions g satisfying the conditions of this theorem were drawn in Figure 10 in the Introduction. The proof follows from Remark 7.

Remark For every $t > 0$ we have the relation

$$\lim_{\varepsilon \to 0} \frac{1}{T(\varepsilon)} E \int_0^{tT(\varepsilon)} |x_\varepsilon(s) - \chi(s/T(\varepsilon)|\, ds = 0.$$

So the transitions from a neighborhood of a stable static state to a neighborhood of another one occurs at essentially *nonrandom times*! These occur periodically, but with a very large period. This phenomenon illustrates stochastic resonance, in which a pattern of an entire system emerges when noise is applied to it, while in the absence of noise only one mode of the system's dynamics emerges.

6.3 Randomly Perturbed Difference Equations

Consider the initial value problem for a difference equation of the form

$$x_{n+1}^\varepsilon - x_n^\varepsilon = \varepsilon \varphi_{n+1}(x_n^\varepsilon, \omega), \quad x_0^\varepsilon = x_0 \tag{6.58}$$

that was introduced and studied in Section 3.3. Assume that there exists a nonrandom $\hat{x} \in R^d$ for which $\varphi_n(\hat{x}, \omega) = 0$ for all $n \in \mathbf{Z}_+$ and all $\omega \in \Omega$. Then $x_n^\varepsilon = \hat{x}$ is the solution of equation (6.58) when $x_0 = \hat{x}$. This solution is said to be *stable with probability* 1 if

$$\lim_{\varepsilon \to 0} \lim_{x_0 \to \hat{x}} P\left\{ \lim_{n \to \infty} x_n^\varepsilon = \hat{x} \right\} = 1. \tag{6.59}$$

If the conditions of the averaging theorem (see Theorem 8 of Chapter 3) are satisfied, we can consider the differential equation

$$\dot{\bar{x}}(t) = \bar{\varphi}(\bar{x}(t)), \tag{6.60}$$

where $\bar{\varphi}(x) = E\varphi_n(x, \omega)$. Since $\bar{\varphi}(\hat{x}) = 0$, the point $\hat{x}$ is a static state for equation (6.60). The main problem we consider here is the following: Let

$\hat{x}$ be an asymptotically stable state for equation (6.60) (this means that $\lim_{t\to\infty} \bar{x}(t) = \hat{x}$ if $|x_0 - \hat{x}|$ is small enough). Under what additional conditions will $\hat{x}$ be stable with probability 1 for the solution of equation (6.58)? We investigate this question in a series of steps of increasing complexity.

6.3.1 Markov Perturbation of a Linear Equation

Let $\{y_n^*,\, n \in \mathbf{Z}_+\}$ be an ergodic Markov chain in a measurable space $(Y, \mathcal{C})$ with transition probability matrix for one step $P(y, C)$ and an ergodic distribution $\rho(dy)$. Let $A(\cdot) : Y \to L(R^d)$ be a measurable function for which $\int \|A(y)\|\, \rho(dy) < \infty$. We consider the solution x_n^ε to the linear difference equation

$$x_{n+1}^\varepsilon - x_n^\varepsilon = \varepsilon A(y_n^*)\, x_n^\varepsilon, \quad x_0^\varepsilon = x_0. \tag{6.61}$$

Set

$$\bar{A} = \int A(y)\, \rho(dy).$$

In this case $\hat{x} = 0$, and the solution $\bar{x} \equiv 0$ of the averaged equation

$$\dot{\bar{x}}(t) = \bar{A}\, \bar{x}(t) \tag{6.62}$$

is asymptotically stable if and only if there exists a positive symmetric matrix $C \in L(R^d)$ for which $C\bar{A} + \bar{A}^* C < 0$ (i.e., $\bar{A}$ is a negative symmetric matrix). Then

$$\left\| e^{t\bar{A}} \right\| \le b e^{-\alpha t}$$

for some positive constants $b > 0$ and $\alpha > 0$.

Theorem 6 *Assume that*

a) $\|A(y)\| \le b_1$ *for* $y \in Y$ *where* $b_1 > 0$ *is a constant,*

b) $\|\exp\{t\bar{A}\}\| \to 0$ *as* $t \to \infty$,

c) *The Markov chain* $\{y_n^*\}$ *is uniformly ergodic.*

Then there exist $\delta > 0$ *and* $\varepsilon_0 > 0$ *for which*

$$P\Big\{ \lim_{n\to\infty} |x_n^\varepsilon| e^{\varepsilon n \delta} = 0 \Big\} = 1$$

for all $\varepsilon < \varepsilon_0$.

Proof It follows from Theorem 8 in Chapter 3 that for any $T > 0$ and $x_0 \in R^d$,

$$\lim_{\varepsilon\to 0} \sup_y E \sup_{n\varepsilon \le T} \left| x_n^\varepsilon - e^{n\varepsilon\bar{A}} x_0 \right| = 0.$$

Since $x_n^\varepsilon = (I + \varepsilon A(y_{n-1}^*)) \cdots (I + \varepsilon A(y_0^*)) x_0$,

$$\lim_{\varepsilon\to 0} \sup_{|x_0|\le 1} \sup_y E \sup_{n\varepsilon\le T} \left| x_n^\varepsilon - e^{n\varepsilon\bar{A}} x_0 \right| = 0.$$

Let T satisfy the condition

$$\|e^{t\bar{A}}\| \le \frac{1}{2} \quad \text{for} \quad t \ge \frac{T}{2}.$$

There exists $\varepsilon_0 > 0$ for which

$$\sup_{|x_0|\le 1} \sup_y E \sup_{n\varepsilon \le T} \left| x_n^\varepsilon - e^{n\varepsilon \bar{A}} x_0 \right| \le \frac{1}{4} \tag{6.63}$$

if $\varepsilon \le \varepsilon_0$. Then

$$\sup_y E|x_n^\varepsilon| \le \frac{3}{4}|x_0| \quad \text{if} \quad \frac{T}{2} \le n\varepsilon \le T.$$

It is easy to see that

$$E\big(|x_{m+n}^\varepsilon|/\mathcal{F}_m\big) \le \frac{3}{4}|x_m^\varepsilon| \quad \text{if} \quad \frac{T}{2} \le n\varepsilon \le T$$

for any m, where $\mathcal{F}_m$ is the σ-algebra generated by $\{y_1^*, \ldots, y_m^*\}$. Therefore, for $l \le n$ and any k,

$$E\big(|x_{l+kn}^\varepsilon|/\mathcal{F}_l\big) \le \left(\frac{3}{4}\right)^k |x_l^\varepsilon|.$$

Note that for $n_1 < n_2$,

$$|x_{n_2}^\varepsilon| \le \prod_{k=n_1}^{n_2-1} \big\|I + \varepsilon A(y_k^*)\big\| \cdot |x_{n_1}^\varepsilon| \le \exp\{(n_2 - n_1)\varepsilon b_1\} \cdot |x_{n_1}^\varepsilon|.$$

So we have for $\alpha > 0$,

$$P\big\{\sup_{kn \le m \le (k+1)n} e^{\varepsilon m \alpha}|x_m^\varepsilon| \ge 1\big\} \le P\big\{|x_{kn}^\varepsilon| \ge e^{-\varepsilon(k+1)n\alpha} e^{-\varepsilon n b_1}\big\}$$

$$\le P\left\{|x_{kn}^\varepsilon| \ge e^{-(k+1)\alpha T} e^{-T b_1}\right\} \le \left(\frac{3}{4}\right)^k e^{k\alpha T} e^{T(\delta + b_1)}.$$

If $\log \frac{3}{4} + \alpha T < 0$, then

$$\sum P\left\{\sup_{kn \le m \le (k+1)n} e^{\varepsilon m \alpha}|x_m^\varepsilon| \ge 1\right\} < \infty$$

and

$$P\left\{\limsup_{m\to\infty} e^{\varepsilon m \alpha}|x_n^\varepsilon| \le 1\right\} = 1.$$

This shows that the conclusions of the theorem hold for any $\delta < \alpha$.

□

Remark 6 *Assume that conditions* (a) *and* (c) *of Theorem 6 hold, and for any* $c > 0$ *and for any* $x_0 \in R^d$,

$$\lim_{\varepsilon \to 0,\, t \to \infty} \sup_{\varepsilon n \geq t} P\{|x_n^\varepsilon| > c\} = 0. \tag{6.64}$$

Then

$$\lim_{t \to \infty} \big\|\exp\{t\bar{A}\}\big\| = 0. \tag{6.65}$$

To prove this statement we note that relation (6.64) implies the following: For any $c > 0$,

$$\lim_{\varepsilon \to 0,\, t \to \infty} \sup_{\varepsilon n \geq t} \sup_{|x_0| \leq 1} P\{|x_n^\varepsilon| > c|x_0|\} = 0, \tag{6.66}$$

which is a consequence of the fact that x_n^ε is a linear function of x_0. Assume that for $\varepsilon < \varepsilon_0$ and $\varepsilon n \geq \frac{T}{2}$,

$$P\left\{|x_n^\varepsilon| > \frac{1}{4}|x_0|\right\} \leq \frac{1}{4}.$$

Together with relation (6.63) this implies that

$$\big\|\exp\{\varepsilon n\bar{A}\}\big\| \leq \frac{1}{2} \quad \text{if } \frac{T}{2} \leq n\varepsilon \leq T,$$

so

$$\big\|\exp\{t\bar{A}\}\big\| \leq \frac{1}{2} \quad \text{if } t \in \left[\frac{T}{2}, T\right].$$

Thus, relation (6.65) is proved.

6.3.2 Stationary Perturbations of a Linear Equation

Let $\{A_n(\omega),\, n = 0, 1, 2, \dots\}$ be an $L(R^d)$-valued stationary sequence. Consider the difference equation

$$x_{n+1}^\varepsilon - x_n^\varepsilon = \varepsilon A_n(\omega) x_n^\varepsilon, \quad x_0^\varepsilon = x_0. \tag{6.67}$$

The solution of this equation can be represented in the form

$$x_{n+1}^\varepsilon = U_{n+1}^\varepsilon(\omega) x_0, \tag{6.68}$$

where

$$U_0^\varepsilon = I, \quad U_n^\varepsilon(\omega) = \big(I + \varepsilon A_n(\omega)\big)\big(I + \varepsilon A_{n-1}(\omega)\big) \cdots \big(I + \varepsilon A_0(\omega)\big). \tag{6.69}$$

Note that $U_n^\varepsilon(\omega)$ also satisfies the difference equation

$$U_{n+1}^\varepsilon(\omega) - U_n^\varepsilon(\omega) = \varepsilon A_n(\omega) U_n^\varepsilon(\omega). \tag{6.70}$$

Assume that $E\|A_n(\omega)\| < \infty$ and denote by $\bar{U}(t)$ the solution of the averaged equation

$$\dot{\bar{U}}(t) = \bar{A}\bar{U}(t), \quad \bar{U}(0) = I. \tag{6.71}$$

It follows from Theorem 8 of Chapter 3 that for any $T > 0$ we have the relation

$$\lim_{\varepsilon \to 0} P\left\{ \sup_{\varepsilon n \le T} \|U_n^\varepsilon(\omega) - \bar{U}(\varepsilon n)\| > \delta \right\} = 0. \tag{6.72}$$

Theorem 7 *Assume that the following conditions are satisfied:*

(1) *The stationary sequence $\{A_n(\omega)\}$ is ergodic.*

(2) *For all $\lambda > 0$, $E\exp\{\lambda\|A_0(\omega)\|\} < \infty$.*

(3) $\lim_{t\to\infty} \|\bar{U}(t)\| = 0$.

Then there exist $\varepsilon_0 > 0$ and $\delta > 0$ for which

$$P\left\{ \sup_n \|U_n^\varepsilon\| e^{\varepsilon n \delta} < \infty \right\} = 1 \quad \textit{for } \varepsilon < \varepsilon_0. \tag{6.73}$$

Proof Note that for $\alpha > 0$,

$$\begin{aligned} E\|U_n^\varepsilon(\omega)\|^\alpha &\le E\prod_{k=0}^n \exp\{\alpha\varepsilon\|A_k(\omega)\|\} \\ &\le \left(\prod_{k=0}^n E\exp\{(n+1)\alpha\varepsilon\|A_k(\omega)\|\}\right)^{\frac{1}{n+1}} \\ &= E\exp\{(n+1)\alpha\varepsilon\|A_0(\omega)\|\}. \end{aligned} \tag{6.74}$$

(Here we used Hölder's inequality.) So

$$\sup_{\varepsilon(n+1)\le T} E\|U_n^\varepsilon(\omega)\|^\alpha \le E\exp\{\alpha T\|A_0(\omega)\|\}. \tag{6.75}$$

Assume that $\|\bar{U}(t)\| < \frac{1}{2}$. Let $\varepsilon n_\varepsilon \le T$ and $\lim_{\varepsilon\to 0}(\varepsilon n_\varepsilon) = T$. Since

$$\lim_{\varepsilon \to 0} E\|U_{n_\varepsilon}^\varepsilon(\omega)\| = \|\bar{U}(T)\| \tag{6.76}$$

because of relations (6.72) and (6.75), there exists ε_0 for which $E\|U_{n_\varepsilon}^\varepsilon(\omega)\| \le \frac{1}{2}$ for $\varepsilon \le \varepsilon_0$.

First we show that

$$P\left\{ \lim_{k\to\infty} \|U_{kn_\varepsilon}^\varepsilon(\omega)\| e^{kn_\varepsilon\varepsilon\delta_1} = 0 \right\} = 1 \tag{6.77}$$

if $\delta_1 < T^{-1}\log 2$. Consider the sequence of random variables

$$\zeta_k^\varepsilon = \|(I + \varepsilon A_{kn_\varepsilon}(\omega))(I + \varepsilon A_{kn_\varepsilon - 1}(\omega)) \cdots (I + \varepsilon A_{(k-1)n_\varepsilon + 1}(\omega))\|. \tag{6.78}$$

It is easy to see that $\{\zeta_k^\varepsilon,\ k=1,2,\dots\}$ is an ergodic stationary sequence and $\zeta_1^\varepsilon = \|U_{n_\varepsilon}^\varepsilon(\omega)\|$. Since

$$\|U_{kn_\varepsilon}^\varepsilon(\omega)\| \le \prod_{i=1}^k \zeta_i^\varepsilon = \exp\left\{\sum_{i=1}^k \log\zeta_i^\varepsilon\right\}$$

and

$$\|U_{kn_\varepsilon}^\varepsilon(\omega)\| e^{kn_\varepsilon\varepsilon\delta_1} \le \exp\left\{\sum_{i=1}^k \log\zeta_i^\varepsilon + kT\delta\right\}$$
$$= \exp\left\{k\left(T\delta_1 + \frac{1}{k}\sum_{i=1}^k \log\zeta_i^\varepsilon\right)\right\},$$

it follows that $E\log\zeta_i^\varepsilon \le -\log 2$, so relation (6.77) is a consequence of the ergodic theorem.

It is easy to see that for $0 \le l \le n_\varepsilon$ we can write the inequality

$$\|U_{kn_\varepsilon+l}^\varepsilon(\omega)\| \le \|U_{kn_\varepsilon}^\varepsilon(\omega)\| v_k^\varepsilon, \tag{6.79}$$

where

$$v_k^\varepsilon = \exp\left\{\varepsilon \sum_{i=kn_\varepsilon+1}^{(k+1)n_\varepsilon} \|A_i(\omega)\|\right\},$$

and $\{v_k^\varepsilon,\ k=0,1,2,\dots\}$ is an ergodic stationary sequence. Let $a<1$. Then with probability 1,

$$\lim_{k\to\infty} a^k v_k^\varepsilon \le \lim_{k\to\infty}(ka^k)\frac{1}{k}\sum_{i=1}^k v_i^\varepsilon = 0,$$

which follows from the ergodic theorem. So for any $\delta_2>0$,

$$P\left\{\lim_{k\to\infty} v_k^\varepsilon \exp\{-kn_\varepsilon\varepsilon\delta_2\} = 0\right\} = 1. \tag{6.80}$$

Relations (6.77), (6.80) and inequality (6.79) imply the statement of the theorem with $\delta = \delta_1 - \delta_2$.

□

6.3.3 Markov Perturbations of Nonlinear Equations

Consider now the difference equation

$$x_{n+1}^\varepsilon - x_n^\varepsilon = \varepsilon\varphi(x_n^\varepsilon, y_n^*), \tag{6.81}$$

where $\{y_n^*\}$ is the same Markov chain as in Section 6.3.1. We assume that the function $\varphi: R^d \times Y \to R^d$ is measurable, and in addition that it satisfies the following conditions:

I. There exists $\bar{x} \in R^d$ for which $\varphi(\bar{x}, y) = 0$ for all $y \in Y$.

II. $\varphi(x, y)$ is continuous in x if $|x - \bar{x}|$ is small enough, and there exists the derivative $\varphi_x(\bar{x}, y) = A(y)$, where $\|A(y)\| \le b$ for all $y \in Y$, where $b > 0$ is a constant.

III. For some $c > 0$

$$|\varphi(x, y) - A(y)(x - \bar{x})| \le c|x - \bar{x}|^2.$$

IV. If $\bar{A} = \int A(y)\,\rho(dy)$, there exists a positive matrix $B \in L(R^d)$ for which $(B\bar{A}x, x) + (\bar{A}x, Bx) < 0$ if $|x| > 0$.

Theorem 8 *Assume that conditions* I–IV *are satisfied and that the Markov chain* $\{y_n^*\}$ *satisfies condition* SMC II *of Section 2.3. Then there exists* $\varepsilon_0 > 0$ *for which*

$$\lim_{x \to \bar{x}} P\left\{\lim_{n \to \infty} x_n^\varepsilon = \bar{x}\right\} = 1 \quad \text{if } \varepsilon < \varepsilon_0.$$

Proof Denote by $P_n(y, C)$ the transition probability of the Markov chain for n steps:

$$P(y_n^* \in C / y_0^*) = P_n(y_0^*, C).$$

It follows from condition SMC II that the function

$$R(y, C) = \sum_{n=1}^{\infty} [P_n(y, C) - \rho(C)]$$

is a signed measure for $C \in \mathcal{C}$ with its variation uniformly bounded in $y \in Y$. Set

$$\hat{R}(y, C) = 1_{\{y \in C\}} + R(y, C).$$

Then

$$\int P(y, dy')\hat{R}(y', C) = \hat{R}(y, C) - 1_{\{y \in C\}} + \rho(C).$$

Define the function

$$\begin{aligned} \psi_\varepsilon(x, y) &= (B(x - \bar{x}), x - \bar{x}) \\ &+ \varepsilon \int \{(BA(y')[x - \bar{x}], x - \bar{x}) + (B[x - \bar{x}], A(y')[x - \bar{x}])\}\hat{R}(y, dy'). \end{aligned} \tag{6.82}$$

Using formula (6.81) we can show that

$$\begin{aligned} &\int \psi_\varepsilon(x + \varepsilon\varphi(x, y), y')P(y, dy') - \psi_\varepsilon(x, y) \\ &\quad = \varepsilon\big(\bar{A}[x - \bar{x}], B[x - \bar{x}]\big) + \varepsilon\big(B\bar{A}[x - \bar{x}], x - \bar{x}\big) \\ &\quad + \varepsilon\big(B[x - \bar{x}], \varphi(x, y) - A(y)[x - \bar{x}]\big) \\ &\quad + \varepsilon\big(B\big[\varphi(x, y) - A(y)[x - \bar{x}]\big], x - \bar{x}\big) + O\big(\varepsilon^2|x - \bar{x}|^2\big). \end{aligned} \tag{6.83}$$

It follows from formula (6.82) that for some $c_1 > 0$ we can write the inequality

$$(1-c_1\varepsilon)\big(B(x-\bar{x}),x-\bar{x}\big) \le \psi_\varepsilon(x,y) \le (1+c_1\varepsilon)\big(B(x-\bar{x}),x-\bar{x}\big). \quad (6.84)$$

Using condition III and relation (6.83) we can find $\delta > 0$ for which

$$\int \psi_\varepsilon(x+\varepsilon\varphi(x,y),y')P(y,dy') - \psi_\varepsilon(x,y) \le -\varepsilon\big(\bar{A}[x-\bar{x}],B[x-\bar{x}]\big) \quad (6.85)$$

if $|x-\bar{x}| \le \delta$.
So taking into account this inequality (6.83) we have the inequality

$$\int \psi_\varepsilon(x+\varepsilon\varphi(x,y),y')P(y,dy') - \psi_\varepsilon(x,y) \le -c_2\varepsilon\psi_\varepsilon(x,y) \quad (6.86)$$

for $|x-\bar{x}| \le \delta$, where $c_2 > 0$ is a constant.
We consider the sequence of random variables

$$\eta_n^\varepsilon = \psi_\varepsilon(x_n^\varepsilon, y_n^*).$$

If $\mathcal{F}_n$ is the σ-algebra generated by $y_0^*,\dots,y_n^*$ then

$$E(\eta_{n+1}^\varepsilon/\mathcal{F}_n) \le (1-c_2\varepsilon)\eta_n^\varepsilon \quad \text{if } |x_n^\varepsilon - \bar{x}| \le \delta, \quad (6.87)$$

which is a consequence of relation (6.86). Set

$$\zeta_n = 1_{\{\sup_{k\le n}|x_k^\varepsilon - \bar{x}| \le \delta\}}. \quad (6.88)$$

Then $\zeta_{n+1} \le \zeta_n$, and ζ_n is an $\mathcal{F}_n$-measurable random variable. It follows from (6.87) that

$$E(\eta_{n+1}^\varepsilon\zeta_{n+1}/\mathcal{F}_n) \le (1-c_2\varepsilon)\eta_n^\varepsilon\zeta_n. \quad (6.89)$$

So $\eta_n^\varepsilon\zeta_n$ is a nonnegative supermartingale, and there exists $\lim_{n\to\infty}\eta_n^\varepsilon\zeta_n$. It is easy to see that

$$P\Big\{\lim_{n\to\infty}\eta_n^\varepsilon\zeta_n = 0\Big\} = 1.$$

Note that

$$\eta_0^\varepsilon\zeta_0 = \psi_\varepsilon(x_0,y_0^*)\,1_{\{|x_0-\bar{x}|\le\delta\}},$$

so

$$P\Big\{\sup_n \eta_n^\varepsilon\zeta_n > \delta_1\Big\} \le \frac{(1+c_1\varepsilon)\big(B(x_0-\bar{x}),x_0-\bar{x}\big)}{\delta_1}$$

(here we use inequality (6.84)). Assume that

$$\sup_n \eta_n^\varepsilon\zeta_n \le \delta_1$$

and $\nu = \sup\{k : \zeta_k = 1\}$. Then

$$\delta_1 \ge \psi_\nu^\varepsilon(x_\nu^\varepsilon, y_\nu^*) \ge (1-c_1\varepsilon)\big(B(x_\nu^\varepsilon-\bar{x}),x_\nu^\varepsilon-\bar{x}\big) \ge (1-c_1\varepsilon)b_1|x_\nu^\varepsilon-\bar{x}|^2,$$

where $b_1 = \|B^{-1}\|^{-1}$. So

$$|x_{\nu+1}^{\varepsilon} - \bar{x}| \le |x_{\nu}^{\varepsilon} - \bar{x}| + \varepsilon|\varphi(x_{\nu}^{\varepsilon}, y_{\nu}^{*})| \le (1 + \varepsilon b_2)|x_{\nu}^{\varepsilon} - \bar{x}|$$
$$\le (1 + \varepsilon b_2)^{\frac{1}{2}}(1 - c_1\varepsilon)^{-\frac{1}{2}} b_1^{-\frac{1}{2}} \delta_1^{\frac{1}{2}}.$$

If $\varepsilon_0 < \frac{1}{c_1}$, then

$$|x_{\nu+1}^{\varepsilon} - \bar{x}| \le \theta_0 \delta_1^{\frac{1}{2}} \quad \text{for } \varepsilon \le \varepsilon_0,$$

where θ_0 is a constant that does not depend on ε. Assume that $\zeta_0 = 1$ and $\theta_0 \delta_1^{\frac{1}{2}} \le \delta$. Then $\nu = +\infty$. So

$$P\left\{\lim_{n\to\infty} \eta_n^{\varepsilon} = 0\right\} \ge 1 - \frac{(1 + c_1\varepsilon)\big(B(x_0 - \bar{x}), x_0 - \bar{x}\big)}{\delta_1},$$

and

$$\lim_{x_0 \to \bar{x}} P\left\{\lim_{n\to\infty} \eta_n^{\varepsilon} = 0\right\} = 1.$$

It follows from inequality (6.84) that

$$P\left\{\lim_{n\to\infty} \eta_n^{\varepsilon} = 0\right\} = P\left\{\lim_{n\to\infty} x_n^{\varepsilon} = \bar{x}\right\}.$$

□

Remark 7 *Inequality* (6.89) *implies that*

$$(1 - c_2\varepsilon)^{-n} \eta_n^{\varepsilon} \zeta_n$$

is a nonnegative supermartingale. So

$$P\left\{\sup_n \eta_n^{\varepsilon} \zeta_n (1 - c_2\varepsilon)^{-n} > h\right\} \le \frac{(1 + c_1\varepsilon)\big(B(x_0 - \bar{x}), x_0 - \bar{x}\big)}{h}.$$

We have proved that $\lim_{x_0 \to \bar{x}} P\{\nu = +\infty\} = 1$. *Therefore,*

$$\lim_{x_0 \to \bar{x}} P\{\lim_{n\to\infty} |x_n^{\varepsilon} - \bar{x}| e^{\varepsilon \alpha n} = 0\} = 1 \tag{6.90}$$

if $1 < e^{2\varepsilon\alpha} < \frac{1}{1 - \varepsilon c_2}$.

Remark 8 *Let an operator* $\hat{G}_{\varepsilon}$ *acting in the space* $B(R^d \times Y)$ *of bounded measurable functions* $f : R^d \times Y \to R$ *be defined by the formula*

$$\hat{G}_{\varepsilon} f(x, y) = \varepsilon^{-1} \int \big[f(x + \varepsilon\varphi(x, y), y') - f(x, y)\big] P(y, dy'). \tag{6.91}$$

Assume that the Markov chain $\{y_n^*\}$ *satisfies condition* SMC II *of Section 2.3, that* $\varphi(x, y)$ *satisfies condition* I, *and that there exists a bounded measurable function* $F_{\varepsilon}(x, y)$ *for which:*

(a) $F_{\varepsilon}(x, y) \ge 0$, $F_{\varepsilon}(\bar{x}, y) = 0$, $F_{\varepsilon}(x, y) \ge \lambda(x)$, *where* $\lambda(x)$ *is a continuous function with* $\lambda(x) > 0$ *if* $x \ne \bar{x}$ *and* $\lambda(\bar{x}) = 0$.

(b) $\hat{G}_\varepsilon F_\varepsilon(x,y) \le -\lambda(x)$ *if* $|x-\bar{x}| \le \delta$, $y \in Y$, *where* $\delta > 0$ *is a constant.*

Then there exists $\varepsilon_0 > 0$ *for which*

$$\lim_{x_0 \to \bar{x}} P\{\lim x_n^\varepsilon = \bar{x}\} = 1 \quad \text{for } \varepsilon < \varepsilon_0.$$

If in addition we have:

(c) *For some constant* c_3,

$$F_\varepsilon(x,y) \le c_3 \lambda(x).$$

Then

$$\lim_{x_0 \to \bar{x}} P\left\{\lim_{n\to\infty} \lambda(x_n^\varepsilon) e^{\varepsilon \alpha n} = 0\right\} = 1$$

for $\alpha < 1/c_3$.

The proof of these statements is based on the formula

$$E\big(F_\varepsilon(x_{n+1}^\varepsilon, y_{n+1}^*)/\mathcal{F}_n\big) = F_\varepsilon(x_n^\varepsilon, y_n^*) + \varepsilon \hat{G}_\varepsilon F_\varepsilon(x_n^\varepsilon, y_n^*)$$

and can be completed in the same way as the proofs of Theorem 8 and Remark 6.

Remark 9 *Assume that the averaged differential equation for the difference equation* (6.81)*; i.e.,*

$$\dot{\bar{x}}(t) = \bar{\varphi}(\bar{x}(t)),$$

has Liapunov function $f_0(x)$ *at the point* $\bar{x}$ *(see Remark* 2 *in Section 6.1), which satisfies conditions* (α), (β) *of Remark* 2. *Let*

$$F_\varepsilon(x,y) = f_0(x) + \varepsilon f_1(x,y), \tag{6.92}$$

where

$$f_1(x,y) = \int \hat{R}(y,dy')\Big(\varphi(x,y'), \frac{df_0}{dx}(x)\Big). \tag{6.93}$$

Then

$$\hat{G}_\varepsilon F_\varepsilon(x,y) = \frac{1}{\varepsilon}\left[f_0(x+\varepsilon\varphi(x,y)) - f_0(x) - \Big(\frac{df_0}{dx}(x), \varphi(x,y)\Big)\right]$$

$$+\Big(\bar{\varphi}(x), \frac{df_0}{dx}(x)\Big) + \int \big[f_1(x+\varepsilon\varphi(x,y),y') - f_1(x,y')\big] P(y,dy'). \tag{6.94}$$

Using this representation for $F_\varepsilon(x,y)$ *and* $\hat{G}_\varepsilon F_\varepsilon(x,y)$, *we can check whether conditions (a), (b), (c) of Remark 8 are fulfilled.*

6.3.4 Stationary Perturbations of Nonlinear Equations

Now we consider the difference equation (6.58), where

$$\{\varphi_n(x, \omega),\ n = 0, \pm 1, \pm 2, \dots\}$$

is a stationary sequence of random functions; this means that the following condition holds:

(d1) for all $m > 0$ and any $x_1, x_2, \dots, x_m \in R^d$ the sequence

$$\{(\varphi_n(x_1, \omega), \dots, \varphi_n(x_m, \omega)),\ n = 0, 1, 2, \dots\} \tag{6.95}$$

is a stationary sequence in $(R^d)^m$.

We assume that the sequence $\{\varphi_n(x, \omega)\}$ satisfies the following additional conditions:

(d2) The sequence (6.95) is ergodic.

(d3) Denote by $\mathcal{F}_n$ the σ-algebra that is generated by the random variables $\{\varphi_k(x, \omega) : x \in R^d,\ k \le n\}$; introduce on the space L_2^φ of all square integrable $\mathcal{F}_\infty$-measurable random variables, the operator θ for which $\theta(\xi_1 \cdot \xi_2) = \theta\xi_1 \cdot \theta\xi_2$ and $\theta\varphi_k(x, \omega) = \varphi_{k+1}(x, \omega)$. We assume that for any $\mathcal{F}_0$-measurable random variable $\xi \in \mathcal{F}_\infty$ with $E\xi = 0$ the series

$$R\xi = \sum_{k=0}^{\infty} E(\theta^k \xi / \mathcal{F}_0) \tag{6.96}$$

converges in L_2^φ.

(d4) There exists a point $\hat{x} \in R^d$ for which $\varphi_k(\hat{x}, \omega) = 0$ for all $\omega \in \Omega$ and $k \ge 0$, and this point $\hat{x}$ is an asymptotically stable equilibrium for the averaged differential equation

$$\dot{\bar{x}}(t) = \bar{\varphi}(\bar{x}(t)).$$

(d5) There exist derivatives

$$\frac{\partial \varphi_k}{\partial x}(\hat{x}, \omega) = A_k(\omega) \quad \text{and} \quad \frac{\partial \bar{\varphi}}{\partial x}(\hat{x}) = \bar{A}, \quad \text{with} \quad EA_k(\omega) = \bar{A},$$

and there exists a positive operator B for which operator $B\bar{A} + \bar{A}^* B$ is negative.

(d6) There exist $c > 0$ and $\delta > 0$ for which

$$P\{\|A_0(\omega)\| + \|R(A_0(\omega) - \bar{A})\| + \sup_{|x - \hat{x}| \le \delta} \|\varphi_0(x, \omega)\| \le c\} = 1,$$

$$\|A_k(\omega)\| \le c, \quad \|\varphi_k(x, \omega) - A_k(\omega)(x - \hat{x})\| \le c|x - \hat{x}|^2 \quad \text{if } |x - \hat{x}| \le \delta,$$

and for every $\beta > 0$ there exists $\gamma > 0$ for which

$$P\left\{ \sup_{|x - \hat{x}| \le \gamma} \left|\varphi_0(x, \omega) - A_0(\omega)(x - \hat{x})\right| \le \beta |x - \hat{x}|^2 \right\} = 1.$$

Theorem 9 *Let conditions* (d1)–(d6) *be satisfied. Then there exist* $\varepsilon_0 > 0$ *and* $\alpha > 0$ *for which*

$$\lim_{x_0 \to \hat{x}} P\left\{\lim_{n\to\infty} |x_n^\varepsilon - \hat{x}|e^{\varepsilon n \alpha} = 0\right\} = 1, \tag{6.97}$$

for $\varepsilon \le \varepsilon_0$.

Proof We define a stationary $L(R^d)$-valued sequence by the formula

$$D_n(\omega) = \theta^n\big(BR[A_0(\omega) - \bar{A}]\big). \tag{6.98}$$

It follows from the definition of the operator R that for all $m \ge 0$,

$$E(\theta^{m+1}R\xi/\mathcal{F}_m) = \theta^m R\xi - \theta^m \xi, \tag{6.99}$$

for any ξ that is an $\mathcal{F}_0$-measurable random variable from L_2^φ. So

$$E(D_{n+1}(\omega)/\mathcal{F}_n) = D_n(\omega) - B[A_n(\omega) - \bar{A}]. \tag{6.100}$$

Next, we consider the sequence

$$\zeta_n = \big(B(x_n^\varepsilon - \hat{x}), x_n^\varepsilon - \hat{x}\big) + \varepsilon\big(D_n(\omega)(x_n^\varepsilon - \hat{x}), x_n^\varepsilon - \hat{x}\big). \tag{6.101}$$

Note that

$$\begin{aligned} E(\zeta_{n+1}/\mathcal{F}_n) &= \zeta_n + 2\varepsilon\big(B(x_n^\varepsilon - \hat{x}), \varphi_n(x_n^\varepsilon,\omega)\big) + \varepsilon^2\big(\varphi_n(x_n^\varepsilon,\omega), \varphi_n(x_n^\varepsilon,\omega)\big) \\ &\quad + \varepsilon E\big[\big(D_{n+1}(\omega)(x_n^\varepsilon - \hat{x}), x_n^\varepsilon - \hat{x}\big)/\mathcal{F}_n\big] - \varepsilon\big(D_{n+1}(\omega)(x_n^\varepsilon - \hat{x}), x_n^\varepsilon - \hat{x}\big) \\ &= 2\varepsilon\big(B\bar{A}(x_n^\varepsilon - \hat{x}), x_n^\varepsilon - \hat{x}\big) + \varepsilon^2\mu_n|x_n^\varepsilon - \hat{x}|^2 \end{aligned}$$

(here we use relation (6.100) and condition (d6)), where the random variable μ_n is bounded by some nonrandom c_1 if $|x_n^\varepsilon - \hat{x}| < \delta$ (δ was introduced in condition (d6)). The remainder of the proof is the same as in Theorem 8.

□

6.3.5 Small Perturbations of a Stable System

Consider equation (6.81) with the function $\varphi(x,y)$ not satisfying condition I of Section 6.3.3. Rather we assume here that the averaged equation has an asymptotically stable state $\bar{x} \in R^d$. We investigate the asymptotic behavior of x_n^ε for small ε and large n in this case.

Theorem 10 *Assume that the Markov chain* $\{y_n^*\}$ *satisfies condition* SMC II *of Section 2.3 and that the functions* $\varphi(x,y)$ *and* $\bar{\varphi}(x)$ *satisfy the following conditions:*

(1) $\varphi(x,y)$ *is a continuous function of* x *for* $|x - \bar{x}| \le \delta$, *where* $\delta > 0$, *there exists the derivative* $\frac{d\varphi}{dx}(\bar{x},y) = A(y)$ *for which*

$$\sup_y\big(\|A(y)\| + \sup_{|x-\bar{x}|\le\delta} |\varphi(x,y)|\big) < \infty,$$

and there exists $c > 0$ for which

$$|\varphi(x,y) - \varphi(\bar{x},y) - A(y)(x-\bar{x})| \le c|x-\bar{x}|^2 \quad \text{if } |x-\bar{x}| \le \delta.$$

(2) $\bar{\varphi}(x) = \int \varphi(x,y)\rho(dy)$ *is such that* $\bar{\varphi}(\bar{x}) = 0$, *and there exists a positive operator* $C \in L(R^d)$ *for which* $C\bar{A} + \bar{A}^*C$ *is a negative operator, where*

$$\bar{A} = \frac{d\bar{\varphi}}{dx}(x)\bigg|_{x=\bar{x}} = \int A(y)\,\rho(dy).$$

Let $N \in \mathbf{Z}_+$, *and define*

$$z_n^{\varepsilon,N} = \frac{x_{N+n}^{\varepsilon} - \bar{x}}{\sqrt{\varepsilon}}, \quad n \ge -N,$$

and

$$z_N^{\varepsilon}(t) = \sum_n 1_{\{n \ge -N\}} 1_{\{\varepsilon n \le t < \varepsilon(n+1)\}} z_n^{\varepsilon,N}, \quad t \in R.$$

Introduce an R^d*-valued Gaussian stationary process* $\psi(t)$

$$\psi(t) = \int_{-\infty}^{t} \exp\{(t-s)\bar{A}\}\,\hat{\Phi}(ds), \tag{6.102}$$

where $\hat{\Phi}(ds)$ *is an* R^d*-valued Gaussian measure with independent values on* R^d, *and*

$$E\hat{\Phi}(ds) = 0,$$

$$E(z, \hat{\Phi}(ds))^2 = (Bz, z)\,ds,$$

where

$$\begin{aligned}(Bz,z) = &\int \big(\varphi(\bar{x},y), z\big)^2 \rho(dy) \\ &+ 2\int\!\!\int \big(\varphi(\bar{x},y), z\big)\big(\varphi(\bar{x},y'), z\big) R(y,dy')\rho(dy).\end{aligned} \tag{6.103}$$

($R(y, C)$ was introduced in Theorem 8.)

Finally, assume that $\varepsilon N \to \infty$, $N = o\big(\frac{1}{\varepsilon}\log\frac{1}{\varepsilon}\big)$ *as* $\varepsilon \to 0$, *and* $x_0^{\varepsilon} = O(\sqrt{\varepsilon})$. *Then the stochastic process* $z_N^{\varepsilon}(t)$ *converges weakly in* C *to the process* $\psi(t)$ *on any finite interval as* $\varepsilon \to 0$.

Proof Consider the stochastic sequences

$$z_n^{\varepsilon} = \frac{x_n^{\varepsilon} - \bar{x}}{\sqrt{\varepsilon}},$$
$$u_n^{\varepsilon} = \sqrt{\varepsilon}\big[\varphi(\bar{x} + \sqrt{\varepsilon}z_n^{\varepsilon}, y_n^*) - \varphi(\bar{x}, y_n^*) - \sqrt{\varepsilon}A(y_n^*)z_n^{\varepsilon}\big],$$

and

$$\hat{z}_n^{\varepsilon} = U^{\varepsilon}(0,n)z_0^{\varepsilon} + \sqrt{\varepsilon}\sum_{k=1}^{n} U^{\varepsilon}(k,n)\varphi(\bar{x}, y_{k-1}^*),$$

where

$$U^\varepsilon(k,n) = (I+\varepsilon A(y^*_{n-1}))\cdots(I+\varepsilon A(y^*_k)) \quad \text{if } k<n, \qquad U^\varepsilon(n,n) = I. \tag{6.104}$$

Then $\{z^\varepsilon_n\}$ and $\{\hat{z}^\varepsilon_n\}$ satisfy the difference equations

$$z^\varepsilon_{n+1} - z^\varepsilon_n = \varepsilon A(y^*_n) z^\varepsilon_n + \sqrt{\varepsilon}\varphi(\bar{x}, y^*_n) + u^*_n,$$
$$\hat{z}^\varepsilon_{n+1} - \hat{z}^\varepsilon_n = \varepsilon A(y^*_n) \hat{z}^\varepsilon_n + \sqrt{\varepsilon}\varphi(\bar{x}, y^*_n),$$

with the same initial conditions

$$z^\varepsilon_0 = \hat{z}^\varepsilon_0 = \frac{x_0 - \bar{x}}{\sqrt{\varepsilon}}.$$

So

$$(z^\varepsilon_n - \hat{z}^\varepsilon_n) = \sum_{k=1}^n U^\varepsilon(k,n) u^\varepsilon_{k-1}$$

and

$$z^\varepsilon_n = U^\varepsilon(0,n)\frac{x_0 - \bar{x}}{\sqrt{\varepsilon}} + \sqrt{\varepsilon}\sum_{k=1}^n U^\varepsilon(k,n)\varphi(\bar{x}, y^*_{k-1}) + \sum_{k=1}^n U^\varepsilon(k,n) u^\varepsilon_{k-1}. \tag{6.105}$$

First, we investigate the limit behavior of the sequence $\{\hat{z}^\varepsilon_n\}$. Note that the operator $(I+\varepsilon A(y))$ is invertible for $\varepsilon > 0$ small enough, because $\sup_y \|A(y)\| < \infty$. So the operators $U^\varepsilon(0,n)$ and $U^\varepsilon(k,n)$ are invertible. Let

$$v^\varepsilon_n = \big[U^\varepsilon(0,n)\big]^{-1}\hat{z}^\varepsilon_n.$$

It follows from (6.104) and the formula

$$U^\varepsilon(0,n) = U^\varepsilon(k,n)U^\varepsilon(0,k)$$

that v^ε_n can be written as

$$v^\varepsilon_n = z^\varepsilon_0 + \sqrt{\varepsilon}\sum_{k=1}^n \big[U^\varepsilon(0,k)\big]^{-1}\varphi(\bar{x}, y^*_{k-1}). \tag{6.106}$$

Set

$$\hat{v}^\varepsilon_n = z^\varepsilon_0 + \sqrt{\varepsilon}\sum_{k=1}^n \bar{V}^\varepsilon_k \varphi(\bar{x}, y^*_{k-1}), \tag{6.107}$$

where

$$\bar{V}^\varepsilon_k = (I + \varepsilon\bar{A})^{-k}. \tag{6.108}$$

Next, we estimate $E|v^\varepsilon_n - \hat{v}^\varepsilon_n|^2$. Set

$$V^\varepsilon_k = \big[U^\varepsilon(0,k)\big]^{-1}.$$

Then

$$\begin{aligned}
v_n^\varepsilon - \hat{v}_n^\varepsilon &= \sqrt{\varepsilon}\sum_{k=1}^n [V_k^\varepsilon - \bar{V}_k^\varepsilon]\varphi(\bar{x}, y_{k-1}^*) \\
&= \sqrt{\varepsilon}\sum_{k=1}^n [V_{k-1}^\varepsilon - \bar{V}_{k-1}^\varepsilon]\varphi(\bar{x}, y_{k-1}^*) \\
&\quad + O\left(\varepsilon^{\frac{3}{2}}\sum_{k=1}^n (\|\bar{V}_{k-1}^\varepsilon\| + \|V_{k-1}^\varepsilon - \bar{V}_{k-1}^\varepsilon\|)\right).
\end{aligned} \tag{6.109}$$

Using the representation

$$V_n^\varepsilon - \bar{V}_n^\varepsilon = \sum_{k<n} V_k^\varepsilon \Delta_k^\varepsilon \bar{V}_{n-k-1}^\varepsilon = \sum_{k<n} (V_k^\varepsilon - \bar{V}_k^\varepsilon)\Delta_k^\varepsilon \bar{V}_{n-k-1}^\varepsilon + \sum_{k<n} \bar{V}_k^\varepsilon \Delta_k^\varepsilon \bar{V}_{n-k-1}^\varepsilon,$$

where

$$\Delta_k^\varepsilon = (I + \varepsilon A(y_k^*))^{-1} - (I + \varepsilon \bar{A})^{-1},$$

and the estimate

$$\|\bar{V}_k^\varepsilon\| \sim \exp\{k\varepsilon\rho\},$$

where $\rho > 0$ is a constant, we can establish the inequalities

$$E\left\|\sum_{k<n} \bar{V}_k^\varepsilon \Delta_k^\varepsilon \bar{V}_{n-k-1}^\varepsilon\right\|^2 \le c_1\left[\varepsilon^2 \exp\{2\varepsilon\rho n\} n + \varepsilon^4 \exp\{4\varepsilon\rho n\} n^2\right]$$

and

$$E\|V_n^\varepsilon - \bar{V}_n^\varepsilon\|^2 \le c_1\left[\varepsilon^2 \exp\{2\varepsilon\rho n\} n + \varepsilon^4 \exp\{4\varepsilon\rho n\} n^2\right]\left[\sum_{k<n} E\|V_n^\varepsilon - \bar{V}_n^\varepsilon\|^2 + n\right], \tag{6.110}$$

where $c_1 > 0$ is a constant.
It follows from relation (6.110) that

$$E\|V_n^\varepsilon - \bar{V}_n^\varepsilon\|^2 \le c_2 n\varepsilon^2 \exp\{2\varepsilon\rho n\} \tag{6.111}$$

if $n\varepsilon^2 \exp\{2\varepsilon\rho n\}$ is small enough.
From relation (6.107) we can obtain that for some constant c_3 the inequality

$$E|v_n^\varepsilon - \hat{v}_n^\varepsilon|^2 \le c_3\varepsilon\left(\sum_{k<n} E\|V_n^\varepsilon - \bar{V}_n^\varepsilon\|^2 + \varepsilon^3 \sum_{k<n} E\|\bar{V}_n^\varepsilon\|^2\right)$$

is satisfied. So

$$E|v_n^\varepsilon - \hat{v}_n^\varepsilon|^2 = O\big(n^2\varepsilon^3 \exp\{2\varepsilon\rho n\}\big) \tag{6.112}$$

if

$$n^2\varepsilon^3 \exp\{2\varepsilon\rho n\} = o(1). \tag{6.113}$$

Note that in the same way that we established relation (6.111) we can prove the relation

$$E\|U^\varepsilon(0,n) - \bar{U}_n^\varepsilon\|^2 = O(n\varepsilon^2), \quad \bar{U}_n^\varepsilon = (I + \varepsilon\bar{A})^n \tag{6.114}$$

(in this case $\rho = 0$ because $\|\bar{U}_n^\varepsilon\| = O(1)$). We can write the representation for $\hat{z}_n^\varepsilon$ as

$$\hat{z}_n^\varepsilon = U_n^\varepsilon v_n^\varepsilon = \bar{U}_n^\varepsilon \hat{v}_n^\varepsilon + \bar{U}_n^\varepsilon(v_n^\varepsilon - \hat{v}_n^\varepsilon) + [U^\varepsilon(0,n) - \bar{U}_n^\varepsilon]v_n^\varepsilon. \tag{6.115}$$

It follows from formula (6.107) that

$$E|\hat{v}_n^\varepsilon|^2 = O(\exp\{2\varepsilon\rho n\});$$

therefore,

$$\hat{z}_n^\varepsilon = \bar{U}_n^\varepsilon \hat{v}_n^\varepsilon + O\big(\|\bar{U}_n^\varepsilon\||v_n^\varepsilon - \hat{v}_n^\varepsilon| + \|U^\varepsilon(0,n) - \bar{U}^\varepsilon\|(|v_n^\varepsilon| + |v_n^\varepsilon - \hat{v}_n^\varepsilon|)$$

and

$$E|\hat{z}_n^\varepsilon - \bar{U}_n^\varepsilon \hat{v}_n^\varepsilon|^2 = O\big(n^2\varepsilon^3 \exp\{2\varepsilon\rho n\} + \sqrt{n}\varepsilon(n^2\varepsilon^3 + 1)^{\frac{1}{2}} \exp\{\varepsilon\rho n\}\big),$$

if relation (6.113) is fulfilled.
Using the representation

$$z_n^\varepsilon = \hat{z}_n^\varepsilon + \sqrt{\varepsilon} U_n^\varepsilon \sum_{k=1}^n V_k^\varepsilon u_{k-1}^\varepsilon$$

and the inequality

$$|u_{k-1}^\varepsilon| \le O\left(\varepsilon^{\frac{3}{2}} |z_n^\varepsilon|^2\right)$$

we can prove that if relation (6.113) is satisfied, then

$$E|z_n^\varepsilon - \bar{U}_n^\varepsilon \hat{v}_n^\varepsilon|^2 = O\big(n^2\varepsilon^3 \exp\{2\varepsilon\rho n\} + \sqrt{n}\varepsilon(n^2\varepsilon^3 + 1)^{\frac{1}{2}} \exp\{\varepsilon\rho n\}\big). \tag{6.116}$$

Assume that $n = o\big(\frac{1}{\varepsilon}\log\frac{1}{\varepsilon}\big)$. Then it follows from relation (6.116) that

$$E|z_n^\varepsilon - \bar{U}_n^\varepsilon \hat{v}_n^\varepsilon|^2 = o(1), \quad |\bar{U}_n^\varepsilon z_0^\varepsilon| = o(1).$$

Therefore, the stochastic process $z_N^\varepsilon(t)$ coincides asymptotically with the stochastic process

$$\tilde{z}_N^\varepsilon(t) = \sum_{k<N+t\varepsilon^{-1}} (I + \varepsilon\bar{A})^{N-k+\frac{t}{\varepsilon}} \varphi(\bar{x}, y_{k-1}^*). \tag{6.117}$$

This process converges weakly in C to the stochastic process $\psi(t)$, because of Theorem 4 in Section 6.3.2 and the fact that

$$(I + \varepsilon\bar{A})^{N-k+t/\varepsilon} \sim \exp\{t\bar{A} + (N-k)\varepsilon\bar{A}\}.$$

□

Remark 10 *Let τ^ε be a stopping time with respect to the filtration $\{\mathcal{F}_n\}$ for which $x_{\tau^\varepsilon}^\varepsilon = O(\sqrt{\varepsilon})$. Then the theorem can be applied to the sequence*

$x^\varepsilon_{\tau^\varepsilon+n}$. *So the point* $\bar{x}$ *is not a stable point for* x^ε_n, *since* $E|x^\varepsilon_n - \bar{x}|^2 \leq \alpha\varepsilon$ *for* n *large enough and some positive constant* $\alpha > 0$. *The behavior of* x^ε_n *in a neighborhood of* $\bar{x}$ *can be described in the following way: First* x^ε_n *approaches* $\bar{x}$ *according to the averaging theorem until the distance* $|x^\varepsilon_n - \bar{x}|$ *becomes of order* $\sqrt{\varepsilon}$. *After this,* x^ε_n *moves in a neighborhood of* $\bar{x}$ *like the stationary Gaussian process that was introduced in the theorem.*

6.4 Convolution Integral Equations

Consider an integral equation of the form

$$x_\varepsilon(t) = \varphi(t) + \int_0^t M\big(t-s, y\big(\frac{t}{\varepsilon}\big)\big)x_\varepsilon(s)\,ds, \tag{6.118}$$

where $\varphi : R_+ \to R$ and $M : R_+ \times Y \to R$ are measurable, locally bounded functions of t, and $y(t)$ is a homogeneous Markov process in the space $(Y, \mathcal{C})$ satisfying SMC II of Section 2.3.
Set

$$\bar{M}(t) = \int M(t,y)\,\rho(dy)$$

and let $\bar{x}(t)$ be the solution of the average integral equation

$$\bar{x}(t) = \varphi(t) + \int_0^t \bar{M}(t-s)\bar{x}(s)\,ds. \tag{6.119}$$

It follows from Theorem 1 of Chapter 3 that $x_\varepsilon(t) - \bar{x}(t) \to 0$ uniformly in t on any finite interval with probability 1.
In this section we investigate the characteristic value

$$\limsup_{t\to\infty} \frac{1}{t} \log|x_\varepsilon(t)|$$

for small $\varepsilon > 0$.
We need some auxiliary results.

Lemma 4 *The solution of equation (6.119) is represented by the formula*

$$\bar{x}(t) = \varphi(t) + \int_0^t \Gamma(t-s)\varphi(s)\,ds, \tag{6.120}$$

where Γ *is given by the resolvent series*

$$\Gamma(t) = \bar{M}(t) + \sum_{n=1}^\infty \int_{0<s_1<\cdots<s_n<t} \bar{M}(t-s_n)\bar{M}(s_n-s_{n-1})\cdots\bar{M}(s_1)\,ds_1\cdots ds_n. \tag{6.121}$$

Proof Let $B(t) = \sup_{s \le t} |\bar{M}(s)|$. Then

$$\Big| \int_{0 < s_1 < \cdots < s_n < t} \bar{M}(t - s_n)\bar{M}(s_n - s_{n-1}) \cdots \bar{M}(s_1)\, ds_1 \cdots ds_n \Big| \le \frac{B^n(t) t^n}{n!}.$$

So the series on the right-hand side of equality (6.121) is absolutely and uniformly convergent on any finite interval. Therefore,

$$\int_0^t \bar{M}(t - s)\Gamma(s)\, ds = \Gamma(t) - \bar{M}(t). \tag{6.122}$$

As a result,

$$\begin{aligned}
&\int_0^t \bar{x}(u)\Gamma(t - u)\, du \\
&= \int_0^t \varphi(u)\Gamma(t - u)\, du + \int_0^t \Gamma(t - u) \int_0^u \bar{M}(u - s)\bar{x}(s)\, ds\, du \\
&= \int_0^t \varphi(u)\Gamma(t - u)\, du + \int_0^t [\Gamma(t - s) - \bar{M}(t - s)]\bar{x}(s)\, ds
\end{aligned}$$

and

$$\int_0^t \varphi(u)\Gamma(t - u)\, du = \int_0^t \bar{M}(t - u)\bar{x}(u)\, du. \tag{6.123}$$

Relation (6.120) is a consequence of relations (6.123) and (6.119).

□

Lemma 5 *Let $\tilde{M}(t, y) = M(t, y) - \bar{M}(t)$ and*

$$L(t, y) = \tilde{M}(t, y) + \int_0^t \Gamma(t - s)\tilde{M}(s, y)\, ds. \tag{6.124}$$

Then the solution $x_\varepsilon(t)$ of equation (6.118) satisfies the equation

$$x_\varepsilon(t) = \bar{x}(t) + \int_0^t L\left(t - s, y\left(\frac{s}{\varepsilon}\right)\right) x_\varepsilon(s)\, ds. \tag{6.125}$$

Proof It follows from (6.118) and (6.119) that

$$x_\varepsilon(t) - \bar{x}(t) = \int_0^t \tilde{M}\left(t - s, y\left(\frac{s}{\varepsilon}\right)\right) x_\varepsilon(s)\, ds + \int_0^t \bar{M}(t - s)[x_\varepsilon(s) - \bar{x}(s)]\, ds.$$

Therefore, using relation (6.122) we obtain

$$\begin{aligned}
&x_\varepsilon(t) - \bar{x}(t) + \int_0^t \Gamma(t - s)[x_\varepsilon(s) - \bar{x}(s)]\, ds \\
&= \int_0^t L\left(t - s, y\left(\frac{s}{\varepsilon}\right)\right) x_\varepsilon(s)\, ds + \int_0^t \Gamma(t - s)[x_\varepsilon(s) - \bar{x}(s)]\, ds.
\end{aligned} \tag{6.126}$$

□

It is easy to see that

$$\int L(t,y)\rho(dy) = 0 \quad \text{for all } t.$$

Lemma 6

$$x_\varepsilon(t) = \bar{x}(t) + \sum_{n=1}^{\infty} \int \cdots \int_{0<s_1<\cdots<s_n<t} L\left(t - s_n, y\left(\frac{s_n}{\varepsilon}\right)\right) \times \cdots \\ \times L\left(s_2 - s_1, y\left(\frac{s_1}{\varepsilon}\right)\right) \bar{x}(s_1)\, ds_1 \cdots ds_n. \tag{6.127}$$

Proof The local boundedness of the function $L(t,y)$ in t ensures that equation (6.124) has a unique solution. If $B_1(t) = \sup_{y\in Y,\, s\le t} |B(t,y)|$, then

$$\left| \int \cdots \int_{0<s_1<.<s_n<t} L(t - s_n, y(\frac{s_n}{\varepsilon})) \cdots L(s_2 - s_1, y(\frac{s_1}{\varepsilon}))\bar{x}(s_1) ds_1 . ds_n \right| \\ \le \frac{B_1^n(t)}{n!} t^n c_t,$$

where $c_t = \sup_{s\le t} |\bar{x}(s)|$. Therefore, the series on the right-hand side of equality (6.127) is uniformly and absolutely convergent. It is easy to check that the sum on the right-hand side of equality (6.127) satisfies equation (6.124).

□

6.4.1 Laplace Transforms and Their Inverses

Nonperturbed convolution equations can be solved using Laplace transformations (see [200]). We describe this technique for equation (6.119): The Laplace transformations for the functions $\varphi(t)$, $\bar{M}(t)$, and $\bar{x}(t)$ are

$$\hat{\varphi}(p) = \int_0^\infty \varphi(t) e^{-pt}\, dt,$$

$$\hat{M}(p) = \int_0^\infty \bar{M}(t) e^{-pt}\, dt,$$

$$\hat{x}(p) = \int_0^\infty \bar{x}(t) e^{-pt}\, dt,$$

respectively, where the complex number p is the transform variable. We assume that there exists $\alpha \in R_+$ for which

$$|\varphi(t)| + |\bar{M}(t)| = O(e^{\alpha t}). \tag{6.128}$$

Then the functions $\hat{\varphi}(p)$ and $\hat{M}(p)$ are analytic in the half-plane $\{p : \mathrm{Re}\, p > \alpha\}$, where Re p is the real part of the complex number p. Let α_1 satisfy the inequality

$$|\hat{M}(p)| < 1 \quad \text{for } \mathrm{Re}\, p \ge \alpha_1,$$

The existence of such α_1 follows from the relations

$$|\hat{M}(p)| \leq \int_0^\infty |\bar{M}(t)| e^{-(\mathrm{Re}\, p)t}\, dt, \qquad \lim_{\lambda\to+\infty} \int_0^\infty |\bar{M}(t)| e^{-\lambda t}\, dt = 0.$$

Since

$$\int_0^\infty e^{-pt} \int_0^t M(t-s)N(s)\, ds = \int_0^\infty e^{-pt} M(t)\, dt \cdot \int_0^\infty e^{-pt} N(t)\, dt$$

if $\int_0^\infty \left|e^{-pt}M(t)\right| dt < \infty$ and $\int_0^\infty \left|e^{-pt}N(t)\right| dt < \infty$, it follows from (6.121) that

$$\int_0^\infty \Gamma(t) e^{-pt}\, dt = \sum_{k=1}^\infty \hat{M}^k(p) = \frac{\hat{M}(p)}{1-\hat{M}(p)},$$

and so

$$\hat{x}(p) = \hat{\varphi}(p) + \frac{\hat{M}(p)}{1-\hat{M}(p)} \hat{\varphi}(p). \tag{6.129}$$

Thus, for convolution equations, it is easy to solve for the Laplace transform of the unknown x.

Let $f : R_+ \to R$ be a continuous function satisfying the relation $f(t) = O(e^{\alpha t})$ for some $\alpha > 0$. Set $\hat{f}(p) = \int_0^\infty e^{-pt} f(t)\, dt$. Then for all t and $\lambda_0 > \alpha$,

$$\int_0^t f(s)\, ds = \lim_{N\to\infty} \frac{1}{2\pi} \int_{-N}^N \frac{\hat{f}(\lambda_0 + iu)}{\lambda_0 + iu} e^{(\lambda_0 - iu)t}\, du \tag{6.130}$$

(see [200, Chapter 2]).

Remark 11 If for some $\lambda_0 > \alpha$,

$$\int_{-\infty}^\infty \left|\hat{f}(\lambda_0 + iu)\right| du < \infty,$$

then

$$f(t) = \frac{1}{2\pi i} \int_{\mathrm{Re}\, p=\lambda_0} \hat{f}(p) e^{p^* t} dp, \tag{6.131}$$

where $p^* = \lambda_0 - iu$ if $p = \lambda_0 + iu$. (This is Cauchy's formula for inverting the Laplace transform of f.)

Corollary 1 Assume that $f(t)$ is a continuous function from R_+ into R and $f(t) = O(e^{\alpha t})$ for some $\alpha > 0$. Set

$$\Delta_h(t) = t\, 1_{\{t\leq h\}} + (2h - t)\, 1_{\{h\leq t\leq 2h\}}.$$

Then

$$\int_0^t \Delta_h(t-s) f(s)\, ds = \frac{1}{2\pi i} \int_{Re\, p=\lambda_0} \frac{1}{p^2}(1 - e^{-ph})^2 \hat{f}(p) e^{p^* t} dp \tag{6.132}$$

for $\lambda_0 > \alpha$. This formula is a consequence of the formula

$$\hat{\Delta}_h(p) = \int_0^\infty \Delta_h(t)e^{-pt}\,dt = \frac{1}{p^2}(1-e^{-ph})^2.$$

We apply the Laplace transformation to investigate the asymptotic properties of the solution of convolution integral equations as $t \to \infty$.

Lemma 7 *Assume that*

$$\bar{M}(t) \geq 0, \quad \int_0^\infty \bar{M}(t)\,dt = +\infty, \quad \bar{M}(t) + |\varphi(t)| = O(e^{\alpha t}),$$

and for all $\lambda > \alpha$,

$$\int_{-\infty}^\infty |\hat{\varphi}(\lambda + iu)|\,du < \infty.$$

Let there exist $\alpha_0 > \alpha$ *for which* $\hat{M}(\alpha_0) = 1$.
Then there exists $\alpha_1 \in (\alpha, \alpha_0)$ *for which*

$$\bar{x}(t) = c_0 e^{\alpha_0 t} + O(e^{\alpha_1 t}),$$

where

$$c_0 = \alpha_0 \hat{\varphi}(\alpha_0)\Big(\int te^{-\alpha_0 t}\bar{M}(t)\,dt\Big)^{-1}.$$

Proof Note that $|\hat{M}(\lambda + iu)| < \hat{M}(\lambda)$ if $u \neq 0$ and $|\hat{M}(\lambda + iu)| \to 0$ as $|u| \to \infty$ uniformly in λ on any finite closed interval $[\gamma, \delta] \subset (\alpha, \infty)$. We can find $\alpha_1 < \alpha_0$ for which the function $\hat{M}(p) - 1$ has only one zero in the region $\operatorname{Re} p \geq \alpha_1$. We consider the function

$$\frac{\hat{\varphi}(p)}{1 - \hat{M}(p)} - \frac{c_0}{p(p - \alpha_0)} = \hat{f}(p).$$

This function is the Laplace transformation of the function

$$f(t) = \bar{x}(t) - \frac{c_0}{\alpha_0}(e^{\alpha_0 t} - 1).$$

The function $\hat{f}(p)$ is analytic in the region $\operatorname{Re} p \geq \alpha_1$. So

$$f(t) = \frac{1}{2\pi i}\int_{-\infty}^\infty e^{pt}\hat{f}(p)dp,$$

since $\int_{-\infty}^\infty |\hat{f}(p)|dp < \infty$ and

$$|f(t)| \leq \frac{1}{2\pi}e^{\alpha_1 t}\int_{-\infty}^\infty |\hat{f}(\alpha_1 + iu)|\,du.$$

This implies the statement of the lemma.

□

Remark 12 *The same result is true if there exists $l \in \mathbf{Z}_+$ for which*

$$\int_{-\infty}^{\infty} |\hat{M}(\lambda + iu)|^l \, du < \infty \quad \textit{for } \lambda > \alpha.$$

To prove this we introduce functions

$$\bar{M}_1(t) = \bar{M}(t), \quad \bar{M}_k(t) = \int_0^t \bar{M}_{k-1}(t-s)\bar{M}(s)\, ds, \quad k = 2, 3, \ldots,$$
$$\bar{x}_l(t) = \sum_{k=l}^{\infty} \int_0^t \varphi(t-s)\bar{M}_k(s)\, ds.$$

Then

$$\bar{x}(t) = \varphi(t) + \sum_{i=1}^{l-1} \int_0^t \varphi(t-s)\bar{M}_i(s)\, ds + \bar{x}_l(t).$$

The conditions of the lemma imply that as $t \to \infty$,

$$\varphi(t) = O(e^{\alpha t}), \quad \bar{M}_i(t) = O(t^{i-1}e^{\alpha t}),$$

and

$$\bar{x}(t) - \bar{x}_l(t) = O(t^{l-1}e^{\alpha t}).$$

Note that

$$\int_0^{\infty} e^{-pt}\bar{x}_l(t)\, dt = \hat{x}_l(p) = \frac{\left(\hat{M}(p)\right)^l \hat{\varphi}(p)}{1 - \hat{M}(p)}.$$

Since

$$\int_{-\infty}^{\infty} |\hat{M}(\lambda + iu)|^l \cdot |\hat{\varphi}(\lambda + iu)|\, du < \infty,$$

we can apply the same calculation to $\hat{x}_l(p)$ as we used in the proof of Lemma 7.

6.4.2 Laplace Transforms of Noisy Kernels

The Laplace transform of $Ex_\varepsilon(t)$ can be calculated. For this we need some notation. Set

$$
\begin{aligned}
R(t, y, dy') &= P(t, y, dy') - \rho(dy'),\\
\hat{L}(p, y) &= \int_0^\infty e^{-pt} L(t, y)\, dt,\\
\hat{L}_1(\varepsilon, p, y) &= \hat{L}(p, y),\\
\hat{L}_{k+1}(\varepsilon, p, y) &= \int_0^\infty \int R(s, y, dy') L(\varepsilon s, y) L_k(\varepsilon, p, y') e^{-\varepsilon ps}\, ds, \quad k > 0,\\
L_1^*(\varepsilon, p, y) &= \hat{x}(p),\\
L_{k+1}^*(\varepsilon, p, y) &= \int_0^\infty \int \bar{x}(\varepsilon s) R(s, y, dy') L_k(\varepsilon, p, y') e^{-\varepsilon ps}\, ds, \quad k > 0,\\
\hat{L}_k(\varepsilon, p) &= \int \hat{L}_k(\varepsilon, p, y) \rho(dy),\\
\mathcal{L}(\varepsilon, p) &= \sum_{k=1}^\infty \varepsilon^k \hat{L}_{k+1}(\varepsilon, p),\\
\mathcal{L}^*(\varepsilon, p, y) &= \sum_{k=0}^\infty \varepsilon^k L_{k+1}^*(\varepsilon, p, y).
\end{aligned}
$$

Note that $\hat{L}_1(\varepsilon, p) = 0$.

Remark 13 *Assume that the conditions of Lemma 7 are satisfied and the constants c, c_1 satisfy the relations*

$$|\bar{x}(t)| + |M(t, y)| \le c e^{\alpha_0 t}, \qquad \int_0^\infty \mathrm{Var} R(t, y, \cdot)\, dt \le c_1.$$

Then for $\mathrm{Re}\, p > \alpha_0$,

$$|\hat{L}_k(\varepsilon, p)| \le \frac{c}{\mathrm{Re}\, p - \alpha_0} (cc_1)^{k-1},$$

$$|L_k^*(\varepsilon, p)| \le \frac{c}{\mathrm{Re}\, p - \alpha_0} (cc_1)^{k-1}.$$

Therefore, for $\mathrm{Re}\, p > \alpha_0$ *and* $\varepsilon < 1/cc_1$ *the functions* $\mathcal{L}(\varepsilon, p)$ *and* $\mathcal{L}^*(\varepsilon, p)$ *are well-defined and analytic in* p, *and they satisfy the inequalities*

$$|\mathcal{L}(\varepsilon, p)| \le \frac{c}{\mathrm{Re}\, p - \alpha_0} \sum_{k=2}^\infty (\varepsilon cc_1)^{k-1} \le \frac{c}{\mathrm{Re}\, p - \alpha_0} \frac{\varepsilon cc_1}{1 - \varepsilon cc_1},$$

and $|\mathcal{L}(\varepsilon, p)| < 1$ *if, in addition,*

$$\frac{c}{\mathrm{Re}\, p - \alpha_0} \frac{\varepsilon cc_1}{1 - \varepsilon cc_1} < 1.$$

Lemma 8 *Assume that the conditions of Lemma 7 are satisfied. Then for any $\hat{\alpha} > \alpha_0$ there exists a number $\hat{\varepsilon}_0 > 0$ for which*

$$\int_0^\infty e^{-pt} Ex_\varepsilon(t)\,dt = \mathcal{L}^*(\varepsilon, p, y_0)\big(1 - \mathcal{L}(\varepsilon, p)\big)^{-1} \tag{6.133}$$

for $\varepsilon < \hat{\varepsilon}_0$ and $\operatorname{Re} p > \hat{\alpha}$, where y_0 is the initial value of the Markov process $y(t)$: $y_0 = y(0)$.

Proof It follows from formula (6.127) that

$$\begin{aligned} Ex_\varepsilon(t) - \bar{x}(t) &= \sum_{n=1}^\infty \int \cdots \int_{0<s_1<\cdots<s_n<t} L(t-s_n, y_n)\cdots L(s_2 - s_1, y_1) \\ &\quad \times P\Big(\frac{s_1}{\varepsilon}, y, dy_1\Big)\cdots P\Big(\frac{s_n - s_{n-1}}{\varepsilon}, y_{n-1}, dy_n\Big)\bar{x}(s_1)\,ds_1\cdots ds_n \\ &= \sum_{n=1}^\infty \int \cdots \int_{0<s_1<\cdots<s_n<t} L(t-s_n, y_n)\cdots L(s_2 - s_1, y_1)\bar{x}(s_1) \\ &\quad \times \prod_{k=1}^n \bigg(R\Big(\frac{s_k - s_{k-1}}{\varepsilon}, y_{k-1}, dy_k\Big) + \rho(dy_k)\bigg)\,ds_1\cdots ds_n, \end{aligned}$$

where $s_0 = 0$. So

$$\begin{aligned} &\int_0^\infty e^{-pt} Ex_\varepsilon(t)\,dt - \hat{x}(p) = \sum_{n=1}^\infty \int \cdots \int \int_0^\infty \cdots \int_0^\infty e^{-p(u_0+\cdots+u_n)}\bar{x}(u_0) \\ &\times \prod_{k=1}^n \bigg[L(u_k, y_k)\bigg(R\Big(\frac{u_{k-1}}{\varepsilon}, y_{k-1}, dy_k\Big) + \rho(dy_k)\bigg)\bigg]\,du_0\cdots du_n \\ &= \sum_{n=1}^\infty \int \cdots \int \int_0^\infty \cdots \int_0^\infty e^{-p(u_0+\cdots+u_n)} \\ &\times \sum_{0<k_1<\cdots<k_l\le n} N^{(n)}_{k_1,\cdots,k_l}(u_0,\ldots,u_n,y_0,\ldots,y_n,dy_1,\ldots,dy_n)\,du_0\cdots du_n, \end{aligned}$$

where

$$\begin{aligned} &N^{(n)}_{k_1,\ldots,k_l}(u_0,\ldots,u_n,y_0,\ldots,y_n,dy_1,\ldots,dy_n) \\ &= \bar{x}(u_0)\prod_{k=1}^n \bigg[L(u_k, y_k)R\Big(\frac{u_{k-1}}{\varepsilon}, y_{k-1}, dy_k\Big)\bigg] \times \prod_{i=1}^l \frac{\rho(dy_{k_i})}{R(u_{k_i-1}/\varepsilon, y_{k_i-1}, dy_{k_i})}. \end{aligned} \tag{6.134}$$

It is easy to see that

$$\begin{aligned} &\int \cdot \int \int_0^\infty \cdot \int_0^\infty e^{-p(u_0+\cdots+u_n)} N^{(n)}_{k_1,\ldots,k_l}(u_0,\ldots,u_n,y_0,\ldots,y_n,dy_1,\ldots,dy_n) \\ &\qquad \times du_0\cdots du_n = \varepsilon^{n-l} L^*_{k_1}(\varepsilon,p,y_0)\hat{L}_{k_2-k_1}(\varepsilon,p)\cdots\hat{L}_{n-k_l+1}(\varepsilon,p). \end{aligned}$$

Therefore,

$$\int_0^\infty e^{-pt} E x_\varepsilon(t)\, dt = \hat{x}(p)$$

$$+ \sum_{n=1}^{\infty} \sum_{\substack{i_1+i_2+\cdots+i_l=n+1 \\ i_1\ge 1, i_2\ge 1,\ldots,i_l\ge 1}} \big(\varepsilon^{i_1-1} L^*_{i_1}(\varepsilon,p), y_0\big)\big(\varepsilon^{i_2-1}\hat{L}_{i_2}(\varepsilon,p)\big)\cdots\big(\varepsilon^{i_l-1}\hat{L}_{i_l}(\varepsilon,p)\big)$$

$$= L^*(\varepsilon,p,y_0) + \sum_{l=2}^{\infty} L^*(\varepsilon,p,y_0)\big(\mathcal{L}(\varepsilon,p)\big)^{l-1}.$$

□

We will extend the analytic function given by the expression on the right-hand side of equation (6.133). For this we need some additional assumptions.

Lemma 9 *Assume that the conditions of Lemma 7 are satisfied and that there exists $\delta > 0$ for which*

$$|R(t,y,C)| \le c_2 e^{-\delta t}. \tag{6.135}$$

Then there exists a number $\varepsilon_0 > 0$ for which

$$\mathcal{L}(\varepsilon,p,y) = \frac{1}{(p-\alpha_0)p} G(\varepsilon,p,y) + \mathcal{K}(\varepsilon,p,y)$$

for $\varepsilon \le \varepsilon_0$, where $G(\varepsilon,p,y)$ and $\mathcal{K}(\varepsilon,p,y)$ are analytic functions in the region $\mathrm{Re}\, p \ge p_1$ *that can be represented by the formulas*

$$G(\varepsilon,p,y) = \sum_{k=1}^{\infty} \varepsilon^{k-1} G_k(\varepsilon,p,y), \tag{6.136}$$

$$\mathcal{K}(\varepsilon,p,y) = \sum_{k=1}^{\infty} \varepsilon^{k-1} \mathcal{K}_k(\varepsilon,p,y), \tag{6.137}$$

where

$$G_1(\varepsilon,p,y) = G(y),$$
$$G_{r+1}(\varepsilon,p,y) = \int_0^\infty \int R(s,y,dy') L(\varepsilon s,y) G_r(\varepsilon,p,y') e^{-\varepsilon p s}\, ds,$$
$$\mathcal{K}_1(\varepsilon,p,y) = \mathcal{K}(p,y),$$
$$\mathcal{K}_{r+1}(\varepsilon,p,y) = \int_0^\infty \int R(s,y,dy') L(\varepsilon s,y) \mathcal{K}_r(\varepsilon,p,y') e^{-\varepsilon p s}\, ds, \quad r > 0,$$

and

$$G(y) = \alpha_0 \int_0^\infty e^{-\alpha_0 t}[M(t,y) - \bar{M}(t)]\, dt \Big/ \int_0^\infty t e^{-\alpha_0 t} \bar{M}(t)\, dt,$$

$$\mathcal{K}(p,y) = L(p,y) - \frac{G(y)}{(p-\alpha_0)p}.$$

Proof $G(y)$ is a bounded function and

$$\mathcal{K}(p,y) = \frac{1}{1-\hat{M}(p)}\int_0^\infty e^{-pt}\tilde{M}(t,y)\,dt - \frac{\alpha_0}{(p-\alpha_0)p}\frac{\int_0^\infty e^{-\alpha_0 t}\tilde{M}(t,y)\,dt}{\int_0^\infty te^{-\alpha_0 t}\bar{M}(t)\,dt}$$

is an analytic function in the region $\operatorname{Re} p \geq \alpha_1$.
Let c_3 satisfy the relation

$$|G(y)| + |\mathcal{K}|(p,y)| \leq c_3 \quad \text{if } \operatorname{Re} p \geq \alpha_1.$$

Then

$$|G_2(\varepsilon,p,y)| \leq c_3 \int_0^\infty e^{-\delta t} c e^{\alpha_0 \varepsilon t} e^{-\varepsilon\alpha_1 t}\,dt = 2c_2 \frac{2\,c_2\,c_3}{\delta+\varepsilon\alpha_1-\alpha_0\varepsilon},$$

and

$$|G_{r+1}(\varepsilon,p,y)| \leq \left(\frac{2c_2}{\delta+\varepsilon\alpha_1-\alpha_0\varepsilon}\right)^r c_3,$$

and in the same way

$$|\mathcal{K}_{r+1}(\varepsilon,p,y)| \leq \left(\frac{2c_2}{\delta+\varepsilon\alpha_1-\alpha_0\varepsilon}\right)^r c_3$$

under the assumption that $\delta - \varepsilon(\alpha_0-\alpha_1) > 0$. If ε_0 satisfies the inequality

$$\varepsilon_0 < \frac{\delta}{2c_2+\alpha_0-\alpha_1},$$

then

$$\frac{2c_2\varepsilon}{\delta-(\alpha_0-\alpha_1)\varepsilon} < 1 \quad \text{for} \quad 0 < \varepsilon \leq \varepsilon_0,$$

and series (6.136) and (6.137) are uniformly convergent in the region Re $p \geq \alpha_1$. This completes the proof.

□

Corollary 2 *Let*

$$G^*(\varepsilon,p,y) = \int_0^\infty \bar{x}(\varepsilon s)R(s,y,dy')G(\varepsilon,p,y')e^{-\varepsilon ps}\,ds,$$

$$\mathcal{K}^*(\varepsilon,p,y) = \int_0^\infty \bar{x}(\varepsilon s)R(s,y,dy')\mathcal{K}(\varepsilon,p,y')e^{-\varepsilon ps}\,ds.$$

Then $G^*(\varepsilon,p,y)$ *and* $\mathcal{K}^*(\varepsilon,p,y)$ *are bounded analytic functions in the region* $\operatorname{Re} p \geq \alpha_1$*, and*

$$\mathcal{L}^*(\varepsilon,p,y) = \hat{x}(p) + \frac{\varepsilon}{p(p-\alpha_0)}G^*(\varepsilon,p,y) + \varepsilon\mathcal{K}^*(\varepsilon,p,y). \tag{6.138}$$

The last formula is a consequence of the following representation:

$$\mathcal{L}^*(\varepsilon,p,y) = \hat{x}(p) + \varepsilon\int_0^\infty \bar{x}(\varepsilon s)R(s,y,dy')\mathcal{L}(\varepsilon,p,y')e^{-\varepsilon ps}\,ds.$$

The proof of the corollary is left to the reader. The next result estimates the expected value of $x_\varepsilon(t)$.

Theorem 11 *Assume that there exists a function $\theta : R \to R_+$ for which*

$$|\hat{\varphi}(\lambda + iu)| + |\hat{M}(\lambda + iu, y)| \le \theta(u) \quad \textit{for } \lambda \ge \alpha_1$$

and $\int \theta(u)\, du < \infty$.
Then for ε small enough there exists α_ε satisfying the relation $\alpha_\varepsilon - \alpha_0 = O(\varepsilon)$ and constants A_ε, B_ε for which

$$Ex_\varepsilon(t) = A_\varepsilon \frac{e^{\alpha_\varepsilon t} - e^{\alpha_0 t}}{\alpha_\varepsilon - \alpha_0} + B_\varepsilon e^{\alpha_0 t} + O(e^{\alpha_1 t}).$$

(The numbers α_0, α_1 were introduced in Lemma 7.)

Proof We will use the inversion formula for the expression on the right-hand side of equality (6.138). Let

$$G(\varepsilon, p) = \int G(\varepsilon, p, y)\rho(dy),$$
$$\mathcal{K}(\varepsilon, p) = \int \mathcal{K}(\varepsilon, p, y)\rho(dy).$$

Then

$$\begin{aligned}(1 - \mathcal{L}(\varepsilon, p))^{-1} &= \left(1 - \frac{G(\varepsilon, p)}{(p - \alpha_0)p} - \mathcal{K}(\varepsilon, p)\right)^{-1} \\ &= (p - \alpha_0)\left(p - \alpha_0 - \frac{G(\varepsilon, p)}{p} - (p - \alpha_0)\mathcal{K}(\varepsilon, p)\right)^{-1}.\end{aligned}$$

It follows from general properties of analytic functions that there exists a unique $\alpha_\varepsilon = \alpha_0 + O(\varepsilon)$ for which

$$\alpha_\varepsilon - \alpha_0 - \frac{G(\varepsilon, p)}{\alpha_\varepsilon} - (\alpha_\varepsilon - \alpha_0)\mathcal{K}(\varepsilon, p) = 0, \tag{6.139}$$

$\alpha_\varepsilon > 0$, and

$$\left| p - \alpha_0 - \frac{G(\varepsilon, p)}{p} - (p - \alpha_0)\mathcal{K}(\varepsilon, p) \right| > 0$$

if $\operatorname{Re} p \ge \alpha_1$ and $p \ne \alpha_\varepsilon$.
Using equation (6.139), we obtain the representation

$$\alpha_\varepsilon = \alpha_0 + \frac{\varepsilon}{\alpha_0} \iint \tilde{M}(0, y)G(y')R(y, dy')\rho(dy) + O(\varepsilon^2). \tag{6.140}$$

Using formula (6.133), Lemma 6, and equation (6.138) we can represent the Laplace transformation for $Ex_\varepsilon(t)$ in the form

$$\int_0^\infty e^{-pt} Ex_\varepsilon(t)\, dt = \frac{\hat{x}(p)}{\mathcal{L}'(\varepsilon, \alpha_\varepsilon)(\alpha_\varepsilon - p)} + \frac{\varepsilon G^*(\varepsilon, \alpha_\varepsilon, y_0)}{p(p-\alpha)\mathcal{L}'(\varepsilon, \alpha_\varepsilon)(\alpha_\varepsilon - p)} + \frac{\varepsilon \mathcal{K}(\varepsilon, p, y_0)}{\mathcal{L}'(\varepsilon, \alpha_\varepsilon)(\alpha_\varepsilon - p)} + \mathcal{H}(\varepsilon, p, y_0),$$

where the function $\mathcal{H}(\varepsilon, p, y_0)$ is analytic and bounded in the region $\operatorname{Re} p \geq \alpha_1$. Since the function

$$\frac{1}{1 - \mathcal{L}(\varepsilon, p)} - \frac{1}{\mathcal{L}'(\varepsilon, \alpha_\varepsilon)(\alpha_\varepsilon - p)}$$

is bounded in the region $\operatorname{Re} p \geq \alpha_1$, we have that

$$|\mathcal{H}(\varepsilon, p, y_0)| = O\biggl(|\hat{x}(p)| + |\mathcal{K}(\varepsilon, p, y_0)| + O\Bigl(\frac{1}{|p|^2}\Bigr)\biggr)$$

for $\operatorname{Re} p \geq \alpha_1$. We can use the inversion formula for the Laplace transform of the function $\mathcal{H}(\varepsilon, p, y_0)$, because under the assumption of the theorem we have that

$$\int |\hat{x}(\alpha_1 + iu)|\, du < \infty \quad \text{and} \quad \int |\mathcal{K}(\varepsilon, \alpha_1 + iu, y_0)|\, du < \infty.$$

Therefore,

$$Ex_\varepsilon(t) = \frac{-1}{\mathcal{L}'(\varepsilon, \alpha_\varepsilon)} \int_0^t \bar{x}(t-s) e^{\alpha_\varepsilon s}\, ds + \frac{\varepsilon G^*(\varepsilon, \alpha_\varepsilon, y)}{\mathcal{L}'(\varepsilon, \alpha_\varepsilon)} \frac{e^{\alpha t} - e^{\alpha_\varepsilon t}}{\alpha - \alpha_\varepsilon} + \frac{\varepsilon}{\mathcal{L}'(\varepsilon, \alpha_\varepsilon)} \int_0^t k_\varepsilon(s, y_0) e^{\alpha_\varepsilon (t-s)}\, ds + O(e^{\alpha_1 t}). \tag{6.141}$$

(We used here the property of Laplace transforms that the Laplace transform of the convolution of two functions is the product of their Laplace transforms.) Here $k_\varepsilon(s, y_0)$ is the function for which

$$\int_0^\infty e^{-pt} k_\varepsilon(t, y_0)\, dt = \mathcal{K}(\varepsilon, p, y_0).$$

□

Next, we estimate $E(x_\varepsilon(t) - \bar{x}(t))^2$. It follows from Lemma 6 that

$$E(x_\varepsilon(t) - \bar{x}(t))^2 = \sum_{l=1}^\infty \sum_{m=1}^\infty \int \cdots \int_{\substack{0<s_1<\cdots<s_l<t \\ 0<u_1<\cdots<u_m<t}} L_t(s_1, \ldots, s_l, u_1, \ldots, u_m, y_1, \ldots, y_{l+m}) \times \prod_{k=1}^{l+m} \Bigl[R\Bigl(\frac{\tau_k}{\varepsilon}, y_{k-1}, dy_k\Bigr) + \rho(dy_k)\Bigr]\, ds_1 \cdots ds_l\, du_1 \cdots du_m, \tag{6.142}$$

where $\tau_1 = v_1$, $\tau_2 = v_2 - v_1, \dots \tau_{l+m} = v_{l+m} - v_{l+m-1}$, and where

$$\begin{aligned}
v_1 &= \min\{s_1, \dots, s_l\} \cup \{u_1, \dots, u_m\}, \\
v_r &= \min[\{s_1, \dots, s_l\} \cup \{u_1, \dots, u_m\} \setminus \{v_1, \dots, v_{r-1}\}],\ r > 1, \\
&L_t(s_1, \dots, s_l, u_1, \dots, u_m, y_1, \dots, y_{l+m}) \\
&= \bar{x}(s_1)L(s_2 - s_1, y_{j_1}) \cdots L(t - s_l, y_{j_l}) \\
&\times \bar{x}(u_1)L(u_2 - u_1, y_{i_1}) \cdots L(t - u_m, y_{i_m}),
\end{aligned}$$

and $v_{j_1} = s_1, \dots, v_{j_l} = s_l$, $v_{i_1} = u_1, \dots, v_{i_m} = u_m$.

Lemma 10 *Assume that the conditions of Lemma 9 are satisfied. Then there exists a constant $b > 0$ for which*

$$\left| \iint L_t(s_1, .., s_l, u_1, .., u_m, y_1, .., y_{l+m}) \prod_{k=1}^{l+m} \left[R\Big(\frac{\tau_k}{\varepsilon}, y_{k-1}, dy_k\Big) + \rho(dy_k) \right] \right|$$

$$\leq b^{l+m} e^{2\alpha_0 t} \sum 1_{\{S \in U_2(l+m)\}} \exp\left\{ -\frac{\delta}{\varepsilon} \sum \tau_k 1_{\{k \in S\}} \right\},$$

where $U_2(n)$ is the set of those subsets S of the set $\{1, \dots, n\}$ for which $n \in S$ and $\{k, k+1\} \cap S$ is not empty for $k = 1, \dots, n-2$.

Proof Since $\int L(t, y)\rho(dy) = 0$, we have for $1 \leq r_1 < r_2 < \cdots < r_\nu \leq l+m$ the relation

$$\begin{aligned}
\int &L_t(s_1, \dots, s_l, u_1, \dots, u_m, y_1, \dots, y_{l+m}) \\
&\times N^{(l+m)}_{r_1, \dots, r_\nu}(\tau_1, \dots, \tau_{l+m}, y_1, \dots, y_{l+m}, dy_1, \dots, dy_{l+m}) = 0
\end{aligned}$$

if $r_\nu = l + m$ or $\min_i(r_{i+1} - r_i) = 1$. This means that the set difference $\{1, \dots, l+m\} \setminus \{r_1, \dots, r_\nu\}$ belongs to $U_2(l+m)$. The function N is defined in relation (6.134).
Using inequality (6.135) and the relation

$$|L(t, y)| \leq c_1 e^{\alpha_0 t},$$

we obtain the inequality

$$\begin{aligned}
&\Big| \int L_t(s_1 \dots, s_l, u_1, \dots, u_m, y_1, \dots, y_{l+m}) \\
&\quad \times N^{(l+m)}_{r_1, \dots, r_\nu}(\tau_1 \dots, \tau_{l+m}, u_1, y_1, \dots, y_{l+m}, dy_1, \dots, dy_{l+m}) \Big| \\
&\leq c_1^{l+m} e^{2\alpha_0 t} c_2^{l+m-\nu} \exp\left\{ -\frac{\delta}{\varepsilon} \Big(\sum_{i=1}^{l+m} \tau_i - \sum_{k=1}^{\nu} \tau_{r_k} \Big) \right\}.
\end{aligned}$$

This completes the proof of Lemma 10.

□

Corollary 3 *For ε small enough we have the inequality*

$$E(x_\varepsilon(t) - \bar{x}(t))^2 \le B_1 e^{2\alpha_0 t}(e^{\varepsilon b_1 t} - 1), \tag{6.143}$$

where B_1 and b_1 are nonnegative constants.

Proof To prove this we can write, using relation (6.140) and Lemma 10,

$$E(x_\varepsilon(t) - \bar{x}(t))^2 \le \sum_{n=2}^{\infty} \sum_{l+m=n} b^{l+m} e^{2\alpha_0 t} \sum_{\frac{n}{2} \le k \le n} \frac{n!}{k!(n-k)!} I_t(n,k),$$

where

$$I_t(n,k) = \int \cdots \int_{s_1+\cdots+s_n \le t} \exp\left\{-\frac{\delta}{\varepsilon}(s_1 + \cdots + s_k)\right\}$$
$$\times 1_{\{s_1 \ge 0, \ldots, s_n \ge 0\}}\, ds_1 \cdots ds_n.$$

We used here the fact that Card $S \ge n/2$ if $S \in U_2(n)$. We have the following estimates:

$$I_t(n,k) = \int_0^t \exp\left\{-\frac{\delta}{\varepsilon}u\right\} \frac{u^{k-1}}{(k-1)!} \frac{(t-u)^{n-k}}{(n-k)!}\, du$$
$$\le \frac{t^{n-k}}{(n-k)!} \int_0^\infty \exp\left\{-\frac{\delta}{\varepsilon}u\right\} \frac{u^{k-1}}{(k-1)!}\, du = \left(\frac{\varepsilon}{\delta}\right)^k \frac{t^{n-k}}{(n-k)!}.$$

Since $\frac{n!}{k!(n-k)!} < 2^n$, we obtain the inequality

$$E(x_\varepsilon(t) - \bar{x}(t))^2 \le \sum_{n=2}^{\infty} n(2b)^n e^{2\alpha_0 t} \sum_{\frac{n}{2} \le k \le n} \left(\frac{\varepsilon}{\delta}\right)^k \frac{t^{n-k}}{(n-k)!}.$$

Let b_2 and B_2 satisfy the inequality

$$n(2b)^n \le B_2 b_2^n.$$

Then

$$E(x_\varepsilon(t) - \bar{x}(t))^2 \le B_2 e^{2\alpha_0 t} \sum_{n=2}^{\infty} b_2^n \sum_{\frac{n}{2} \le k \le n} \left(\frac{\varepsilon}{\delta}\right)^k \frac{t^{n-k}}{(n-k)!}$$

$$\le B_2 e^{2\alpha_0 t} \sum_{l=1}^{\infty} \sum_{k=l}^{\infty} b_2^{k+l} \left(\frac{\varepsilon}{\delta}\right)^k \frac{t^l}{l!}$$

$$= B_2 e^{2\alpha_0 t} \sum_{l=1}^{\infty} \sum_{k=l}^{\infty} \frac{1}{1 - b_2 \frac{\varepsilon}{\delta}} b_2^{2l} \left(\frac{\varepsilon}{\delta}\right)^l \frac{t^l}{l!} = \frac{B_2}{1 - \varepsilon \frac{b_2}{\delta}} e^{2\alpha_0 t} \left(\exp\left\{\frac{\varepsilon t b_2^2}{\delta}\right\} - 1\right)$$

if $\varepsilon < \delta/b_2$.

□

Finally, we estimate the growth of solutions to the perturbed equation. We first establish some auxiliary results.

Lemma 11 *Assume that* $\varphi(t) > 0$, $M(t, y) \geq 0$, *and*

$$\lim_{h \to 0} \sup_t \left[\left| \frac{\varphi(t+h) - \varphi(t)}{\varphi(t)} \right| + \sup_y \left| \frac{M(t+h, y) - M(t, y)}{M(t, y)} \right| \right] = 0. \quad (6.144)$$

Then

$$\lim_{h \to 0} \sup_t \frac{|x_\varepsilon(t+h) - x_\varepsilon(t)|}{x_\varepsilon(t)} = 0.$$

Proof Set

$$\rho_h = \sup_t \left[\left| \frac{\varphi(t+h) - \varphi(t)}{\varphi(t)} \right| + \sup_y \left| \frac{M(t+h, y)_{M}(t, y)}{M(t, y)} \right| \right],$$

$$C_h = \sup_{y,\, s \leq h} |M(s, y)|.$$

Then

$$\begin{aligned} |x_\varepsilon(t+h) - x_\varepsilon(t)| &= \Bigg| \varphi(t+h) - \varphi(t) \\ &\quad + \int_0^t \left[M\left(t+h-s, y\left(\frac{s}{\varepsilon}\right)\right) - M\left(t-s, y\left(\frac{s}{\varepsilon}\right)\right) \right] x_\varepsilon(s)\, ds \\ &\quad + \int_t^{t+h} M\left(t+h-s, y\left(\frac{s}{\varepsilon}\right)\right) x_\varepsilon(s)\, ds \Bigg| \\ &\leq \rho_h \varphi(t) + \rho_h \int_0^t M\left(t-s, y\left(\frac{s}{\varepsilon}\right)\right) x_\varepsilon(s)\, ds + C_h \left| \int_t^{t+h} x_\varepsilon(s)\, ds \right| \\ &\leq [\rho_h + hC_h] x_\varepsilon(t) + C_h \int_0^h |x_\varepsilon(t+u) - x_\varepsilon(t)|\, du. \end{aligned}$$

This implies the inequality

$$|x_\varepsilon(t+h) - x_\varepsilon(t)| \leq [\rho_h + hC_h] \exp\{hC_h\} x_\varepsilon(t).$$

(Note that under the assumptions of the lemma $x_\varepsilon(t)$ is nonnegative.) The statement of the lemma follows, since $\lim_{h \to 0} \rho_h = 0$.

□

Theorem 12 *Assume that the functions* $M(t, y)$ *and* $\varphi(t)$ *satisfy the following conditions:*

(1) $\varphi(t) \geq 0$ *and* $M(t, y) \geq 0$.

(2) *Relation (6.144) is satisfied.*

(3) $M(t, y) + \varphi(t) = O(e^{\alpha t})$, *where* $\alpha \geq 0$.

(4) *There exists* $\alpha_0 > \alpha$ *for which*

$$\int_0^\infty e^{-\alpha_0 t} \bar{M}(t)\, dt = 1.$$

(5) *There exist a measurable function* $\theta : R \to R_+$ *with* $\int_{-\infty}^{\infty} \theta(u)\, du < \infty$ *and* $\alpha_1 \in (\alpha, \alpha_0)$ *for which*

$$|\hat{\varphi}(\lambda + iu)| + |\hat{M}(\lambda + iu, y)| \leq \theta(u), \quad u \in R,\ y \in Y,$$

if $\lambda \geq \alpha_1$.

Then there exist $\varepsilon_0 > 0$ *and* $\gamma > 0$ *such that*

$$P\left\{\limsup_{t\to\infty} \frac{1}{t} \log x_\varepsilon(t) > \alpha_0 + \gamma\varepsilon\right\} = 0 \quad \textit{for } \varepsilon < \varepsilon_0.$$

Proof Using Lemma 11 we need only prove that for all $h > 0$,

$$P\left\{\limsup_{k\to\infty} \frac{1}{kh} \log x_\varepsilon(kh) > \alpha_0 + \gamma\varepsilon\right\} = 0 \quad \text{for } \varepsilon < \varepsilon_0.$$

Note that it follows from Corollary 3 and Lemma 7 that

$$\begin{aligned} P\big\{x_\varepsilon(kh) > \exp\{(\alpha_0 + \gamma\varepsilon)kh\}\big\} &\leq E x_\varepsilon^2(kh) \cdot \exp\{-2kh(\alpha_0 + \gamma\varepsilon)\} \\ &= O\big(\exp\{(2\alpha_0 + \varepsilon b_1)kh - 2kh(\alpha_0 + \gamma\varepsilon)\}\big). \end{aligned}$$

If $2\gamma > b_1$, then

$$\sum P\big\{x_\varepsilon(kh) > \exp\{(\alpha_0 + \gamma\varepsilon)kh\}\big\} < \infty.$$

□

7

Markov Chains with Random Transition Probabilities

In this chapter we consider a Markov chain $\{\xi_k,\ k \in \mathbf{Z}_+\}$ having a finite state space but random transition probabilities. Of interest here is the asymptotic behavior of the chain and its transition probabilities as $n \to \infty$ under the assumption that the transition probabilities of the Markov chain are close to those of a homogeneous Markov chain having nonrandom transition probabilities.

By a Markov chain with random transition probabilities we mean the following: Denote by $\mathcal{P}_r$ the set of $r \times r$ stochastic matrices, i.e., matrices

$$M = (\mu_{ij})_{i,j\in\overline{1,r}},$$

having the following properties:

(a) $\mu_{ij} \geq 0$;

(b) $\sum_{j=1}^{r} \mu_{ij} = 1$.

We consider a $\mathcal{P}_r$-valued stochastic process $\{P_k(\omega),\ k \in \mathbf{Z}_+\}$. Let $\mathcal{E}$ be the σ-algebra generated by the stochastic process $\{P_k(\omega),\ k \in \mathbf{Z}_+\}$.

A stochastic process $\{\xi_n, n \in \mathbf{Z}_+\}$ with values in $\overline{1,r}$ is called a Markov chain with transition probabilities $\{P_n(\omega)\}$ if for any $k, n \in \mathbf{Z}_+$, and $i_0, \ldots, i_n \in \overline{1,r}$, the following relation is satisfied:

$$P\{\xi_k(\omega) = i_0, \ldots, \xi_{k+n}(\omega) = i_n/\mathcal{E}\} = P\{\xi_k(\omega) = i_0/\mathcal{E}\} \prod_{l=0}^{n-1} p_{k+l}(\omega, i_l, i_{l+1}), \tag{7.1}$$

where $p_n(\omega, i, j)$ are the elements of the matrix $P_n(\omega)$.

The stochastic process $\{P_n(\omega),\ n \in \mathbf{Z}_+\}$ is called a *random environment*, and $\{\xi_n\}$ is called a *Markov chain in a random environment.*

7.1 Stationary Random Environment

A stationary random environment is defined by a stationary $\mathcal{P}_r$-valued stochastic process $\{P_n(\omega),\ n \in \mathbf{Z}_+\}$. It is convenient for our construction to consider the process extended from $\mathbf{Z}_+$ to $\mathbf{Z}$; any stationary stochastic process on $\mathbf{Z}_+$ can be extended as a stationary process to $\mathbf{Z}$.
We use the following notation for the transition probability from step l to step n:

$$P_{l,n}(\omega) = P_l(\omega)P_{l+1}(\omega)\cdots P_{n-1}(\omega).$$

Denote by $\mathcal{D}_r$ the space of row vectors $\vec{a} = (a_1, \ldots, a_r)$ with the norm

$$|\vec{a}|_1 = \sum_{k=1}^{r} |a_k|,$$

and denote by $\mathcal{C}_r$ the space of column vectors

$$c^{\uparrow} = \begin{pmatrix} c^1 \\ \vdots \\ c^r \end{pmatrix}$$

with the norm

$$|c^{\uparrow}|^1 = \max_k |c^k|.$$

Let $\mathcal{D}_r^1$ be the subset of probability distributions in $\mathcal{D}_r$: If $a \in \mathcal{D}_r^1$, then $a_i \geq 0$ and $\sum_{i=1}^r a_i = 1$.

Definitions (1) *A $\mathcal{D}_r^1$-valued stationary stochastic process $\{\vec{d}_k(\omega),\ k \in \mathbf{Z}\}$ is called a stationary distribution for the stationary random environment $\{P_n(\omega),\ n \in \mathbf{Z}\}$ if*

$$\{(\vec{d}_k(\omega), P_k(\omega)),\ k \in \mathbf{Z}\}$$

is a stationary $\mathcal{D}_r^1 \times \mathcal{P}_r$-valued process and the relation

$$\vec{d}_{k+1}(\omega) = \vec{d}_k(\omega)P_k(\omega)$$

holds for all $k \in \mathbf{Z}$.
(2) *A $\mathcal{C}_r$-valued stochastic process $\{c_k^{\uparrow}(\omega),\ k \in \mathbf{Z}\}$ is called a stationary vector for the stationary random environment $\{P_n(\omega),\ n \in \mathbf{Z}\}$ if*

$$\{(c_k^{\uparrow}(\omega), P_k(\omega)),\ k \in \mathbf{Z}\}$$

is a stationary $\mathcal{C}_r \times \mathcal{P}_r$-valued process and the relation

$$c_k^{\uparrow}(\omega) = P_k(\omega)c_{k+1}^{\uparrow}(\omega)$$

holds for all $k \in \mathbf{Z}$.

Remark 1 *If $P\{c_k^\uparrow(\omega) = e^\uparrow\} = 1$, where $e^\uparrow = (1, \ldots, 1)^T$, then $\{c_k^\uparrow(\omega)\}$ is a stationary vector for any random environment.*

Theorem 1 *For any stationary random environment $\{P_k(\omega),\ k \in \mathbf{Z}\}$ there exists a stationary distribution $\{\vec{d}_k(\omega),\ k \in \mathbf{Z}\}$.*

Proof Let $\vec{a} \in \mathcal{D}_r^1$. For $N > 0$ introduce the $\mathcal{D}_r^1$-valued stochastic process

$$\vec{a}_k(N, \omega) = \frac{1}{N} \sum_{l=k-N+1}^{k} \vec{a}\, P_{l,k}(\omega), \quad k \in \mathbf{Z}.$$

It is easy to see that $\{\vec{a}_k(N, \omega), P_k(\omega)\}$ is a $\mathcal{D}_r^1 \times \mathcal{P}_r$-valued stationary stochastic process. It satisfies the relation

$$\vec{a}_{k+1}(N, \omega) - \vec{a}_k(N, \omega) = \frac{1}{N}(\vec{a}P_k(\omega) - \vec{a}P_{k-N+1,k+1}(\omega)). \tag{7.2}$$

Denote by Q_N the probability measure on the space $(\mathcal{D}_r^1 \times \mathcal{P}_r)^{\mathbf{Z}}$ that is the distribution of the stochastic process $\{(\vec{a}_k(N, \omega), P_k(\omega)),\ k \in \mathbf{Z}\}$. Since $\mathcal{D}_r^1 \times \mathcal{P}_r$ is compact (in the natural topology as a subset of $R^r \times R^{r^2}$), the set of measures $\{Q_N,\ N \in \mathbf{Z},\ N > 0\}$ is compact with respect to weak convergence of measures. Let $\bar{Q}$ be a limit point of this set. We consider some probability space $\{\Omega', \mathcal{F}', P'\}$ and a stochastic process $\{(\vec{b}_k(\omega'), P_k^*(\omega'),\ k \in \mathbf{Z}\}$ that has the distribution $\bar{Q}$. Then $\{(\vec{b}_k(\omega'), P_k^*(\omega')),\ k \in \mathbf{Z}\}$ is a stationary process for which

$$\begin{aligned} &P\{|\vec{b}_{k+1}(\omega') - \vec{b}_k(\omega')P_k^*(\omega')|_1 > \varepsilon\} \\ &\quad = \lim_{N\to\infty} P\Big\{\frac{1}{N}|\vec{a}P_k(\omega) - \vec{a}P_{k-N+1,k+1}(\omega)| > \varepsilon\Big\} = 0 \end{aligned}$$

for all $\varepsilon > 0$, because of relation (7.2).
Thus

$$\vec{b}_{k+1}(\omega') = \vec{b}_k(\omega')P_k^*(\omega'). \tag{7.3}$$

Denote by $\mathcal{I}'$ the σ-algebra generated by $\{P_k(\omega'),\ k \in \mathbf{Z}\}$. Then

$$E(\vec{b}_{k+1}(\omega')/\mathcal{I}') = E(\vec{b}_k(\omega')/\mathcal{I}')P_k^*(\omega'). \tag{7.4}$$

It is easy to see that the distribution of $\{P_k^*(\omega'),\ k \in \mathbf{Z}\}$ is the same as $\{P_k(\omega),\ k \in \mathbf{Z}\}$. Besides,

$$\big\{\big(E(\vec{b}_k(\omega')/\mathcal{I}'), P_k^*(\omega')\big),\ k \in \mathbf{Z}\big\} \tag{7.5}$$

is a stationary stochastic process. Denote by $F_k : \mathcal{P}_r^{\mathbf{Z}} \to \mathcal{D}_r^1$ a measurable function for which $F_k\big(\{P_k^*(\omega),\ k \in \mathbf{Z}\}\big) = E(\vec{b}_k(\omega')/\mathcal{I}')$.
Set

$$\vec{d}_k(\omega) = F_k\big(\{P_k(\omega),\ k \in \mathbf{Z}\}\big).$$

Then the distribution of the process $\{(\vec{d}_k(\omega), P_k(\omega)),\ k \in \mathbf{Z}\}$ coincides with the distribution of the process in (7.5). This and relation (7.3) imply that $\{\vec{d}_k(\omega)\}$ is a stationary distribution for $\{P_k(\omega)\}$.

□

It is known that an ergodic homogeneous Markov chain has a unique stationary distribution. We will investigate under what conditions a stationary random environment $\{P_k(\omega),\ k \in \mathbf{Z}\}$ has a unique stationary distribution. Denote by $\mathcal{D}_r^0$ the subset of vectors $\vec{a} = (a_1, \ldots, a_r) \in \mathcal{D}_r$ for which $\sum_{k=1}^r a_k = 0$. Let $\Pi \in \mathcal{P}_r$ and define

$$\|\Pi\|_0 = \sup\{|\vec{a}\Pi|_1 : \vec{a} \in \mathcal{D}_r^0,\ |\vec{a}|_1 = 1\}. \tag{7.6}$$

Note that $\mathcal{D}_r^0$ is a linear subspace in $\mathcal{D}_r$, and it is an invariant subspace for each $\Pi \in \mathcal{P}_r$; i.e., $\vec{a}\Pi \in \mathcal{D}_r^0$ for all $\vec{a} \in \mathcal{D}_r^0$.
The norm $\|\cdot\|_0$ is useful for investigation of ergodic properties of finite Markov chains. Following are some of its properties:

I. $\|\Pi\|_0 \le 1$ for any $\Pi \in \mathcal{P}_r$.

II. $\|\Pi_1\Pi_2\|_0 \le \|\Pi_1\|_0 \|\Pi_2\|_0$, for any $\Pi_1,\ \Pi_2 \in \mathcal{P}_r$.

III. If $\|\Pi\|_0 = 0$ and $\Pi \in \mathcal{P}_r$, then all rows of the matrix Π coincide, i.e., $\pi_{1j} = \pi_{ij}$ for $i = 2, \ldots, r$, $j = 1, \ldots, r$, where π_{ij}, $i, j \in \overline{1, r}$, are the elements of the matrix Π.

IV. . If $\|\Pi\|_0 < 1$ and $\Pi \in \mathcal{P}_r$, then there exists a unique vector $\vec{a} \in \mathcal{D}_r^1$ for which $\vec{a}\Pi = \vec{a}$, and

$$\lim_{n\to\infty} \Pi^n = \tilde{\Pi},$$

where $\tilde{\Pi} \in \mathcal{P}_r$, $\|\tilde{\Pi}\|_0 = 0$, and all rows of the matrix $\tilde{\Pi}$ coincide with the vector $\vec{a}$.

V. If $\|\Pi\|_0 = 1$ and $\Pi \in \mathcal{P}_r$, then there exists $k \le r$ and $\vec{a} \in \mathcal{D}_r^0$ for which $\vec{a} = \vec{a}\Pi^k$. If this relation is true for some k and $\vec{a} \in \mathcal{D}_r^0$, then $\|\Pi\|_0 = 1$.

VI. For all $\Pi \in \mathcal{P}_r$ there exists the limit

$$\lim_{n\to\infty} n^{-1} \sum_{k=1}^{n} \Pi^k.$$

Denote this limit by $\tilde{\Pi}$. Then $\|\tilde{\Pi}\|_0 = 0$ if and only if

$$\left\| r^{-1} \sum_{i=1}^{r} \Pi^i \right\|_0 < 1;$$

in this case the rows of the matrix $\tilde{\Pi}$ are equal to the vector $\vec{a}$, which is the unique solution of the equation

$$\vec{a} = \vec{a}\left(r^{-1}\sum_{i=1}^{r}\Pi^{i}\right), \quad \vec{a} \in \mathcal{D}_r^1.$$

VII. Let $A, B, \Pi \in \mathcal{P}_r$. Then

$$\|A\Pi - B\Pi\|_1 \leq 2\|\Pi\|_0.$$

VIII. Set $\alpha(\Pi) = -\log \|\Pi\|_0$, $\Pi \in \mathcal{P}_r$. (So, if $\|\Pi\|_0 = 0$, then $\alpha(\Pi) = +\infty$.) Then for any matrices $\Pi_1, \ldots, \Pi_n \in \mathcal{P}_r$ we have that

$$\alpha(\Pi_1\Pi_2\cdots\Pi_n) \geq \sum_{k=1}^{n}\alpha(\Pi_k).$$

The following statement is a variant for stationary processes of the ergodic theorem for Markov chains in random environments.

Theorem 2 *Let $\{\xi_n(\omega),\ n \in \mathbf{Z}\}$ be a Markov chain in a random environment $\{P_n(\omega),\ n \in \mathbf{Z}\}$. Assume that $E\alpha(P_1(\omega)) > 0$ and that $\{P_n(\omega),\ n \in \mathbf{Z}\}$ is an ergodic stationary process. Then the following statements hold:*
(1) There exists a matrix $\tilde{\Pi} \in \mathcal{P}_r$ with $\|\tilde{\Pi}\|_0 = 0$ for which

$$\lim_{n\to\infty} E\, P_{0,n}(\omega) = \tilde{\Pi}.$$

(2) There exists a unique stationary distribution $\{\vec{d}_n(\omega),\ n \in \mathbf{Z}\}$ that satisfies the relation

$$\lim_{n\to\infty} |\vec{d}_n(\omega) - \vec{a}P_{0,n}(\omega)|_1 = 0$$

almost surely for all $\vec{a} \in \mathcal{D}_r^1$.
(3) $\lim_{n\to\infty} n^{-1}\sum_{k=1}^{n} P_{0,k}(\omega) = \tilde{\Pi}$ almost surely.
(4) Let $f(i_1, \ldots, i_m)$ be a function from $\overline{1,r}^{\,m}$ into R. Then

$$P\left\{\lim_{n\to\infty} n^{-1}\sum_{k=0}^{n-1} f\left(\xi_{k+1}(\omega), \ldots, \xi_{k+m}(\omega)\right) = \lambda_f\right\} = 1,$$

where

$$\lambda_f = E\sum_{i_1,\ldots,i_m} \pi_1(i_1,\omega)p_1(i_1,i_2,\omega)\cdots p_{m-1}(i_{m-1},i_m,\omega)f(i_1,\ldots,i_m)$$

and $\vec{d}_1(\omega) = (\pi_1(1,\omega), \ldots, \pi_r(1,\omega))$, $p_k(i,j,,\omega)$ are the elements of the matrix $P_k(\omega)$.

Proof First we prove that there exists a limit

$$\lim_{n\to\infty} P_{-n,k}(\omega) = \tilde{P}_k(\omega) \tag{7.7}$$

for all $k \in \mathbf{Z}$ almost surely in ω. It follows from property VII that

$$\|P_{-n,k}(\omega) - P_{-l,k}(\omega)\| \leq 2\|P_{-l,k}(\omega)\|_0$$

for $-n < -l \leq -k$. Property VIII implies the inequality

$$\|P_{-l,k}(\omega)\|_0 \leq \exp\Big\{-\sum_{i=-l}^{k-1} \alpha(P_i(\omega))\Big\}. \tag{7.8}$$

Since

$$\frac{1}{l}\sum_{i=-l}^{k-1} \alpha(P_i(\omega)) \to E\alpha(P_0(\omega))$$

with probability 1, $\lim_{l\to\infty} \|P_{-l,k}(\omega)\|_0 = 0$. The existence of the limit on the left–hand side of (7.7) is proved. From (7.8) we see that $\|\tilde{P}_k(\omega)\|_0 \to 0$. Denote by $\vec{d}_k(\omega)$ the vector representing the rows of the matrix $\tilde{P}_k(\omega)$ (the norms are the same). Since $\tilde{P}_{k+1}(\omega) = \tilde{P}_k(\omega)P_k(\omega)$, we have $\vec{d}_{k+1}(\omega) = \vec{d}_k(\omega)P_k(\omega)$, so $\{\vec{d}_k(\omega),\, k \in \mathbf{Z}\}$ is a stationary distribution.
If $\{\vec{\pi}_k(\omega),\, k \in \mathbf{Z}\}$ is an arbitrary stationary distribution, then

$$|\vec{d}_k(\omega) - \vec{\pi}_k(\omega)|_1 \leq \|P_{-l,k}(\omega)\|_0 |\vec{d}_{k-l}(\omega) - \vec{\pi}_{-l}(\omega)|_1 \leq 2\|P_{-l,k}(\omega)\|_0 \to 0$$

as $l \to \infty$. With this, we have proved the uniqueness of a stationary distribution.
In the same way

$$|\vec{d}_n(\omega) - \vec{a}P_{a,n}(\omega)|_1 \leq 2\|P_{0,n}(\omega)\|_0 \to 0$$

almost surely as $n \to \infty$. Statement (2) is therefore proved. Statement (1) follows from the relations

$$\lim_{n\to\infty} EP_{0,n}(\omega) = \lim_{n\to\infty} EP_{-n,0}(\omega) = E\tilde{P}_0(\omega).$$

To prove statement (3) we use the inequalities

$$\|P_{0,k}(\omega) - \tilde{P}_k(\omega)\|_1 \leq 2\|P_{0,k}(\omega)\|_0 \tag{7.9}$$

and

$$\Big\|n^{-1}\sum_{k=1}^{n}\big(P_{0,k}(\omega) - \tilde{P}_k(\omega)\big)\Big\|_1 \leq \frac{2}{n}\sum_{k=1}^{n} \|P_{0,k}(\omega)\|_0.$$

Therefore,

$$\lim_{n\to\infty}\Big\|n^{-1}\sum_{k=1}^{n}\big(P_{0,k}(\omega) - \tilde{P}_k(\omega)\big)\Big\|_0 = 0$$

almost surely, since

$$\lim_{n\to\infty}\Big\|n^{-1}\sum_{k=1}^{n} P_{0,k}(\omega)\big)\Big\|_0 = \lim_{n\to\infty} \|P_{0,n}(\omega)\|_0 = 0.$$

The ergodicity of the process $\{P_k(\omega),\, k \in \mathbf{Z}\}$ implies the relation

$$P\Big\{\lim n^{-1}\Big(\sum_{k=1}^{n} \tilde{P}_k(\omega)\Big) = E\tilde{P}_0(\omega)\Big\} = 1.$$

Proof of statement (4)

First, we assume that $P\{\xi_0(\omega) = i/\mathcal{E}\} = \pi_0(i,\omega)$. Then $\{(\xi_n(\omega), P_n(\omega))\}$ for $n > 0$ is a stationary $\overline{1,r} \times \mathcal{P}_r$-valued process. This follows from the relation

$$\begin{aligned} &E\big(f(\xi_{k+1}(\omega),\dots,\xi_{k+m}(\omega))/\mathcal{E}\big)\\ &= \sum_{i_1,\dots,i_m} \pi_{k+1}(i_1,\omega)p_{k+1}(i_1,i_2,\omega)\cdots p_{k+m-1}(i_{m-1},i_m,\omega)f(i_1,\dots,i_m)\\ &= \Phi_f(\vec{d}_{k+1}(\omega), P_{k+1}(\omega),\dots,P_{k+m-1}(\omega)),\end{aligned}$$

where Φ_f is a bounded measurable function from $\mathcal{D}_p \times (\mathcal{P}_p)^{m-1}$ and from the fact that $\{\vec{d}_k(\omega), P_k(\omega)\}$ is a stationary process. This implies the existence of the limit

$$\lim_{n\to\infty} \frac{1}{n}\sum_{k=1}^{n} f(\xi_{k+1}(\omega),\dots,\xi_{k+m}(\omega)) = \tilde{f}(\omega).$$

It is easy to see that $E\tilde{f}(\omega) = \lambda_f$.

Denote by Ξ_k the σ-algebra generated by $\{\xi_0(\omega),\dots,\xi_k(\omega)\}$. Then

$$E\big(f(\xi_{k+1}(\omega),.,\xi_{k+m}(\omega))/\Xi_k \vee \mathcal{E}\big) = \Phi_f(\xi_{k+1}(\omega), P_{k+1}(\omega),.,P_{k+m-1}(\omega))$$

$$= \sum_{i_1,\dots,i_m} \delta(i_1,\xi_{k+1}(\omega))p_{k+1}(i_1,i_2,\omega)\cdots p_{k+m-1}(i_{m-1},i_m,\omega)f(i_1,\dots,i_m),$$

and for $j < k+1$,

$$\begin{aligned}&E\big(f(\xi_{k+1}(\omega),\dots,\xi_{k+m}(\omega))/\Xi_j \vee \mathcal{E}\big)\\ &= \sum_{i,i_1} \delta(i,\xi_j(\omega))p_j(i,i_1,\omega)\Phi_f\big(i_1, P_{k+1}(\omega),\dots,P_{k+m-1}(\omega)\big),\end{aligned}$$

where $\delta(\alpha,\beta) = 1$ if $\alpha = \beta \in \overline{1,r}$, but $\delta(\alpha,\beta) = 0$ if $\alpha \neq \beta$, $\alpha,\beta \in \overline{1,r}$.

Set

$$\Phi_f(k+1,\omega) = \sum_i \Phi_f\left(i, P_{k+1}(\omega),\dots,P_{k+m-1}(\omega)\right).$$

It follows from inequality (7.9) that

$$E\big(f(\xi_{k+1}(\omega),\dots,\xi_{k+m}(\omega))/\Xi_j \vee \mathcal{E}\big) = \Phi_f(k+1,\omega) + O\big(\|P_{j,k+1}(\omega)\|_0\big).$$

Let $k+m < l$. Then

$$\begin{aligned}&E\big(f(\xi_{k+1}(\omega),\dots,\xi_{k+m}(\omega))f(\xi_{l+1}(\omega),\dots,\xi_{l+m}(\omega))/\mathcal{E}\big)\\ &= E\big[E\big(f(\xi_{k+1}(\omega),\dots,\xi_{k+m}(\omega))f(\xi_{l+1}(\omega),\dots,\xi_{l+m}(\omega))/\mathcal{E} \vee \Xi_{k+m}\big)/\mathcal{E}\big]\\ &= \Phi_f(k+1,\omega)\Phi_f(l+1,\omega) + O\big(\|P_{k+m,l+1}(\omega)\|_0\big).\end{aligned}$$

Therefore,

$$E(\tilde{f}^2(\omega)/\mathcal{E}) = \lim_{n\to\infty} \frac{1}{n^2}\Big(\sum_{k=1}^{n} \Phi_f(k,\omega)\Big)^2 + O\Big(\frac{1}{n^2}\sum_{k+m<l} \|P_{k+m,l+1}(\omega)\|_0\Big) + O\Big(\frac{1}{n}\Big).$$

Since $\{P_k(\omega),\ k \in \mathbf{Z}\}$ is an ergodic process, we have that

$$\lim_{n\to\infty} \frac{1}{n}\sum_{k=1}^{n} \Phi_f(k,\omega) = \lambda_f,$$

and

$$E(\tilde{f}^2(\omega)/\mathcal{E}) = \lambda_f^2, \quad P\{\tilde{f}(\omega) = \lambda_f\} = 1.$$

Statement (4) is proved if the distribution of $\xi_k(\omega)$ is the stationary distribution for $\{P_k(\omega)\}$. Let the conditional distribution of $\xi_0(\omega)$ with respect to $\mathcal{E}$ be arbitrary, say denote it by $\vec{p}_0(\omega)$. Let

$$A_k = \Big\{\omega : \lim_{n\to\infty} \frac{1}{n}\sum_{j=k}^{n+k-1} f\big(\xi_j(\omega),\ldots,\xi_{j+m-1}(\omega)\big) = \lambda_f\Big\},\ k = 0,1,2,\ldots.$$

Note that $A_k = A_0$ for all k and that statement (4) is equivalent to the relation $P\{A_0\} = 1$. Denote by $\vec{p}_k(\omega)$ the conditional distribution of $\xi_k(\omega)$ with respect to $\mathcal{E}$. Denote by $Q_k(\omega)$ the conditional distribution of $\{\xi_k(\omega),\xi_{k+1}(\omega),\ldots\}$ with respect to $\mathcal{E}$ if $\xi_k(\omega)$ has the conditional distribution $\vec{p}_k(\omega)$, and denote by $Q'_k(\omega)$ the same conditional distribution if $\xi_k(\omega)$ has the conditional distribution $\vec{d}_k(\omega)$. Then

$$\operatorname{Var}\big(Q_k(\omega) - Q'_k(\omega)\big) \le 2|\vec{d}_k(\omega) - \vec{p}_k(\omega)|_1.$$

Therefore,

$$Q(\omega, A_k) \ge 1 - 2|\vec{d}_k(\omega) - \vec{p}_k(\omega)|_1,$$

since it has been proved that $Q'_k(\omega, A_k) = 1$. This implies the relation

$$P\{A_0\} = P\{A_k\} = EQ(\omega, A_k) \ge 1 - 2E|\vec{d}_k(\omega) - \vec{p}_k(\omega)|_1.$$

This completes the proof of statement (4).
Statement (2) implies that

$$\lim_{k\to\infty} E|\vec{d}_k(\omega) - \vec{p}_k(\omega)|_1 = 0.$$

This completes the proof of the theorem.

□

Now we consider the general ergodic properties of the compound stationary process $\{(\xi_k(\omega), P_k(\omega)), k \in \mathbf{Z}_+\}$ under the assumption that the conditional distribution of $\xi_n(\omega)$ with respect to the σ-algebra $\mathcal{E}$ is determined by a stationary distribution $\{\vec{d}_k(\omega),\ k \in \mathbf{Z}\}$.

Denote by Λ the set $\overline{1,r}^{\mathbf{Z}_+}$ with the cylindrical σ-algebra C_Λ; the elements of Λ will be denoted by $\lambda = (\xi_0, \xi_1, \dots)$, where $\xi_k \in \overline{1,r}$. We introduce a shift mapping from Λ onto Λ by τ:

$$\tau\lambda = (\xi_1, \xi_2, \dots) \quad \text{if} \quad \lambda = (\xi_0, \xi_1, \dots).$$

Let $\tilde{P}_{\vec{d}}$ be the probability measure on $\Omega \times \Lambda$ for which

$$\tilde{P}_{\vec{d}}(A \times \{\lambda : \xi_0 = i_0, \dots, \xi_n = i_n\}) = E_P 1_A d_0^{i_0}(\omega) \prod_{k=1}^{n} P_{k-1}(i_{k-1}, i_k, \omega). \tag{7.10}$$

Let T be the measurable mapping of Ω into Ω for which $P_k(\omega) = T^k\omega$, $k \in \mathbf{Z}$. Denote by $\mathcal{I}$ the σ-algebra of those measurable subsets $B \in \mathcal{E} \otimes C_\Lambda$ for which

$$\int |1_B(\omega, \lambda) - 1_B(T\omega, \tau\lambda)| \tilde{P}_{\vec{d}}(d\omega, d\lambda) = 0.$$

Next, we establish a lemma.

Lemma 1 *Assume that the random environment $\{P_k(\omega),\ k \in \mathbf{Z}\}$ is ergodic. Then for any $B \in \mathcal{I}$ there exists a random subset $S_B(\omega) \subset \overline{1,r}$ satisfying the following conditions:*
(a) *for all n*

$$1_B(\omega, \lambda) = 1_{\{\xi_n \in S_B(T^n\omega)\}};$$

(b)

$$E_{\tilde{P}}(1_B(\omega, \lambda)/\mathcal{E}) = \sum 1_{\{i \in S_B(\omega)\}} d_0^i(\omega) \quad \textit{is a constant.}$$

Here $E_{\tilde{P}}$ is the expectation with respect to the measure $\tilde{P}_{\vec{d}}$.

Proof Let $\theta = 1_B(\omega, \lambda)$,

$$\varphi_B(\xi_0, \omega) = E_{\tilde{P}}(\theta/\Xi_0 \vee \mathcal{E}),$$

where Ξ_k is the σ-algebra in Λ generated by $\xi_0, \xi_1, \dots, \xi_k$. It follows from the Markov property that

$$E_{\tilde{P}}(\theta/\Xi_n \vee \mathcal{E}) = \varphi_B(\xi_n, T^n\omega),$$

since

$$1_B(\omega, \lambda) = 1_B(T^n\omega, \tau^n\lambda).$$

Note that

$$\theta = \lim_{n\to\infty} E_{\tilde{P}}(\theta/\mathcal{E} \vee \Xi_n) = \lim_{n\to\infty} \varphi(\xi_n, T^n\omega).$$

Therefore,

$$\lim_{n\to\infty} E_{\tilde{P}}|\varphi(\xi_n, T^n\omega) - \varphi(\xi_{n+1}, T^{n+1}\omega)| = 0.$$

Since

$$E_{\tilde{P}}|\varphi(\xi_0, \omega) - \varphi(\xi_1, T^1\omega)| = E_{\tilde{P}}|\varphi(\xi_n, T^n\omega) - \varphi(\xi_{n+1}, T^{n+1}\omega)|,$$

we have

$$\varphi(\xi_0, \omega) = \varphi(\xi_1, T\omega) = \varphi(\xi_n, T^n\omega)$$

for all $n \in \mathbf{Z}$, and $1_B(\omega, \lambda) = \varphi(\xi_0, \omega)$.
Denote by $S(\omega)$ the set of those $i \in \overline{1,r}$ for which $\varphi(\xi_0, \omega) = 1$ if $i \in S(\omega)$. Then $S(\omega)$ satisfies condition (a) of the lemma. Statement (b) is a consequence of (a) and the ergodicity of $\{P_k(\omega)\}$.
This completes the proof of the lemma.

□

Remark 2 *The stochastic process*

$$E_{\tilde{P}}(\theta/\mathcal{E}) = \sum 1_{\{i\in S_B(T^n\omega)\}} d_n^i(\omega)$$

is invariant with respect to the mapping T, so it is a constant.

Two direct corollaries to this result follow.

Corollary 1 *The σ-algebra $\mathcal{I}$ is an atomic one: Assume that $E_{\tilde{P}}\theta > 0$. Then $E_{\tilde{P}}(\theta/\mathcal{E}) = E_{\tilde{P}}\theta$,*

$$P\{S_B(\omega) = \emptyset\} = 0, \quad \textit{and} \quad E_{\tilde{P}}(\theta/\mathcal{E}) \geq \min_i d_0^i(\omega) > 0.$$

Corollary 2 *The elements $p_n(i, j, \omega)$ of the matrix $P_n(\omega)$ satisfy the identities*

$$\sum p_n(i, j, \omega) 1_{\{j\in S_B(T^{n+1}\omega)\}} = 1 \quad \textit{for all} \quad i \in S_B(T^n\omega),$$
$$\sum p_0(i, j, \omega) 1_{\{j\in S_B(T^{n+1}\omega)\}} = 0 \quad \textit{for all} \quad i \in \overline{1,r} \setminus S_B(T^n\omega).$$

Corollary 3 *If*

$$c = \sum 1_{\{i\in S_B(\omega)\}} d_0^i(\omega), \quad 0 < c < 1,$$

then the vector stochastic processes

$$\vec{p}_n(\omega) = \frac{1}{c}\left(d_n^1(\omega) 1_{\{1\in S_B(T^n\omega)\}}, \ldots, d_n^r(\omega) 1_{\{r\in S_B(T^n\omega)\}}\right)$$

and

$$\vec{p}_n^{\,*}(\omega) = \frac{1}{1-c}\left(\vec{d}_n(\omega) - c\vec{p}_n(\omega)\right)$$

are stationary distributions for the random environment $\{P_n(\omega),\ n \in \mathbf{Z}\}$.

To prove this it suffices to verify that

$$\vec{p}_n(\omega)P_n(\omega) = \vec{p}_{n+1}(\omega),$$

which is a consequence of Corollary 2.

Remark 3 *It is easy to see that*

$$\|\vec{p}_n(\omega) - \vec{p}_n^{\,*}(\omega)\|_1 = 1.$$

Since

$$\big(\vec{p}_n(\omega) - \vec{p}_n^{\,*}(\omega)\big)P_n(\omega) = \big(\vec{p}_{n+1}(\omega) - \vec{p}_{n+1}^{\,*}(\omega)\big),$$

$\|P_n(\omega)\|_0 = 1.$

Definition 3 *A stationary distribution* $\{\vec{d}_n(\omega),\ n \in \mathbf{Z}\}$ *is called* irreducible *if for any stationary distribution* $\{\vec{p}_n(\omega),\ n \in \mathbf{Z}\}$ *such that* $p_n^i(\omega) = 0$ *for all* i *for which* $d_n^i(\omega) = 0$, *we have* $\vec{p}_n(\omega) = \vec{d}_n(\omega)$.

Theorem 3 *Let the random environment* $\{P_n(\omega),\ n \in \mathbf{Z}\}$ *be ergodic. The stationary process* $\{(\xi_k(\omega), P_k(\omega)),\ k \in \mathbf{Z}_+\}$ *with the distribution* $\tilde{P}_{\vec{d}}$ *given by formula (7.10) is ergodic if and only if the stationary distribution* $\{\vec{d}_n(\omega),\ n \in \mathbf{Z}_+\}$ *is irreducible.*

Proof It follows from Corollary 3 that if the σ-algebra $\mathcal{I}$ of the invariant subsets for the process $\{(\xi_k(\omega), P_k(\omega)),\ k \in \mathbf{Z}_+\}$ is not trivial, then there exists a stationary distribution $\vec{p}_n(\omega)$ satisfying the following conditions:

(1) $\vec{p}_n(\omega) \neq \vec{d}_n(\omega)$;

(2) $p_n^i(\omega) = 0$ if $d_n^i(\omega) = 0$ for all $i \in \overline{1,r}$, $n \in \mathbf{Z}_+$, $\omega \in \Omega$.

Assume that there exists a stationary distribution $\{\vec{p}_n(\omega),\ n \in \mathbf{Z}_+\}$ satisfying conditions (1) and (2). Set

$$S(\omega) = \{i : p_0^i(\omega) > 0\}$$

and

$$\theta(\lambda, \omega) = 1_{\{\xi_0 \in S(\omega)\}}.$$

It is easy to check that $\theta(\tau\lambda, T\omega) = \theta(\lambda, \omega)$. So $\theta(\lambda, \omega)$ is a nontrivial invariant function for the stochastic process $\{(\xi_n(\omega), P_n(\omega)),\ n \in \mathbf{Z}_+\}$. This means that the σ-algebra $\mathcal{I}$ is not trivial, and this completes the proof of the theorem.

□

Remark 4 *Assume that $B_1, \dots, B_s$ are all the atoms of the σ-algebra $\mathcal{I}$ with respect to the measure $\tilde{P}_{\vec{d}}$. Then the stationary distributions:*

$$\{\vec{d}_n(i,\omega),\, n \in \mathbf{Z}_+\}, \quad i = 1, \dots, s,$$

where

$$d_n^j(i,\omega) = \frac{1}{c_i} d_n^j(\omega) 1_{\{j \in S_{B_i}(T^n\omega)\}}, \quad c_i = \sum d_n^j(\omega) 1_{\{j \in S_{B_i}(T^n\omega)\}},$$

are irreducible.

Corollary 4 *If $\{\vec{d}_n(\omega), n \in \mathbf{Z}_+\}$ is the unique stationary distribution for an ergodic random environment $\{P_n(\omega),\, n \in \mathbf{Z}\}$, then the process $\{(\xi_n(\omega), P_n(\omega)),\, n \in \mathbf{Z}_+\}$ is ergodic.*

7.2 Weakly Random Environments

In this section we consider random environments of the form

$$P_n(\omega) = P_n^\varepsilon(\omega) = P_0 + \varepsilon Q_n(\omega), \quad n \in \mathbf{Z}, \tag{7.11}$$

where P_0 is a nonrandom matrix from $\mathcal{P}_r$, $\varepsilon > 0$ is a small parameter, and $Q_n(\omega)$ for fixed n and ω is a matrix from the space $\mathcal{Q}_r$ of matrices of order $r \times r$ with rows from $\mathcal{D}_r^0$. We refer to such a random environment as weakly random.

We assume that $\{Q_n(\omega),\, n \in \mathbf{Z}\}$ is a $\mathcal{Q}_r$-valued stationary process for which

$$P\{Q_n(\omega) \geq -cP_0\} = 1, \tag{7.12}$$

for some $c > 0$. (If A and B are matrices, then the relation $A \geq B$ means that all $a_{ij} \geq b_{ij}$ where a_{ij} are the elements of A and b_{ij} are elements of B.) The relation (7.12) implies that $P_n^\varepsilon(\omega) \in \mathcal{P}_r$ for $\varepsilon > 0$ small enough. The random environment $\{P_n(\omega),\, n \in \mathbf{Z}_+\}$ can be treated as a random perturbation of the transition probability P_0.

We assume that the stationary process $\{Q_n(\omega),\, n \in \mathbf{Z}\}$ is ergodic and satisfies the following ergodic mixing condition:

EMC Denote by $\mathcal{E}_l^k$ the σ-algebra generated by $\{Q_n(\omega),\, n \in \overline{k,l}\}$. Then there exists a sequence $\{\mu_l,\, l \in \mathbf{Z}_+\}$ for which $\mu_l \to 0$ and

$$|E\xi_1\xi_2 - E\xi_1 E\xi_2| \leq \mu_l$$

if ξ_1 is an $\mathcal{E}_n^{-\infty}$–measurable random variable and ξ_2 is a $\mathcal{E}_\infty^{n+l}$-measurable random variable, $|\xi_1| \leq 1$, $|\xi_2| \leq 1$.

If

$$E\alpha(P_0 + \varepsilon Q_0(\omega)) > 0 \quad \text{for } \varepsilon \text{ small enough}, \tag{7.13}$$

then Theorem 2 implies the existence of the limit

$$\tilde{\Pi}_\varepsilon = \lim_{n\to\infty} E \prod_{k=0}^{n} (P_0 + \varepsilon Q_k(\omega)) \tag{7.14}$$

and the perturbed transition probabilities satisfy the relation

$$\lim_{n\to\infty} \frac{1}{n} \sum_{k=1}^{n} P^\varepsilon_{0,k}(\omega) = \tilde{\Pi}_\varepsilon, \tag{7.15}$$

where for $k < l$,

$$P^\varepsilon_{k,l}(\omega) = P^\varepsilon_k(\omega) \cdots P^\varepsilon_{l-1}(\omega). \tag{7.16}$$

Set $\bar{Q} = EQ_0(\omega)$. Then

$$\|P_0 + \varepsilon\bar{Q}\|_0 \le E\|P_0 + \varepsilon Q_0(\omega)\|_0 < 1.$$

It follows from property IV of $\|\cdot\|_0$ that there exists the limit

$$\lim_{n\to\infty} (P_0 + \varepsilon\bar{Q})^n = \Pi_\varepsilon.$$

Theorem 4 (1) *Let relation (7.13) hold for all small $\varepsilon > 0$, and let the stationary process $\{Q_n(\omega),\ n \in \mathbf{Z}\}$ be ergodic. Suppose that $\alpha(P_0) > 0$. Then*

$$\|\tilde{\Pi}_\varepsilon - \Pi_\varepsilon\|_1 = O(\varepsilon^2).$$

(2) *Let* EMC *be satisfied and $\alpha(P_0) = 0$. Then*

$$\lim_{\varepsilon\to 0} \|\tilde{\Pi}_\varepsilon - \Pi_\varepsilon\|_1 = 0.$$

Proof (1) We use the representation

$$EP^\varepsilon_{0,n}(\omega) = P_0^n + \sum_{k\ge 1} \varepsilon^k \sum_{0\le i_1<i_2<\cdots<i_k<n} W(i_1,\ldots,i_k),$$

where

$$W(i_1,\ldots,i_k) = EP_0^{i_1} Q_{i_1}(\omega) P_0^{i_2-i_1-1} Q_{i_2}(\omega) \cdots Q_{i_k}(\omega) P_0^{n-1-i_k}.$$

Note that $Q_i(\omega)\Pi = 0$ if $\|\Pi\|_0 = 0$, since the rows of the matrix $Q_i(\omega)$ are from $\mathcal{D}^0_r$. Let

$$\Pi = \lim_{n\to\infty} P_0^n.$$

Then $\|\Pi\|_0 = 0$, since $\alpha(P_0) > 0$. Let $R_i = P^i - \Pi$. Then

$$W(i_1,\ldots,i_k) = P_0^{i_1} Q_{i_1}(\omega) R_{i_2-i_1-1} \cdots Q_{i_k} R_{n-1-i_k}.$$

There exist constants c and $0 < \theta < 1$ for which $\|R_i\| \le c\theta^i$. Since $\|Q_n(\omega)\| \le 2$ for $\varepsilon < 1$, we have that

$$\|W(i_1,\dots,i_k)\|_1 \le c^{k-1}2^k\theta^{n-i},$$

$$\sum_{0\le i_1<i_2<\cdots<i_k<n-1} W(i_1,\dots,i_k) \le c^{k-1}2^k\left(\frac{1}{1-\theta}\right)^k,$$

and

$$EP^{\varepsilon}_{0,n}(\omega) = P_0^n + \varepsilon\sum_{i=0}^{n-1} W(i) + O(\varepsilon^2).$$

In the same way we can obtain the relation

$$(P_0+\varepsilon\bar{Q})^n = P_0 + \varepsilon\sum_{i=0}^{n-1}\bar{W}(i) + O(\varepsilon^2),$$

where

$$\bar{W}(i) = P_0^i\bar{Q}P_0^{n-1-i}.$$

It is easy to see that $\bar{W}(i) = W(i)$. So

$$\|(P_0+\varepsilon\bar{Q})^n - EP^{\varepsilon}_{0,n}(\omega)\|_1 \le c_1\varepsilon^2,$$

where the constant c_1 does not depend on n. This relation implies statement (1) of the theorem.
(2) Let $n, l, r, \in \mathbf{Z}_+$. It follows from EMC that

$$\left\|E\prod_{k=0}^{s-1}P^{\varepsilon}_{k(n+l),k(n+l)+n}(\omega) - \left(EP^{\varepsilon}_{0,n}(\omega)\right)^s\right\|_1 \le c_2 s\mu_l, \tag{7.17}$$

where c_2 is a constant that does not depend on n, l, s.
Let $\tilde{Q}_k(\omega) = Q_k(\omega) - \bar{Q}$. Then

$$\|P^{\varepsilon}_{k,k+l}(\omega) - (P_0+\varepsilon\bar{Q})^l\|_1 \le l\varepsilon b_1, \tag{7.18}$$

where b_1 is a constant for which $\|\tilde{Q}_k(\omega)\| \le b_1$. Relations (7.17) and (7.18) imply the inequality

$$\left\|EP^{\varepsilon}_{0,nr}(\omega) - \left(EP^{\varepsilon}_{0,n}(\omega)\right)^s\right\|_1 \le c_3 s(\mu_l + s\varepsilon). \tag{7.19}$$

Set

$$\tilde{W}(i_1,\dots,i_k) = E(P_0+\varepsilon\bar{Q})^{i_1}\tilde{Q}_{i_1}(\omega)(P_0+\varepsilon\bar{Q})^{i_2-i_1-1}\cdots\tilde{Q}_{i_k}(\omega)(P_0+\varepsilon\bar{Q})^{n-1-i_k}. \tag{7.20}$$

Then

$$EP^{\varepsilon}_{0,n}(\omega) - \left(P_0+\varepsilon\bar{Q}\right)^n = \sum_{k\ge 2}\varepsilon^k\sum_{0\le i_1<\cdots<i_k<n}\tilde{W}(i_1,\dots,i_k).$$

Since $\|\tilde{W}(i_1, \ldots, i_k)\| \le b_1^k$, we have that

$$\left\|EP^{\varepsilon}_{0,n}(\omega)-(P_0+\varepsilon\bar{Q})^n\right\|_1 \le \sum_{k\ge 2} b_1^k \,\frac{n(n-1)\cdots(n-k+1)}{k!} \le e^{n\varepsilon b_1}-1-n\varepsilon b_1.$$

Let $\varepsilon n = \lambda$. Then the last inequality and (7.19) imply the relation

$$\left\|EP^{\varepsilon}_{0,ns}(\omega) - (P_0+\varepsilon\bar{Q})^{ns}\right\|_1 \le c_3 s\left(\mu_l + \lambda\frac{l}{n}\right) + s(\varepsilon^{b_1\lambda} - 1 - b_1\lambda). \quad (7.21)$$

In the same way in which inequality (7.19) was obtained, we can obtain the following one: For $m > ns$,

$$\left\|EP^{\varepsilon}_{0,m}(\omega) - EP^{\varepsilon}_{0,m-ns}(\omega)EP^{\varepsilon}_{0,ns}(\omega)\right\|_1 \le c_3\left(\mu_l + \lambda\frac{l}{n}\right). \quad (7.22)$$

Property VII implies that

$$\left\|EP^{\varepsilon}_{0,m-ns}(\omega)EP^{\varepsilon}_{0,ns}(\omega) - EP^{\varepsilon}_{0,ns}(\omega)\right\|_1 \le 2\left\|EP^{\varepsilon}_{0,ns}(\omega)\right\|_0. \quad (7.23)$$

In the same way we can write

$$\left\|(P_0+\varepsilon\bar{Q})^m - (P_0+\varepsilon\bar{Q})^{ns}\right\|_1 \le 2\left\|(P_0+\varepsilon\bar{Q})^{ns}\right\|_0. \quad (7.24)$$

Since

$$\left\|EP^{\varepsilon}_{0,ns}(\omega)\right\|_0 \le 2\left\|(P_0+\varepsilon\bar{Q})^{ns}\right\|_0 + \left\|EP^{\varepsilon}_{0,ns} - (P_0+\varepsilon\bar{Q})^{ns}\right\|_1,$$

inequalities (7.21)–(7.24) imply that $\|EP^{\varepsilon}_{0,m} - (P_0+\varepsilon\bar{Q})^m\|_1$

$$\le c_3(r+1)\,(\mu_l + \lambda l/n) + s(\varepsilon^{b_1\lambda} - 1 - b_1\lambda) + 4\left\|(P_0+\varepsilon\bar{Q})^{ns}\right\|_0.$$

There exists $\beta > 0$ for which

$$\alpha(P_0+\varepsilon\bar{Q}) \ge \beta\varepsilon,$$

so

$$\left\|(P_0+\varepsilon\bar{Q})^{ns}\right\|_0 \le e^{-ns\beta\varepsilon} = e^{-\lambda s\beta}.$$

Thus,

$$\limsup_{m\to\infty}\left\|EP^{\varepsilon}_{0,m} - (P_0+\varepsilon\bar{Q})^m\right\|_1 \le c_3(s+1)\left(\mu_l + \lambda\frac{l}{n}\right) + \frac{s\lambda^2 b_1^2}{2}e^{b_1\lambda} + e^{-\lambda s\beta};$$

and for $n\to\infty$, $\varepsilon n\to\lambda$, $\varepsilon\to 0$, we have for fixed s and l,

$$\limsup_{\varepsilon\to 0, m\to\infty}\left\|EP^{\varepsilon}_{0,m} - (P_0+\varepsilon\bar{Q})^m\right\|_1 \le c_3(S+1)\mu_l + \frac{s\lambda^2 b_1^2}{2}e^{b_1\lambda} + e^{-\lambda s\beta}.$$

The expression on the right-hand side of the inequality tends to zero if $l\to\infty$, $\lambda s\to\infty$, $\lambda^2 s\to 0$. This completes the proof of the theorem.

□

Next, we investigate $\lim_{\varepsilon\to 0}\Pi_\varepsilon$. For this purpose we need some additional facts about matrices $P_0 \in \mathcal{P}_r$.

Definitions (1) $P_0 \in \mathcal{P}_r$ *is a* reducible matrix *if there exist two different vectors* $\vec{d}_1$, $\vec{d}_2$ *from* $\mathcal{D}_r^1$ *for which*

$$\vec{d}_i P_0 = \vec{d}_i, \quad i = 1, 2.$$

(2) $P_0 \in \mathcal{P}_r$ *is an* aperiodic matrix *if there exists* $\lim_{n\to\infty} P_0^n$.
(3) *The least number* $d \in \mathbf{Z}_+$ *for which there exists* $\lim_{n\to\infty} P_0^{nd}$ *is called the* period *of the matrix* P_0.

Remark 5 (a) *If* P_0 *is a reducible matrix, then* $\|P_0\|_0 = 1$.
(b) *If* P_0 *is an aperiodic matrix, then* $d = 1$. *If* $d > 1$, *then* $\|P_0\|_0 = 1$ *because of the relation* $\|P_0^d\|_0 = 1$.
(c) *For any matrix* $P_0 \in \mathcal{P}_r$ *there exists a period.*
(d) *If* P_0 *is an aperiodic matrix, then the matrix* $\Pi_0 = \lim_{n\to\infty} P_0^n$ *satisfies the relations*

$$\Pi_0 P_0 = P_0 \Pi_0 = \Pi_0, \quad \Pi_0^2 = \Pi_0.$$

(e) *If* d *is the period of the matrix* P_0, *and* $\Pi_0 = \lim_{n\to\infty} P_0^{nd}$, *then the matrices* $\Pi_0, \Pi_0 P_0, \dots, \Pi_0 P_0^{d-1}$ *are distinct, and*

$$\Pi_0 P_0^d = P_0^d \Pi_0 = \Pi_0, \quad \Pi_0^2 = \Pi_0.$$

The proof of these statements can be found in [54, p.117, Theorem 12].

Theorem 5 *Suppose that the conditions of Theorem* 4 *are satisfied. Then*
(1) *If* $\alpha(P_0) > 0$, *then*

$$\lim_{\varepsilon\to 0} \Pi_\varepsilon = \Pi_0 = \lim_{n\to\infty} P_0^n.$$

(2) *If* $\alpha(P_0) = 0$, *but* P_0 *is an aperiodic matrix, then*

$$\lim_{\varepsilon\to 0} \Pi_\varepsilon = \lim_{t\to\infty} \Pi_0 \exp\{t \Pi_0 \bar{Q} \Pi_0\},$$

where $\Pi_0 = \lim_{n\to\infty} P_0^n$.
(3) *If* $d > 1$, *then*

$$\lim_{\varepsilon\to 0} \Pi_\varepsilon = \lim_{t\to\infty} \exp\{tR\},$$

where $\Pi_0 = \lim_{n\to\infty} P_0^{nd}$,

$$R = \sum_{i=0}^{d-1} \Pi_0 P_0^i \bar{Q} P_0^{d-i-1} \Pi_0.$$

Proof We use the relation

$$\left(P_0 + \varepsilon\bar{Q}\right)^n = P_0^n + \sum_{k\geq 1} \varepsilon^k \sum_{0\leq i_1<\cdots<i_k<n} P_0^{i_1}\bar{Q}\cdots P_0^{i_k - i_{k-1}-1}\bar{Q} P_0^{n-i_k-1}. \tag{7.25}$$

Note that

$$\left\|(P_0 + \varepsilon\bar{Q})^n - \Pi_\varepsilon\right\|_1 \leq 2\|(P_0 + \varepsilon\bar{Q})^n\|_0 \leq 2\exp\{-n\alpha(P_0 + \varepsilon\bar{Q})\}. \tag{7.26}$$

So

$$\|P_0^n - \Pi_\varepsilon\| \le 2\exp\{-n\alpha(P_0 + \varepsilon\bar{Q})\} + \sum_{k\ge 1} \frac{(\varepsilon n)^k}{k!} b_1^k, \tag{7.27}$$

where $b_1 > 0$ is a constant for which $\|\bar{B}\|_1 \le b_1$. If $\alpha(P_0) > 0$, then $\alpha(P_0 + \varepsilon\bar{Q}) > \frac{1}{2}\alpha(P_0)$ for $\varepsilon > 0$ small enough. Therefore, relation (7.27) implies statement (1).
Now we consider the proof of statement (2). A sequence $\gamma_k \to 0$ exists for which $\gamma_{k+1} \le \gamma_k$ and for which

$$\|P_0^k - \Pi_0\| \le \gamma_k.$$

For any $l \in \mathbf{Z}_+$ we have the inequality

$$\Big\| \sum_{0\le i_1<\cdots<i_k<n} P_0^{i_1}\bar{Q}\cdots P_0^{i_k-i_{k-1}-1}\bar{Q}P_0^{n-i_k-1}$$

$$\times\big(1 - 1_{\{i_1\ge l,\ldots,i_k-i_{k-1}-1\ge l, n-i_k-1\ge l\}}\big)\Big\|_1 \le \frac{n^{k-1}l}{(k-1)!} b_1^k.$$

This implies the following estimate:

$$\Big\| \sum_{0\le i_1<\cdots<i_k<n} \big(P_0^{i_1}\bar{Q}\cdots P_0^{i_k-i_{k-1}-1}\bar{Q}P_0^{n-i_k-1} - \Pi_0\bar{Q}\cdots\Pi_0\bar{Q}\Pi_0\big)\Big\|$$

$$\le \frac{2n^{k-1}lb_1^k}{(k-1)!} + \frac{2^{k-1}n^k b_1^k \gamma_l}{k!}. \tag{7.28}$$

It follows from relations (7.25) and (7.28) that

$$(P_0 + \varepsilon\bar{B})^n = \Pi_0 \exp\{\varepsilon n \Pi_0\bar{B}\Pi_0\} + O\Big(\Big(\frac{l}{n} + \gamma_l\Big)\exp\{2\varepsilon n b_1\}\Big).$$

Taking into account relation (7.26), we can write

$$\big\|\Pi_\varepsilon - \Pi_0 \exp\{\varepsilon n\Pi_0\bar{B}\Pi_0\}\big\|_1 \le 2\exp\{-\varepsilon n\beta\} + O\left(\left(\frac{l}{n} + \gamma_l\right)\exp\{2\varepsilon n b_1\}\right), \tag{7.29}$$

where β is the same as in the proof of Theorem 4.
Let $\varepsilon \to 0$, $n \to \infty$, $l \to \infty$, $l/n \to 0$, and $\varepsilon n \to t$. Then

$$\limsup_{\varepsilon\to 0}\big\|\Pi_\varepsilon - \Pi_0\exp\{\varepsilon n\Pi_0\bar{B}\Pi_0\}\big\|_1 \le 2\exp\{-t\beta\}. \tag{7.30}$$

This relation implies statement (2).
To prove statement (3) we use the representation

$$(P_0 + \varepsilon\bar{B})^d = \big(P_0^d + \varepsilon R + O(\varepsilon^2)\big).$$

It is easy to check that

$$\big\|(P_0 + \varepsilon\bar{B})^{nd} - (P_0^d + \varepsilon R)^n\big\|_1 = O(n\varepsilon^2).$$

Since P_0^d is an aperiodic matrix, statement (3) follows from statement (2). □

Remark 6 *Let*

$$\hat{\Pi}_0 = \lim_{t\to\infty} \Pi_0 \exp\{t\hat{R}\},$$

where $\hat{R} = \Pi_0 \bar{Q} \Pi_0$ *in case* (2) *and* $\hat{R} = R$ *in case* (3).
Note that under the conditions of the theorem, $\Pi_0 \exp\{t\hat{R}\} \in \mathcal{P}_r$ *and* $\|\Pi_0 \exp\{t\hat{R}\}\|_0 < 1$. *The matrix* $\hat{\Pi}_0$ *satisfies the relation*

$$\hat{\Pi}_0 \Pi_0 \exp\{t\hat{R}\} = \hat{\Pi}_0 \quad \textit{for all } t > 0,$$

where $\hat{\Pi}_0 \in \mathcal{P}_r$, $\|\hat{\Pi}_0\|_0 = 0$, *and* $\hat{\Pi}_0$ *is the unique solution of the equation*

$$\hat{\Pi}_0 \hat{R} = \hat{\Pi}_0, \quad \hat{\Pi}_0 \in \mathcal{P}_r.$$

Thus, the limit matrix $\hat{\Pi}_0$ is one having identical rows.

7.3 Markov Processes with Randomly Perturbed Transition Probabilities

Let $\{P_{s,t}(\omega),\ -\infty < s \le t < \infty\}$ be a $\mathcal{P}_r$-valued random function satisfying the following conditions:

(i) $P_{s,t}(\omega)$ is right continuous in s, t,

(ii) $P_{t,t}(\omega) = I$,

(iii) $P_{s,t}(\omega) P_{t,u}(\omega) = P_{s,u}(\omega)$ if $s < t < u$.

We will call this random function a *random environment.*
Denote by $\mathcal{E}$ the σ-algebra generated by the random variables

$$\{P_{s,t}(\omega),\ -\infty < s \le t < \infty\}. \tag{7.31}$$

Consider an $\overline{1,r}$-valued stochastic process $\{\xi(t,\omega),\ t \in R\}$, and let Ξ_t be the σ–algebra generated by the random variables $\{\xi(s,\omega),\ s \le t\}$. Here $\xi(t,\omega)$ is called a Markov process in the random environment (7.31) if

$$P\{\xi(t,\omega) = j/\mathcal{E} \vee \Xi_s\} = p_{s,t}(\xi(s,\omega), j, \omega), \tag{7.32}$$

where $p_{s,t}(i,j,\omega)$, for $i, j \in \overline{1,r}$, are the elements of the matrix $P_{s,t}(\omega)$.

7.3.1 Stationary Random Environments

We assume that the random environment (7.31) satisfies an additional condition:

(iv) For all $h > 0$ the stochastic process $P_{t,t+h}(\omega)$ is a stationary–in–t $\mathcal{P}_r$-valued stochastic process.

Such a random environment is called stationary. Here we will consider random environments satisfying a stronger condition:

(v) The limit

$$A(t,\omega) = \lim_{h\to 0} \frac{1}{h}\left(P_{t,t+h}(\omega) - I\right) \tag{7.33}$$

exists in probability and $\{A(t,\omega),\ t \in R\}$ is a $\tilde{\mathcal{Q}}_r$-valued stationary process, where $\tilde{\mathcal{Q}}_r$ is the space of those $r \times r$ matrices Q for which $I + \varepsilon Q \in \mathcal{P}_r$ for $\varepsilon > 0$, $\varepsilon < \|Q\|_1$, and $E\|A(t,\omega)\|_1 < \infty$.

It is easy to check that condition (v) implies the following differential equations for $P_{s,t}(\omega)$:

$$\frac{\partial P_{s,t}}{\partial s}(\omega) = A(s,\omega)P_{s,t}(\omega), \tag{7.34}$$

$$\frac{\partial P_{s,t}}{\partial t}(\omega) = P_{s,t}(\omega)A(t,\omega). \tag{7.35}$$

Definition *An $\mathcal{E}$-measurable $\mathcal{D}_r^1$-valued stochastic process $\vec{d}(t,\omega)$ is called a stationary distribution for the random environment (7.31) if it satisfies the following properties:*
(a) *$\{(A(s,\omega),\vec{d}(s,\omega)),\ s \in R\}$ is a $\tilde{\mathcal{Q}}_r \times \mathcal{D}_r^1$-valued stationary process.*
(b) *For $s < t$,*

$$\vec{d}(t,\omega) = \vec{d}(s,\omega)P_{s,t}(\omega).$$

It follows from (7.33) that a stationary distribution $\vec{d}(t,\omega)$ satisfies the differential equation

$$\frac{d\vec{d}}{dt}(t,\omega) = \vec{d}(t,\omega)A(t,\omega). \tag{7.36}$$

Lemma 2 *Assume that a $\mathcal{D}_r^1$-valued stationary process $\vec{d}(t,\omega)$ having property (a) satisfies relation (7.36). Then $\vec{d}(t,\omega)$ is a stationary distribution.*

Proof Relation (7.36) implies that

$$\begin{aligned}\vec{d}(t,\omega) &= \vec{d}(s,\omega) + \int_s^t \vec{d}(u,\omega)A(u,\omega)du \\ &= \int_s^t [\vec{d}(u,\omega) - \vec{d}(s,\omega)P_{s,u}(\omega)]A(u,\omega)du + \vec{d}(s,\omega) \\ &\quad + \vec{d}(s,\omega)\int_s^t P_{s,u}(\omega)A(u,\omega)du \\ &= \int_s^t [\vec{d}(u,\omega) - \vec{d}(s,\omega)P_{s,u}(\omega)]A(u,\omega)du + \vec{d}(s,\omega)P_{s,t}(\omega),\end{aligned}$$

and

$$\vec{d}(t,\omega) - \vec{d}(s,\omega)P_{s,t}(\omega) = \int_s^t [\vec{d}(u,\omega) - \vec{d}(s,\omega)P_{s,u}(\omega)]A(u,\omega)du. \quad (7.37)$$

Set

$$\rho(t) = \left\|\vec{d}(t,\omega) - \vec{d}(s,\omega)P_{s,t}(\omega)\right\|, \quad F(t) = \int_s^t \|A(u,\omega)\|_1 du.$$

Then $\rho(t) \leq 2$, $F(t)$ is an increasing continuous function, and

$$\rho(t) \leq \int_s^t \rho(u)dF(u), \quad t > s.$$

This implies that $\rho(t) = 0$.

□

Lemma 3 *If condition* (v) *is satisfied, then there exists a stationary distribution.*

Proof Let $\vec{a} \in \mathcal{D}_r^1$. For $T > 0$ set

$$\vec{d}_T(t,\omega) = \frac{1}{T}\int_{t-T}^t \vec{a}P_{s,t}(\omega)ds. \quad (7.38)$$

It is easy to check that the process $\{(\vec{d}_T(\omega), A(t,\omega)),\ t \in R\}$ is stationary and $\vec{d}_T(t,\omega)$ satisfies the differential equation

$$\frac{d\,\vec{d}_T}{dt}(t,\omega) = \vec{d}_T(t,\omega)A(t,\omega) + \frac{1}{T}\big(\vec{a} - \vec{a}P_{t-T,t}(\omega)\big).$$

We can choose a sequence for which the joint distributions of the processes $\vec{d}_{T_n}(t,\omega)$, $A(t,\omega)$ converge to the joint distributions of some pair of processes $\tilde{d}(t,\omega)$ and $\tilde{A}(t,\omega)$. The remainder of the proof is the same as in Theorem 1.

□

Lemma 4 *Suppose*
(1) *Let* $C \in \tilde{\mathcal{Q}}_r$, *then* $e^{tC} \in \mathcal{P}_r$ *for all* $t > 0$.

(2) *For all* $C \in \tilde{Q}_r$ *there exists the nonnegative limit*

$$\hat{\alpha}(C) = \lim_{t \to 0} \frac{1}{t} \alpha(e^{tC}). \tag{7.39}$$

If $\|e^{tC}\|_0 < 1$ *for all* t *small enough, then* $\hat{\alpha}(C) > 0$.

Proof (1) $e^{tC} = \lim_{n \to \infty} \big(I + \frac{t}{n} C\big)^n$, and $I + \frac{t}{n} C \in \mathcal{P}_r$ for t small enough. (2) Let $P \in \mathcal{P}_r$. Then $\|P\|_0 < 1$ if and only if 1 is a simple eigenvalue of the matrix P and all other eigenvalues λ satisfy the relation $|\lambda| < 1$. Since the matrix C satisfies the condition $\|e^{tC}\| < 1$ for all $t > 0$, the number 0 is a simple eigenvalue for the matrix C, and all other eigenvalues λ of C satisfy the relation $\operatorname{Re} \lambda < 0$. The proof of statement (2) follows from the Jordan representation of the matrix e^{tC}.

□

As a direct corollary of this result, we have the following result:

Corollary 5

$$\|P_{s,t}(\omega)\|_0 \le \exp\Big\{ - \int_s^t \hat{\alpha}\big(A(u,\omega)\big)\, du \Big\}. \tag{7.40}$$

Theorem 6 *Assume that* $\{A(t,\omega),\, t \in R\}$ *is an ergodic stationary process and*

$$E\hat{\alpha}\big(A(u,\omega)\big) > 0.$$

Then the following statements are valid:

(1) *There exists a unique stationary distribution* $\{\vec{d}(t,\omega),\, t \in R\}$, *and for all* $\vec{a} \in \mathcal{D}^1_r$,

$$\lim_{t \to \infty} \big| \vec{d}(t,\omega) - \vec{a} P_{0,t}(\omega) \big|_1 = 0$$

almost surely.

(2) *There exists the limit*

$$\lim_{t \to \infty} E P_{0,t}(\omega) = \tilde{\Pi}.$$

(3)

$$\lim_{T \to \infty} \frac{1}{T} \int_0^T P_{0,t}(\omega)\, dt = \tilde{\Pi} \quad \textit{almost surely.}$$

These statements can be proved in the same way as statements (1), (2), (3) of Theorem 2, and the details are left to the reader.

7.3.2 Ergodic Theorem for Markov Processes in Random Environments

Let $\{\vec{d}(t,\omega),\ t \in R\}$ be a stationary process. We consider the stochastic process $\{(\xi(t), A(t,\omega)),\ t \in R\}$ in the space $\overline{1,r} \times \tilde{\mathcal{Q}}_r$ for which

$$\begin{aligned} &P\{\xi(t_1) = i_1, \dots, \xi(t_k) = i_k/\mathcal{E}\} \\ &= d^{i_1}(t_1,\omega) p_{t_1,t_2}(i_1, i_2, \omega) \cdots p_{t_{k-1},t_k}(i_{k-1}, i_k, \omega), \end{aligned} \tag{7.41}$$

where $\mathcal{E}$ is the σ-algebra generated by the process $\{A(t,\omega),\ t \in R\}$; $d^i(t,\omega)$ are the elements of the vector $\vec{d}(t,\omega)$; $p_{s,t}(i,j,\omega)$ are the elements of the matrix $P_{s,t}(\omega)$; and $P_{s,t}(\omega)$ is connected with the process $A(t,\omega)$ by equations (7.34) and (7.35).
We consider the probability space Ω of all $\tilde{\mathcal{Q}}_r$-valued functions $\omega = Q(\cdot)$ with the cylindrical σ-algebra in this space, and the distribution of the process $A(t,\omega)$ defines the space's measure P. Denote by Y the space of $\overline{1,r}$-valued functions $y(t)$ defined on R and denote by $\tilde{P}_{\vec{d}}$ the measure on $Y \times \Omega$ that is the distribution of the stochastic process $\{(\xi(t), A(t,\omega)),\ t \in R\}$ satisfying equation (7.41). This is a stationary process.

Definition *A stationary distribution* $\{\vec{d}(t,\omega),\ t \in R\}$ *is called* irreducible *if for any stationary distribution* $\{\vec{p}(t,\omega),\ t \in R\}$ *satisfying the condition* $p^i(t,\omega) = 0$ *for all* i *for which* $d^i(t,\omega) = 0$, *we have* $\vec{p}(t,\omega) = \vec{d}(t,\omega)$ *(where* $p^i(\cdot)$ *are the coordinates of the vector* $\vec{p}(\cdot)$*).*

Theorem 7 *If the stationary process* $\{A(t,\omega),\ t \in R\}$ *is ergodic and the stationary distribution* $\vec{d}(t,\omega)$ *is irreducible, then the compound stationary process* $\{(\vec{d}(t,\omega), A(t,\omega)),\ t \in R\}$ *is ergodic.*

The proof of the theorem is based on the description of the σ-algebra $\mathcal{I}$ of invariant sets for the compound process: For every $B \in \mathcal{I}$ there exists a random subset $S_B(\omega) \subset \overline{1,r}$ satisfying the following conditions:

(a) For all $t \in R$

$$1_B(y(\cdot),\omega) = 1_{\{y(t) \in S_B(T^t\omega)\}}$$

where $T^t\omega(s) = \omega(s+t)$.

(b)

$$E_{\tilde{P}}\big(1_B(\xi(\cdot),\omega)/\mathcal{E}\big) = \sum 1_{\{i \in S_B(\omega)\}} d^i(0,\omega),$$

where $E_{\tilde{P}}$ is the expectation with respect to measure $\tilde{P}_{\vec{d}}$.

This can be proved in the same way as in Lemma 1. The remainder of the proof is an obvious modification of the proof of Theorem 3. The details are left to the reader.

Corollary 6 (Ergodic theorem for a Markov process in a random environment). *Let the conditions of Theorem 7 be satisfied, and let $g(y,\omega)$ be a function for which*

$$E\sum_{i=1}^{r}|g(i,\omega)|d^i(0,\omega)<\infty.$$

Then

$$\lim_{t\to\infty}\frac{1}{t}\int_0^t g(\xi(s),T^s\omega)\,ds=E\sum_{i-1}^{r}g(i,\omega)d^i(0,\omega)$$

with probability 1.

7.3.3 Markov Process in a Weakly Random Environment

Let $\xi(t)$ be a homogeneous Markov process with phase space $\overline{1,r}$. Its transition probabilities $p_{ij}(t)$ satisfy the differential equations

$$\frac{d}{dt}P(t)=AP(t)=P(t)A,$$

where $P(t)$ is the matrix with elements $p_{ij}(t)$ and

$$A=\lim_{t\to 0}\frac{1}{t}(P(t)-I) \tag{7.42}$$

is a matrix from $\tilde{\mathcal{Q}}_r$.
We will next consider ergodic properties of a Markov process in a random environment generated by a $\tilde{\mathcal{Q}}_r$-valued stationary process of the form

$$A^\varepsilon(t,\omega)=A+\varepsilon B(t,\omega)$$

where $A\in\tilde{\mathcal{Q}}_r$, $\{B(t,\omega),\ t\in R\}$ is a $\tilde{\mathcal{Q}}_r$-valued bounded stationary process and $\varepsilon>0$ is sufficiently small.
Denote by $P^\varepsilon_{s,t}(\omega)$ the solutions of the differential equations

$$\frac{\partial P^\varepsilon_{s,t}(\omega)}{\partial s}=(A+\varepsilon B(s,\omega))P^\varepsilon_{s,t}(\omega),\quad s<t,\ P^\varepsilon_{t,t}(\omega)=I. \tag{7.43}$$

They also satisfy the "forward" equations

$$\frac{\partial P^\varepsilon_{s,t}}{\partial t}(\omega)=P^\varepsilon_{s,t}(\omega)(A+\varepsilon B(t,\omega)),\quad t>s. \tag{7.44}$$

We will assume that

$$E\hat{\alpha}(A+\varepsilon B(t,\omega))>0\quad \text{for } \varepsilon \text{ small enough.} \tag{7.45}$$

Let $\bar{B}=EB(t,\omega)$. Then $\hat{\alpha}(A+\varepsilon\bar{B})>0$.
Let the $\mathcal{P}_r$-valued function $\bar{P}^\varepsilon(t)$ satisfy the differential equation

$$\frac{d\bar{P}^\varepsilon}{dt}(t)=(A+\varepsilon\bar{B})\bar{P}^\varepsilon(t) \tag{7.46}$$

with the initial condition $\bar{P}^\varepsilon(0) = I$. Set $\tilde{B}(t,\omega) = B(t,\omega) - \bar{B}$.
With these assumptions and notation, we have the following lemma.

Lemma 5 *Let $t_0 < t$. Then*

$$P^\varepsilon_{t_0,t}(\omega) = \bar{P}^\varepsilon(t - t_0)$$

$$+ \sum_{k=1}^{\infty} \varepsilon^k \int \cdots \int_{t_0 < s_1 < \cdots < s_k < t} \bar{P}^\varepsilon(s_1 - t)\tilde{W}^\varepsilon(s_1, \ldots, s_k)\bar{P}^\varepsilon(t - s_k) ds_1 \cdots ds_k, \tag{7.47}$$

where

$$\tilde{W}^\varepsilon(s_1, \ldots, s_k) = \tilde{B}(s_1, \omega)\bar{P}^\varepsilon(s_2 - s_1)\tilde{B}(s_2, \omega) \cdots \bar{P}^\varepsilon(s_k - s_{k-1})\tilde{B}(s_k, \omega). \tag{7.48}$$

These formulas are consequences of equation (7.43).

Remark 7 *If $P(t)$ is the solution of equation (7.42), then*

$$P^\varepsilon_{t_0,t}(\omega) = P(t)$$

$$+ \sum_{k=1}^{\infty} \varepsilon^k \int \cdots \int_{t_0 < s_1 < \cdots < s_k < t} P(s_1 - t)W(s_1, \ldots, s_k)P(t - s_k) ds_1 \cdots ds_k, \tag{7.49}$$

where

$$W(s_1, \ldots, s_k) = B(s_1, \omega)P(s_2 - s_1)B(s_2, \omega) \cdots P(s_k - s_{k-1})B(s_k, \omega). \tag{7.50}$$

Using these representations for $P^\varepsilon_{s,t}(\omega)$, we can obtain the following extension of Theorem 4 to continuous–time processes:

Theorem 8 *Let condition (7.45) hold.*
(1) *If $\hat{\alpha}(A) > 0$, then*

$$\limsup_{t\to\infty} \|EP^\varepsilon_{0,t}(\omega) - \bar{P}^\varepsilon(t)\| = O(\varepsilon^2).$$

(2) *If the stochastic process $B(t,\omega)$ satisfies the condition*
EMC

$$|E\xi_1\xi_2 - E\xi_1 E\xi_2| \le \mu(t), \quad t > 0,$$

where ξ_1 is an $\mathcal{E}^{-\infty}_s$–measurable and ξ_2 is an $\mathcal{E}^{s+t}_\infty$-measurable random variable, $|\xi_1| \le 1$, $|\xi_2| \le 1$, and $\mu(t) \to 0$ as $t \to \infty$ (here $\mathcal{E}^s_t$ is the σ-algebra generated by random variables $\{B(u,\omega),\, u \in [s,t]\}$), then

$$\lim_{\varepsilon\to 0} \limsup_{t\to\infty} \|EP^\varepsilon_{0,t}(\omega) - \bar{P}^\varepsilon(t)\| = 0.$$

Remark 8 *There exists the limit*

$$\Pi_0 = \lim_{t\to\infty} P(t).$$

If $\hat{\alpha}(A) > 0$, then

$$\lim_{\varepsilon\to 0}\lim_{t\to\infty} \bar{P}^{\varepsilon}(t) = \Pi_0.$$

If $\hat{\alpha}(A) = 0$, then

$$\lim_{\varepsilon\to 0}\lim_{t\to\infty} \bar{P}^{\varepsilon}(t) = \lim_{t\to\infty} \Pi_0 \exp\{tR\},$$

where $R = \Pi_0 \bar{B} \Pi_0$.

If condition (7.45) holds, then $\hat{\alpha}(R) > 0$ and

$$\left\| \lim_{t\to\infty} \Pi_0 \exp\{tR\} \right\|_0 = 0.$$

The proof of these statements is similar to the proof of Theorem 5.

8

Randomly Perturbed Mechanical Systems

An important aspect of mechanical systems is that they often experience bifurcations. For example, in a conservative system, as the energy increases through a local maximum of the potential energy, the system can pass through a saddle–saddle connection and go from having two possible oscillations to having only one. On the other hand, when dissipation is present, local minima of the potential function become stable equilibria. For example, in the case of a two–well potential function, which is the basis of binary quantum–mechanical devices, movement between the two states due to random fluctuations is of great interest. Such systems are difficult to study. We present in this chapter a method for studying these kinds of behaviors in two–dimensional systems that is based on graphs.
Next, we consider random perturbations of oscillatory linear systems, and at the end of this chapter we consider random perturbations of rigid–body motions.

8.1 Conservative Systems with Two Degrees of Freedom

We consider in detail the case of a conservative system with two degrees of freedom. First, we study aspects of the nonrandom problem by formulating a description of the system's dynamics using a potential function graph. Next, we study the same problem, but now perturbed by random noise when the system is on graph edges and not near a graph knot. Finally, we investigate the behavior of the system when it is near a knot where bifurcations can occur.

8.1.1 Conservative Systems

The position of a particle of unit mass moving in a straight line in a frictionless medium is described by the differential equation

$$\ddot{x}(t) = -\frac{\partial U}{\partial x}(x(t)), \tag{8.1}$$

where $x(t)$ is the position of the particle at time t, $\dot{x} = dx/dt$ is its velocity, and $U(x)$ is a smooth function from R into R that describes the potential energy of the system.

Suppose that $U(x)$ satisfies the following conditions:

($\mathcal{P}$) $U(x) \to \infty$ as $|x| \to \infty$, its second derivative $U''(x) = \partial^2 U/\partial x^2$ exists and is continuous, the set of extremals $\{x : U'(x) = 0\}$ is finite, and $U''(x) \neq 0$ if $U'(x) = 0$.

Equation (8.1) can be rewritten in terms of phase space variables as a first–order system of two differential equations:

$$\frac{d}{dt}\begin{pmatrix} x(t) \\ \dot{x}(t) \end{pmatrix} = \begin{pmatrix} \dot{x}(t) \\ -U'(x(t)) \end{pmatrix}, \tag{8.2}$$

where the plane $\{(x, \dot{x}) : x \in R, \dot{x} \in R\}$ is called the phase space of the system. Following are some properties of this system.

Equation (8.2) has a first integral, namely, the energy of the system

$$\mathcal{E}(t) \equiv \mathcal{E}(x, \dot{x}) = U(x) + \frac{\dot{x}^2}{2}. \tag{8.3}$$

So

$$U\big(x(t)\big) + \frac{\big(\dot{x}(t)\big)^2}{2} = c, \tag{8.4}$$

for any solution $x(t)$, where the constant c is determined by the initial conditions of the system.

Let $L_c = \{(x, \dot{x}) : \mathcal{E}(x, \dot{x}) = c\}$ denote the level set of energy c. If $c \geq u_0 = \inf_x U(x)$, then the set L_c is not empty, and it has a finite number of connected components. Any connected component of L_c is called an orbit of the system. Orbits can be described as follows: We consider the set

$$\Delta_c = \{x : U(x) \leq c\}.$$

Then there exists a finite number of closed intervals $[\alpha_1^c, \beta_1^c]$, $[\alpha_2^c, \beta_2^c]$,..., $[\alpha_k^c, \beta_k^c]$, $\beta_1^c < \alpha_2^c \ldots \beta_{k-1}^c < \alpha_k^c$, for which

$$\Delta_c = \bigcup_i [\alpha_i^c, \beta_i^c].$$

The closed curve

$$\frac{\dot{x}^2}{2} + U(x) = c, \quad x \in [\alpha_i^c, \beta_i^c], \tag{8.5}$$

is an orbit of the system. This orbit is *regular* if $\alpha_i^c < \beta_i^c$ and $U(x) < c$ for $x \in (\alpha_i^c, \beta_i^c)$. The motion of the particle along a regular orbit can be determined by solving the differential equations

$$\frac{dx(t)}{dt} = \pm\sqrt{2c - 2U\big(x(t)\big)}, \tag{8.6}$$

where we choose the sign $+$ for the motion from α_i^c to β_i^c and the negative sign $-$ for the backward motion. This motion is periodic with period

$$T = 2\int_{\alpha_i^c}^{\beta_i^c} \frac{dx}{\sqrt{2c - 2U(x)}}. \tag{8.7}$$

Moreover, for any continuous function $F(x, \dot{x})$ the average of F along an orbit exists:

$$\begin{aligned} &\lim_{t\to\infty} \frac{1}{t}\int_0^t F\big(x(s), \dot{x}(s)\big)ds \\ &= \frac{1}{2T}\int_{\alpha_i^c}^{\beta_i^c} \frac{F(x, \sqrt{2c - 2U(x)}) + F(x, -\sqrt{2c - 2U(x)})}{\sqrt{2c - 2U(x)}} dx. \end{aligned} \tag{8.8}$$

If an orbit is not regular, it is called *singular*. In this case, either $\alpha_i^c = \beta_i^c$ and α_i^c is a local minimum of $U(x)$, or the orbit includes several rest points

$$\alpha_i^c < \gamma_1 < \cdots < \gamma_l < \beta_i^c,$$

where $U(\gamma_j) = c$ for $j = 1, \ldots, l$, and $U(x) < c$ for $x \in (\alpha_i^c, \beta_i^c)$, $x \neq \gamma_j$, $j = 1, \ldots, l$. The points $\gamma_1, \ldots, \gamma_l$ are local maxima of $U(x)$. In the last case, the orbit is the union of $l + 1$ loops that are determined by equation (8.5) on the intervals $[\alpha_i^c, \gamma_1], [\gamma_1, \gamma_2], \ldots, [\gamma_l, \beta_i^c]$. If the initial conditions of the system are $x = \alpha_i^c$, $\dot{x} = 0$, and $U'(\alpha_i^c) = 0$, then $x(t) = \alpha_i^c$ for all t and α_i^c is a state of stable equilibrium. If $x \in [\alpha_i^c, \beta_i^c]$ and $\gamma_1 < \gamma_2 < \cdots < \gamma_l$ are local maxima of the function $U(x)$ in the interval $[\alpha_i^c, \beta_i^c]$, then $x(t) \to \gamma_i$ as $t \to \infty$, where the number $i = 1, 2, \ldots, l$, depends on the initial position of the system (we assume that the initial conditions $(x, \dot{x})$ satisfy relation (8.5)).

Remark 1 If for given c there exists a singular orbit, then c is a local minimum or a local maximum of the function $U(x)$. So there exists only a finite number of singular orbits.

Denote by $O_1(c), \ldots, O_{k_c}(c)$ the orbits of energy c, that is, for which $\mathcal{E}(x, \dot{x}) = c$. These are enumerated in such a way that their projections onto the x-axis are intervals with increasing left endpoints.

The singular values of U make up the set $U_s = \{u_0, u_1, \ldots, u_r\} = \{U(x) : U'(x) = 0\}$. These are labeled so that $u_0 < u_1 < \cdots < u_r$. Then $O_i(c)$ is a regular orbit if $c \in (u_0, \infty) \setminus U_s$. The number of orbits of energy c, k_c, is continuous on the set $(u_0, \infty) \setminus U_s$, and the set–valued functions $O_i(c)$ are continuous in the metric generated by the distance

$$d(O_i(c), O_i(c')) = \sup_{(\tilde{x}, \dot{\tilde{x}}) \in O_i(c')} \inf_{(x, \dot{x}) \in O_i(c)} (|x - \tilde{x}| + |\dot{x} - \dot{\tilde{x}}|)$$

if $c \in (u_0, \infty) \setminus U_s$.
Moreover, these functions are right-continuous at $c \in U_s$. If $O_i(c)$ is defined on the interval $c \in [u_{l-1}, u_l)$ and $O_i(u_l)$ is a regular orbit, then the function $O_i(c)$ can be extended as a continuous function on the interval $[u_{l-1}, u_l)$. Note that the left limit $O_i(u_{l+1}^-)$ exists in the metric d; this limit can be a regular orbit, and in this case $O_i(c)$ can be extended as a continuous function on the interval $[u_{l+1}, u_{l+2})$. Another possibility is that $O_i(u_l^-)$ is one of the loops of a singular orbit. If a singular orbit $O_j(u_l)$ has k loops, then there exist k continuous functions $O_{i_1}(c), \ldots, O_{i_k}(c)$ for which $O_{i_1}(u_l^-), \ldots, O_{i_k}(u_l^-)$ are different loops of the orbit $O_j(u_l)$. For a regular orbit $O_i(c)$, there exists a maximal interval on which the function $O_i(c)$ is continuous; this interval is one of the intervals $[u_r, \infty)$ or $[u_k, u_l)$, $k \le l < j \le r$.
Now consider all continuous functions on their maximal intervals of continuity, which we denote by $O_1^*(c), O_2^*(c), \ldots, O_m^*(c)$, and let $I_1, I_2, \ldots, I_m$ be the domains of these functions. Note that it can happen that $I_k = I_l$, but we consider these intervals as being different because the functions $O_k^*(c)$ and $O_l^*(c)$ are different.
The set $G = \{(c, k) : c \in I_k, k = 1, \ldots, m\}$ is said to be the *orbit graph* of the system. There exists a one-to-one correspondence between G and the set of all orbits of the system: Each orbit can be represented in a unique way in the form

$$O = O_k^*(c), \; c \in I_k.$$

The *edges* of the graph are the subsets of G having the form $E_k = \{(c, k), \, c \in I_k\}$ for $k = 1, \ldots, m$. Let $u_k^* = \inf I_k$. The points (u_k^*, k) for $k = 1, \ldots, m$ are called the *vertices* of the graph. The edge E_i is connected to the edge E_k if $\sup I_i = u_k^*$ and $O_i^*(u_k^*-) \subset O_k^*(u_k^*)$. In this case, we will say that the vertex (u_k^*, k) is the *origin* of the edge E_i. A vertex (u_k^*, k) is called an *end* if there are no edges for which it is the origin. In this case, we will call the edge E_k also an end. A vertex that is not an end is called a *knot*.

Example 1 Suppose that the function $U(x)$ has two local minima at the points $x_1 < x_2$ and a local maximum at the point $x_3 \in (x_1, x_2)$, $U(x_i) = u_i^*$, such that $u_1^* < u_3^*$, $u_2^* < u_3^*$. Then the graph G has the edges $\{(c, 1), \, c \in I_1\}$, $\{(c, 2), \, c \in I_2\}$, and $\{(c, 3), \, c \in I_3\}$, where $I_1 = [u_3^*, \infty)$, $I_2 = [u_1^*, u_3^*)$, and

$I_3 = [u_2^*, u_3^*)$. The vertices are the singular orbits: A knot

$$\left\{ (x, \dot{x}) : \frac{\dot{x}^2}{2} + U(x) = u_3^*, \; x \in [\alpha, \beta] \right\}$$

and the two ends

$$\{(x_1, u_1^*)\}, \quad \{(x_2, u_2^*)\},$$

where $\alpha < x_3 < \beta$, $U(\alpha) = U(\beta) = u_3^*$. Figure 8.1 describes the potential function graph and the orbits for $U = 2.5x^4/4 - 0.5x^2/2 + 0.5x$.

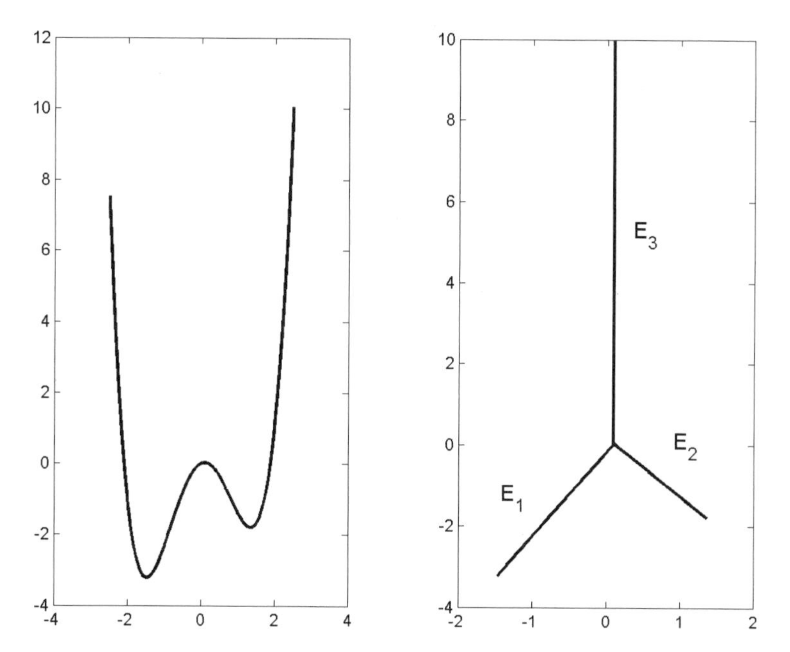

Figure 8.1. Potential function graph of the system having $U(x) = 2.5x^4/4 - 5x^2/2 + 0.5x$. Left shows the potential energy function U. Right shows the potential energy graph associated with U. The extremals are at $x = -1.4618, x = 1.3613$, and $x = 0.1005$, and we denote by E_1 the edge to the lower left, E_3 the edge going straight up, and E_2 the edge to the lower right.

Remark 2 If an orbit is regular, then a small change of the energy of the particle leaves the orbit on the same edge. But with a small increase in energy of a particle on a singular orbit, the particle leaves the orbit on the same edge, but the new orbit is regular. After a small decrease of the energy of a particle (if that is possible) on a singular orbit, the new orbit belongs to one of the edges whose origin is the vertex where the singular orbit is situated. If a singular orbit contains only one point, then a decrease of the energy of the particle may not be possible.

Denote by $C(G)$ the space of all bounded continuous functions from the graph space G into R. Suppose that $\Psi(u,k)$ for $u \in R$, $k \in \{1,\dots,m\}$ and $(u,k) \in G$ is a continuous function. Then $\Psi(u,k)$ is continuous in u on the interval I_k for fixed k, and

$$\Psi(u_k^*, k) = \Psi(v_{l-}^*, l)$$

for all l for which the vertex (u_k^*, k) is the origin of the edge E_l, where $v_l^* = \sup I_l$. If the derivative with respect to u, $\Psi'(u,k)$, exists for $u \in I_k$, $k \in \{1,2,\dots,m\}$, and if

$$\Psi'(u_k^*, k) = \Psi'(v_{l-}^*, l)$$

for all k and l for which the vertex (u_k^*, k) is the origin of the edge E_l, then $\Psi(u,k) \in C^{(1)}(G)$. In the same way, we can define the derivative spaces $C^{(r)}(G)$, $r = 1, 2, \dots$.

Remark 3 A function $\nu(x,\dot{x}) : R^2 \to \{1,\dots,m\}$ exists for which $\nu(x,\dot{x}) = k$ if $(x,\dot{x}) \in O$ and the orbit O is in E_k. So for any $\Psi \in C(G)$ the function

$$\Psi(\mathcal{E}(x,\dot{x}), \nu(x,\dot{x})) \in C(R^2),$$

and it is easy to check that if $\Psi \in C^{(r)}(G)$, we have that

$$\Psi(\mathcal{E}(x,\dot{x}), \nu(x,\dot{x})) \in C^{(r)}(R^2).$$

8.1.2 Randomly Perturbed Conservative Systems

Consider now the solution to the randomly perturbed differential equation

$$\ddot{x}_\varepsilon(t) = -U_x\left(x_\varepsilon(t), y\left(\frac{t}{\varepsilon}\right)\right), \tag{8.9}$$

where the potential energy is described by a measurable function $U(x,y)$ of y, and $y(t)$ is a homogeneous Markov process taking values in $(Y, \mathcal{C})$, which is a measurable space called the *noise space*. We assume that $y(t)$ satisfies condition SMC II of Section 2.3, and we denote by $P(t,y,\mathcal{C})$ the transition probability function and by $\rho(dy)$ the ergodic distribution of the process. Suppose that $U(x,y)$ satisfies the following additional conditions:

($\mathcal{PP}$) The partial derivatives $U_x(x,y)$, $U_{xx}(x,y)$, and $U_{xxx}(x,y)$ in x exist as continuous bounded functions, and the function

$$U(x) = \int U(x,y)\rho(dy) \tag{8.10}$$

satisfies condition ($\mathcal{P}$) of Section 8.1.1.

With this condition, equation (8.1) is the averaged equation for equation (8.9). We will investigate the behavior of the solution to equation (8.9) using the averaging theorems from Section 4.2, the diffusion approximation for first integrals done in Section 5.1.5, and the fact that system (8.2) has the first integral $\mathcal{E}(x, \dot{x})$.

First, we transform equation (8.9) into a system of differential equations of the first order:

$$\frac{d}{dt}\begin{pmatrix} x_\varepsilon(t) \\ \dot{x}_\varepsilon(t) \end{pmatrix} = \begin{pmatrix} \dot{x}_\varepsilon(t) \\ F\big(x_\varepsilon(t), y_\varepsilon(t)\big) \end{pmatrix}, \tag{8.11}$$

where as before, we use the notation

$$\dot{x}_\varepsilon(t) = \frac{dx_\varepsilon}{dt}(t), \quad F(x, y) = -U_x(x, y), \quad y_\varepsilon(t) = y\Big(\frac{t}{\varepsilon}\Big).$$

Remark 4 If $x(t)$ is the solution to equation (8.1) satisfying the initial conditions $x(0) = x_0$, $\dot{x}(0) = \dot{x}_0$ and if $x_\varepsilon(t)$ is the solution to equation (8.9) satisfying the same initial conditions, then for any t_0,

$$P\Big\{\limsup_{\varepsilon \to 0} \sup_{t \le t_0} \big(|x_\varepsilon(t) - x(t)| + |\dot{x}_\varepsilon(t) - \dot{x}(t)|\big) = 0\Big\} = 1.$$

This follows from Theorem 6 of Chapter 3.

Next, we consider the stochastic process defined by the energy function

$$\hat{z}_\varepsilon(t) = \mathcal{E}\big(x_\varepsilon(t), \dot{x}_\varepsilon(t)\big). \tag{8.12}$$

Lemma 1 *We have*

$$\frac{d\hat{z}_\varepsilon(t)}{dt} = \tilde{F}\big(x_\varepsilon(t), y_\varepsilon(t)\big)\dot{x}_\varepsilon(t), \tag{8.13}$$

where $\tilde{F}(x, y) = F(x, y) - F(x)$ *and* $F(x) = -U'(x)$. *In particular,* $\tilde{F}(x, y) = -U_x(x, y) + U_x(x)$.

The proof can be obtained by direct calculation and is left to the reader. The following lemma is comparable to integration by parts in nonrandom calculus.

Lemma 2 *Let* $\phi(x, \dot{x}, z, y)$ *be a measurable function from* $R^3 \times Y$ *into* R *that is bounded and continuous in* $x, \dot{x}, z$ *and*

$$\int \phi(x, \dot{x}, z, y)\rho(dy) = 0.$$

Suppose that the derivatives $\phi_x, \phi_{\dot{x}}, \phi_z$ *exist and are bounded and continuous in* $x, \dot{x}, z$. *Set*

$$R\phi(x, \dot{x}, z, y) = \int \phi(x, \dot{x}, z, y')R(y, dy'),$$

where

$$R(y, C) = \int_0^\infty \big(P(t, y, C) - \rho(C)\big)dt$$

(see conditions SMC II in Section 2.3). Let $\mathcal{F}_t^\varepsilon$ be the σ-algebra generated by the history of y, that is, by the sets $\{y_\varepsilon(s), s \le t\}$. Then for $0 \le t_1 < t_2$ we have the relation

$$
\begin{aligned}
&E\Big(\int_{t_1}^{t_2} \phi(x_\varepsilon(s), \dot{x}_\varepsilon(s), \hat{z}_\varepsilon(s), y_\varepsilon(s))ds \Big/ \mathcal{F}_{t_1}^\varepsilon\Big) \\
&= \varepsilon E\Big(R\phi(x_\varepsilon(t_1), \dot{x}_\varepsilon(t_1), \hat{z}_\varepsilon(t_1), y_\varepsilon(t_1)) - R\phi(x_\varepsilon(t_2), \dot{x}_\varepsilon(t_2), \hat{z}_\varepsilon(t_2), y_\varepsilon(t_2)) \\
&\quad + \varepsilon \int_{t_1}^{t} \phi_1(x_\varepsilon(s), \dot{x}_\varepsilon(s), \hat{z}_\varepsilon(s), y_\varepsilon(s))ds \Big/ \mathcal{F}_{t_1}^\varepsilon\Big),
\end{aligned}
\tag{8.14}
$$

where

$$
\begin{aligned}
\phi_1(x, \dot{x}, z, y) &= \left[\frac{\partial R\phi}{\partial x}(x, \dot{x}, z, y) - \frac{\partial R\phi}{\partial z}(x, \dot{x}, z, y)F(x)\right]\dot{x} \\
&\quad + \left[\frac{\partial R\phi}{\partial \dot{x}}(x, \dot{x}, z, y) + \frac{\partial R\phi}{\partial z}(x, \dot{x}, z, y)\dot{x}\right]\tilde{F}(x, y).
\end{aligned}
$$

Proof The proof is accomplished by applying formula (5.7) to the function $g(x, \dot{x}, y) = R\phi\left(x, \dot{x}, \frac{1}{2}\dot{x}^2 + U(x), y\right)$ and taking into account Lemma 1 of Chapter 5.

□

Remark 5 Suppose that the function $\phi(x, \dot{x}, z, y)$ satisfies the condition

$$\sup_{|x|\le C, |\dot{x}|\le C} \sup_{y\in Y} (|\phi_x(x, \dot{x}, z, y)| + |\phi_{\dot{x}}(x, \dot{x}, z, y)| + |\phi_z(x, \dot{x}, z, y)|) < \infty$$

for any $C > 0$. Then formula (8.14) is true with

$$t_2 = \tau_c \wedge t,$$

where

$$\tau_c = \inf\{s : |x_\varepsilon(s)| + |\dot{x}_\varepsilon(s)| \ge c\}.$$

Lemma 3 *A constant $c_1 > 0$ exists for which*

$$E\hat{z}_\varepsilon^2(t) \le (\hat{z}_\varepsilon^2(0) + c_1\varepsilon)\exp\{c_1\varepsilon t\}.$$

Proof Lemma 1 implies the relation

$$\hat{z}_\varepsilon^2(t) = \hat{z}_\varepsilon^2(0) + \int_0^t \phi(x_\varepsilon(s), \dot{x}_\varepsilon(s), \hat{z}_\varepsilon(s), y_\varepsilon(s))\, ds,$$

where

$$\phi(x, \dot{x}, z, y) = 2\tilde{F}(x, y)\dot{x}z.$$

So, using the notation of Lemma 2, we have

$$\phi_1(x,\dot{x},z,y) = 2\left[\frac{\partial}{\partial x}R\tilde{F}(x,y)\dot{x}z - \dot{x}R\tilde{F}(x,y)F(x)\right]\dot{x} + 2[R\tilde{F}(x,y)z + R\tilde{F}(x,y)\dot{x}^2]\tilde{F}(x,y).$$

It follows from Lemma 2 and Remark 5 that

$$\begin{aligned} E\hat{z}_\varepsilon^2(t\wedge\tau_c) &= \hat{z}_\varepsilon^2(0) + \varepsilon 2R\tilde{F}(x_\varepsilon(0),y_\varepsilon(0))\dot{x}_\varepsilon(0)\hat{z}_\varepsilon(0) \\ &\quad - \varepsilon E_2 R\tilde{F}(x_\varepsilon(t\wedge\tau_c),y_\varepsilon(t\wedge\tau_c))\dot{x}_\varepsilon(t\wedge\tau_c)\hat{z}_\varepsilon(t\wedge\tau_c) \\ &\quad + \varepsilon E\int_0^{t\wedge\tau_c}\phi_1(x_\varepsilon(s),\dot{x}_\varepsilon(s),\hat{z}_\varepsilon(s),y_\varepsilon(s))\,ds \end{aligned} \tag{8.15}$$

for all $t > 0$, where τ_c is introduced in Remark 5. Note that

$$\dot{x}^2 \le 2z - 2\inf U(x).$$

So for some constant $c_2 > 0$,

$$R\tilde{F}(x,y)\dot{x}z \le c_2(1+z^2)$$

and

$$\phi_1(x,\dot{x},z,y) \le c_2(1+z^2).$$

This and (8.15) imply that a constant c_1 exists for which

$$E\hat{z}_\varepsilon^2(t\wedge\tau_c) \le \hat{z}_\varepsilon^2(0) + c_1\varepsilon + c_1\varepsilon E\int_0^{t\wedge\tau_c}\hat{z}_\varepsilon^2(s)\,ds. \tag{8.16}$$

So

$$E\hat{z}_\varepsilon^2(t\wedge\tau_c) \le (\hat{z}_\varepsilon^2(0) + c_1\varepsilon)\exp\{c_1\varepsilon t\}.$$

□

Remark 6 In the same way we can prove that

$$E(\hat{z}_\varepsilon(t) - \hat{z}_\varepsilon(0))^2 = O(\varepsilon t).$$

Denote by $z_\varepsilon(t)$ the G-valued stochastic process that is determined by the pair $(\hat{z}_\varepsilon(t),\theta_\varepsilon(t))$, where $\theta_\varepsilon(t) = \nu(x_\varepsilon(t),\dot{x}_\varepsilon(t))$ and $\nu(x,\dot{x})$ is introduced in Remark 3. This means that if $\theta_\varepsilon(t) = k$, then $z_\varepsilon(t)$ is considered to be an element of the edge E_k. If $\varphi \in C(G)$, then a function $f_k \in C(I_k)$ exists for which

$$\varphi(z_\varepsilon(t)) = \sum_k f_k(\hat{z}_\varepsilon(t))1_{\{\theta_\varepsilon(t)=k\}}. \tag{8.17}$$

Lemma 4 *Assume* $\varphi \in C^{(3)}(G)$. *Set*

$$\hat{F}(x) = \iint \tilde{F}(x,y')\tilde{F}(x,y)R(y,dy')\rho(dy). \tag{8.18}$$

Then for $0 \le t_1 < t_2$,

$$E(\varphi(z_\varepsilon(t_2)) - \varphi(z_\varepsilon(t_1)) \big/ \mathcal{F}^\varepsilon_{t_1})$$
$$= \varepsilon E\bigg(\int_{t_1}^{t_2} [\varphi''(z_\varepsilon(s))\dot{x}_\varepsilon^2(s) + \varphi'(z_\varepsilon(s))]\hat{F}(x_\varepsilon(s))\, ds \Big/ \mathcal{F}^\varepsilon_{t_1} \bigg) \tag{8.19}$$
$$+ O(\varepsilon + \varepsilon^2(t_2 - t_1)).$$

Proof Suppose $\varphi(c,k) = f_k(c)$ if $(c,k) \in E_k$, $k = 1, \ldots, m$, $f_k \in C^{(3)}(I_k)$. Then $\varphi(z_\varepsilon(t))$ is represented by formula (8.17) and

$$\frac{d\varphi}{dt}(z_\varepsilon(t)) = \sum_k f'_k(\hat{z}_\varepsilon(t))\tilde{F}(x_\varepsilon(t), y_\varepsilon(t))\dot{x}_\varepsilon(t) 1_{\{\theta_\varepsilon(t)=k\}}$$
$$= \varphi'(z_\varepsilon(t))\, \tilde{F}(x_\varepsilon(t), y_\varepsilon(t))\, \dot{x}_\varepsilon(t).$$

Set

$$\phi(x, \dot{x}, z, y) = \phi'(z)\tilde{F}(x,y)\dot{x},$$
$$\phi_1(x, \dot{x}, z, y) = \phi'(z) \left[\tilde{F}(x,y)R\tilde{F}(x,y) + \dot{x}^2 \frac{\partial}{\partial x} R\tilde{F}(x,y) \right]$$
$$+ \varphi''(z)[\dot{x}^2 \tilde{F}(x,y)R\tilde{F}(x,y) + \dot{x}^2 F(x)R\tilde{F}(x,y)].$$

Using Lemma 2 we can write

$$E(\varphi(z_\varepsilon(t_2)) - \varphi(z_\varepsilon(t_1)) \big/ \mathcal{F}^\varepsilon_{t_1})$$
$$= E\bigg(\int_{t_1}^{t_2} \phi(x_\varepsilon(s), \dot{x}_\varepsilon(s), z_\varepsilon(s), y_\varepsilon(s))\, ds \Big/ \mathcal{F}^\varepsilon_{t_1} \bigg)$$
$$= \varepsilon E\bigg(\int_{t_1}^{t_2} \phi_1(x_\varepsilon(s), \dot{x}_\varepsilon(s), z_\varepsilon(s), y_\varepsilon(s))\, ds \Big/ \mathcal{F}^\varepsilon_{t_1} \bigg) + O(\varepsilon). \tag{8.20}$$

Define

$$\tilde{\phi}_1(x, \dot{x}, z, y) = \phi_1(x, \dot{x}, z, y) - \int \phi_1(x, \dot{x}, z, y)\rho(dy).$$

Then $\int \tilde{\phi}_1(x, \dot{x}, z, y)\rho(dy) = 0$, and Lemma 2 can be applied to the function $\tilde{\phi}_1$. So

$$E\bigg(\int_{t_1}^{t_2} \tilde{\phi}_1(x_\varepsilon(s), \dot{x}_\varepsilon(s), z_\varepsilon(s), y_\varepsilon(s))\, ds \Big/ \mathcal{F}^\varepsilon_{t_1})$$
$$= \varepsilon E\bigg(\int_{t_1}^{t_2} \phi_2(x_\varepsilon(s), \dot{x}_\varepsilon(s), z_\varepsilon(s), y_\varepsilon(s))\, ds \Big/ \mathcal{F}^\varepsilon_{t_1} \bigg) + O(\varepsilon),$$

where

$$\begin{aligned}\phi_2(x, \dot{x}, z, y) &= \left[\frac{\partial}{\partial x} R\tilde{\phi_1}(x, \dot{x}, z, y) - \frac{\partial}{\partial z} R\tilde{\phi_1}(x, \dot{x}, z, y) F(x)\right] \dot{x} \\ &\quad + \left[\frac{\partial}{\partial \dot{x}} R\tilde{\phi_1}(x, \dot{x}, z, y) + \frac{\partial}{\partial z} R\tilde{\phi_1}(x, \dot{x}, z, y)\dot{x}\right] \tilde{F}(x, y).\end{aligned}$$

It is easy to see that under the conditions on φ and $U(x, y)$, the function ϕ_2 is bounded. So

$$\begin{aligned}&E\left(\int_{t_1}^{t_2} \phi_1(x_\varepsilon(s), \dot{x}_\varepsilon(s), z_\varepsilon(s), y_\varepsilon(s))\, ds \Big/ \mathcal{F}_{t_1}^\varepsilon\right) \\ &= E\left(\int_{t_1}^{t_2} \bar{\phi}_1(x_\varepsilon(s), \dot{x}_\varepsilon(s), z_\varepsilon(s))\, ds \Big/ \mathcal{F}_{t_1}^\varepsilon\right) + O(\varepsilon) + O(\varepsilon(t_2 - t_1)),\end{aligned} \tag{8.21}$$

where

$$\bar{\phi}_1(x, \dot{x}, z) = \int \phi_1(x, \dot{x}, z, y)\rho(dy) = (\varphi''(z)\dot{x}^2 + \varphi'(z))\hat{F}(x). \tag{8.22}$$

Substituting relations (8.21) and (8.22) in formula (8.20), we obtain formula (8.19).

□

Let $z \in G$ be a regular point, i.e., the corresponding orbit is regular and $z \in E_k$. Denote by $T(z)$ the period of the function $x(t)$ for which $\mathcal{E}(x(0), \dot{x}(0)) = c$, where $z = (c, k)$. Introduce the functions

$$a(z) = \frac{1}{T(z)} \int_0^{T(z)} \hat{F}(x(t))dt, \tag{8.23}$$

and

$$b(z) = \frac{2}{T(z)} \int_0^{T(z)} \dot{x}^2(t)\hat{F}(x(t))dt. \tag{8.24}$$

For $\varphi \in C^{(2)}(G)$ we define an operator L by

$$L\varphi(z) = a(z)\varphi'(z) + \frac{1}{2}b(z)\varphi''(z). \tag{8.25}$$

Lemma 5 *Suppose that $z_\varepsilon(s)$ is a regular point of the graph G. Then*

$$\left|E\left(\int_s^{s+T(z_\varepsilon(s))} (L\varphi(z_\varepsilon(u)) - [\varphi''(z_\varepsilon(u))\dot{x}_\varepsilon^2(u) + \varphi'(z_\varepsilon(u))]\hat{F}(x_\varepsilon(u)))\, du \Big/ \mathcal{F}_s^\varepsilon\right)\right|$$

$$= O\left(\sqrt{\varepsilon}(T(z_\varepsilon(s)) + T^{3/2}(z_\varepsilon(s)))\right).$$

Proof It follows from Theorem 4 and formula (4.35) that

$$E\Big([\varphi''(z_\varepsilon(u))\dot{x}_\varepsilon^2(u) + \varphi'(z_\varepsilon(u))]\hat{F}(x_\varepsilon(u)) - [\varphi''(z_\varepsilon(s))(\dot{x}(u))^2 + \varphi'(z_\varepsilon(s))]\hat{F}(x^*(u))/\mathcal{F}_s^\varepsilon\Big) = O(\sqrt{\varepsilon})$$

for $u \in (s, s+T(z_\varepsilon(s)))$; here $x^*(u)$ is the solution to system (8.2) on the interval $(s, s+T(z_\varepsilon(s))$ with initial conditions

$$x^*(s) = x_\varepsilon(s), \qquad \dot{x}^*(s) = \dot{x}_\varepsilon(s).$$

So

$$E\left(\int_s^{s+T(z_\varepsilon(s))} [\varphi''(z_\varepsilon(u))\dot{x}_\varepsilon^2(u) + \varphi'(z_\varepsilon(u))]\hat{F}(x_\varepsilon(u))\,du \Big/ \mathcal{F}_s^\varepsilon\right)$$
$$= E(L\varphi(z_\varepsilon(s))T(z_\varepsilon(s))/\mathcal{F}_s^\varepsilon) + O(\sqrt{\varepsilon}\,T(z_\varepsilon(s))).$$

Since the function $L\varphi$ has a bounded derivative, it follows from Remark 6 that for $u > s$,

$$E(L\varphi(z_\varepsilon(u)) - L\varphi(z_\varepsilon(s))/\mathcal{F}_s^\varepsilon) = O(\sqrt{\varepsilon(u-s)}).$$

Therefore,

$$E\left(L\varphi(z_\varepsilon(s))T(z_\varepsilon(s)) - \int_0^{T(z_\varepsilon(s))} L\varphi(z_\varepsilon(u))\,du \Big/ \mathcal{F}_s^\varepsilon\right) = O(\sqrt{\varepsilon}\,T^{3/2}(z_\varepsilon(s))).$$

This completes the proof of the lemma.

□

Denote by L_k the differential operator given by the formula

$$L_k\varphi(u) = a(u)\varphi'(u) + \frac{1}{2}b(u)\varphi''(u)$$

on the space $C^{(2)}(I_k)$. Denote by $z_k(t)$ the diffusion process on the interval I_k with absorbing boundaries and generator L_k. The next theorem describes the local behavior of the randomly perturbed system before it reaches a vertex of the graph G.
Let V be the subset of all vertices of the graph G. Set

$$\zeta_\varepsilon = \inf\{t : z_\varepsilon(t) \in V\}.$$

Theorem 1 *Let $(x_\varepsilon(t), \dot{x}_\varepsilon(t))$ be the solution of Equation (8.11) that satisfies the initial conditions $x_\varepsilon(0) = x_0$, $\dot{x}_\varepsilon(0) = \dot{x}_0$ for all $\varepsilon > 0$. Let $\nu(x_0, \dot{x}_0) = k$ and $\mathcal{E}(x_0, \dot{x}_0) = c$ for $(c, k) \in G \setminus V$. Suppose that the Markov process $y(t)$ satisfies condition SMC II of Section 2.3 and $U(x, y)$ satisfies condition $(\mathcal{PP})$. Then the stochastic process $\{\hat{z}_\varepsilon(t/\varepsilon), t/\varepsilon < \zeta_\varepsilon\}$ converges weakly in $\mathcal{C}$ as $\varepsilon \to 0$ to a diffusion process $z_k(t)$ satisfying the initial conditions $z_k(0) = c$.*

Proof Introduce stopping times

$$\zeta_\varepsilon^\delta = \sup\left\{t : \hat{z}_\varepsilon(s) \in \left[u_k^* + \delta, \left(v_k^* \wedge \frac{1}{\delta}\right) - \delta\right], s \le t\right\},$$

$$\zeta^\delta = \sup\left\{t : z_k(s) \in \left[u_k^* + \delta, \left(v_k^* \wedge \frac{1}{\delta}\right) - \delta\right], s \le t\right\},$$

where $u_k^* = \inf I_k$ and $v_k^* = \sup I_k$, (possibly $v_k^* = +\infty$). To prove the theorem, it suffices to prove that the stochastic process $\hat{z}_\varepsilon(t \wedge \zeta_\varepsilon^\delta/\varepsilon)$ converges weakly in C to the stochastic process

$$z_1(t \wedge \zeta^\delta).$$

Note that $T(u)$ is a continuous function on a closed and bounded interval $[u_k^* + \delta, (v_k^* \wedge \frac{1}{\delta}) - \delta]$, so that a constant T_δ exists for which $T(u) \le T_\delta$ if $u \in [u_k^* + \delta, (v_k^* \wedge \frac{1}{\delta}) - \delta]$. Set

$$\phi_\varepsilon(u) = (\varphi''(\hat{z}_\varepsilon(u))\, \dot{x}_\varepsilon^2(u) + \varphi'(\hat{z}_\varepsilon(u)))\hat{F}(x_\varepsilon(u)),$$

where $\varphi \in C^{(3)}(I_k)$. Then using the relation

$$|T(z_\varepsilon(s)) - T(z_\varepsilon(u))| \le c|\hat{z}_\varepsilon(s) - \hat{z}_\varepsilon(u)|$$

if $s \le \zeta_\varepsilon^\delta$, $u \le \zeta_\varepsilon^\delta$ where the constant c depends on δ, and using Remark 6, we can prove that

$$E\left(\int_{t_{1/\varepsilon}}^{t_{2/\varepsilon}} \phi_\varepsilon(u \wedge \zeta_\varepsilon^\delta)\, du \Big/ \mathcal{F}^\varepsilon_{t_{1/\varepsilon}\wedge\zeta_\varepsilon^\delta}\right)$$
$$= E\left(\int_{t_{1/\varepsilon}}^{t_{2/\varepsilon}} \frac{1}{T(z_\varepsilon(s \wedge \zeta_\varepsilon^\delta))} \int_s^{s+T(z_\varepsilon(s\wedge\zeta_\varepsilon^\delta))} \phi_\varepsilon(u)\, du\, ds \Big/ \mathcal{F}^\varepsilon_{t_{1/\varepsilon}\wedge\zeta_\varepsilon^\delta}\right)$$
$$+ O\left(\sqrt{\varepsilon}\, \frac{t_2 - t_1}{\varepsilon}\right).$$

Now using Lemmas 5 and 4 we can write

$$E\left(\varphi(\hat{z}((t_2/\varepsilon) \wedge \zeta_\varepsilon^\delta)) - \varphi(\hat{z}((t_1/\varepsilon) \wedge \zeta_\varepsilon^\delta))/\mathcal{F}_{\zeta_\varepsilon^\delta \wedge (t_1/\varepsilon)}\right)$$
$$= E\left(\int_{t_1}^{t_2} L\varphi(\hat{z}((s/\varepsilon) \wedge \zeta_\varepsilon^\delta))\, ds/\mathcal{F}_{(t_1/\varepsilon)\wedge\zeta_\varepsilon^\delta}\right) + O(\sqrt{\varepsilon}).$$

The weak convergence of the process $\hat{z}_\varepsilon((t/\varepsilon) \wedge \zeta_\varepsilon^\delta)$ to the process $z_1(t \wedge \zeta^\delta)$ follows from Theorem 2 and Remark 7 of Chapter 2. Using Remark 6 we can prove the compactness of the family of stochastic processes $\{\hat{z}(t/\varepsilon \wedge \zeta_\varepsilon^\delta), \varepsilon > 0\}$ for all $\delta > 0$ in the same way as in Theorem 5 of Chapter 5.

□

Now we consider the behavior of the process $\hat{z}_\varepsilon(t)$ in a neighborhood of an end. It will be the same as in the case where the function $U(x)$ has only one minimum and the graph G consists of one edge. Assume that $U(0) = 0$, $U(x) > 0$ if $x \neq 0$, and that $U''(0) = 2q$ where $q > 0$. Then $U(x) \approx qx^2$ and $F(x) \approx -2qx$ as $x \to 0$. It can be shown that the asymptotic behavior of the system in a neighborhood of the point $x = 0$, $\dot{x} = 0$, is the same as the system with $U(x) = qx^2$, for which

$$x(t) = (q^{-1}E_0)^{1/2} \cos \sqrt{2q}\,(t + \varphi),$$
$$\dot{x}(t) = -(2E_0)^{1/2} \sin \sqrt{2q}\,(t + \varphi),$$

where $E_0 = \frac{1}{2}\dot{x}_0^2 + qx_0^2$ is the initial energy of the system and φ is the initial phase of the system. Then the period $T(z)$ does not depend on z and equals

$$T = \frac{2\pi}{\sqrt{2q}}.$$

Note (see Lemma 4) that

$$\hat{F}(x) = q_0 + q_1 x + q_2 x^2 + O(x^3),$$

where

$$q_0 = \iint F(0,y)F(0,y')R(y,dy')\rho(dy),$$
$$q_2 = \iint F_x(0,y)F_x(0,y')R(y,dy')\rho(dy),$$

and

$$q_1 = \iint F(0,y)F_x(0,y')R(y,dy')\rho(dy).$$

Then using formulas (8.23), (8.24), we can write

$$a(z) \approx q_0,\ b(z) \approx 2q_0 z \quad \text{for } q_0 > 0,$$

and

$$a(z) \approx \frac{q_2}{2q} z, \quad b(z) \approx \frac{q_2}{2q} z^2,$$

as $z \to 0$.

Introduce the function

$$V(z) = \exp\left\{ - \int_{z_0}^{z} 2a(u)b^{-1}(u)\, du \right\}.$$

Then $V(z) \approx c/z$ as $z \to 0$ if $q_0 > 0$, c is a constant. So

$$I_1 = \int_0^{z_0} V(z)dz = +\infty.$$

(This integral was introduced in Remark 4 of Chapter 5.) This means that the point $z = 0$ is not accessible for the limit diffusion process with diffusion coefficients $a(z)$ and $b(z)$.

In the case $q_0 = 0$ we have

$$V(z) \approx \frac{c}{z^2}, \quad \text{so } I_1 = \infty,$$

and we have the same statement. We formulate these results as a theorem, which we state without further proof.

Theorem 2 *Let $U(x)$ have a unique minimum, say $\inf_x U(x) = 0$, and suppose that the conditions of Theorem 1 are satisfied. Assume that $\hat{F}(0) + \hat{F}''(0) > 0$. Then the stochastic process $\hat{z}_\varepsilon(t/\varepsilon)$ converges weakly in C as $\varepsilon \to 0$ on the interval $(0, \infty)$ to a diffusion process $z(t)$ with the generator L given by formula (8.25) and initial value $z(0) = E_0$.*

Corollary 1

$$\lim_{\varepsilon \to 0} P\{\varepsilon\, \zeta_\varepsilon < t\} = 0$$

for all $t > 0$.

This follows from the relation

$$P\{\zeta < t\} = 0 \quad \text{for all } t > 0.$$

Example 2 Suppose that $U(x, y) = q(y)x^2/2$, where $q(y)$ is a bounded measurable function from Y into R and $q = \int q(y)\rho(dy) > 0$. The averaged system

$$\frac{\partial \dot{x}}{dt} = -qx,$$
$$\frac{\partial x}{dt} = \dot{x},$$

has first integral $\frac{1}{2}\dot{x}^2 + qx^2 = \mathcal{E}(x, \dot{x})$, and the function

$$z_\varepsilon(t) = \mathcal{E}\left(x\left(\frac{t}{\varepsilon}\right), \dot{x}\left(\frac{t}{\varepsilon}\right)\right)$$

converges weakly on the interval $[0, \infty)$ to the diffusion process $z(t)$ having generator

$$L\theta(z) = \frac{\hat{q}}{q}\left(z^2\theta''(z) + 2z\theta'(z)\right),$$

where

$$\hat{q} = \iint q(y)q(y')\rho(dy)R(y, dy').$$

We get the coefficients of the differential operator L from formulas (8.23) and (8.24).

To describe the motion of the perturbed system we introduce new variables (r, φ) that are connected to $(x, \dot{x})$ by relations

$$x = r \sin \sqrt{q}\varphi, \quad \dot{x} = r\sqrt{q} \cos \sqrt{q}\varphi. \tag{8.26}$$

Then the motion of the averaged system is determined by the equations

$$\dot{r}(t) = 0, \quad \dot{\varphi}(t) = 1.$$

Let $r_\varepsilon(t)$ and $\varphi_\varepsilon(t)$ satisfy the relation

$$x_\varepsilon(t) = r_\varepsilon(t) \sin \sqrt{q}\,\varphi_\varepsilon(t), \quad \dot{x}_\varepsilon(t) = r_\varepsilon(t)\sqrt{q}\, \cos \sqrt{q}\,\varphi_\varepsilon(t). \tag{8.27}$$

Then

$$\big(\sin \sqrt{q}\,\varphi_\varepsilon(t)\big)\frac{dr_\varepsilon}{dt}(t) + \big(\sqrt{q}\,r_\varepsilon(t) \cos \sqrt{q}\,\varphi_\varepsilon(t)\big)\frac{d\varphi_\varepsilon}{dt}(t) = \sqrt{q}\,r_\varepsilon(t) \cos \sqrt{q}\,\varphi_\varepsilon(t),$$

$$\big(\cos \sqrt{q}\,\varphi_\varepsilon(t)\big)\frac{dr_\varepsilon}{dt}(t) - \big(\sqrt{q}\,r_\varepsilon(t) \sin \sqrt{q}\,\varphi_\varepsilon(t)\big)\frac{d\varphi_\varepsilon}{dt}(t) = -\sqrt{q}\,r_\varepsilon(t) \sin \sqrt{q}\,\varphi_\varepsilon(t).$$

From these relations we can obtain the following equations for the perturbed system:

$$\begin{aligned} \sqrt{q}\,\dot{r}_\varepsilon(t) &= \tilde{q}\,\big(y\big(\frac{t}{\varepsilon}\big)\big)\, r_\varepsilon(t)\, \sin 2\sqrt{q}\,\varphi_\varepsilon(t), \\ q\,\dot{\varphi}_\varepsilon(t) &= q - \tilde{q}\big(y\big(\frac{t}{\varepsilon}\big)\big)\, 2\, \sin^2 \sqrt{q}\,\varphi_\varepsilon(t), \end{aligned} \tag{8.28}$$

where $\tilde{q} = (q - q(y))/2$.
Note that

$$\mathcal{E}(x, \dot{x}) = \frac{1}{2}\dot{x}^2 + \frac{1}{2}qx^2 = \frac{1}{2}r^2.$$

So the asymptotic behavior of $r_\varepsilon(t/\varepsilon)$ is the same as the asymptotic behavior of the function $\sqrt{2q^{-1}z_\varepsilon(t)}$.
Set $\varphi_\varepsilon(t) = t + \psi_\varepsilon(t)$ and $\log r_\varepsilon(t) = \alpha_\varepsilon(t)$. Then

$$\begin{aligned} \sqrt{q}\,\dot{\alpha}_\varepsilon(t) &= \tilde{q}\big(y\big(\frac{t}{\varepsilon}\big)\big)\, \sin 2\sqrt{q}\,(t + \psi_\varepsilon(t)), \\ q\,\dot{\psi}_\varepsilon(t) &= -\tilde{q}\big(y\big(\frac{t}{\varepsilon}\big)\big)\, 2\, \sin^2 \sqrt{q}\,(t + \psi_\varepsilon(t)). \end{aligned} \tag{8.29}$$

We can apply Theorem 3, Chapter 5 to the system (8.29). This theorem implies that the stochastic process $(\alpha_\varepsilon(t/\varepsilon), \tilde{\varphi}_\varepsilon(t/\varepsilon))$ converges weakly in C to the stochastic process

$$\big(c_0 w_1(t), c_0 w_2(t)\big),$$

where $w_1(t)$, $w_2(t)$ are independent Wiener processes, $Ew_k(t) = 0$, $Ew_k^2(t) = t$, and

$$c_0 = \left(\frac{1}{2}\iint q(y)q(y')\rho(dy)R(y, dy')\right)^{1/2}.$$

So we have the following asymptotic representation for the process $x_\varepsilon(t)$:

$$x_\varepsilon(t) \approx r_0 \exp\{c_0\, w_1(\varepsilon t)\} \sin \sqrt{q}\,(\varphi_0 + t + c_0\, w_2(\varepsilon t)). \tag{8.30}$$

8.1.3 Behavior of the Perturbed System near a Knot

Consider a system with two potential wells, for which the orbit graph has three edges, two ends, and one knot. This means that the function $U(x)$ has two minima and one maximum. Let minima be situated at the points $x = -1$, $x = 1$, and the maximum be at point $x = 0$. Assume $U(0) = 0$, and set $u_{-1} = U(-1)$, $u_1 = U(1)$. Denote the edges of the graph G by E_{-1}, E_0, E_1, where $E_\theta = \{(c, \theta), c \in I_\theta\}$ for $\theta \in \{-1, 0, 1\}$, and the corresponding energy intervals are $I_{-1} = [u_{-1}, 0)$, $I_0 = [0, +\infty)$, $I_1 = [u_1, 0)$, respectively. The function $\nu(x, \dot{x})$ is determined as follows: $\nu(x, \dot{x}) = -1$ if $\mathcal{E}(x, \dot{x}) \in I_{-1}$, $x < 0$; $\nu(x, \dot{x}) = 1$ if $\mathcal{E}(x, \dot{x}) \in I_1$, $x > 0$; and $\nu(x, \dot{x}) = 0$ if $\mathcal{E}(x, \dot{x}) \in I_0$. The orbit corresponding to the knot in the phase plane is determined by the relation $\dot{x}^2 + 2U(x) = 0$; the point $(x, \dot{x}) = (0, 0)$ is an unstable point of equilibrium; and the unperturbed system tends to this point as $t \to \infty$. We need some results concerning the behavior of the unperturbed system in a neighborhood of the point $(x, \dot{x}) = (0, 0)$.
Let $E_0 = \mathcal{E}(x_0, \dot{x}_0) = c$, where $|c|$ is small enough, $x_0 = 0$ if $c > 0$, and $\dot{x}_0 = 0$ if $c < 0$. Let $x(t)$ be the position of the system at time t. Set $T_{-1}(c) = T(c, -1)$, $T_1(c) = T(c, 1)$ if $c > 0$, and $T(c, -1) = \{\inf\{t > 0 : x(t) = 0\}$ if $\dot{x}_0 < 0$, $T(c, 1) = \inf\{t > 0 : x(t) = 0\}$ if $\dot{x}_0 > 0$.
Then in the same way as in formula (8.7) we can prove that for $c > 0$,

$$T_{-1}(c) = 2\int_{\alpha_0^c}^{0} \frac{dx}{\sqrt{2c - 2U(x)}}, \quad T_1(c) = 2\int_0^{\beta_0^c} \frac{dx}{\sqrt{2c - 2U(x)}}, \tag{8.31}$$

where $U(\alpha_0^c) = c$, $U(\beta_0^c) = c$, $\alpha_0^c < 0 < \beta_0^c$. If $\alpha_{-1}^c < -1 < \beta_{-1}^c < 0 < \alpha_1^c < 1 < \beta_1^c$ for $c < 0$ and

$$U(\alpha_{-1}^c) = U(\beta_{-1}^c) = U(\alpha_1^c) = U(\beta_1^c) = c,$$

then $T_{-1}(c)$ and $T_1(c)$ are determined by formula (8.7). It follows from formula (8.31) that

$$T_\theta(c) \approx \frac{1}{2}\left(-U''(0)\right)^{-1/2} \log\frac{1}{|c|}, \quad \theta = -1, 1. \tag{8.32}$$

Now we consider the perturbed system with the same initial conditions x_0, $\dot{x}_0$. Introduce the sequence of stopping times $\{\tau_n^\varepsilon, n = 0, 1, 2, \ldots\}$ by the relations: $\tau_0^\varepsilon = 0$,

$$\tau_{n+1}^\varepsilon = \min\left\{s > \tau_n^\varepsilon : \max_{\tau_n^\varepsilon \le u \le s} |x_\epsilon(u)| \ge 1, |x_\varepsilon(s)| < 1, |x_\varepsilon(s)| \wedge |\dot{x}_\varepsilon(s)| = 0\right\}.$$

Set

$$\begin{aligned}\eta_n^\varepsilon &= \mathcal{E}(x_\varepsilon(\tau_k^\varepsilon), \dot{x}_\varepsilon(\tau_k^\varepsilon)) - \mathcal{E}(x_\varepsilon(\tau_{k-1}^\varepsilon), \dot{x}_\varepsilon(\tau_{k-1}^\varepsilon)) \\ &= \int_{\tau_{k-1}^\varepsilon}^{\tau_k^\varepsilon} \tilde{F}(x_\varepsilon(s), y_\varepsilon(s))\, ds, \\ \zeta_n^\varepsilon &= \sum_{k<n} \eta_k^\varepsilon = \mathcal{E}(x_\varepsilon(\tau_n^\varepsilon), \dot{x}_\varepsilon(\tau_n^\varepsilon)) - \mathcal{E}(x_0, \dot{x}_0), \\ \theta_n^\varepsilon &= \operatorname{sign} \dot{x}_\varepsilon(\tau_n^\varepsilon) + \operatorname{sign} x_\varepsilon(\tau_n^\varepsilon).\end{aligned} \tag{8.33}$$

(We assume that $\operatorname{sign} 0 = 0$.) Note that $\theta_n^\varepsilon = 0$ if $\dot{x}_\varepsilon(\tau_n^\varepsilon) = 0$ and $x_\varepsilon(\tau_n^\varepsilon) = 0$. We will consider the systems for that $\theta_n^\varepsilon \neq 0$ for all n with probability that tends to 1 as $\varepsilon \to 0$.
It is easy to see that the pair of random variables $(\zeta_n^\varepsilon, \theta_n^\varepsilon)$ determines the position of the system on the orbit graph at time τ_n^ε: If $\zeta_n^\varepsilon > 0$, then $\nu(x_\varepsilon(\tau_n^\varepsilon), \dot{x}_\varepsilon(\tau_n^\varepsilon)) = 0$. If $\zeta_n^\varepsilon < 0$, then $\nu(\dot{x}_\varepsilon(\tau_n^\varepsilon), \dot{x}_\varepsilon(\tau_n^\varepsilon)) = \theta_n^\varepsilon$. We will investigate the sequence

$$\{(\zeta_n^\varepsilon, \theta_n^\varepsilon), \quad n = 0, 1, \ldots\}$$

in the space $R \times \Theta$, where $\Theta = \{-1, 0, 1\}$. We reformulate Theorem 5 of Chapter 4 for the system determined by equations (8.11):

Lemma 6 *Set*

$$\dot{X}_\varepsilon(t) = \varepsilon^{-1/2}(\dot{x}_\varepsilon(t) - \dot{x}(t)). \tag{8.34}$$

The stochastic process $\dot{X}_\varepsilon(t)$ converges weakly in C as $\varepsilon \to 0$ to the stochastic process $\dot{X}(t)$ that is the solution to the integral equation

$$\dot{X}(t) = W(t) + \int_0^t \left(\int_0^s F_x(x(u))\, du \right) \dot{X}(s)\, ds, \tag{8.35}$$

where $W(t)$ is the Gaussian stochastic process with independent increments for which

$$EW(t) = 0, \quad EW^2(t) = 2\int_0^t \hat{F}(x(s))\, ds, \tag{8.36}$$

where $\hat{F}$ was introduced in Lemma 4.

Lemma 7 *The conditional distribution of η_{k+1}^ε with respect to the σ-algebra $\mathcal{F}_{\tau_k^\varepsilon}^\varepsilon$ is asymptotically Gaussian as $\varepsilon \to 0$ with*

$$E(\eta_{k+1}^\varepsilon / \mathcal{F}_{\tau_k^\varepsilon}^\varepsilon) = \varepsilon a_{\theta_k^\varepsilon}(\zeta_{\tau_k^\varepsilon}^\varepsilon) + o(\varepsilon)$$

and

$$E((\eta_{k+1}^\varepsilon)^2 / \mathcal{F}_{\tau_k^\varepsilon}^\varepsilon) = \varepsilon b_{\theta_k^\varepsilon}(\zeta_{\tau_k^\varepsilon}^\varepsilon) + o(\varepsilon),$$

where

$$a_\theta(z) = \int_{\alpha_\theta^z}^{\beta_\theta^z} (2z - 2U(x))^{-1/2} \hat{F}(x)\, dx$$

and

$$b_\theta(z) = \int_{\alpha_\theta^z}^{\beta_\theta^z} (2z - 2U(x))^{1/2} \hat{F}(x)\, ds,$$

for $z < 0$*, and*

$$a_{-1}(z) = \int_{\alpha_0^z}^{0} (2z - 2U(x))^{-1/2} \hat{F}(x)\, dx$$

$$a_1(z) = \int_0^{\beta_0^z} (2z - w2U(x))^{-1/2} \hat{F}(x)\, dx$$

and

$$b_{-1}(z) = \int_{\alpha_0^z}^{0} (2z - 2U(x))^{1/2} \hat{F}(x)\, dx$$

$$b_1(z) = \int_0^{\beta_0^z} (2z - 2U(x))^{1/2} \hat{F}(x)\, dx$$

for $z > 0$.

The proof follows from Lemma 6 and the representation for η_{k+1}^ε.

Remark 7 It follows from Lemma 7 that the joint distribution of the random variables

$$\left\{\left(\frac{1}{\sqrt{\varepsilon}}(\zeta_k^\varepsilon - E_0), \theta_k^\varepsilon\right), k = 1, 2, \ldots, n\right\}$$

converges to the joint distribution of the random variables $\{(\hat{\zeta}_k, \hat{\theta}_k), k = 1, 2, \ldots, n\}$ as $\varepsilon \to 0$, $E_0 \to 0$, and $\sqrt{\varepsilon}\log E_0 \to 0$ for all n, where $(\hat{\zeta}_k, \hat{\theta}_k)$ is the Markov chain in the space $R \times \{-1, 1\}$ that is determined as follows. Let $\{\gamma_k(\theta), k = 1, 2, \ldots, \theta = -1, 1\}$ be independent Gaussian random variables for which

$$E\gamma_k(\theta) = 0, \quad E(\gamma_k(\theta))^2 = b_\theta, \quad b_\theta = b_\theta(0).$$

Then

$$\hat{\zeta}_{k+1} = \hat{\zeta}_k + \gamma_{k+1}(\hat{\theta}_k), \quad \hat{\theta}_{k+1} = -(\operatorname{sign}\zeta_{k+1})\hat{\theta}_k,$$

and $\hat{\zeta}_0 = 0$.
We will consider this Markov chain with an arbitrary initial state $(\hat{\zeta}_0, \hat{\theta}_0)$. We define an operator S on the space of bounded measurable functions $f(x, \theta) : R \times \{-1, 1\} \to R$ by

$$Sf(x, \theta) = \int [f(x + u, \theta)1_{(x+u<0)} + f(x_u, -\theta)1_{(x+u>0)}] g_\theta(u)\, du, \quad (8.37)$$

where $g_\theta(u)$ is the density distribution of the random variable $\gamma_k(\theta)$.

Denote by $P_n(z_0, \theta_0, \theta, x)$ the density of the conditional probability

$$P\{\hat{\zeta}_n < x, \hat{\theta}_n = \theta \,/\, \hat{\zeta}_0 = z_0, \hat{\theta}_0 = \theta_0\}.$$

Then

$$P_n(z_0, \theta_0, \theta, x) = \int [P_{n-1}(z_0, \theta_0, \theta, x-u) g_\theta(u) 1_{(x-u<0)} + P_{n-1}(z_0, \theta_0, -\theta, x-u) g_{-\theta}(u) 1_{(x-u>0)}]. \tag{8.38}$$

It follows from formula (8.38) that

$$P_n(z_0, \theta_0, \theta, x) = \sum_{\theta_k \in \{-1,1\}, k=1,\ldots,n-1} \iint \mathcal{D}_n(\theta_0, \theta_1, \ldots, \theta_n, z_0, u_1, \ldots, u_n, x)\, du_1 \cdots du_n, \tag{8.39}$$

where

$$\begin{aligned}\mathcal{D}_n(\theta_0, \ldots, \theta_n, z_0, u_1, \ldots, u_n, x) &= g_{\theta_0}(u_1 - z_0) g_{\theta_1}(u_2 - u_1) \ldots g_{\theta_n}(x - u_n) \\ &\quad \times 1_{\{(\theta_1/\theta_0)u_1<0, \ldots, (\theta_n/\theta_{n-1})u_n<0, (\theta/\theta_n)(x-u_n)<0\}}.\end{aligned} \tag{8.40}$$

Lemma 8 *For $\delta > 0$, $c > 1$, and $n \in \mathbf{Z}_+$,*

$$E\frac{1}{n}\sum_{k=1}^{n} 1_{\{|\hat{\zeta}_k| \le \delta\sqrt{n}\}} \log\left(\frac{\sqrt{n}}{|\hat{\zeta}_k|} \vee c\right) = O(\delta). \tag{8.41}$$

Proof Using formulas (8.39), (8.40) one can prove that a constant $A > 0$ exists depending on z_0 such that

$$P_n(z_0, \theta_0, \theta, x) \le An^{-1/2}.$$

So

$$E1_{\{|\hat{\zeta}_k| \le \delta\}} \log\left(\frac{\sqrt{n}}{|\hat{\zeta}_k|} \vee c\right) \le A\frac{1}{\sqrt{k}} \int_{-n^{1/2}\delta}^{n^{1/2}\delta} \log\left(\frac{\sqrt{n}}{|u|} \vee c\right) du,$$

and the expression on the left–hand side of equality (8.41) is less than

$$A\frac{1}{\sqrt{n}} \int_{-n^{1/2}\delta}^{n^{1/2}\delta} \log\left(\frac{\sqrt{n}}{|u|} \vee c\right) du = O(\delta).$$

This completes the proof of the lemma. □

Corollary 2 Assume that $\varepsilon n = O(1)$. Then

$$E\sum_{k=1}^{n} \varepsilon(\tau_k^\varepsilon - \tau_{k-1}^\varepsilon) 1_{\{|\zeta_{k-1}^\varepsilon| \le \delta\}} = O(\delta). \tag{8.42}$$

Proof It follows from relation (8.32) that

$$E(\tau_k^\varepsilon - \tau_{k-1}^\varepsilon / \mathcal{F}^\varepsilon_{\tau^\varepsilon_{k-1}}) = O\left(\log\left(\frac{1}{|\zeta^\varepsilon_{k-1}|} \vee c \right)\right).$$

So relation (8.42) is a consequence of Lemmas 7 and 8.

□

Corollary 3

$$E \int_0^{t/\varepsilon} 1_{\{|\hat{z}_\varepsilon(s)| \le \delta\}}\, ds = O(t\delta). \tag{8.43}$$

The reader can show that this relation follows from (8.42).

Lemma 9 *If $\varphi \in C^{(3)}(G)$ and $0 \le t_1 < t_2$, then*

$$\begin{aligned}\limsup_{\varepsilon \to 0} E\Bigg| E\Bigg(\varphi\left(z_\varepsilon\left(\frac{t_2}{\varepsilon}\right)\right) - \varphi\left(z_\varepsilon\left(\frac{t_1}{\varepsilon}\right)\right) \\ - \int_{t_1}^{t_2} L\varphi\left(z_\varepsilon\left(\frac{s}{\varepsilon}\right)\right) ds \Big/ \mathcal{F}^\varepsilon_{t_1/\varepsilon}\Bigg)\Bigg| = 0,\end{aligned} \tag{8.44}$$

where $z_\varepsilon(t)$ is the stochastic process introduced in Remark 6, and the differential operator L is defined by formula (8.25).

Proof It follows from Lemma 4 that it suffices to prove the relation

$$\limsup_{\varepsilon \to 0} E\left| \int_0^t \left(L\varphi\left(z_\varepsilon\left(\frac{s}{\varepsilon}\right)\right) - \mathcal{L}\varphi\left(z_\varepsilon\left(\frac{s}{\varepsilon}\right)\right)\right) ds\right| = 0 \tag{8.45}$$

for all $t > 0$, where

$$\mathcal{L}\varphi(z_\varepsilon(s)) = [\varphi''(z_\varepsilon(s))\dot{x}_\varepsilon^2(s) + \varphi'(z_\varepsilon(s))]\hat{F}(x_\varepsilon(s)).$$

Let $\mathcal{X}_\delta(x)$ be a function from $C^{(3)}(R)$ defined by the relations: $\mathcal{X}_\delta(x) = 1$ if $|x| > \delta$, $\mathcal{X}_\delta(x) = 0$ if $|x| \le \delta/2$. Then using the proof of Theorem 1 we can write that for any $\delta > 0$,

$$\limsup_{\varepsilon \to 0} E\left| \int_0^t \left[L\varphi\left(z_\varepsilon\left(\frac{s}{\varepsilon}\right)\right) - \mathcal{L}\varphi\left(z_\varepsilon\left(\frac{s}{\varepsilon}\right)\right)\right] \mathcal{X}_\delta(\hat{z}_\varepsilon(s))\, ds\right| = 0.$$

Formula (8.43) implies that

$$\limsup_{\varepsilon \to 0} E\left| \int_0^t \left[L\varphi\left(z_\varepsilon\left(\frac{s}{\varepsilon}\right)\right) - \mathcal{L}\left(\varphi\left(z_\varepsilon\left(\frac{s}{\varepsilon}\right)\right)\right)\right] (1 - \mathcal{X}_\delta(\hat{z}_\varepsilon(s)))\, ds\right| = O(\delta),$$

so

$$\limsup_{\varepsilon \to 0} E\left| \int_0^t \left[L\varphi\left(z_\varepsilon\left(\frac{s}{\varepsilon}\right)\right) - \mathcal{L}\varphi\left(z_\varepsilon\left(\frac{s}{\varepsilon}\right)\right)\right] ds\right| = O(\delta)$$

for all $\delta > 0$.

This completes the proof of the lemma.

□

Remark 8 The Markov chain $\{(\hat{\zeta}_n, \hat{\theta}_n), n = 0, 1, \dots\}$ has an invariant measure π on the space $R \times \{-1, 1\}$ that is defined by the relation

$$\pi (A \times \{\theta\}) = l(A), \quad A \in \mathcal{B}(R), \quad \theta \in \{-1, 1\},$$

where $l(A)$ is the Lebesgue measure on R. We can show that the Markov chain is Harris recurrent with respect to the measure π. This follows from the fact that the random walk is Harris recurrent in R if the distribution of one step ξ satisfies the conditions $E\xi^2 < \infty$, $E\xi = 0$, and the density of the distribution exists with respect to the Lebesgue measure (see [167], Section 1.3.1).

It follows from Theorem 4 of Chapter 1 that with probability 1,

$$\lim_{n\to\infty} \left(E \sum_{k=1}^{n} f(\hat{\zeta}_k, \hat{\theta}_k) \Big/ E \sum_{k=1}^{n} G(\hat{\zeta}_k) \right) = \frac{\int (f(x,-1) + f(x,1))dx}{2 \int G(x)dx} \tag{8.46}$$

for any measurable functions $f(x, \theta)$ and $G(x)$ for which
(1) $\int (|f(x,-1)| + |f(x,1)|)\, dx < \infty$,
(2) $G(x) > 0$, $\int G(x)\, dx < \infty$,
and
(3) $\sum G(\hat{\zeta}_n) = +\infty$ with probability 1.

It follows from the Harris recurrence that (3) holds for all positive functions. We define the function

$$\Psi(l_{-1}, l_0, l_1, x, \theta) = \sum_{\theta'} [l_{\theta'} x 1_{(x<0)} + l_0 x 1_{(x>0)}] 1_{(\theta=\theta')}.$$

This function is independent of θ for $x > 0$, so it can be considered as a function on the graph G. It is linear on any edge of the graph, and it is equal to $l_\theta x$ on the edge E_θ.

It follows that

$$\begin{aligned} &S\Psi(l_{-1}, l_0, l_1, x, \theta) - \Psi(l_{-1}, l_0, l_1, x, \theta) = \mathcal{X}(l_{-1}, l_0, l_1, x, \theta) \\ &= \sum_{\theta'} (l_{\theta'} - l_0)[-b_{\theta'} g_{\theta'}(x) + x(G_{\theta'}(-x) - 1_{(x<0)})] 1_{(\theta=\theta')}, \end{aligned}$$

where $G_\theta(x) = \int_{-\infty}^{x} g_\theta(u)\, du$.

It is easy to see that

$$\int \mathcal{X}(l_{-1}, l_0, l_1, x, \theta)\, dx = \frac{1}{2}(l_0 - l_\theta) b_\theta.$$

Note that $\Psi(-1, 1, -1, x, \theta) = |x|$,

$$\int \mathcal{X}(-1, 1, x, \theta)\, dx = b_\theta.$$

We have the representation

$$E(\Psi(l_{-1},l_0,l_1,\hat{\zeta}_n,\hat{\theta}_n)-\Psi(l_{-1},l_0,l_1,\hat{\zeta}_0,\hat{\theta}_0)\Big/\mathcal{F}_0^{\varepsilon})$$

$$=E\biggl(\sum_{k=0}^{n-1}\mathcal{X}(l_{-1},l_0,l_1,\hat{\zeta}_k,\hat{\theta}_k)\Big/\mathcal{F}_0^{\varepsilon}\biggr). \tag{8.47}$$

Using formulas (8.46) and (8.47) we can obtain the relation

$$\begin{aligned}\lim_{n\to\infty}&\frac{E(\Psi(l_{-1},l_0,l_1,\hat{\zeta}_n,\hat{\theta}_n)-\Psi(l_{-1},l_0,l_1,\hat{\zeta}_0,\hat{\theta}_0))}{E|\hat{\zeta}_n|}\\&=\frac{(l_0-l_1)b_{-1}+(l_0-l_1)b_1}{2(b_{-1}+b_1)}.\end{aligned} \tag{8.48}$$

Now we prove the main result concerning the asymptotic behavior of the process $z_\varepsilon(\frac{t}{\varepsilon})$ on the graph G as $\varepsilon\to 0$. Set

$$P_{-1}=\frac{b_{-1}}{2(b_{-1}+b_1)},\quad P_1=\frac{b_1}{2(b_{-1}+b_1)},\quad P_0=\frac{1}{2},$$

and note that

$$P_{-1}+P_0+P_1=1.$$

Let $\mathbf{C}(P_{-1},P,P_1,G)$ be the set of functions $f(z):G\to R$ satisfying the conditions

(i) $f(z)$ is continuous, $f'(z)$, $f''(z)$, $f'''(z)$ are defined at $z\neq z_0=(0,0)$, they are bounded and continuous for $z\neq 0$, and $f''(z)$ and $f'''(z)$ have limits as $z\to z_0$,

(ii) $f'(x,\theta)$ has a limit as $x\to 0$ and

$$-f'(0-,-1)P_{-1}+f'(0+,0)P_0-f'(0-,1)P_1=0. \tag{8.49}$$

Note that

$$\Psi(l_{-1},l_0,l_1,x,\theta)\in\mathbf{C}(P_{-1},P_0,P_1,G)\quad\text{if}\quad -l_{-1}P_{-1}+l_0P_0-l_1P_1=0.$$

Theorem 3 *For all $\varphi\in\mathbf{C}(P_{-1},P_0,P_1,G)$ relation (8.44) holds for all $0\le t_1<t_2$.*

Proof It is easy to see that for $\varphi\in\mathbf{C}(P_{-1},P_0,P_1,G)$ constants l_{-1},l_0,l_1 exist for which

$$\varphi^*(x,\theta)=\varphi(x,\theta)-\Psi(l_{-1},l_0,l_1,x,\theta)\in C^{(3)}(G)$$

and $-l_{-1}P_{-1}+l_0P_0-l_1P_1=0$. Therefore, it suffices to prove that relation (8.44) is satisfied for the function

$$\Psi(x,\theta)=\Psi(l_{-1},l_0,l_1,x,\theta).$$

Fix δ_1 and n_1 and introduce the sequence of stopping times $\{\lambda_k^\varepsilon, k = 0, 1, 2, \dots\}$, which are determined by the relations

$$\lambda_0^\varepsilon = \{\inf \tau_k^\varepsilon : |\zeta_k^\varepsilon| < \delta_1\}, \quad \lambda_1^\varepsilon = \tau_{k_0+n_1}^\varepsilon \quad \text{if} \quad \lambda_0^\varepsilon = \tau_{k_0}^\varepsilon,$$
$$\lambda_{2l}^\varepsilon = \{\inf \tau_k^\varepsilon > \lambda_{2l-1}^\varepsilon : |\zeta_k^\varepsilon| < \delta_1\}, \quad \lambda_{2l+1}^\varepsilon = \tau_{k_0+n_1}^\varepsilon \text{ if } \lambda_{2l}^\varepsilon = \tau_{k_0}^\varepsilon.$$

It follows that

$$\begin{aligned} &E(\Psi(z_\varepsilon(\lambda_{2l}^\varepsilon)) - \Psi(z_\varepsilon(\lambda_{2l-1}^\varepsilon))/\mathcal{F}_{\lambda_{2l-1}^\varepsilon}^\varepsilon) \\ &= E\biggl(\varepsilon \sum a_\theta(\zeta_k^\varepsilon) 1_{\{\lambda_{2l-1}^\varepsilon \le \zeta_k^\varepsilon < \lambda_{2l-1}^\varepsilon\}} \Big/ \mathcal{F}_{\lambda_{2l-1}^\varepsilon}^\varepsilon\biggr) \\ &\quad + O(\varepsilon^{1/2} \exp\{-c\delta_1^2 \varepsilon^{-1}\}), \quad \theta = \theta_{k_0}^\varepsilon \text{ if } \lambda_{2l-1}^\varepsilon = \tau_{k_0}^\varepsilon, \end{aligned} \tag{8.50}$$

where $c > 0$ is a constant.
To obtain this estimate we use the fact that for $\tau_k^\varepsilon \in [\lambda_{2l-1}^\varepsilon, \lambda_{2l}^\varepsilon)$, θ_k^ε is the same, $|\zeta_k^\varepsilon| > \delta_1$, and $\Psi(\zeta_k^\varepsilon, \theta_k^\varepsilon) = \lambda_\theta \zeta_k^\varepsilon$.
Now we consider

$$E(\Psi(z_\varepsilon(\lambda_{2l+1}^\varepsilon)) - \Psi(z_\varepsilon(\lambda_{2l}^\varepsilon))/\mathcal{F}_{\lambda_{2\lambda}^\varepsilon}^\varepsilon).$$

We may assume that $|z_\varepsilon(0)| < \delta_1$ and consider the case $l = 0$. Using Remark 7 we can prove that

$$\lim_{\substack{\varepsilon \to 0 \\ \varepsilon n \to 0}} \sup_{k \le n} \left| \frac{E|\zeta_k^\varepsilon|}{\sqrt{\varepsilon} E|\hat{\zeta}_k|} - 1 \right| = 0,$$

$$\lim_{\substack{\varepsilon \to 0 \\ \varepsilon n \to 0}} \sup_{k \le n} \left| \frac{E\Psi(\zeta_k^\varepsilon) - \sqrt{\varepsilon} E\Psi(\hat{\zeta}_k)}{\sqrt{\varepsilon} E|\hat{\zeta}_k|} \right| = 0,$$

if $\sqrt{\varepsilon} \log|\zeta_0^\varepsilon| \to 0$.
It follows from (8.48) that

$$|E\Psi(z_\varepsilon(\lambda_1^\varepsilon)) - \Psi(z_\varepsilon(0))| \le \mu(\varepsilon, n_1, \delta_1) E|z_\varepsilon(\lambda_1^\varepsilon)|,$$

where $\mu(\varepsilon, n_1, \delta_1) \to 0$ as $\delta_1 \to 0$, $\varepsilon \to 0$, $\varepsilon n_1 \to 0$ and $\sqrt{\varepsilon} \log|\zeta_0^\varepsilon| \to 0$, taking into account that the expression on the right-hand side of equality (8.48) is equal to 0. Note that

$$|z_\varepsilon(\lambda_{2l}^\varepsilon)| = \delta_1 + O(\sqrt{\varepsilon})$$

with probability

$$O\Bigl(\frac{1}{\sqrt{\varepsilon}} \exp\Bigl\{-\frac{\delta_1^2}{\varepsilon}\Bigr\}\Bigr).$$

Moreover,

$$E|z_\varepsilon(\lambda_1^\varepsilon)| = O(\sqrt{E|z_\varepsilon(\lambda_1^\varepsilon)|^2}) = O(\sqrt{\varepsilon n_1}).$$

This implies the inequality

$$\begin{aligned} & E|E(\Psi(z_\varepsilon(\lambda_{2l+1}^\varepsilon)) - \Psi(z_\varepsilon(\lambda_{2l}^\varepsilon))/\mathcal{F}_{\lambda_{2l}^\varepsilon}^\varepsilon)| \\ & \le A\mu(\varepsilon, n_1, \delta_1)\varepsilon^{1/2} n_1^{1/2} + B\varepsilon^{1/2} \exp\left\{ -\frac{\delta_1^2}{\varepsilon} \right\}, \end{aligned} \tag{8.51}$$

where A, B are some constants.

Relations (8.50) and (8.51) imply that

$$E\left|E\left(\psi(z_\varepsilon(\lambda_{k_2}^\varepsilon)) - \Psi(z_\varepsilon(\lambda_{k_1}^\varepsilon)) - \varepsilon \int_{\lambda_{k_1}^\varepsilon}^{\lambda_{k_2}^\varepsilon} L\Psi(z_\varepsilon(s))\, ds \Big/ \mathcal{F}_{\lambda_{k_1}^\varepsilon}^\varepsilon\right)\right|$$

$$\le A_1 \mu_1(\varepsilon, n_1, \delta_1)\varepsilon^{1/2} n_1^{1/2} E(k_2 - k_1), \tag{8.52}$$

where A_1 is a constant, $0 < k_1 < k_2$ are integer-valued random variables for which $\lambda_{k_1}^\varepsilon, \lambda_{k_2}^\varepsilon$ are stopping times, and

$$\mu_1(\varepsilon, n_1, \delta_1) = \mu(\varepsilon, n_1, \delta_1) + n_1^{-1/2}.$$

Let $k_i = \inf\{k : \lambda_k^\varepsilon \le t_i/\varepsilon\}$, $i = 1, 2$. Then $E|z_\varepsilon(\lambda_{k_2}^\varepsilon) - z_2(\frac{1}{\varepsilon} t_i)| \to 0$ as $\varepsilon \to 0$ for $i = 1, 2$. Relation (8.52) implies that the proof of the theorem is a consequence of the relation

$$\varepsilon^{1/2} n_1^{1/2} E(k_2 - k_1) = O(1). \tag{8.53}$$

To establish this, we need to estimate $E(\lambda_{k+1}^\varepsilon - \lambda_k^\varepsilon)$. We have that $E(\lambda_{2l+1}^\varepsilon - \lambda_{2l}^\varepsilon) = \lambda_{2l+1}^\varepsilon - \lambda_{2l}^\varepsilon = n_1$. To estimate $\lambda_{2l}^\varepsilon - \lambda_{2l-1}^\varepsilon$ we take into account that $z^\varepsilon(t)$ is on the same edge for $t \in [\lambda_{2l-1}^\varepsilon, \lambda_{2l}^\varepsilon]$. Using Remark 7 and the approximation of $\{\sqrt{\varepsilon}\, \hat{\zeta}_k, k = 1, 2, \ldots, n\}$ for $|\hat{\zeta}_0| > 0$ by a Wiener process, and using the relation

$$E\tau_a \wedge c \approx c_1 a \quad \text{if} \quad a \to 0$$

for any $c > 0$, where $a > 0$ and $\tau_a = \inf\{t : w(t) = a\}$, ($w(t)$ is the approximating Wiener process, and c_1 does not depend on c), we can show that

$$\begin{aligned} E(\lambda_{2l}^\varepsilon - \lambda_{2l-1}^\varepsilon/\mathcal{F}_{2l-1}^\varepsilon) &\ge \frac{1}{\varepsilon} c_2 E\left(\left|z_\varepsilon(\lambda_{2l-1}^\varepsilon)\right|/\mathcal{F}_{2l-1}^\varepsilon\right) \\ &\approx c_2 \frac{1}{\sqrt{\varepsilon}} E|\hat{\zeta}_{n_1}| \ge c_3 \frac{1}{\sqrt{\varepsilon}} \sqrt{n_1}, \end{aligned}$$

where c_2, c_3 are some positive constants. Note that

$$\frac{t_2 - t_1}{\varepsilon} \ge \lambda_{k_2 - 1}^\varepsilon \lambda_{k_1}^\varepsilon \ge \sum_{i=k_1+1}^{k_2 - 1} (\lambda_i^\varepsilon - \lambda_{i-1}^\varepsilon)$$

and

$$\frac{t_2 - t_1}{\varepsilon} \ge E \sum_{i=k_1+1}^{k_2-1} (\lambda_i^\varepsilon - \lambda_{i-1}^\varepsilon) \ge c_3 \frac{1}{\sqrt{\varepsilon}} \sqrt{n_1} E(k_2 - k_1 - 2).$$

This inequality implies (8.53), and it completes the proof of Theorem 3. □

Now we consider the system for which the function $U(x)$ has $r \geq 2$ maxima and $r+1$ minima, $U'(x) = 0$ for $x = x_k$, $k = 1, \ldots, 2r+1$, $x_1 < x_2 < \cdots < x_{2r+1}$, and $U(x_2) = \cdots = U(x_{2r}) = 0$ are identical maximum values. In this case, the graph G has $r+2$ edges and $r+2$ vertices; the vertex corresponding to the singular orbit

$$\{(x, \dot{x}) : \mathcal{E}(x, \dot{x}) = 0, x \in [\alpha, \beta]\},$$

where $\alpha < x_1 < x_{2r+1} < \beta$ are such that $U(\alpha) = U(\beta) = 0$, is a knot, that is the origin of $r+1$ edges. Denote them by $E_1, \ldots, E_{r+1}$. The orbit O belongs to E_l if for $(x, \dot{x}) \in O$ we have $\mathcal{E}(x, \dot{x}) < 0$, $x \in (x_{2k-2}, x_{2k})$, $x_0 = \alpha$, $x_{2r+2} = \beta$. Denote by E_0 the edge for which $\mathcal{E}(x, \dot{x}) \geq 0$.
To investigate the behavior of the perturbed systems if $\mathcal{E}(x_\theta, \dot{x}_\theta)$ is small enough, we consider the sequence of stopping times $\{\tau_k^\varepsilon, k = 0, \ldots\}$, $\tau_0^\varepsilon = 0$. Suppose τ_k^ε is determined, $z_\varepsilon(\tau_k^\varepsilon) \in E_l$, $l > 0$. Then

$$\begin{aligned}\tau_{k+1}^\varepsilon = \inf\{s > \tau_k^\varepsilon : &|x_\varepsilon(s) - x_\varepsilon(\tau_k^\varepsilon)| > |x_{2l-1} - x_\varepsilon(\tau_k^\varepsilon)|,\\ &|\dot{x}_\varepsilon(s)| \wedge |x_\varepsilon(s) - x_{2l-2}| \wedge |x_\varepsilon(s) - x_{2l}| = 0\}.\end{aligned}$$

If $z_\varepsilon(\tau_k^\varepsilon) > 0$ and $x_\varepsilon(\tau_k^\varepsilon) = x_{2l}$, $l = 1, \ldots, r$, then

$$\tau_{k+1}^\varepsilon = \inf\{s > \tau_k^\varepsilon : |x_\varepsilon(s) - x_{2l+2\mathrm{sign}\dot{x}(\tau_k^\varepsilon)}| \wedge |\dot{x}_\varepsilon(s)| = 0\}$$

if $z_\varepsilon(\tau_k^\varepsilon) > 0$ and $\dot{x}_\varepsilon(\tau_k^\varepsilon) = 0$, then

$$\tau_{k+1}^\varepsilon = \inf\{s > \tau_k^\varepsilon : |x_\varepsilon(s) - x_2| \wedge |\dot{x}_\varepsilon(s)| = 0\} \quad \text{for} \quad x_\varepsilon(\tau_k^\varepsilon) < x_1$$

and

$$\tau_{k+1}^\varepsilon = \inf\{s > \tau_k^\varepsilon : |x_\varepsilon(s) - x_{2r}| \wedge |\dot{x}_\varepsilon(s)| = 0\} \quad \text{for} \quad x_\varepsilon(\tau_k^\varepsilon) > x_{2r+1}.$$

For $z \geq 0$ define the functions

$$\alpha_0(z) = \inf\{x : U(x) = z\}, \quad \beta_0(z) = \sup\{x : U(x) = z\},$$

and for $z < 0$,

$$\begin{aligned}\alpha_k(z) &= \inf\{x \in (x_{2k-2}, x_{2k}) : U(x) = z\},\\ \beta_k(z) &= \sup\{x \in (x_{2k-2}, x_{2k}) : U(x) = z\},\end{aligned}$$

for $k = 1, \ldots, r+1$. Set $\Delta_1 = (-\infty, x_2]$ and $\Delta_2 = [x_2, x_4], \ldots, \Delta_{r+1} = [x_{2r}, \infty)$. Define the sequence of random variables

$$\{(\zeta_k^\varepsilon, \theta_k^\varepsilon, \sigma_k^\varepsilon), k = 0, 1, \ldots\}$$

in the space $R \times \{1, 2, \ldots, r+1\} \times \{-1, 1\}$, where

$$\begin{aligned}&\zeta_k^\varepsilon = \mathcal{E}(x_\varepsilon(\tau_k^\varepsilon), \dot{x}_\varepsilon(\tau_k^\varepsilon)),\\ &\theta_k^\varepsilon = l, l \in \{1, \ldots, r+1\} \text{ if } x_\varepsilon(t) \in \Delta_l \ \text{ for } t \in [\tau_k^\varepsilon, \tau_{k+1}^\varepsilon]\\ &\sigma_k^\varepsilon = \mathrm{sign}(x_\varepsilon(\tau_{k+1}^\varepsilon) - x_\varepsilon(\tau_k^\varepsilon)).\end{aligned}$$

Set $\eta_{k+1}^{\varepsilon} = \zeta_{k+1}^{\varepsilon} - \zeta_{k}^{\varepsilon}$.
Using Lemma 6 we can prove that the conditional distribution of η_{k+1}^{ε} with respect to the σ-algebra $\mathcal{F}_{\tau_k^\varepsilon}^{\varepsilon}$ is asymptotically Gaussian with

$$E(\eta_{k+1}^{\varepsilon}/\mathcal{F}_{\tau_k^\varepsilon}^{\varepsilon}) = \varepsilon\, a_{\theta_k^\varepsilon}(\zeta_{\tau_k^\varepsilon}^{\varepsilon}) + o(\varepsilon)$$

and

$$E((\eta_{k+1}^{\varepsilon})^2/\mathcal{F}_{\tau_k^\varepsilon}^{\varepsilon}) = \varepsilon\, b_{\theta_k^\varepsilon}(\zeta_{\tau_k^\varepsilon}^{\varepsilon}) + o(\varepsilon),$$

where

$$a_\theta(z) = \frac{1}{2}\int_{\alpha_\theta(z)}^{\beta_\theta(z)} (2z - 2U(x))^{-1/2} \hat{F}(x)\, dx,$$

$$\beta_\theta(z) = \frac{1}{2}\int_{\alpha_\theta(z)}^{\beta_\theta(z)} (2z - 2U(x))^{1/2} \hat{F}(x)\, dx, \quad \theta = 1, \ldots, r+1,$$

and for $z > 0$,

$$\alpha_1(z) = \alpha_0(z),\ \alpha_2(z) = x_2,\ \ldots,\ \alpha_{r+1}(z) = x_{2r},$$

$$\beta_1(z) = x_2,\ \ldots,\ \beta_r(z) = x_{2r},\ \beta_{r+1}(z) = \beta_0(z).$$

We consider $r+1$ independent sequences $\{\gamma_k(\theta), k = 1, 2, \ldots\}$, for $\theta \in \{1, \ldots, r+1\}$, of independent Gaussian random variables for which

$$E\gamma_k(\theta) = 0, \quad E(\gamma_k(\theta))^2 = b_\theta = b_\theta(0).$$

Define the sequence of $R \times \{1, \ldots, r+1\} \times \{-1, 1\}$ random variables

$$\{(\hat{\zeta}_k, \hat{\theta}_k, \hat{\sigma}_k), k = 0, 1, 2, \ldots\}, \tag{8.54}$$

where $\hat{\zeta}_0 \in R$, $\hat{\theta}_0 \in \{1, \ldots, r+1\}$, $\hat{\sigma}_0 \in \{-1, 1\}$ are given and

$$\begin{aligned}
&\hat{\zeta}_{k+1} = \hat{\zeta}_k + \gamma_{k+1}(\hat{\theta}_k),\\
&\hat{\theta}_{k+1} = \hat{\theta}_k \ \text{ if } \hat{\zeta}_k < 0, \hat{\theta}_{k+1} = \hat{\theta}_k + \sigma_k \ \text{ if } \ \hat{\zeta}_k \geq 0 \text{ and } \hat{\theta}_k \in [2, r],\\
&\hat{\theta}_{k+1} = r+1 \ \text{ if } \ \hat{\theta}_k = r+1, \hat{\sigma}_k = 1, \hat{\theta}_{k+1} = r \ \text{ if } \ \hat{\theta}_k = r+1, \hat{\sigma}_k = -1\\
&\hat{\theta}_{k+1} = 1 \ \text{ if } \ \hat{\theta}_k = 1, \hat{\sigma}_k = -1, \hat{\theta}_{k+1} = 2 \ \text{ if } \ \hat{\theta}_k = 1, \hat{\sigma}_k = 2,\\
&\hat{\sigma}_{k+1} = -\hat{\sigma}_k \ \text{ if } \ \hat{\zeta}_k < 0, \hat{\sigma}_{k+1} = -\hat{\sigma}_k \ \text{ if } \ \hat{\zeta}_k > 0, \hat{\theta}_k = 1, \hat{\sigma}_k = -1,\\
&\hat{\sigma}_{k+1} = -\hat{\sigma}_k \ \text{ if } \ \hat{\zeta}_k \geq 0, \hat{\theta}_k = r+1, \hat{\sigma}_k = 1,\\
&\hat{\sigma}_{k+1} = \hat{\sigma}_k \ \text{ if } \ \hat{\zeta}_k \geq 0, \hat{\theta}_k \in [2, r], \hat{\sigma}_{k+1} = \hat{\sigma}_k \ \text{ if } \ \zeta_k > 0, \hat{\theta}_k = 1,\\
&\quad \hat{\sigma}_k = 1, \hat{\sigma}_{k+1} = \hat{\sigma}_k \ \text{ if } \ \hat{\zeta}_k \geq 0, \theta_k = r+1, \hat{\sigma}_k = -1.
\end{aligned}$$

With these preliminaries, we have the following lemma.

Lemma 10 *Let $\mathcal{E}(x_0, \dot{x}_0) = E_0 \to 0$ and $\sqrt{\varepsilon} \log \frac{1}{|E_0|} \to 0$. Then for any n, the joint distribution of the random variables*

$$\left\{ \left(\frac{1}{\sqrt{\varepsilon}} (\zeta_k^\varepsilon - E_0, \theta_k^\varepsilon, \sigma_k^\varepsilon) \right), k = 1, \dots, n \right\}$$

converges to the distribution of the random variables

$$\{(\hat{\zeta}_k, \hat{\theta}_k, \hat{\sigma}_k), k = 1, \dots, n\}.$$

The proof is a direct consequence of the asymptotic behavior of the random variable η_k^ε.

Remark 9 The sequence given by (8.54) is a homogeneous Markov chain, it is Harris recurrent, and it has an invariant measure π on the space $R \times \{1, \dots, r+1\} \times \{-1, 1\}$ that is defined by the relation

$$\pi(A \times \{\theta\} \times \{\sigma\}) = l(A), \quad \theta \in \{1, \dots, r+1\}, \quad \sigma \in \{-1, 1\};$$

see Remark 8.

Using this statement we can prove the following result in the same way as Theorem 3.

Theorem 4 *Let $q_0 = 1/2$,*

$$q_\theta = \frac{b_\theta}{2 \sum_{l=1}^{r+1} b_l}, \qquad \theta = 1, 2, \dots, r+1.$$

Denote by $C(q_0, q_1, \dots, q_{r+1}, G)$ the set of functions $\varphi \in C(G)$ for which the derivatives $\varphi'(x, \theta), \varphi''(x, \theta), \varphi'''(x, \theta)$ exist for $x \neq 0$, and $\theta = 0, 1, \dots, r+1$. Suppose that the following limits exist:

$$\begin{aligned} &\varphi'(0+, 0), \varphi'(0-, \theta), \\ &\varphi''(0+, 0), \varphi''(0-, \theta), \\ &\varphi'''(0+, 0), \varphi'''(0-, \theta), \quad \theta = 1, 2, \dots, r+1, \end{aligned}$$

and

$$q_0 \, \varphi'(0+, 0) = \sum_{l=1}^{r+1} q_l \, \varphi'(0-, l),$$

$$\varphi''(0+, 0) = \varphi''(0-, \theta), \quad \varphi'''(0+, 0) = \varphi'''(0-, \theta), \quad \theta = 1, 2, \dots, r+1.$$

Then for all $\varphi \in C(q_0, \dots, q_{r+1}, G)$ relation (8.44) holds for all $0 \le t_1 < t_2$.

This statement can be extended to the randomly perturbed system considered in Section 8.1.2. Let $z_i = (u_i, \theta_i)$, $i = 1, \dots, K$, be all the knots to the graph G and define $\Theta_i \subset \Theta$ as the subset of those $\theta \in \Theta$ for which (u_i, θ_i)

is the origin of the edge E_θ. Let

$$q_{\theta_i}(i) = \frac{1}{2}, \quad q_\theta(i) = \frac{1}{2} b_\theta \Big(\sum_{\theta' \in \Theta_i} b_{\theta'} \Big)^{-1}, \quad i = 1, \ldots, K, \quad \theta \in \Theta_i.$$

Denote by $C(G, q_\theta(i), i = 1, \ldots, K, \theta \in \Theta_i)$ the set of functions $\varphi \in C(G)$ for which the derivatives $\varphi'(z)$, $\varphi''(z)$, and $\varphi'''(z)$ exist for $z \neq z_i$ for all $i = 1, \ldots, K$. Moreover, the limits $\varphi'(u_i+, \theta_i), \varphi'(u_i-, \theta_i)$, $\lim_{z \to z_i} \varphi''(z)$, and $\lim_{z \to z_i} \varphi'''(z)$ exist for $i = 1, \ldots, k$, and

$$q_{\theta_i}(i)\, \varphi'(u_i+, \theta_i) = \sum_{\theta' \in \Theta_i} q_{\theta'(i)}\, \varphi'(u_i-, \theta'), \quad i = 1, \ldots, k. \tag{8.55}$$

Theorem 5 *For all $\varphi \in C(G, q_\theta(i), i = 1, \ldots, k, \theta \in \Theta_i)$, relation (8.44) holds for all $0 \le t_1 < t_2$.*

The proof is the same as the proof of Theorem 3, and it is not presented here.

8.1.4 Diffusion Processes on Graphs

We consider a homogeneous Markov process with values in G, which we denote by $z(t) = (\zeta(t), \theta(t))$, for $\zeta(t) \in R$ and $\theta(t) \in \Theta$. Suppose that the following conditions are satisfied:

(1) $z(t)$ is continuous in G; this means that $\zeta(t)$ is continuous as a real–valued function and $\theta(t)$ is a constant on any interval (t_1, t_2) if $z(t) \in I_\theta$ for all $t \in (t_1, t_2)$.

(2) Set

$$\tau_\theta = \sup\{t : z(s) \in E_\theta \text{ for } s < t\}$$

and let $z_0 \in E_\theta$. Then $\zeta(t)$ on the interval $[0, \tau_\theta)$ is the diffusion process on the interval I_θ with the generator L^θ for which

$$L^\theta f(x) = a(x, \theta) f'(x) + \frac{1}{2} b(x, \theta) f''(x),$$

for $f \in C^{(2)}(I_\theta)$, $a(x, \theta)$, and $b(x, \theta)$, which are continuous functions on the open interval $I_\theta \setminus \{(u_\theta, \theta)\}$, where $u_\theta = \inf I_\theta$ and $b(x, \theta) > 0$.

(3) Denote by P_z the probability under the condition $z(0) = z$. Then for $i = 1, \ldots, k$, the limit

$$\lim_{c \to 0} P_{z_i}\{\theta_\tau c = \theta\} = q_\theta(i), \quad \theta \in \Theta_i \times \bigcup \{\theta_i\},$$

exists, where

$$\tau^c = \inf\{t : |\zeta(t) - u_i| = c\},$$

$$q_{\theta_i}(i) + \sum_{\theta \in \Theta_i} q_\theta(i) = 1.$$

(4) For any $z \in G$,

$$P_z\left\{ \int_0^\infty 1_{\{z(s)\in V\}}\, ds = 0 \right\} = 1;$$

here $V \subset G$ is the subset of all vertices.

The next condition is related to the behavior of the process $\zeta(t)$ on the boundaries of the intervals I_θ. Let $I_\theta = (u_\theta, v_\theta)$.
For fixed $s_0 \in I_\theta$, set

$$V_\theta(x) = \exp\left\{ - \int_{s_0}^x 2a(u,\theta)b^{-1}(u,\theta)\, du, \right\},$$

$$A_\theta = \int_{u_\theta}^{s_0} V_\theta(x)\, dx, \quad A^*_\theta = \int_{s_0}^{v_\theta} V_\theta(x)\, dx,$$

and

$$B_\theta = \iint_{u_\theta < s < u < s_0} \frac{V_\theta(s)}{b_\theta(u,\theta)V_\theta(u)}\, ds\, du,$$

$$B^*_\theta = \iint_{s_0 < u < s < v_\theta} \frac{V_\theta(s)}{b(u,\theta)V_\theta(u)}\, du\, ds.$$

(5) If (u_θ, θ) is an end, then $A_\theta = +\infty$; if (u_θ, θ) is a knot, then

$$A_\theta < \infty, B_\theta < \infty; \quad \text{if} \quad v_\theta < \infty, \quad \text{then} \quad A^*_\theta < \infty, B^*_\theta < \infty.$$

This condition means that any end is an accessible boundary for the process $\zeta(t)$, and any knot is a regular boundary for the diffusion processes on the edges for which this knot is a boundary.
Denote by $\mathcal{D}$ the set of functions $f(z)$ from $C(G)$ for which the derivatives $f'(z), f''(z)$ exist if $z \in G \setminus V$, they are bounded and continuous, and the limits $\lim_{z\to\bar z} f''(z)$, $\bar z \in V^{(k)}$, $f'(u_i - \theta)$, $\theta \in \Theta_i$, $f'(u_i+, \theta_i)$ exist, for $i = 1, 2, \ldots, K$. Moreover, $V^{(k)} = \{(u_i, \theta_i), i = 1, \ldots, K\}$ is the set of all the knots, and relation (8.55) holds with $f(z)$ instead of $\varphi(z)$.

Lemma 11 *Let $f \in \mathcal{D}$. Then for $0 \le t_1 < t_2$,*

$$E\left(\int_{t_1}^{t_2} L^{\theta(s)} f(\zeta(s), \theta(s))\, ds \Big/ \mathcal{F}_0 \right) = E(f(z(t)) - f(z(0))/\mathcal{F}_{t_1}), \quad (8.56)$$

where $\mathcal{F}_{t_1}$ is the σ-algebra generated by $\{z(s), s \le t\}$.

Proof It follows from condition (2) that relation (8.56) holds for a function $f(z) \in \mathcal{D}$ for which $f(z) = 0$ in a neighborhood of any knot z_i, $i = 1, 2, \ldots, K$. Condition (3) implies that for a function $\hat f(z)$ for which

$$\hat f(x, \theta) = l_\theta(x - u_i), \ \theta \in \Theta \cup \{\theta_i\}, \ |x - u_i| \le \delta, \quad i = 1, \ldots, K,$$

where $\delta > 0$ is small enough, we have that

$$E\left(\hat{f}(z(\tau^c)) - f(z_0) - \int_0^{\tau_c} L^{\theta_s}\hat{f}(z(s)) \Big/ \mathcal{F}_0\right) = o(c)$$

because $E\tau^c = \theta(c)$ if $c < \delta$. Introduce the sequence of stopping times $\rho_0 = 0$,

$$\rho_1 = \inf\{s : z(s) \in V^{(k)}\}, \qquad \rho_2 = \inf\{s > \rho_1 : |\zeta(s) - \zeta(\rho_1)| = c\},$$
$$\rho_3 = \inf\{s > \rho_2 : z(s) \in V^{(k)}\}, \quad \rho_4 = \inf\{s > \rho_3 : |\zeta(s) - \zeta(\rho_3)| = c\},$$

and so on. We will have

$$E\left(f(z(\rho_{2k+1})) - f(z(\rho_{2k})) - \int_{\rho_{2k}}^{\rho_{2k+1}} L^{\theta(s)} f(z(s))\, ds \Big/ \mathcal{F}_{\rho_{2k}}\right) = 0,$$
$$E\left(f(z(\rho_{2k})) - f(z(\rho_{2k+1})) - \int_{\rho_{2k-1}}^{\rho_{2k}} L^{\theta(s)} f(z(s))\, ds \Big/ \mathcal{F}_{\rho_{2k}}\right) = o(c).$$

These relations are valid for the stopping time $\rho_{2k+1} \wedge t$. This implies the proof of the lemma.

□

Remark 10 Assume that $z(t) = (\zeta(t), \theta(t))$ is a right continuous stochastic process for which relation (8.56) is fulfilled for all $f \in \mathcal{D}$ and the coefficients $a(x, \theta)$, $b(x, \theta)$ satisfy condition (5). Then $z(t)$ is a continuous strong Markov process satisfying conditions (2), (3), (4). First, we note that the stochastic process

$$f(z(t)) - \int_0^t L^{\theta(s)} f(z(s))\, ds$$

is a martingale, so relation (8.56) is true if t_1, t_2 are stopping times. This implies condition (2) because of the martingale characterization of one-dimensional Markov processes (see Section 2.2). So $z(t)$ is continuous if $z(t) \in G \setminus V^{(k)}$. Let $\hat{f}(x, \theta)$ be the function constructed in the proof of Lemma 11. We can prove that $\hat{f}(z(t))$ is a continuous function. This implies that $z(t)$ is continuous if $z(t) \in V^{(k)}$.

Let z_i be a knot. Denote by ρ_0 the stopping time

$$\rho_0 = \inf\{s : z(s) = z_i\}$$

and let

$$\rho_1 = \inf\{s > \rho_0 : |\zeta(s) - \zeta(u_i)| = c\}.$$

Then

$$E\left(\hat{f}(\rho_1) - \hat{f}(\rho_0) - \int_{\rho_0}^{\rho_1} L^{\theta}\hat{f}(z(s))\, ds \Big/ \mathcal{F}_{\rho_0}\right) = 0,$$

and as a result,

$$\sum_{\theta\in\Theta_i\cup\{\theta_i\}} P\{\theta(\rho_1)=\theta|\mathcal{F}_{\rho_0}\}l_\theta = O(E(\rho_1-\rho_0/\mathcal{F}_{\rho_0})).$$

Since $E(\rho_1-\rho_0/\mathcal{F}_{\rho_0})\to 0$ as $c\to 0$ we see that condition (3) is fulfilled. To prove condition (4) we consider the function $f_i^\delta(z)\in\mathcal{D}$ that satisfies the relation

$$L^\theta f_i^\delta(x,\theta)=h_i^\delta(x,\theta),\quad \delta>0, \theta\in\Theta_i\cup\{\theta_i\},$$

$h_i^\delta(x,\theta)=0$ if $\dot{x}\in I_\theta$, $|x-u_i|\geq\delta$, $h_i^\delta(x,\theta)=1$ if $x\in I_\theta$, $|x-u_i|\leq\frac{1}{2}\delta$, $h_i^\delta(x,\theta)\geq 0$, and $\frac{d}{dx}h_i^\delta(x,\theta)$ is continuous in I_θ. It follows from condition (5) that

$$|f_i^\delta(x,\theta)|=c_\delta|x-u_i|,\quad \text{where}\quad c_\delta\to 0\quad \text{as}\quad \delta\to 0.$$

Relation (8.56) implies the inequality

$$E\int_0^t 1_{\{|\zeta(s)-u_i|\leq\frac{1}{2}\delta\}}\,ds = O(c_\delta E|\zeta(s)-u_i|).$$

This implies condition (4).
Next, we prove that the initial value z_0, the coefficients $a(x,\theta)$ and $b(x,\theta)$, and the constants $q_\theta(i)$ determine the distribution of the process $z(t)$. We consider the graph G with one knot; denote it by $(u_0,0)$. First, we calculate the function

$$R_\lambda f(z_0)=E\int_0^\infty e^{-\lambda t}f(z(t))\,dt,\quad z_0=z(0),\quad \lambda>0,\quad f\in\mathcal{D}.$$

We will use the same stopping times as in Lemma 11. Since ρ_k depends on $c>0$, we use the notation $\{\rho_k^c, k=0,1,\ldots\}$. Then

$$R_\lambda f(z_0)=E\sum_{k=0}^\infty\int_{\rho_k^c}^{\rho_{k+1}^c} e^{-\lambda t}f(z(t))\,dt.$$

It follows from condition (4) that

$$\lim_{c\to 0} E\sum_{k=0}^\infty\left|\int_{\rho_{2k+1}^c}^{\rho_{2k+2}^c} e^{-\lambda t}f(z(t))\,dt\right|=0,$$

so

$$R_\lambda f(z_0)=\lim_{c\to 0} E\sum_{k=0}^\infty\int_{\rho_{2k}^c}^{\rho_{2k+1}^c} e^{-\lambda t}f(z(t))\,dt. \tag{8.57}$$

Denote by $x^\theta(t)$ the diffusion process on the interval I_θ with absorption on the boundary. Let

$$\tau_0^\theta=\inf\{s : x^\theta(s)=u_0\}$$

and

$$R_\lambda^\theta \varphi(x) = E\Big(\int_0^{\tau_0^\theta} e^{-\lambda s} \varphi(x^\theta(s))\, ds \Big/ x^\theta(0) = x \Big), \quad x \in I_\theta,$$

$\varphi \in C^{(2)}(I_\theta)$. Then

$$E \int_0^{\rho_1^c} e^{-\lambda t} f(z(t))\, dt = R_\lambda^{\theta_0} \varphi(x_0, \theta_0),$$

where $(x_0, \theta_0) = z_0$, and

$$E\Big(\int_{\rho_{2k}^c}^{\rho_{2k+1}^c} e^{-\lambda t} f(z(t))\, dt \Big/ \mathcal{F}_{\rho_{2k}^c} \Big) = e^{-\lambda \rho_{2k}^c} R_\lambda^{\theta(\rho_{2k}^c)} (f(c(\theta(\rho_{2k}^c)), \theta(\rho_{2k}^c))),$$

where $c(0) = c$ and $c(\theta) = -c$ for $\theta = 1, \ldots, r+1$. This and condition (3) imply the relation

$$\begin{aligned} &E\Big(\int_{\rho_{2k}^c}^{\rho_{2k+1}^c} e^{-\lambda t} f(z(t))\, dt \Big/ \mathcal{F}_{\rho_{2k-1}^c} \Big) \\ &\quad = e^{-\lambda \rho_{2k-1}^c} \Big[q_0 R_\lambda^0(c, 0) + \sum_{\theta=1}^{r+1} q_\theta R_\lambda^\theta(-c, \theta) + o(c) \Big]. \end{aligned} \tag{8.58}$$

Set

$$G_\lambda^\theta(x) = E(e^{-\lambda \tau_0^\theta} / x^\theta(0) = x).$$

Then

$$E\big(e^{-\lambda \rho_{2k-1}^c} / \mathcal{F}_{\rho_{2k-3}^c}\big) = e^{-\lambda \rho_{2k-3}^c} \Big[q_0\, G_\lambda^0(c, 0) + \sum_{\theta=1}^{r+1} q_\theta\, G_\lambda^\theta(-c, \theta) + o(c) \Big] \tag{8.59}$$

if $2k - 3 > 0$.

It follows from formulas (8.57), (8.58), and (8.59) that $R_\lambda f(z_0)$ is determined by the functions $R_\lambda^\theta \varphi$, $G_\lambda^\theta(x)$ and $\{q_\theta, \theta = 0, \ldots, r+1\}$. Since

$$E\Big(\int_t^\infty e^{-\lambda s} f(x(s))\, ds \Big/ \mathcal{F}_t \Big) = R_\lambda f(z(t)),$$

the conditional distribution of $z(s)$ for $s > t$ with respect to the σ-algebra $\mathcal{F}_t$ depends on $z(t)$ only. This means that $z(t)$ is a Markov process.

Remark 11 Let $z_n(t)$, $n = 1, 2, \ldots$, be a sequence of G-valued continuous processes for which the functions $a(z)$ and $b(z)$ exist and satisfy condition (5), and let the constants $\{q_\theta(i), i = 1, \ldots, K, \theta = \Theta_i \cup \{\theta_i\}\}$ be such that

$$\lim_{n \to \infty} E \Big| E\Big(f(z_n(t_2)) - f(z_n(t_1)) - \int_{t_1}^{t_2} L^{\theta(s)} f(z(s))\, ds \Big/ \mathcal{F}_{t_1}^n \Big) \Big| = 0 \tag{8.60}$$

for all $0 \leq t_1 < t_2$, $f \in \mathcal{D}$, where $\mathcal{F}_t^n$ is the σ-algebra generated by $\{z(s), s \leq t\}$. Then the sequence $z_n(t)$ converges weakly to the diffusion Markov process $z(t)$ on G satisfying conditions (1)–(4). If the sequence $\{z_n(t)\}$ is compact in $C(G)$, then $z_n(t)$ converges to $z(t)$ weakly in C. This statement can be proved in the same way as Theorem 2 and Remark 7 of Chapter 2.
Now we can formulate the main result concerning the asymptotic behavior of a perturbed system.

Theorem 6 *Assume that the coefficients $a(x,\theta), b(x,\theta)$ satisfy condition 5). Then the stochastic processes $\{z_\varepsilon(t/\varepsilon), \varepsilon > 0\}$ converge weakly in C to the diffusion process $z(t)$ on the graph G that is determined by conditions 2)–4).*

Proof The weak convergence follows from Theorem 5 and Remark 10. Using Theorem 5 we can prove that

$$\limsup_{\varepsilon\to 0} E\left(1 - \exp\{-(\zeta_\varepsilon(t+h) - \zeta_\varepsilon(t))^2\}\right) = O(h), \tag{8.61}$$

where $z_\varepsilon(t/\varepsilon) = (z_\varepsilon(t), \theta_\varepsilon(t))$. This implies the compactness of the family $\{z_\varepsilon(t/\varepsilon)\}$ as $\varepsilon \to 0$. In this way, the theorem is proved.

□

8.1.5 Simulation of a Two-Well Potential Problem

To illustrate the dynamics of a system like that described in the preceding, we consider the system

$$\ddot{x} = -(2.5\,x^3 - 5\,x + 0.5) + 10\,y\left(\frac{t}{\varepsilon}\right),$$

where y is a jump process that is uniformly distributed on the interval $[-5, 5]$ having stopping times that are exponentially distributed. The solution beginning at the point $(1.5, 0)$ is shown in Figure 8.2.

8.2 Linear Oscillating Conservative Systems

In this section we consider linear problems in greater detail.

8.2.1 Free Linear Oscillating Conservative Systems

Consider a system with phase space $R^m \times R^m$. We denote the states of the system by $(x, \dot{x})$, where $x \in R^m$, $\dot{x} \in R^m$; $x = (x_1, \ldots, x_m)$ is treated as the position of the system and $\dot{x} = (\dot{x}_1, \ldots, \dot{x}_m)$ as its velocity. Assume

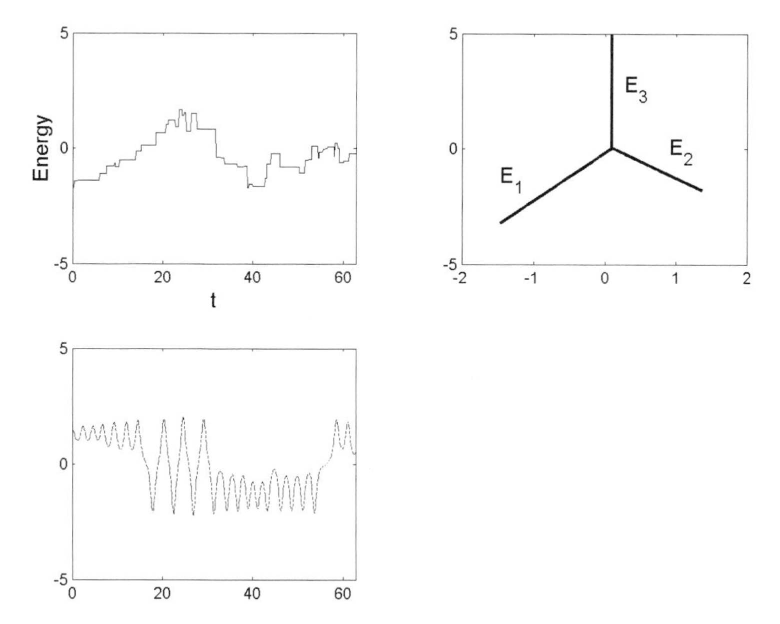

Figure 8.2. Upper left: Energy, $\mathcal{E}(x(t), \dot{x}(t))$, as a function of time. Upper right: Potential energy graph. Lower left: The trajectory of $x(t)$. This shows that x first moves on E_2, then E_3, then E_1, and finally on E_2. Here the simulation is for $0 \leq t \leq 100\pi$.

that the potential energy of the system is a quadratic form in x,

$$U(x) = \frac{1}{2}(\Lambda x, x),$$

and its kinetic energy is of the form

$$T(\dot{x}) = \frac{1}{2}(\dot{x}, \dot{x}).$$

The linear operator $\Lambda \in L(R^m)$ is assumed to be symmetric and positive. Without loss of generality, we assume that Λ has canonical form in the natural basis in R^m, so

$$(\Lambda x, x) = \sum_{k=1}^{m} \lambda_k x_k^2,$$

where $\lambda_k > 0$, $k = 1, \ldots, m$. We assume additionally that $0 < \lambda_1 \leq \lambda_2 \leq \cdots \leq \lambda_m$. Let $(x(t), \dot{x}(t))$ be the state of the system at instant $t \in R_+$. The motion of the system is determined by the equation

$$\frac{d\dot{x}}{dt}(t) = -\Lambda x(t) \tag{8.62}$$

and initial values $x(0)$, $\dot{x}(0)$. (We must also take into account the relation $\dot{x}(t) = \frac{dx}{dt}(t)$.) Under the assumptions on Λ, the solution of equation (8.62) can be represented in the form

$$\begin{aligned} x_k(t) &= a_k \cos(\lambda_k t + \theta_k), \\ \dot{x}_k(t) &= -a_k \lambda_k \sin(\lambda_k t + \theta_k), \quad k = 1, \ldots, m, \end{aligned} \tag{8.63}$$

where the amplitude $a_k \geq 0$ and the phase deviation $\theta_k \in (-\pi, \pi]$ are defined by the initial conditions

$$x_k(0) = a_k \cos\theta_k, \quad \dot{x}_k(0) = -a_k \lambda_k \sin\theta_k, \quad k = 1, \ldots, m.$$

Note that the functions

$$\mathcal{E}_k(x, \dot{x}) = \frac{1}{2}\lambda_k^2 x_k^2 + \frac{1}{2}\dot{x}_k^2, \quad k = 1, \ldots, m,$$

are first integrals for equation (8.62).
It follows from formula (8.63) that the asymptotic behavior of average values of continuous functions along trajectories of the system is described as follows:

Statement *Let $\Phi(x, \dot{x})$ be a continuous function from $R^m \times R^m \to R$. Then there exists a limit*

$$\lim_{T\to\infty} \frac{1}{T} \int_0^T \Phi(x(t), \dot{x}(t))\, dt = A_m[\Phi, \lambda_1, \ldots, \lambda_m, a_1, \ldots, a_m, \theta_1, \ldots, \theta_m], \tag{8.64}$$

where $A_m[\Phi, \ldots]$ is a continuous nonnegative linear function in Φ satisfying the inequality

$$\begin{aligned} &A_m[\Phi, \lambda_1, \ldots, \lambda_m, a_1, \ldots, a_m, \theta_1, \ldots, \theta_m] \\ &\quad \leq \sup\{|\Phi(x, \dot{x})| : |x|^2 \leq |a|^2,\ |\dot{x}|^2 \leq |\Lambda a|^2\}, \end{aligned}$$

where $a = (a_1, \ldots, a_m) \in R^m$.

The function $A_m[\Phi, \lambda_1, \ldots, \lambda_m, a_1, \ldots, a_m, \theta_1, \ldots, \theta_m]$ can be calculated in the following way. There exist $r \in \mathbf{Z}_+$, $r \leq m$, and positive numbers $\delta_1, \ldots, \delta_r$ such that for all $k \leq m$ there exist $n_{k1}, \ldots, n_{kr} \in \mathbf{Z}$ for which

$$\lambda_k = \sum_{j=1}^{r} n_{kj}\delta_j,$$

and for all $n_1, \ldots, n_r$ we have the inequality $|\sum_{j=1}^r n_j \delta_j| > 0$ whenever $\sum |n_j| > 0$. The function $A_m[\Phi, \ldots]$ is expressed in terms of $\delta_1, \ldots, \delta_r$ by the formula

$$\begin{aligned} A_m[\Phi, \lambda_1, \ldots, \lambda_m, a_1, \ldots, a_m, \theta_1, \ldots, \theta_m] &= \frac{\delta_1 \cdots \delta_r}{(2\pi)^r} \\ \times \int_0^{\frac{2\pi}{\delta_1}} \cdots \int_0^{\frac{2\pi}{\delta_r}} \Phi\big(X(\delta_1, s_1, \ldots, \delta_r, s_r), \dot{X}(\delta_1, s_1, \ldots, \delta_r, s_r)\big)\, ds_1 \cdots ds_r& \end{aligned} \tag{8.65}$$

where

$$X_k(\delta_1, s_1, \ldots, \delta_r, s_r) = a_k \cos\Big(\sum_{j=1}^{r} n_{kj}\delta_j s_j + \theta_k\Big),$$

$$\dot{X}_k(\delta_1, s_1, \ldots, \delta_r, s_r) = -\lambda_k a_k \sin\Big(\sum_{j=1}^{r} n_{kj}\delta_j s_j + \theta_k\Big).$$

Remark 12 Using the representation

$$X_k(\delta_1, s_1, \ldots, \delta_r, s_r) = a_k \cos\Big(\sum_{j=1}^{r} n_{kj}\delta_j \big(s_j + \frac{\theta_k}{\lambda_k}\big)\Big)$$

and the analogous representation of $\dot{X}_k$ using formula (8.65) we can prove that $A_m[\Phi, \lambda_1, \ldots, \lambda_m, a_1, \ldots, a_m, \theta_1, \ldots, \theta_m]$ as a function of $\theta_1, \ldots, \theta_m$ depends only on

$$\frac{\theta_2}{\lambda_2} - \frac{\theta_1}{\lambda_1}, \ldots, \frac{\theta_m}{\lambda_m} - \frac{\theta_1}{\lambda_1}.$$

8.2.2 Randomly Perturbed Linear Oscillating Systems

Random perturbations of linear systems take the form

$$\begin{aligned} \frac{d\dot{x}_\varepsilon}{dt}(t) &= -\Lambda x_\varepsilon(t) + b\left(x_\varepsilon(t), \dot{x}_\varepsilon(t), y\left(\frac{t}{\varepsilon}\right)\right) \\ \frac{dx_\varepsilon}{dt}(t) &= \dot{x}_\varepsilon(t), \end{aligned} \tag{8.66}$$

where the function $b : R^m \times R^m \times Y \to R^m$ is bounded, measurable, and continuous in x and $\dot{x}$, and $y(t)$ is a Markov process in Y satisfying condition SMC II of Section 2.3.

Additionally, we assume that the function $b(x, \dot{x}, y)$ satisfies the Lipschitz condition

$$|b(x, \dot{x}, y) - b(x_1, \dot{x}_1, y)| \le L(|x - x_1| + |\dot{x} - \dot{x}_1|),$$

and

$$\int |b(x, \dot{x}, y)|\rho(dy) < \infty \quad \text{and} \quad \int b(x, \dot{x}, y)\rho(dy) = 0.$$

Then the averaged equation for equation (8.66) is equation (8.62). It follows from Theorem 5 of Chapter 3 and the statement in Section 8.1.1 that there exists T_ε such that $T_\varepsilon \to \infty$, $\varepsilon T_\varepsilon \to 0$ for which any continuous function

$\Phi(x, \dot{x})$ has

$$\begin{aligned}&\frac{1}{T_\varepsilon}\int_t^{t+T_\varepsilon} \Phi(x_\varepsilon(s), \dot{x}_\varepsilon(s))\, ds \\ &\approx A\left[\Phi, \lambda_1, \dots, \lambda_m, z_1^\varepsilon(t), \dots, z_m^\varepsilon(t), \frac{1}{\lambda_1}\theta_1^\varepsilon(t), \dots, \frac{1}{\lambda_m}\theta_m^\varepsilon(t)\right],\end{aligned} \tag{8.67}$$

where the stochastic processes $z_1^\varepsilon(t), \dots, z_m^\varepsilon(t)$, $\theta_1^\varepsilon(t), \dots, \theta_m^\varepsilon(t)$, are defined by the relations

$$\begin{aligned}x_{\varepsilon k}(t) &= z_k^\varepsilon(t)\cos\lambda_k\theta_k^\varepsilon(t),\\ \dot{x}_{\varepsilon k}(t) &= -\lambda_k z_k^\varepsilon(t)\sin\lambda_k\theta_k^\varepsilon(t), \quad k = 1, \dots m,\end{aligned}$$

where $x_{\varepsilon k}$ and $\dot{x}_{\varepsilon k}$ are the components of the vectors x_ε and $\dot{x}_\varepsilon$. These relations together with equation (8.66) imply the validity of the following system of differential equations for z_k^ε and $\theta_k^\varepsilon(t)$, $k = 1, \dots, m$:

$$\begin{aligned}\frac{dz_k^\varepsilon}{dt}(t) &= -b_k\left(x_\varepsilon(t), \dot{x}_\varepsilon(t), y\left(\frac{t}{\varepsilon}\right)\right)\frac{\sin\lambda_k\theta_k^\varepsilon(t)}{\lambda_k^2 z_k^\varepsilon(t)},\\ \frac{d\theta_k^\varepsilon}{dt}(t) &= 1 - b_k\left(x_\varepsilon(t), \dot{x}_\varepsilon(t), y\left(\frac{t}{\varepsilon}\right)\right)\frac{\cos\lambda_k\theta_k^\varepsilon(t)}{\lambda_k^2 z_k^\varepsilon(t)}, \quad k = 1, \dots, m,\end{aligned}$$

where $b_k(x, \dot{x}, y)$ are the coordinates of the vector $b(x, \dot{x}, y)$.
Note that

$$z_k^\varepsilon(t) = \sqrt{x_{\varepsilon k}^2(t) + \frac{1}{\lambda_k^2}\dot{x}_{\varepsilon k}^2(t)} = \sqrt{\frac{2\mathcal{E}_k(x_\varepsilon(t), \dot{x}_\varepsilon(t))}{\lambda_k^2}}.$$

Set

$$\begin{aligned}\hat{z}_k^\varepsilon(t) &= \mathcal{E}_k\big(x_\varepsilon(t), \dot{x}_\varepsilon(t)\big), \quad k = 1, 2, \dots, m,\\ \hat{\theta}_k^\varepsilon(t) &= \theta_k^\varepsilon(t) - \theta_1^\varepsilon(t), \quad k = 1, 2, \dots, m.\end{aligned}$$

These functions satisfy the system of differential equations

$$\frac{d\hat{z}_k^\varepsilon}{dt}(t) = b_k\big(x_\varepsilon(t), \dot{x}_\varepsilon(t), y\big(\frac{t}{\varepsilon}\big)\big)\dot{x}_{\varepsilon k}(t), \quad k = 1, \dots, m, \tag{8.68}$$

and

$$\begin{aligned}\frac{d\hat{\theta}_k^\varepsilon}{dt}(t) = \frac{1}{2}\Bigg(&\frac{x_{\varepsilon 1}(t)b_1\big(x_\varepsilon(t), \dot{x}_\varepsilon(t), y\big(\frac{t}{\varepsilon}\big)\big)}{\hat{z}_1^\varepsilon(t)}\\ &- \frac{x_{\varepsilon k}(t)b_k\big(x_\varepsilon(t), \dot{x}_\varepsilon(t), y\big(\frac{t}{\varepsilon}\big)\big)}{\hat{z}_k^\varepsilon(t)}\Bigg), \quad k = 2, \dots, m-1.\end{aligned} \tag{8.69}$$

Note that Remark 12 implies that the expression on the right-hand side of relation (8.67) can be represented as

$$\hat{A}_m\left[\Phi, \lambda_1, \dots, \lambda_m, \left(\frac{2\hat{z}_1^\varepsilon(t)}{\lambda_1^2}\right)^{1/2}, \dots, \left(\frac{2\hat{z}_m^\varepsilon(t)}{\lambda_m^2}\right)^{1/2}, \frac{1}{\lambda_1}\theta_1^\varepsilon(t), \dots, \frac{1}{\lambda_m}\theta_m^\varepsilon(t)\right]$$

$$= B_m\big[\Phi, \hat{z}_1^\varepsilon(t), \dots, \hat{z}_m^\varepsilon(t), \hat{\theta}_2^\varepsilon(t), \dots, \hat{\theta}_m^\varepsilon(t)\big]. \tag{8.70}$$

(Here we consider $\lambda_1, \dots, \lambda_m$ as given and unchanging, so we omit them as the arguments of the function B_m.)
We consider the diffusion approximation for the stochastic process

$$\left(\hat{z}_1^\varepsilon\left(\frac{t}{\varepsilon}\right), \dots, \hat{z}_k^\varepsilon\left(\frac{t}{\varepsilon}\right), \hat{\theta}_2^\varepsilon\left(\frac{t}{\varepsilon}\right), \dots, \hat{\theta}_m^\varepsilon\left(\frac{t}{\varepsilon}\right)\right).$$

Introduce the functions

$$\begin{aligned}
\beta_k(x, \dot{x}) &= \sum_{j=1}^m \iint \frac{\partial b_k}{\partial x_j}(x, \dot{x}, y') b_j(x, \dot{x}, y') \rho(dy) R(y, dy'), \\
\alpha_{kl}(x, \dot{x}) &= \iint b_k(x, \dot{x}, y') b_l(x, \dot{x}, y') \rho(dy) R(y, dy'), \\
G_k(x, \dot{x}) &= \dot{x}_k \beta_k(x, \dot{x}) + \alpha_k(x, \dot{x}), \\
G_{kl}(x, \dot{x}) &= \alpha_{kl}(x, \dot{x}) + \alpha_{lk}(x, \dot{x}), \\
H_k(x, \dot{x}) &= x_k \beta_k(x, \dot{x}), \\
H_{kl}(x, \dot{x}) &= x_k x_l \big(\alpha_{kl}(x, \dot{x}) + \alpha_{lk}(x, \dot{x})\big), \\
\hat{H}_{kl}(x, \dot{x}) &= x_k \big(\alpha_{kl}(x, \dot{x}) + \alpha_{lk}(x, \dot{x})\big).
\end{aligned}$$

Denote by $\tilde{z}^\varepsilon(t)$ the stochastic process in R^m with coordinates $\tilde{z}_k^\varepsilon(t) = \hat{z}_k^\varepsilon(t/\varepsilon)$ for $k = 1, \dots, m$, and by $\tilde{\theta}^\varepsilon(t)$ the stochastic process in R^{m-1} with coordinates

$$\tilde{\theta}_l^\varepsilon(t) = \hat{\theta}_{l+1}^\varepsilon(t/\varepsilon), \quad l = 1, \dots, m-1.$$

With these preliminaries, we have the following result.

Theorem 7 *The compound stochastic process $(\tilde{z}^\varepsilon(t), \tilde{\theta}^\varepsilon(t))$ converges weakly to the diffusion process $(\tilde{z}(t), \tilde{\theta}(t))$ in $R^m \times R^{m-1}$ having generator L that is defined on functions $F(z, \theta)$ from $C^{(2)}(R^{2m-1})$ by the relation*

$$\begin{aligned}
LF(z, \theta) &= \sum_{k=1}^m a_k(z, \theta) \frac{\partial F}{\partial z_k}(z, \theta) + \sum_{l=1}^{m-1} \tilde{a}_l(z, \theta) \frac{\partial F}{\partial \theta_l}(z, \theta) \\
&+ \frac{1}{2} \sum_{k,j=1}^m b_{kj}(z, \theta) \frac{\partial^2 F}{\partial z_k \partial z_j}(z, \theta) + \frac{1}{2} \sum_{l,i=1}^{m-1} \tilde{b}_{li}(z, \theta) \frac{\partial^2 F}{\partial \theta_l \partial \theta_i}(z, \theta) \qquad (8.71) \\
&+ \frac{1}{2} \sum_{k=1}^m \sum_{l=1}^{m-1} c_{kl}(z, \theta) \frac{\partial^2 F}{\partial z_k \partial \theta_l}(z, \theta),
\end{aligned}$$

where

$$\begin{aligned}
a_k(z,\theta) &= \tilde{B}_m[G_k, z, \theta], \quad k \in \overline{1,m}, \\
\tilde{a}_l(z,\theta) &= \tilde{B}_m[H_1, z, \theta](2z_1)^{-1} - \tilde{B}_m[H_{l+1}, z, \theta](2z_{l+1})^{-1}, \quad l \in \overline{1,m-1}, \\
b_{kj}(z,\theta) &= \tilde{B}_m[G_{kj}, z, \theta], \quad k,j \in \overline{1,m}, \\
\tilde{b}_{li}(z,\theta) &= \frac{1}{4}\big(\tilde{B}_m[H_{11}, z, \theta]z_1^{-2} + \tilde{B}_m[H_{l+1,i+1}, z, \theta](z_{l+1}z_{i+1})^{-1} \\
&\quad - \tilde{B}_m[H_{1,l+1}, z, \theta](z_1 z_{l+1})^{-1} - \tilde{B}_m[H_{1,i+1}, z, \theta](z_1 z_{i+1})^{-1}\big), \\
&\qquad l, i \in \overline{1, m-1},
\end{aligned}$$

$$c_{kl}(z,\theta) = \frac{1}{2}\big(\tilde{B}_m[\hat{H}_{1k}, z, \theta]z_1^{-1} - \tilde{B}_m[H_{l+1,k}, z, \theta]z_{l+1}^{-1}\big),$$
$$\text{for } k \in \overline{1,m},\ l \in \overline{1,m-1},$$

and

$$\tilde{B}_m[\Phi, z, \theta] = B_m[\Phi, z_1, \ldots, z_m, \theta_1, \ldots, \theta_m],$$

where B_m is defined by relation (8.70).

Proof Let $F(z,\theta)$ satisfy the conditions of the theorem. Then using equations (8.68) and (8.69) we can write, for $t_1 < t_2$,

$$\begin{aligned}
&E\left(F(\tilde{z}^\varepsilon(t_2), \tilde{\theta}^\varepsilon(t_2)) - F(\tilde{z}^\varepsilon(t_1), \tilde{\theta}^\varepsilon(t_1))/\mathcal{F}^\varepsilon_{t_1/\varepsilon}\right) \\
&= E\Bigg(\int_{t_1/\varepsilon}^{t_2/\varepsilon}\Bigg[\sum_{k=1}^m \frac{\partial F}{\partial z_k}(\tilde{z}^\varepsilon(\varepsilon s), \tilde{\theta}^\varepsilon(\varepsilon s)) b_k\left(x_\varepsilon(s), \dot{x}_\varepsilon(s), y\left(\frac{s}{\varepsilon}\right)\right)\dot{x}_{\varepsilon k}(s) \\
&+ \sum_{k=1}^{m-1}\frac{\partial F}{\partial \theta_l}(\tilde{z}^\varepsilon(\varepsilon s), \tilde{\theta}^\varepsilon(\varepsilon s))\Bigg(x_{\varepsilon 1}(s) b_1\left(x_\varepsilon(s), \dot{x}_\varepsilon(s), y\left(\frac{s}{\varepsilon}\right)\right)\left(2z_1^\varepsilon\left(\frac{s}{\varepsilon}\right)\right)^{-1} \\
&- x_{\varepsilon k}(s) b_{l+1}\left(x_\varepsilon(s), \dot{x}_\varepsilon(s), y\left(\frac{s}{\varepsilon}\right)\right)\left(2z_{l+1}^\varepsilon\left(\frac{s}{\varepsilon}\right)\right)^{-1}\Bigg)\Bigg] ds \Big/ \mathcal{F}^\varepsilon_{t_1/\varepsilon}\Bigg).
\end{aligned}$$

Applying formula (5.7) to the last integral, we can obtain the formula

$$\begin{aligned}
&E\big(F(\tilde{z}^\varepsilon(t_2), \tilde{\theta}^\varepsilon(t_2)) - F(\tilde{z}^\varepsilon(t_1), \tilde{\theta}^\varepsilon(t_1))/\mathcal{F}^\varepsilon_{t_1/\varepsilon}\big) \\
&= \varepsilon E\int_{t_1/\varepsilon}^{t_2/\varepsilon} \hat{F}\big(\tilde{z}^\varepsilon(\varepsilon s), \tilde{\theta}^\varepsilon(\varepsilon s), x_\varepsilon(s), \dot{x}_\varepsilon(s)\big)\, ds + O(\varepsilon),
\end{aligned} \tag{8.72}$$

where

$$\hat{F}(z,\theta,x,\dot{x}) = \sum_{k=1}^{m} G_k(x,\dot{x})\frac{\partial F}{\partial z_k}(z,\theta)$$

$$+\sum_{l=1}^{m-1}\left[H_1(x,\dot{x})(2z_1)^{-1} - H_{l+1}(x,\dot{x})(2z_{l+1})^{-1}\right]\frac{\partial F}{\partial \theta_l}(z,\theta) + \frac{1}{2}\sum_{k,j=1}^{m} G_{kj}(x,\dot{x})$$

$$\times \frac{\partial^2 F}{\partial z_k \partial z_j}(z,\theta) + \frac{1}{8}\sum_{l,i=1}^{m-1}\left(H_{11}(x,\dot{x})z_1^{-2} + H_{l+1,i+1}(x,\dot{x})(z_{l+1}z_{i+1})^{-1}\right.$$

$$\left. - H_{1,l+1}(x,\dot{x})(z_1 z_{l+1})^{-1} - H_{1,i+1}(x,\dot{x})(z_1 z_{i+1})^{-1}\right)\frac{\partial^2 F}{\partial \theta_1 \partial \theta_i}(z,\theta)$$

$$+\frac{1}{4}\sum_{k=1}^{m}\sum_{l=1}^{m}\left(H_{1k}(x,\dot{x})z_1^{-1} - H_{l+1,k}(x,\dot{x})z_{l+1}^{-1}\right)\frac{\partial^2 F}{\partial z_k \partial \theta_l}(z,\theta).$$

The proof can be completed by applying relation (8.67) to equation (8.72). (See proof of Theorem 6 in Chapter 5.)

□

8.3 A Rigid Body with a Fixed Point

A rigid body is a system of point masses in R^3 for which all distances between points are constant. Let $x_1, \dots, x_r$ be the positions of the points of the rigid body and let $m_1, \dots, m_r$ be their masses. We consider motion of rigid bodies for which there exists a fixed point. We assume that the fixed point is O, the origin of R^3. If $x_1(t), \dots, x_r(t)$ are the positions of the points of the rigid body at time t, then $|x_i(t)|$, $i = 1, 2, \dots, r$, and $|x_i(t) - x_j(t)|$, $i, j \in \overline{1, r}$, are constant in time.
The *inertia operator* I of the rigid body is a positive symmetric operator defined by

$$(Iz, z) = \sum_{i=1}^{r} m_i (x_i, z)^2, \quad z \in R^3.$$

Let $x_i(0) = x_i$. The position of the moving rigid body in R^3 at time t can be determined by the orthogonal operator B_t in R^3 satisfying the relation

$$x_i(t) = B_t x_i, \quad i = 1, 2, \dots, r.$$

Denote by I_t the inertia operator of the rigid body at time t. Then

$$I_t = B_t I B_t^*.$$

(Here B^* is the adjoint operator to B.)

The *angular velocity* ω_t of the moving rigid body is the vector in R^3 defined by

$$v_i(t) = [\omega_t, x_i(t)], \quad i = 1, \dots, r,$$

where $v_i(t) = \dot{x}_i(t)$ is the velocity of the point x_i and $[a, b]$ is the vector product of vectors a and b from R^3.
The *angular momentum vector* m_t of the rigid body at time t is defined by

$$m_t = \sum_{i=1}^{r} m_i [x_i(t), v_i(t)],$$

so

$$m_t = I_t \omega_t.$$

The kinetic energy of the moving rigid body at time t, say T_t, is defined by the relation

$$T_t = \frac{1}{2}(I_t \omega_t, \omega_t).$$

Introduce the operator $\Omega_t \in L(R^3)$ for which

$$\Omega_t x = [x, \omega_t].$$

Then the operator function B_t satisfies the differential equation

$$\frac{dB_t}{dt} = B_t \Omega_t.$$

Denote by $f_i(t)$ the force acting on the point x_i at time t. Then

$$n(t) = \sum_{i=1}^{r} [f_i(t), x_i(t)]$$

is the moment of forces acting on the rigid body. The motion of the rigid body is determined by the equation

$$\frac{dm_t}{dt} = n_t. \tag{8.73}$$

8.3.1 *Motion of a Rigid Body around a Fixed Point*

In this case $n_t = 0$ for all t, so $m_t = \text{const}$. The kinetic energy of the rigid body is constant, too. To describe the motion we introduce the inertia ellipsoid of the rigid body defined by the equation

$$(Iz, z) = 1.$$

For the moving body the inertia ellipsoid at time t is determined by the equation

$$(I_t z, z) = 1.$$

Denote by Π the plane

$$\Pi = \left\{ x \in R^3 : (x, m) = \frac{(m, \omega)}{\sqrt{2T}} \right\},$$

where m is the initial angular momentum vector, ω is the angular velocity, and T is the kinetic energy of the rigid body at the initial time.
Poinsot's theorem states that the ellipsoid of inertia rolls without slipping along the plane Π. (See [192, p.375].)
There exists a closed curve on the inertia ellipsoid along which it rolls on the plane Π. Denote by α the angle on which the rigid body is turned around the vector m during one rotation of the point of tangency between the inertia ellipsoid and Π. If $\alpha/2\pi$ is a rational number, then the motion of the rigid body is periodic. If $\alpha/2\pi$ is an irrational number, then $B_{t_1} \neq B_{t_2}$ for $t_1 \neq t_2$.
The ergodic properties of the motion are described by the following theorem.

Theorem 8 *Let $g(m, B)$ be a continuous function from $R^3 \times L(R^3)$ into R. There exists the limit*

$$\lim_{u \to \infty} \frac{1}{u} \int_0^u g(m_s, B_s)\, ds = A(g, m, T, \alpha),$$

where m and T are the angular momentum vector and the kinetic energy of the rigid body, respectively, and $\alpha \in [0, 2\pi]$ is defined above. The limit $A(g, m, T, \alpha)$ does not depend on α if α is an irrational number.

The proof of this theorem is a consequence of Poinsot's theorem mentioned earlier and ergodic theory.

Remark 13 Set

$$\bar{A}(g, m, T) = \frac{1}{2\pi} \int_0^{2\pi} A(g, m, T, \alpha)\, d\alpha.$$

Then

$$\lim_{u \to \infty} \frac{1}{u} \int_0^u g(m_s, B_s)\, ds = \bar{A}(g, m, T)$$

for almost all initial values (m_0, B_0) with respect to the product of the Lebesgue measure in R^3 and the Haar measure on the group $O(R^3)$ of the orthogonal operators in R^3.

8.3.2 Analysis of Randomly Perturbed Motions

Consider now the motion of a rigid body that is determined by the equation

$$\frac{dm_\varepsilon}{dt}(t) = N\left(m_\varepsilon(t), T_\varepsilon(t), B_\varepsilon(t), y\left(\frac{t}{\varepsilon}\right)\right), \tag{8.74}$$

where $m_\varepsilon(t)$ is the angular momentum vector of the rigid body, $T_\varepsilon(t)$ is its kinetic energy, and $B_\varepsilon(t)$ is the orthogonal operator of its position at time t. It is assumed that the Markov process $y(t)$ in a measurable space Y satisfies condition SMC II of Section 2.3. Here $N(m, T, B, y)$ is a function from $R^3 \times R_+ \times O(R^3) \times Y$ into R^3; which represents the moment of the forces acting on the rigid body. It is assumed that the function $N(m, T, B, y)$ is measurable in y and has bounded continuous derivatives

$$\frac{\partial N}{\partial m}, \quad \frac{\partial N}{\partial T}, \quad \frac{\partial N}{\partial B}.$$

with respect to m, T, B. Finally, we assume that the distribution of $N(m, T, B, y(t))$ is absolutely continuous with respect to the Lebesgue measure in R^3.

Remark 14 The function $T_\varepsilon(t)$ can be expressed in terms of $m_\varepsilon(t)$ and $B_\varepsilon(t)$ by the formula

$$T_\varepsilon(t) = \frac{1}{2}\left(B_\varepsilon(t) I^{-1} B_\varepsilon^*(t) m_\varepsilon(t), m_\varepsilon(t)\right),$$

so $T_\varepsilon(t)$ satisfies the differential equation

$$\frac{dT_\varepsilon}{dt}(t) = \left(B_\varepsilon(t) I^{-1} B_\varepsilon^*(t) m_\varepsilon(t), N\left(m_\varepsilon(t), T_\varepsilon(t), B_\varepsilon(t), y\left(\frac{t}{\varepsilon}\right)\right)\right). \tag{8.75}$$

The last equation is a consequence of relation

$$\left(\frac{d}{dt}(B_t I^{-1} B_t^*) m_t, m_t\right) = 0,$$

which is true for any motion.

Let

$$\int N(m, T, B, y)\, \rho(dy) = 0. \tag{8.76}$$

Introduce the functions

$$a_1(m,T,B) = \iint \left(\left[\frac{\partial N}{\partial m}(m,T,B,y') \right]^* N(m,T,B,y) \right.$$
$$\left. + \left(BI^{-1}B^*m, N(m,T,B,y)\right) \frac{\partial N}{\partial T} N(m,T,B,y') \right) R(y,dy')\rho(dy),$$
$$a_0(m,T,B) = \iint \left[\left(\frac{\partial N}{\partial m}(m,T,B,y') BI^{-1}B^*m, N(m,T,B,y) \right) \right.$$
$$\left. + \left(BI^{-1}B^*m, \frac{\partial N}{\partial T}(m,T,B,y') \right) \left(BI^{-1}B^*m, N(m,T,B,y)\right) \right]$$
$$\times R(y,dy')\rho(dy),$$
$$b_2(m,T,B) = 2 \iint N(m,T,B,y') \otimes N(m,T,B,y) R(y,dy')\rho(dy),$$
$$b_1(m,T,B) = 2 \iint \left(B^{**}m, N(m,T,B,y)\right) N(m,T,B,y') R(y,dy')\rho(dy),$$

where $B^{**} = BI^{-1}B^*$, and

$$b_0(m,T,B) = 2 \iint \left(BI^{-1}B^*m, N(m,T,B,y')\right)$$
$$\times \left(BI^{-1}B^*m, N(m,T,B,y)\right) R(y,dy')\rho(dy).$$

Note that $b_2 \in L_2(R^3)$, a_1, $b_1 \in R^3$, a_0, $b_0 \in R$, for a, $b \in R^3$, and we denote by $a \otimes b$ the operator for which

$$a \otimes b\, x = (a,x)b, \quad x \in R^3.$$

Set

$$\bar{a}_k(m,T) = \bar{A}(a_k,m,T), \quad k = 0,1,$$
$$\bar{b}_k(m,T) = \bar{A}(b_k,m,T), \quad k = 0,1,2.$$

Denote by L the differential operator that is defined on functions $F(m,T)$ from $C^{(2)}(R^3 \times R_+)$ by the relation

$$LF(m,T) = \frac{1}{2}\mathrm{Tr}(F_{mm})b_2 + (b_1, F_{mT}) + \frac{1}{2} b_0 F_{TT} + (a_1, F_m) + a_0 F_T. \quad (8.77)$$

With this notation we have the following result.

Theorem 9 *Assume that all previous conditions and relation (8.76) are satisfied. Then the stochastic process $(\tilde{m}_\varepsilon(t), \tilde{T}_\varepsilon(t))$, where*

$$\tilde{m}_\varepsilon(t) = m_\varepsilon\left(\frac{t}{\varepsilon}\right), \quad \tilde{T}_\varepsilon(t) = T_\varepsilon\left(\frac{t}{\varepsilon}\right),$$

converges weakly in C to the Markov diffusion process $(\hat{m}(t), \hat{T}(t))$ in $R^3 \times R_+$ having generator L and initial conditions $\hat{m}(0) = m$ and $\hat{T}(0) = T$, where m and T are the angular momentum vector and the kinetic energy of the rigid body at the initial time, respectively.

Proof Using equations (8.74) and (8.75) and the techniques developed in Section 5.1.5, we can obtain the representation of

$$E\big(F(m_\varepsilon(t_2), T_\varepsilon(t_2)) - F(m_\varepsilon(t_1), T_\varepsilon(t_1))/\mathcal{F}^\varepsilon_{t_1}\big)$$

in terms of the functions $a_1(m, T, B)$, $a_0(m, T, B)$, $b_2(m, T, B)$, and $b_1(m, T, B)$.
The remainder of the proof is based on Theorem 8 and is completed in the same way as in Theorem 5 of Chapter 5.
The details are left to the reader.

□

Remark 15 The motion of the rigid body on intervals of time on the order of $o(1/\varepsilon)$ can be described by the motion of the averaged system as determined using Poinsot's theorem.
Since the form of the inertia ellipsoid does not change in time, we need only describe the change in time of the plane. That is determined by the angular momentum vector $m_\varepsilon(t)$ and the kinetic energy $T_\varepsilon(t)$. Theorem 9 describes their evolution as $\varepsilon \to 0$.

9

Dynamical Systems on a Torus

In this chapter we consider a dynamical system on a two–dimensional torus. A general theory for such systems was developed by A. Poincaré and A. Denjoy, and it has numerous applications in science and engineering. In particular, this theory has important applications to phase-locked loop electronic circuits that are considered in the next chapter. This theory also plays an important role in the Kolmogorov-Arnol'd-Moser theory of oscillatory solutions to nonlinear systems. In the first part of the chapter we describe the theory as it was developed for flows that have no random elements. In the second part, we consider random perturbations of the same problems.

9.1 Theory of Rotation Numbers

Consider the system of differential equations

$$\begin{aligned} \dot{x}(t) &= a\left(x(t), y(t)\right), \\ \dot{y}(t) &= b\left(x(t), y(t)\right), \end{aligned} \tag{9.1}$$

where $x(t)$ and $y(t)$ are functions from R into R, and $a(x, y)$ and $b(x, y)$ are functions from R^2 into R that have the following properties:

(a) $a(x, y)$, $b(x, y)$ are continuous with continuous derivatives a_x, a_y, b_x, b_y, and $a(x, y) > 0$.

(b) They are periodic functions in x and y with period 1:

$$a(x,y) = a(x+1,y) = a(x,y+1),$$
$$b(x,y) = b(x+1,y) = b(x,y+1),$$

for all x, y.

We will denote by $[x]$ the integer part of a real number x, and set $\{x\} = x - [x]$.
The torus is obtained by the mapping $\tau : R^2 \to [0,1)^2$, $\tau(x,y) = (\{x\}, \{y\})$.
Because of property (a) we can consider system (9.1) as a differential equation on the torus

$$T^{(2)} = \mathbb{C}^2,$$

where $\mathbb{C}$ is the circle in the plane of radius $1/(2\pi)$.
We will describe certain properties of the solutions of equations (9.1) in the plane as well as their representations on the torus $T^{(2)}$.
Note that $x(t)$ for $t > 0$ satisfies the property

$$x(t) \geq x(0) + \alpha t,$$

where

$$\alpha = \inf\{a(x,y) : x \in [0,1),\ y \in [0,1)\} > 0,$$

and $x(t)$ is a strictly increasing function. The success of the theory rests on the fact that the system (9.1) can be converted to an equivalent first–order differential equation, as shown in the next lemma.

Lemma 1 *Denote by $\Phi(x, x_0, y_0)$ the solution of the equation*

$$\frac{d\Phi}{dx}(x, x_0, y_0) = c\,(x, \Phi(x, x_0, y_0)) \tag{9.2}$$

satisfying the initial condition

$$\Phi(x_0, x_0, y_0) = y_0,$$

where

$$c(x,y) = b(x,y)a^{-1}(x,y). \tag{9.3}$$

Then the solution of the system (9.1) satisfying the initial condition

$$x(t_0) = x_0,\ y(t_0) = y_0$$

satisfies the relation

$$y(t) = \Phi\big(x(t), x_0, y_0\big). \tag{9.4}$$

The proof follows from (9.1) and the definition of $c(x,y)$.

Remark 1 *The function $c(x,y)$ defined by (9.3) satisfies the following properties:*

(a′) $c(x,y)$ is a continuous function with continuous derivatives $c_x(x,y)$, $c_y(x,y)$.

(b′) $c(x,y)$ is a periodic function in x and y with period 1.

For any fixed $x \in R$ we consider the mapping $\Phi_x : R \to R$ that is defined by the relation

$$\Phi_x(y) = \Phi(x, 0, y). \tag{9.5}$$

(Note that the subscript here does not indicate a derivative.) Following are some properties of this mapping:

(i) $\Phi_x(y)$ is a continuous function in x and y with continuous derivatives $\frac{\partial}{\partial x}\Phi_x(y)$, $\frac{\partial}{\partial y}\Phi_x(y)$.

(ii) $\Phi_x(y+1) = \Phi_x(y) + 1$ (this is called the circle mapping property).

(iii) $\Phi_x(y) < \Phi_x(y')$ for all x if $y < y'$, so there exists the inverse mapping $\Phi_x^{-1}(y)$ for which $\Phi_x(\Phi_x^{-1}(y)) = y$.

(iv)

$$\Phi(x, x_0, y_0) = \Phi_x\big(\Phi_{x_0}^{-1}(y_0)\big). \tag{9.6}$$

Note that properties (ii), (iii), (iv) are consequences of Φ being the solution to equation (9.2).

9.1.1 *Existence of the Rotation Number*

Theorem 1 *There exists a limit*

$$\theta = \lim_{x\to\infty} \frac{1}{x}\Phi(x, x_0, y). \tag{9.7}$$

This limit does not depend on x_0 or on y, and it exists uniformly in $|x_0| \le C$, $|y| \le C$ for any $C > 0$. The limit θ is referred to as the rotation number *for system (9.1).*

Proof Set $\mathbf{\Phi} = \Phi_1$; $\mathbf{\Phi}$ is a mapping of R into R, it is invertible, and $\mathbf{\Phi}^k$ denotes the kth power of this mapping for all integers k.
We will say that $\mathbf{\Phi}$ is periodic if there exist integers n and k and $y_0 \in R$ for which

$$\mathbf{\Phi}^n(y_0) = y_0 + k.$$

Note that for all x, x_0, y we have the relation

$$\Phi(x, x_0, y) = \Phi_x\big(\Phi_{x_0}^{-1}(y)\big) = \Phi_{\{x\}}\left(\mathbf{\Phi}^{[x]}\big(\Phi_{x_0}^{-1}(y)\big)\right).$$

There exists a constant K for which

$$|\Phi_x(y) - y| < K \quad \text{for } x \in [0,1] \text{ for all } y \in R.$$

Therefore, to prove the theorem we have to prove that uniformly for $|y| \le c$,

$$\lim_{m\to\infty} \frac{1}{m} \mathbf{\Phi}^m(y) = \theta. \tag{9.8}$$

If $\mathbf{\Phi}$ is periodic and

$$y_0 - k_0 \le y \le y_0 + k_0$$

for some $k_0 \in \mathbf{Z}_+$, then

$$\mathbf{\Phi}^{nl}(y_0 - k_0) = y_0 - k_0 + lk \le \mathbf{\Phi}^{nl}(y) \le y_0 + k_0 + lk = \mathbf{\Phi}^{nl}(y_0 + k_0).$$

So

$$\frac{y_0 - k_0}{nl} + \frac{k}{n} \le \frac{1}{nl}\mathbf{\Phi}^{nl}(y) \le \frac{y_0 + k_0}{nl} + \frac{k}{n}$$

and

$$\lim_{l\to\infty} \frac{1}{nl}\mathbf{\Phi}^{nl}(y) = \frac{k}{n} \quad \text{uniformly in } y \in [y_0 - k_0, y_0 + k_0].$$

This implies that (9.8) holds uniformly in $|y| \le c$ with $\theta = k/n$.
Assume that $\mathbf{\Phi}$ is not periodic. Then for any n and k the expression $\mathbf{\Phi}^n(y) - y - k$ is either positive for all y or negative for all y. Denote by Θ_+ the set of all rational numbers k/n for which

$$\mathbf{\Phi}^n(y) > y + k \quad \text{for all } y \in R.$$

Assume that $k/n \in \Theta_+$ and $k_1/n_1 > k/n$. Then $k_1/n_1 \in \Theta_+$, because in this case

$$\mathbf{\Phi}^{n_1}(y) > y + k_1.$$

We have

$$\mathbf{\Phi}^{nn_1}(y) > y + nk_1 > y + n_1 k, \tag{9.9}$$

but the relation $\mathbf{\Phi}^n(y) < y + k$ implies the inequality

$$\mathbf{\Phi}^{nn_1}(y) < y + n_1 k,$$

which contradicts (9.9). The set Θ_+ is bounded from below by the number

$$\beta = \inf_{y\in[0,1]} (\Phi(y) - y),$$

and the rational number $[\beta]$ does not belong to Θ_+. Set

$$\theta = \inf \Theta_+.$$

It is easy to show that for $\frac{k}{n} > \theta$,

$$\limsup_{m\to\infty} \frac{1}{m}\mathbf{\Phi}^m(y) \le \frac{k}{n} \quad \text{uniformly in } y \in [-c, c],$$

and for $k/n < \theta$,

$$\liminf_{m\to\infty} \frac{1}{m}\mathbf{\Phi}^m(y) \ge \frac{k}{n} \quad \text{uniformly in } y \in [-c, c].$$

This completes the proof of the theorem.

□

Remark 2 *Assume that $\theta = k_1/n_1$ and that the least common divisor of k_1 and n_1 is 1. Then there exists y_0 for which*

$$\mathbf{\Phi}^{n_1}(y_0) = y_0 + k_1;$$

i.e., $\mathbf{\Phi}$ is periodic.

9.1.2 Purely Periodic Systems

System (9.1) is called periodic if the mapping $\mathbf{\Phi}$ is periodic. Assume that $\mathbf{\Phi}^{n_1}(y_0) = y_0 + k_1$. Consider the solution of system (9.1) in this case with the initial condition $x(0) = 0$, $y(0) = y_0$. Then

$$y(t) = \Phi_{x(t)}(y_0).$$

Since $x(t)$ is an increasing function and $x(t) \to +\infty$ as $t \to +\infty$, there exists t_0 for which $x(t_0) = n_1$. Then

$$y(t_0) = \Phi_{n_1}(y_0) = \mathbf{\Phi}^{n_1}(y_0) = y_0 + k_1.$$

Since we have $\{x(t_0)\} = \{x(0)\}$ and $\{y(t_0)\} = \{y(0)\}$, the solution $(x(t), y(t))$ considered on the torus $T^{(2)}$ is periodic with period t_0. System (9.1) is called *purely periodic* if the solution of the system is periodic on the torus $T^{(2)}$ for any initial conditions, that is, every solution is periodic. The main result concerning purely periodic systems is presented in the following theorem.

Theorem 2 *If (9.1) is purely periodic, there exist functions $\Psi_1(x, y)$, $\Psi_2(x, y)$ such that*

(1) they are continuous in R^2 and have continuous derivatives

$$\frac{\partial}{\partial x}\Psi_k(x, y), \quad \frac{\partial}{\partial y}\Psi_k(x, y), \quad k = 1, 2,$$

(2) $\Psi_k(x, y) = \Psi_k(x + 1, y) = \Psi_k(x, y + 1)$, $k = 1, 2$,

(3) $\Psi_1^2(x, y) + \Psi_2^2(x, y) = 1$,

(4)

$$a(x, y)\frac{\partial}{\partial x}\Psi_k(x, y) + b(x, y)\frac{\partial}{\partial y}\Psi_k(x, y) = 0, \quad k = 1, 2. \tag{9.10}$$

The last equation means that $\Psi_k(x, y)$, $k = 1, 2$, are first integrals for system (9.1).

Proof The function $\Psi_0(x, y) = \Phi_x^{-1}(y)$ satisfies condition (1) for $k = 0$ and $\Psi_0(x, y) = \Psi_0(x, y) + 1$. It is easy to see that

$$y = \Psi_0\big(x, \Phi_x(y)\big).$$

Therefore,

$$\begin{aligned} 0 &= \frac{\partial}{\partial x}\Psi_0\big(x, \Phi_x(y)\big) + \frac{\partial}{\partial y}\Psi_0\big(x, \Phi_x(y)\big)\frac{\partial}{\partial x}\Phi_x(y) \\ &= \frac{\partial}{\partial x}\Psi_0\big(x, \Phi_x(y)\big) + c\big(x, \Phi_x(y)\big)\frac{\partial}{\partial y}\Psi_0\big(x, \Phi_x(y)\big). \end{aligned}$$

This implies that $\Psi_0(x, y)$ satisfies equation (9.10) for $k = 0$.
Now we set

$$\Psi_1(x, y) = \cos 2\pi\Psi_0(x, y), \quad \Psi_2(x, y) = \sin 2\pi\Psi_0(x, y).$$

The functions Ψ_1, Ψ_2 satisfy properties (1)–(4).
This completes the proof of Theorem 2.

□

Remark 3 *Let $(x(t), y(t))$ be the solution of system (9.1) for some initial conditions. Then $\Psi_k(x(t), y(t)) \equiv$ constant for $k = 1, 2$.*

Next we consider a periodic system. Set

$$\varphi(y) = \Phi^{n_1}(y) - k_1.$$

We assume that for some $y_0 \in [0, 1)$ we have $\varphi(y_0) = y_0$. So the function $\varphi(y)$ maps the interval $[y_0, y_0 + 1]$ into itself, the function $\varphi(y)$ is strictly increasing, $\varphi(y_0) = y_0$, and $\varphi(y_0 + 1) = \varphi(y_0) + 1$.
Denote by Π_0 the set of fixed points, i.e., all $y \in [y_0, y_0+1]$ for which $\varphi(y) = y$. Let $[y_0, y_0 + 1] \setminus \Pi_0 = \cup_k\Delta_k$, where $\Delta_k = (\alpha_k, \beta_k)$, and $\Delta_i \cap \Delta_j = \varnothing$ if $i \neq j$.

Theorem 3 *If (9.1) is periodic, denote by φ^n the nth power of the mapping $\varphi : [y_0, y_0 + 1] \to [y_0, y_0 + 1]$. If $\varphi(y) - y > 0$ for $y \in \Delta_k$, then $\lim_{n\to\infty} \varphi^n(y) = \beta_k$ for all $y \in \Delta_k$; if $\varphi(y) - y < 0$ for $y \in \Delta_k$, then $\lim_{n\to\infty} \varphi^n(y) = \alpha_k$ for all $y \in \Delta_k$.*

Proof It is easy to see that $|\varphi(y) - y| > 0$ for $y \in \Delta_k$. If $\varphi(y) > y$, then $\varphi^{n+1}(y) > \varphi^n(y)$ and there exists

$$\lim_{n\to\infty} \varphi^n(y) = \gamma.$$

Then $\varphi(\gamma) = \gamma$. This means that $\gamma = \beta_k$. In the same way we can prove the second statement of the theorem.

□

9.1.3 Ergodic Systems

System (9.1) is called *ergodic* if the set $\{\{\Phi(n, 0, y)\}, n = 0, 1, 2, \dots\}$ is dense on the interval $[0, 1]$, where $\{\Phi\} = \Phi - [\Phi]$, for all $y \in [0, 1)$. (Note

that for a periodic system the set $\{\{\Phi(n,0,y)\},\ n=0,1,2,\dots\}$ is a finite set if $\Phi(n_1,0,y)=y+k_1$.)
So for ergodic systems the rotation number θ is an irrational number. The inverse statement is true for systems satisfying some additional conditions regarding the smoothness of the coefficients.

Theorem 4 *Assume that θ is an irrational number and that $\frac{\partial c}{\partial y}(x,y)$ is a continuous function of bounded variation. Then system (9.1) is ergodic.*

(The proof of this statement is in [16, p.409].)
It follows from general ergodic theorems that under the conditions of Theorem 4 there exists a probability measure $m(dy)$ such that for any continuous function $h(y): R \to R$ satisfying the condition $h(y+1)=h(y)$ there exists

$$\lim_{n\to\infty}\frac{1}{n}\sum_{k=1}^{n} h\big(\Phi^k(y_0)\big)=\int_0^1 h(y)\,m(dy) \tag{9.11}$$

uniformly in $y_0\in[0,1]$.
This result implies the following statement concerning the ergodic distribution for system (9.1).

Theorem 5 *Let the conditions of Theorem 4 be satisfied. Then for any continuous function $h(x,y)$ satisfying the property*

$$h(x,y)=h(x+1,y)=h(x,y+1)$$

there exists the limit

$$\lim_{t\to\infty}\frac{1}{t}\int_0^t h\big(x(s),y(s)\big)ds=\int_0^1\int_0^1 h\big(u,\Phi_u(y)\big)a^{-1}\big(u,\Phi_u(y)\big)\,du\,m(dy) \times\left[\int_0^1\int_0^1 a^{-1}\big(u,\Phi_u(y)\big)\,du\,m(dy)\right]^{-1} \tag{9.12}$$

uniformly in $x(0)\in[0,1]$, $y(0)\in[0,1]$, where $(x(t),y(t))$ is the solution of the system (9.1).

Proof The proof follows from the representation

$$\int_0^t h\big(x(s),y(s)\big)ds=\int_0^{x(t)}\frac{h\big(u,\Phi_u(y)\big)}{a\big(u,\Phi_u(y)\big)}du=\sum_{k=0}^{[x(t)]}\int_k^{k+1}\frac{h\big(u,\Phi_u(y)\big)}{a\big(u,\Phi_u(y)\big)}du$$
$$=\sum_{k=0}^{[x(t)]}\int_0^1\frac{h\big(u,\Phi_u(y)\big)}{a\big(u,\Phi_u(y)\big)}du,$$

which holds up to order $O(1)$, and formula (9.11).

□

The following result concerning the representation of the solution to (9.2) in the ergodic case will be useful.

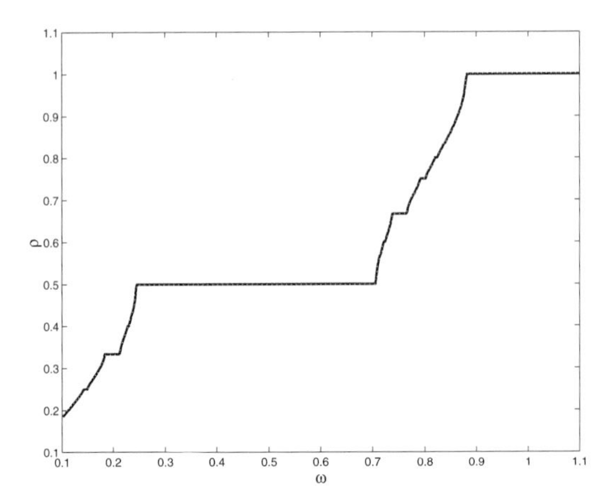

Figure 9.1. Simulation of the rotation number for system (9.13). Note the plateau regions where the frequency response of the system is constant over a range of ω values. This exhibits the phenomenon of phase–locking, which we discuss in Chapter 10.

Theorem 6 *(Bohl's Theorem). Under the conditions of Theorem 4 there exists a continuous function* $\chi(x, y) : R^2 \to R$ *for which*

$$\chi(x, y) = \chi(x + 1, y) = \chi(x, y + 1),$$

and the function

$$G(y) = y + \chi(0, y)$$

is strictly increasing for $y \in R$. *Moreover,*

$$\Phi_x(y) = g(y) + \theta x + \chi(x, g(y) + \theta x))$$

where $g(y) = G^{-1}(y)$, *i.e.,* $G(g(y)) = y$.

This theorem establishes the near-identity transformation that is widely used in the theory of nonlinear oscillations, and the proof of the theorem is in [16, p.414].

9.1.4 Simulation of Rotation Numbers

Consider the system

$$\begin{aligned} \dot{x} &= \omega - 0.1(\sin(x - y) + 2\sin(2x - y)), \\ \dot{y} &= 1 + 0.1(\sin(x - y) + \sin(2x - y)). \end{aligned} \tag{9.13}$$

For each value of $\omega \in [0.1, 1.1]$, this system has a rotation number, say $\rho(\omega)$. Figure 9.1 shows the rotation number calculated for this system for each of 500 values of ω in this interval.

9.2 Randomly Perturbed Torus Flows

Let $(Z, \mathcal{Z})$ be a measurable space, and let $z(t,\omega)$ be a homogeneous Markov process in this space with transition probability function

$$P(t, z, C) = P\{z(t,\omega) \in C/z(0,\omega) = z\}, \quad C \in \mathcal{Z}, \ z \in Z.$$

We assume that $z(t,\omega)$ is a uniformly ergodic process with an ergodic distribution $\rho(dz)$. This stochastic process will describe random noise in the system moving on the torus $T^{(2)}$.

The system is defined by the differential equations

$$\begin{aligned} \dot{x}_\varepsilon(t) &= a\left(x_\varepsilon(t), y_\varepsilon(t), z\left(\frac{t}{\varepsilon}, \omega\right)\right), \\ \dot{y}_\varepsilon(t) &= b\left(x_\varepsilon(t), y_\varepsilon(t), z\left(\frac{t}{\varepsilon}, \omega\right)\right), \end{aligned} \tag{9.14}$$

where $(x_\varepsilon(t), y_\varepsilon(t))$ are R-valued random functions that define the state of the system, and the functions $a(x, y, z)$, $b(x, y, z)$ satisfy the following conditions:

(1) They are defined on $R^2 \times Z$ and take values in R.

(2) They are measurable with respect to $\mathcal{B}(R^2) \otimes \mathcal{Z}$; for fixed z they are continuous in x, y; and they have continuous and bounded derivatives, $a_x(x, y, z)$, $a_y(x, y, z)$, $b_x(x, y, z)$, $b_y(x, y, z)$.

(3) For fixed z they are periodic in x and y with

$$a(x, y, z) \equiv a(x + 1, y, z) \equiv a(x, y + 1, z),$$

$$b(x, y, z) \equiv b(x + 1, y, z) \equiv b(x, y + 1, z),$$

for all x, y, and z.

Under these conditions system (9.14) has a unique solution for any initial conditions $x_\varepsilon(0) = x_0$, $y_\varepsilon(0) = y_0$.

Recall that the $R^2 \times Z$-valued stochastic process

$$\left(x_\varepsilon(t), y_\varepsilon(t), z\left(\frac{t}{\varepsilon}, \omega\right)\right)$$

is a homogeneous Markov process.

As usual, we will investigate the asymptotic behavior of the solution to system (9.14) as $t \to \infty$ and $\varepsilon \to 0$. In particular, we will investigate the ratio $y_\varepsilon(t)/x_\varepsilon(t)$ for large t and small ε.

Set

$$\bar{a}(x, y) = \int a(x, y, z)\rho(dz)$$

and

$$\bar{b}(x,y) = \int b(x,y,z)\rho(dz).$$

We assume in addition that $\bar{a}(x,y) > 0$. Then the functions $\bar{a}(x,y)$, $\bar{b}(x,y)$ satisfy conditions (1)–(3) in Section 9.1. The system of differential equations

$$\begin{aligned} \dot{\bar{x}}(t) &= \bar{a}\big(\bar{x}(t),\bar{y}(t)\big), \\ \dot{\bar{y}}(t) &= \bar{b}\big(\bar{x}(t),\bar{y}(t)\big), \end{aligned} \tag{9.15}$$

determines the averaged system for (9.14), and it can be studied using the methods developed in Section 9.1.

We will use some statements that were proved earlier for general randomly perturbed differential equations, which we summarize next.

I. Let $(x_\varepsilon(t), y_\varepsilon(t))$ be the solution to system (9.14), and let $\bar{z}(t) = (\bar{x}(t),\bar{y}(t))$ be the solution to system (9.15). Assume that $x_\varepsilon(0) = \bar{x}(0)$, $y_\varepsilon(0) = \bar{y}(0)$. Then for any $T > 0$,

$$P\Big\{\lim_{\varepsilon\to 0}\sup_{t\le T}\big(|x_\varepsilon(t)-\bar{x}(t)| + |y_\varepsilon(t)-\bar{y}(t)|\big) = 0\Big\} = 1.$$

This statement is a consequence of Theorem 5 of Chapter 3.

II. Suppose additionally that the Markov process $\bar{z}(t)$ satisfies condition SMC II of Section 2.3. Then under the conditions $x_\varepsilon(0) = \bar{x}(0)$, $y_\varepsilon(0) = \bar{y}(0)$, the two–dimensional process $(\tilde{x}_\varepsilon(t), \tilde{y}_\varepsilon(t))$, where

$$\tilde{x}_\varepsilon(t) = \varepsilon^{-1/2}(x_\varepsilon(t) - \bar{x}(t)), \quad \tilde{y}_\varepsilon(t) = \varepsilon^{-1/2}(y_\varepsilon(t) - \bar{y}(t)),$$

converges weakly to a two–dimensional Gaussian process $(\tilde{x}(t), \tilde{y}(t))$, which is determined by solving the system of integral equations

$$\tilde{x}(t) = \int_0^t \big[\bar{a}_x(\bar{x}(s),\bar{y}(s))\tilde{x}(s) + \bar{a}_y(\bar{x}(s),\bar{y}(s))\tilde{y}(s)\big]ds + v_1(t),$$

$$\tilde{y}(t) = \int_0^t \big[\bar{b}_x(\bar{x}(s),\bar{y}(s))\tilde{x}(s) + \bar{b}_y(\bar{x}(s),\bar{y}(s))\tilde{y}(s)\big]ds + v_2(t),$$

where $(v_1(t), v_2(t))$ is a two–dimensional Gaussian process with independent increments for which $Ev_1(t) = 0$, $Ev_2(t) = 0$, and

$$\begin{aligned} E\big(\alpha_1 v_1(t) + \alpha_2 v_2(t)\big)^2 = \int_0^t \iint &\big[\alpha_1 a(\bar{x}(s),\bar{y}(s),z) + \alpha_2 b(\bar{x}(s),\bar{y}(s),z)\big] \\ &\times \big[\alpha_1 a(\bar{x}(s),\bar{y}(s),z') + \alpha_2 b(\bar{x}(s),\bar{y}(s),z')\big] R(z,dz')\rho(dz)dt. \end{aligned}$$

This follows from Theorem 5 of Chapter 4.

9.2.1 Rotation Number in the Presence of Noise

Here we describe the behavior of the ratio $y_\varepsilon(t)/x_\varepsilon(t)$ as $t \to \infty$ and $\varepsilon \to 0$ for solutions of (9.14). First, we show that $x_\varepsilon(t) \to \infty$ with probability 1. Next, we show that the limit of $y_\varepsilon(t)/x_\varepsilon(t)$ can be calculated in terms of an ergodic measure on the torus. Lastly, we estimate the difference between this limit and the rotation number for the averaged system (9.15).

Lemma 2 *There exists $\varepsilon_0 > 0$ for which*

$$P\{x_\varepsilon(t) \to +\infty \quad \text{as } t \to +\infty\} = 1$$

for all $\varepsilon < \varepsilon_0$.

Proof Set

$$A(\varepsilon, t, x, y, z) = E\big(x_\varepsilon(t)/x_\varepsilon(0) = x, y_\varepsilon(0) = y, z(0) = z\big) - x.$$

Then

$$A(\varepsilon, t_1, x, y, z) = A(\varepsilon, t_1, x+1, y, z) = A(\varepsilon, t_1, x, y+1, z),$$

and $A(\varepsilon, t_1, x, y, z)$, $\frac{\partial}{\partial x}A(\varepsilon, t_1, x, y, z)$, $\frac{\partial}{\partial y}A(\varepsilon, t_1, x, y, z)$ are bounded and continuous functions of (t_1, x, y) over any bounded region. To prove this we consider the functions

$$X_\varepsilon(t, x, y), \quad Y_\varepsilon(t, x, y)$$

that solve system (9.14) and satisfy the initial conditions

$$X_\varepsilon(0, x, y) = x, \quad Y_\varepsilon(0, x, y) = y.$$

It is easy to see that $\frac{\partial}{\partial x}X_\varepsilon(t, x, y)$, $\frac{\partial}{\partial y}X_\varepsilon(t, x, y)$, $\frac{\partial}{\partial x}Y_\varepsilon(t, x, y)$, and $\frac{\partial}{\partial y}Y_\varepsilon(t, x, y)$ satisfy the system of linear ordinary differential equations

$$\begin{aligned}\frac{d}{dt}\frac{\partial}{\partial x}X_\varepsilon(t, x, y) &= a_x\left(X_\varepsilon, Y_\varepsilon, z\left(\frac{t}{\varepsilon}\right)\right)\frac{\partial}{\partial x}X_\varepsilon(t, x, y)\\ &\quad + a_y\left(X_\varepsilon, Y_\varepsilon, z\left(\frac{t}{\varepsilon}\right)\right)\frac{\partial}{\partial x}Y_\varepsilon(t, x, y),\end{aligned}$$

$$\begin{aligned}\frac{d}{dt}\frac{\partial}{\partial x}Y_\varepsilon(t, x, y) &= b_x\left(X_\varepsilon, Y_\varepsilon, z\left(\frac{t}{\varepsilon}\right)\right)\frac{\partial}{\partial x}X_\varepsilon(t, x, y)\\ &\quad + b_y\left(X_\varepsilon, Y_\varepsilon, z\left(\frac{t}{\varepsilon}\right)\right)\frac{\partial}{\partial x}Y_\varepsilon(t, x, y),\end{aligned}$$

$$\begin{aligned}\frac{d}{dt}\frac{\partial}{\partial y}X_\varepsilon(t, x, y) &= a_x\left(X_\varepsilon, Y_\varepsilon, z\left(\frac{t}{\varepsilon}\right)\right)\frac{\partial}{\partial y}X_\varepsilon(t, x, y)\\ &\quad + a_y\left(X_\varepsilon, Y_\varepsilon, z\left(\frac{t}{\varepsilon}\right)\right)\frac{\partial}{\partial y}Y_\varepsilon(t, x, y),\end{aligned}$$

and

$$\frac{d}{dt}\frac{\partial}{\partial y}Y_\varepsilon(t,x,y) = b_x\left(X_\varepsilon, Y_\varepsilon, z\left(\frac{t}{\varepsilon}\right)\right)\frac{\partial}{\partial y}X_\varepsilon(t,x,y) + b_y\left(X_\varepsilon, Y_\varepsilon, z\left(\frac{t}{\varepsilon}\right)\right)\frac{\partial}{\partial y}Y_\varepsilon(t,x,y)$$

with the initial conditions

$$\frac{\partial}{\partial x}X_\varepsilon(0,x,y) = 1, \quad \frac{\partial}{\partial y}X_\varepsilon(0,x,y) = 0,$$

$$\frac{\partial}{\partial x}Y_\varepsilon(0,x,y) = 0, \quad \frac{\partial}{\partial y}Y_\varepsilon(0,x,y) = 1.$$

This implies that the functions $\frac{\partial}{\partial x}X_\varepsilon(t,x,y)$, $\frac{\partial}{\partial y}X_\varepsilon(t,x,y)$, $\frac{\partial}{\partial x}Y_\varepsilon(t,x,y)$, and $\frac{\partial}{\partial y}Y_\varepsilon(t,x,y)$ are continuous in (t,x,y) and are bounded by a nonrandom constant over any bounded region. Note that

$$A(\varepsilon,t,x,y,z) = \int_0^t E\left[a\left(X_\varepsilon(s,x,y), Y_\varepsilon(s,x,y), z\left(\frac{s}{\varepsilon}\right)\right)/z(0) = z\right]ds$$

and

$$\frac{\partial A}{\partial x}(\varepsilon,t,x,y,z) = \int_0^t E\left[a_x\left(X_\varepsilon(s,x,y), Y_\varepsilon(s,x,y), z\left(\frac{s}{\varepsilon}\right)\right)\frac{\partial}{\partial x}X_\varepsilon(s,x,y) + a_y\left(X_\varepsilon(s,x,y), Y_\varepsilon(s,x,y), z\left(\frac{s}{\varepsilon}\right)\right)\frac{\partial}{\partial x}Y_\varepsilon(s,x,y)/z(0) = z\right]ds,$$

and $\frac{\partial A}{\partial y}(\varepsilon,t,x,y,z)$ can be represented in the same way. These formulas establish the properties of the functions $A(\varepsilon,t,x,y,z)$.
It follows from the averaging theorem that

$$\lim_{\varepsilon\to 0} A(\varepsilon,t,x,y,z) = \int_0^t \bar{a}\big(X(s,x,y), Y(s,x,y)\big)ds$$

uniformly in x, y, z for each $t > 0$, where $(X(s,x,y), Y(s,x,y))$ is the solution to system (9.15) with initial conditions

$$X(0,x,y) = x, \quad Y(0,x,y) = y.$$

Since $\bar{a}(x,y) > \delta$, there exist $\varepsilon_0 > 0$ and t_1 for which

$$\inf_{\varepsilon\le\varepsilon_0,\, x,y,z} A(\varepsilon,t_1,x,y,z) = \delta_1 > 0.$$

Then

$$\frac{x_\varepsilon(nt_1)}{n} = \frac{1}{n}\sum_{k=0}^{n-1}[x_\varepsilon((k+1)t_1) - x_\varepsilon(kt_1)]$$

$$= \frac{1}{n}\sum_{k=0}^{n-1}\left[x_\varepsilon((k+1)t_1) - x_\varepsilon(kt_1) - A\left(\varepsilon, t_1, x_\varepsilon(kt_1), y_\varepsilon(kt_1), z\left(\frac{kt_1}{\varepsilon}\right)\right)\right]$$

$$+ \frac{1}{n}\sum_{k=0}^{n-1} A\left(\varepsilon, t_1, x_\varepsilon(kt_1), y_\varepsilon(kt_1), z\left(\frac{kt_1}{\varepsilon}\right)\right).$$

If $\varepsilon \le \varepsilon_0$, then

$$\frac{x_\varepsilon(nt_1)}{n} \ge \delta + \frac{1}{n}\sum_{k=0}^{n-1}, \xi(\varepsilon, kt_1),$$

where

$$\xi(\varepsilon, kt_1) = x_\varepsilon((k+1)t_1) - x_\varepsilon(kt_1) - E\big(x_\varepsilon((k+1)t_1) - x_\varepsilon(kt_1)/\mathcal{F}^\varepsilon_{kt_1}\big),$$

where $\mathcal{F}^\varepsilon_t$ is the σ- algebra generated by $\{z(s/\varepsilon),\ s \le t\}$.
It is easy to see that $E(\xi(\varepsilon, kt_1)/\mathcal{F}^\varepsilon_{kt_1}) = 0$ and

$$E\big((\xi(\varepsilon, kt_1))^2/\mathcal{F}^\varepsilon_{kt_1}\big) \le c.$$

To complete the proof we prove the following statement, which will also be useful later.

Lemma 3 *Let* $\{\xi_k\}$ *be a martingale with respect to the discrete time filtration* $\{\mathcal{F}_k\}, E\xi_k = 0$, *and* $E(\xi_k^2/\mathcal{F}_{k-1}) \le c$. *Then*

$$\frac{1}{n}\sum_{k=1}^{n}\xi_k \to 0 \quad \textit{with probability } 1.$$

Proof of Lemma 3: Set

$$\eta_m = \sup_{n \le 2^m}\left|\sum_{k=1}^{n}\xi_k\right|.$$

Then for $2^{m-1} < n \le 2^m$,

$$\frac{1}{n}\left|\sum_{k=1}^{n}\xi_k\right| \le \frac{1}{2^{m-1}}\eta_m \tag{9.16}$$

and

$$P\left\{\frac{1}{2^{m-1}}\eta_m \ge a\right\} \le P\left\{\sup_{n\le 2^m}\left|\sum_{k=1}^{n}\xi_k\right| \ge 2^{m-1}a\right\} \le \frac{2^m c}{(2^{m-1}a)^2}.$$

So

$$\sum_m P\left\{\frac{1}{2^{m-1}}\eta_m \ge q^m\right\} \le 4c\sum_m \frac{1}{2^m q^{2m}} < \infty$$

if $0 < q < 1$ and $2q^2 > 1$. This implies that

$$P\Big\{\lim_{m\to\infty} 2^{-m}\eta_m = 0\Big\} = 1,$$

and completes the proof of Lemma 3.

□

We return to the proof of Lemma 2. Since

$$P\Big\{\lim \frac{1}{n}\sum_{k=0}^{n-1}\xi(\varepsilon, kt_1) = 0\Big\} = 1,$$

we have that

$$P\Big\{\liminf_{n\to\infty}\frac{x_\varepsilon(nt_1)}{n} \geq \delta\Big\} = 1.$$

If $t \in [nt_1, (n+1)t_1]$, then $|x_\varepsilon(nt_1) - x_\varepsilon(t)| \leq c_1$, where c_1 is a constant. Therefore,

$$P\Big\{\liminf_{t\to\infty}\frac{x_\varepsilon(t)}{t} \geq \delta\Big\} = 1.$$

This completes the proof of Lemma 2.

□

We define a Markov process in the space $T^{(2)} \times Z$ by

$$\left(\{x_\varepsilon(t)\}, \{y_\varepsilon(t)\}, z\left(\frac{t}{\varepsilon}\right)\right), \tag{9.17}$$

which we assume to be an ergodic process for $\varepsilon > 0$ with ergodic distribution $m_\varepsilon(dx, dy, dz)$. This means that for all measurable functions $F(x, y, z) : R^2 \times Z \to R$ for which $F(x, y, z) \equiv F(x+1, y, z) \equiv F(x, y+1, z)$ and

$$\int |F(\{x\}, \{y\}, z)| m_\varepsilon(dx, dy, dz) < \infty,$$

then with probability 1,

$$\lim_{t\to\infty}\frac{1}{t}\int_0^t F\left(x_\varepsilon(s), y_\varepsilon(s), z\left(\frac{s}{\varepsilon}\right)\right) ds = \int F(\{x\}, \{y\}, z) m_\varepsilon(dx, dy, dz). \tag{9.18}$$

Formula (9.18) implies that

$$\lim_{t\to\infty}\frac{y_\varepsilon(t)}{x_\varepsilon(t)} = \lim_{t\to\infty}\left(y_\varepsilon(0)+\int_0^t b\left(x_\varepsilon(s),y_\varepsilon(s),z\left(\frac{s}{\varepsilon}\right)\right)ds\right)$$
$$\times\left(x_\varepsilon(0)+\int_0^t a\left(x_\varepsilon(s),y_\varepsilon(s),z\left(\frac{s}{\varepsilon}\right)\right)ds\right)^{-1}$$
$$=\lim_{t\to\infty}\left(\int_0^t b\left(x_\varepsilon(s),y_\varepsilon(s),z\left(\frac{s}{\varepsilon}\right)\right)ds\right)\left(\int_0^t a\left(x_\varepsilon(s),y_\varepsilon(s),z\left(\frac{s}{\varepsilon}\right)\right)ds\right)^{-1}$$
$$=\int b(\{x\},\{y\},z)m_\varepsilon(dx,dy,dz)\Big/\int a(\{x\},\{y\},z)m_\varepsilon(dx,dy,dz).$$

So the following statement has been proved.

Theorem 7 *Let the Markov process (9.17) be ergodic with the ergodic distribution* $m_\varepsilon(dx,dy,dz)$. *Then with probability* 1,

$$\lim_{t\to\infty}\frac{y_\varepsilon(t)}{x_\varepsilon(t)}=\frac{\int b(\{x\},\{y\},z)m_\varepsilon(dx,dy,dz)}{\int a(\{x\},\{y\},z)m_\varepsilon(dx,dy,dz)}. \tag{9.19}$$

Remark 4 *It follows from Lemma 2 that*

$$\int a(\{x\},\{y\},z)m_\varepsilon(dx,dy,dz)>0$$

for $\varepsilon<\varepsilon_0$, *where* $\varepsilon_0>0$ *is the number whose existence is proved there.*

We consider now the ergodic property of the Markov process (9.17) under the further assumption that $z(t)$ is a jump Markov process. In this case there exist a bounded measurable positive function $\lambda(z)$ and a transition probability in Z, say $Q(z,dz')$, such that the generator G of the Markov process $z(t)$ is represented in the form

$$Gh(z)=\lambda(z)\times\int[h(z')-h(z)]Q(z,dz') \tag{9.20}$$

for any integrable function h.
Note that the ergodic distribution $\rho(dz)$ of the process $z(t)$ satisfies the relation

$$\int_C\lambda(z)\rho(dz)=\int\lambda(z)\rho(dz)Q(z,C). \tag{9.21}$$

The process $z(t,\omega)$ can be represented in the form

$$z(t,\omega)=\sum_n z_n(\omega)1_{\{\sum_{k=0}^{n-1}\eta_k(\omega)\lambda^{-1}(z_k(\omega))\le t<\sum_{k=0}^{n}\eta_k(\omega)\lambda^{-1}(z_k(\omega))\}}, \tag{9.22}$$

where $\{z_k(\omega),\ k=0,1,2,\dots\}$ is a sequence of Z-valued random variables, and $\{\eta_k(\omega)\}$, $k=0,1,\dots$, is a sequence of R-valued random variables. In addition, $\{z_k(\omega)\}$ is the Markov chain in $(Z, \mathcal{Z})$ with transition probability $Q(z, C)$, and

$$P\left[\eta_k(\omega) > t\lambda(z_k(\omega))/\eta_0(\omega), \ldots, \eta_{k-1}(\omega), z_0(\omega), \ldots, z_k(\omega)\right] = e^{-t}.$$

Denote by $S_z(t, x, y)$ the function from $Z \times R_+ \times T^{(2)}$ into $T^{(2)}$ that is defined by

$$S_z(t, x, y) = (\{X^z(t, x, y)\}, \{Y^z(t, x, y)\}),$$

where $(X^z(t, x, y),\ Y^z(t, x, y))$ is the solution of the system of differential equations

$$\frac{d}{dt} X^z(t, x, y) = a\big(X^z(t, x, y), Y^z(t, x, y), z\big),$$

$$\frac{d}{dt} Y^z(t, x, y) = b\big(X^z(t, x, y), Y^z(t, x, y), z\big),$$

satisfying the initial conditions $X^z(0, x, y) = x$, $Y^z(0, x, y) = y$.
Set $\tau_k = \eta_k(\omega)\lambda^{-1}(z_k(\omega))$, and let $X(t, x, y, \omega)$, $Y(t, x, y, \omega)$ be the solution of system (9.14) for $\varepsilon = 1$ with the initial conditions $X(0, x, y) = x$, $Y(0, x, y) = y$. Denote by $\{u_n(\omega),\ n = 0, 1, 2, \ldots\}$ the sequence of $T^{(2)}$-valued random variables that are defined by

$$u_n(\omega) = \left(\left\{X\left(\sum_{k=0}^{n-1} \tau_k, x, y, \omega\right)\right\}, \left\{Y\left(\sum_{k=0}^{n-1} \tau_k, x, y, \omega\right)\right\}\right). \tag{9.23}$$

Then

$$u_n(\omega) = S_{z_{n-1}(\omega)}(\tau_{n-1}, u_{n-1}(\omega)), \tag{9.24}$$

where

$$z_n(\omega) = z\left(\sum_{k=0}^{n-1} \tau_k, \omega\right). \tag{9.25}$$

It is easy to see that $\{(u_n(\omega), z_n(\omega)),\ n = 0, 1, 2, \ldots\}$ is a homogeneous Markov chain in the space $T^{(2)} \times Z$.

Lemma 4 *Assume that the following conditions are fulfilled:*

(1) $|a(x, y, z)b(x, y, z') - a(x, y, z')b(x, y, z)| > 0$, $a^2(x, y, z) + b^2(x, y, z) > 0$ *for all* x, y, $z \neq z'$

(2) $\rho((z)) < 1$ *for every* $z \in Z$. *Here* (z) *is a singleton containing the point* z.

Then

(a) the Markov chain $\{(u_n(\omega), z_n(\omega)),\ n = 0, 1, 2, \ldots\}$ *is Harris recurrent with respect to the measure*

$$Q_2(dx, dy)\rho(dz) \tag{9.26}$$

on the torus $T^{(2)}$, where Q_2 is Lebesgue measure on $[0,1] \times [0,1]$,

(b) there exists a unique invariant probability measure for this Markov chain, and it is absolutely continuous with respect to measure (9.26).

Proof The function

$$\Delta(x, y, z, z') = a(x, y, z)b(x, y, z') - a(x, y, z')b(x, y, z)$$

is continuous in x, y, so

$$\inf_{0 \le x \le 1,\, 0 \le y \le 1} |\Delta(x, y, z, z')| > 0.$$

Therefore, for fixed z, z' the two systems of curves $\{U_u^z(t),\, u \in T^{(2)}\}$ and $\{U_u^{z'}(t),\, u \in T^{(2)}\}$ can be used as local coordinates, where

$$U_u^z = \big((X^z(t, x, y)),\, (Y^z(t, x, y))\big) \quad \text{if } u = ((x), (y)).$$

This means that for any point $u_0 \in T^{(2)}$, say $u_0 = (\{x_0\}, \{y_0\})$, we can find such $\varepsilon > 0$ and δ_0 that for $x \in (x_0 - \varepsilon, x_0 + \varepsilon)$ and $y \in (y_0 - \varepsilon, y_0 + \varepsilon)$ there are numbers t_1, $|t_1| < \delta_0$, and t_2, $|t_2| < \delta_0$, for which

$$u = (\{x\}, \{y\}) = S_{z'}(t_2, S_z(t_1, u)).$$

Moreover, t_1, t_2 satisfying these conditions are unique. This implies that for $u \in T^{(2)}$ there exists an open set $G_u \subset T^{(2)}$, $u \in G_u'$ for which

$$P\{u_2(\omega) \in A/u_0(\omega) = u,\, z_0(\omega) = z\} \ge c(z)Q_2(A \cap G_u), \qquad (9.27)$$

where $c(z) > 0$. Using this inequality we can prove that for any measurable set C we have

$$\sum_n E\big(1_{\{(u_n(\omega), z_n(\omega)) \in C\}}/u_{n-2}(\omega), z_{n-2}(\omega)\big)$$

$$\ge \sum c(z_{n-2}(\omega))1_{\{z_n(\omega) \in C'\}}Q_2(C_{z_{n-2}(\omega)} \cap C_{u_{n-2}}(\omega)) = +\infty,$$

where $C_z = \{u : (u, z) \in C\}$ and $C' = \{z : C_z \ne \varnothing\}$.
This implies statement (a) of the lemma. The existence and uniqueness of the invariant measure is a consequence of the compactness of $T^{(2)}$ and the continuity of the transition probability in x and y.
This completes the proof of Lemma 4.

□

The next theorem describes the behavior of the ratio $y_\varepsilon(t)/x_\varepsilon(t)$ for a randomly perturbed system.

Theorem 8 *Assume that the averaged system (9.15) has rotation number θ:*

$$\lim_{t \to \infty} \bar{y}(t)/\bar{x}(t) = \theta \qquad (9.28)$$

uniformly with respect to $x(0) \in [0, 1]$.

Then for any $\delta > 0$,

$$\lim_{\varepsilon\to 0} P\{\limsup_{t\to\infty} |y_\varepsilon(t)/x_\varepsilon(t) - \theta| < \delta\} = 1. \tag{9.29}$$

Proof It follows from (9.28) that for any $\delta_1 > 0$ there exists a time $t = a > 0$ for which

$$\sup_{x_0,y_0} \left|\theta - (Y(a,x_0,y_0) - y_0)/(X(a,x_0,y_0) - x_0)\right| < \delta_1, \tag{9.30}$$

where $(X(t,x,y), Y(t,x,y))$ is the solution to system (9.15) with the initial conditions

$$X(0,x,y) = x, \quad Y(0,x,y) = y.$$

For any $n \in \mathbf{Z}_+$ we have

$$y_\varepsilon(na)/na = \frac{1}{na}\left(y_0 + \sum_{k=1}^{n-1}[y_\varepsilon(ka) - y_\varepsilon((k-1)a)]\right)$$

$$= \frac{1}{na}\left(y_0 + \sum_{k=1}^{n-1}\delta\tilde{y}_\varepsilon(ka)\right) + \frac{1}{na}\sum_{k=1}^{n-1} E\big(y_\varepsilon(ka) - y_\varepsilon((k-1)a)/\mathcal{F}^\varepsilon_{(k-1)a}\big),$$

where

$$\delta\tilde{y}_\varepsilon(ka) = y_\varepsilon(ka) - y_\varepsilon((k-1)a) - E\big(y_\varepsilon(ka) - y_\varepsilon((k-1)a)/\mathcal{F}^\varepsilon_{(k-1)a}\big).$$

It follows from Lemma 3 that

$$P\left\{\lim_{n\to\infty}\frac{1}{n}\sum_{k=1}^{n}\delta\tilde{y}_\varepsilon(ka) = 0\right\} = 1.$$

It follows from statement I that

$$\begin{aligned} &E\big(y_\varepsilon(ka) - y_\varepsilon((k-1)a)/\mathcal{F}^\varepsilon_{(k-1)a}\big)\\ &= E\big(Y_\varepsilon(a, x_\varepsilon((k-1)a), y_\varepsilon((k-1)a)) - y_\varepsilon((k-1)a)/\mathcal{F}^\varepsilon_{(k-1)a}\big)\\ &= Y(a, x_\varepsilon((k-1)a), y_\varepsilon((k-1)a)) - y_\varepsilon)((k-1)a) + O(\alpha(\varepsilon,a)), \end{aligned}$$

where $(X_\varepsilon(t,x,y), Y_\varepsilon(t,x,y))$ is the solution of system (9.14), and where the gauge function $\alpha(\varepsilon,a)$ goes to 0 for all $a > 0$.
So

$$\frac{y_\varepsilon(na)}{na} = \frac{1}{na}\sum_{k<n}[Y(a, x_\varepsilon(ka), y_\varepsilon(ka)) - y_\varepsilon(ka)] + O(\alpha(\varepsilon,a)) + o(1).$$

In the same way we can obtain the representation

$$\frac{x_\varepsilon(na)}{na} = \frac{1}{na}\sum_{k<n}[X(a, x_\varepsilon(ka), y_\varepsilon(ka)) - x_\varepsilon(ka)] + O(\alpha(\varepsilon,a)) + o(1).$$

Note that

$$\left|\frac{Y(a, x_\varepsilon(ka), y_\varepsilon(ka)) - y_\varepsilon(ka)}{X(a, x_\varepsilon(ka), y_\varepsilon(ka)) - x_\varepsilon(ka)} - \theta\right| < \delta_1$$

because of relation (9.30).
Lemma 2 implies the relation $x_\varepsilon(na) \to +\infty$ with probability 1. So

$$\frac{y_\varepsilon(na)}{x_\varepsilon(na)} = \theta + O(\alpha(\varepsilon, a0 + \delta_1) + o(1).$$

Assume that $t \in [na, (n+1)a]$. Then

$$|y_\varepsilon(t) - y_\varepsilon(na)| \leq c_1 a,$$

$$|x_\varepsilon(t) - x_\varepsilon(na)| \leq c_2 a,$$

and

$$\frac{y_\varepsilon(t)}{x_\varepsilon(t)} = \theta + O\left(\alpha(\varepsilon, a) + \delta_1 + \frac{a}{x_\varepsilon(na)}\right) + o(1),$$

where $P\{\lim_{n\to\infty} o(1) = 0\} = 1$.
This completes the proof of Theorem 8.

□

Remark 5 Assume that the Markov process $(\{x_\varepsilon(t)\}, \{y_\varepsilon(t)\}, z(t/\varepsilon))$, for $t \geq 0$, is ergodic in $T^{(2)} \times Z$ for all $\varepsilon > 0$. Then with probability 1 there exists a limit

$$\lim_{t\to\infty} y_\varepsilon(t)/x_\varepsilon(t) = \theta_\varepsilon$$

for sufficiently small $\varepsilon > 0$ for which $\lim_{t\to\infty} x_\varepsilon(t) = +\infty$. It follows from Theorem 8 that $\theta_\varepsilon \to \theta$ as $\varepsilon \to 0$.

9.2.2 Simulation of Rotation Numbers with Noise

The simulation in Figure 9.2 shows the rotation numbers, calculated as in Figure 9.1, but for each of ten sample paths for ω in the system

$$\dot{x} = \omega(t/\varepsilon) - 0.1(\sin(x - y) + 2\sin(2x - y)),$$
$$\dot{y} = 1 + 0.1(\sin(x - y) + \sin(2x - y)).$$

In the simulation here, $\omega(t/\varepsilon) = 10(z(t/\varepsilon) - 0.5) + \bar{\omega}$, where the stopping times for the jump process z are exponentially distributed and its values are uniformly distributed on $[0, 1]$.

9.2.3 Randomly Perturbed Purely Periodic Systems

We will consider here the systems defined by equations (9.14) for which the averaged system defined by equations (9.15) is purely periodic. Denote by $\Psi_1(x, y)$ and $\Psi_2(x, y)$ the first integrals for system (9.15), the existence of which was proved in Theorem 2. These functions satisfy conditions (1)–(3)

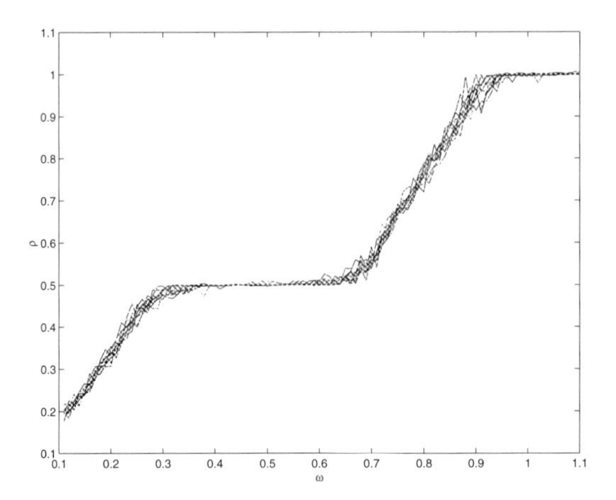

Figure 9.2. Simulation of rotation numbers for a torus flow with noise. It is surprising that even in the presence of such strong noise, the general pattern of phase locking persists.

of Theorem 2 and the following partial differential equations:

$$\bar{a}(x,y)\frac{\partial}{\partial x}\Psi_k(x,y) + \bar{b}(x,y)\frac{\partial}{\partial y}\Psi_k(x,y) = 0, \tag{9.31}$$

for $k = 1, 2$. We assume that $\Psi_1(x,y)$ and $\Psi_2(x,y)$ are constructed in the same way as in Theorem 2: $\Psi_1(x,y) = \cos 2\pi\Psi_0(x,y)$, $\Psi_2(x,y) = \sin 2\pi\Psi_0(x,y)$, and

$$\Psi_0(x,y) = \Phi_x^{-1}(y),$$

where $\Phi_x(y)$ is the solution to the equation

$$\frac{d}{dx}\Phi_x(y) = \bar{c}(x, \Phi_x(y)), \quad \Phi_0(y) = y, \tag{9.32}$$

and $\bar{c}(x,y) = \bar{b}(x,y)/\bar{a}(x,y)$.
Assume that $\theta = k/l$ and $\Phi_l(y) = y+k$ for all y. Denote by O_y, for $y \in [0,1]$, the orbit on the torus $T^{(2)}$ corresponding to the solution of equation (9.32),

$$O_y = \big\{\{x\}, \{\Phi_x(y)\}, \, x \in [0,l]\big\}.$$

Note that O_y is the orbit for the solution of equations (9.15) with initial conditions $\bar{x}(0) = 0$, $\bar{y}(0) = y$. Now we denote by $m_y(dx, dy)$ the ergodic distribution on the orbit O_y: For any continuous function $f(x,y)$ satisfying the conditions $f(x,y) \equiv f(x+1,y) \equiv f(x, y+1)$,

$$\int_{T^{(2)}} f(x', y') m_y(dx', dy') = \lim_{t\to\infty} \frac{1}{t}\int_0^t f\big(\bar{x}(s), \bar{y}(s)\big) ds,$$

where $(\bar{x}(s), \bar{y}(s))$ is the solution to equation (9.15) with the orbit O_y. In the same way as formula (9.12) was obtained, we can prove that

$$\int_{T^{(2)}} f(x', y') m_y(dx', dy') \\ = \int_0^l \frac{f(u, \Phi_u(y))}{\bar{a}((u, \Phi_u(y))} du \cdot \left[\int_0^l [\bar{a}((u, \Phi_u(y))]^{-1} du \right]^{-1}. \tag{9.33}$$

For the stochastic processes

$$\psi_\varepsilon^k(t) = \Psi_k\left(x_\varepsilon\left(\frac{t}{\varepsilon}\right), y_\varepsilon\left(\frac{t}{\varepsilon}\right)\right), \quad k = 1, 2,$$

we can use the theorem on diffusion approximations for first integrals (Chapter 5, Theorem 6). We summarize these results in the following theorem.

Theorem 9 *The stochastic process $(\psi_\varepsilon^1(t), \psi_\varepsilon^2(t))$ converges weakly in C to the diffusion process $(\hat{\psi}^1(t), \hat{\psi}^2(t))$ whose generator $\hat{L}$ is defined as follows: For sufficiently smooth functions $H(\varphi, \psi) : R^2 \to R$, then*

$$\hat{L}H(\varphi, \psi) = \hat{a}_1(\varphi, \psi) \frac{\partial H}{\partial \varphi}(\varphi, \psi) + \hat{a}_2(\varphi, \psi) \frac{\partial H}{\partial \psi}(\varphi, \psi) \\ + \frac{1}{2}\Big(\hat{b}_{11}(\varphi, \psi) \frac{\partial^2 H}{\partial \varphi^2}(\varphi, \psi) + 2\hat{b}_{12}(\varphi, \psi) \frac{\partial^2 H}{\partial \varphi \partial \psi}(\varphi, \psi) \\ + \hat{b}_{22}(\varphi, \psi) \frac{\partial^2 H}{\partial \psi^2}(\varphi, \psi) \Big), \tag{9.34}$$

where the coefficients of the differential operators are defined by the relations

$$\hat{L}H(\varphi, \psi) = \iiiint \Big\{ a(x, y, z') \frac{\partial}{\partial x} \Big[a(x, y, z) \frac{\partial H}{\partial x}(\Psi_1(x, y), \Psi_2(x, y)) \\ + b(x, y, z) \frac{\partial H}{\partial y}(\Psi_1(x, y), \Psi_2(x, y)) \Big] \\ + b(x, y, z') \frac{\partial}{\partial y} \Big[a(x, y, z) \frac{\partial H}{\partial x}(\Psi_1(x, y), \Psi_2(x, y)) \\ + b(x, y, z) \frac{\partial H}{\partial y}(\Psi_1(x, y), \Psi_2(x, y)) \Big] \Big\} R(z, dz') \rho(dz) m_{y(\varphi, \psi)}(dx, dy), \tag{9.35}$$

where $y(\varphi, \psi) \in [0, 1)$ is defined by the equations $\varphi = \cos(2\pi y(\varphi, \psi))$, $\psi = \sin(2\pi y(\varphi, \psi))$.

Remark 6 In particular,

$$\hat{a}_k(\varphi.\psi) = \iiiint \Big\{ a(x,y,z') \frac{\partial}{\partial x} \Big[a(x,y,z)\, dx\, \Psi_k(x,y)$$

$$+ b(x,y,z)\, dy\, \Psi_k(x,y) \Big] + b(x,y,z') \frac{\partial}{\partial y} \Big[a(x,y,z) \frac{\partial}{\partial x} \Psi_k(x,y)$$

$$+ b(x,y,z) \frac{\partial}{\partial y} \Psi_k(x,y) \Big] \Big\} R(z,dz') \rho(dz) m_{y(\varphi,\psi)}(dx,dy),$$

and

$$\begin{aligned} &\hat{b}_{11}(\varphi,\psi)\lambda_1^2 + 2\hat{b}_{12}(\varphi,\psi)\lambda_1\lambda_2 + \hat{b}_{22}(\varphi,\psi)\lambda_2^2 \\ &= \iiiint \Big\{ a(x,y,z') \frac{\partial}{\partial x} \Big[\lambda_1 \Psi_1(x,y) + \lambda_2 \Psi_2(x,y) \Big] \\ &\quad + b(x,y,z') \frac{\partial}{\partial y} \Big[\lambda_1 \Psi_1(x,y) + \lambda_2 \Psi_2(x,y) \Big] \Big\} \\ &\quad \times \Big\{ a(x,y,z) \frac{\partial}{\partial x} \Big[\lambda_1 \Psi_1(x,y) + \lambda_2 \Psi_2(x,y) \Big] \\ &\quad + b(x,y,) \frac{\partial}{\partial y} \Big[\lambda_1 \Psi_1(x,y) + \lambda_2 \Psi_2(x,y) \Big] \Big\} \\ &\quad \times \mathcal{D}R(z,dz') \rho(dz) m_{y(\varphi,\psi)}(dx,dy). \end{aligned}$$

Remark 7 It is easy to see that $[\Psi_1(t)]^2 + [\Psi_2(t)]^2 = 1$. Consider the complex-valued stochastic process

$$\zeta(t) = \Psi_1(t) + i\Psi_2(t),$$

which is a diffusion process on the circle $|z| = 1$ in the complex plane. Set $\theta(t) = \arg \zeta(t)$, so $\zeta(t) = \exp\{i\theta(t)\}$. Assume that the stochastic process $\theta(t)$ is continuous on the real line. Then it is a homogeneous diffusion process, and its diffusion coefficients are periodic with period 2π. Denote the diffusion coefficients for $\theta(t)$ by $\alpha(\theta)$ and $\beta(\theta)$. Then

$$\begin{aligned} \hat{a}_1(\cos\theta, \sin\theta) &= -\sin\theta \cdot \alpha(\theta) - \frac{1}{2}\cos\theta \cdot \beta(\theta), \\ \hat{a}_2(\cos\theta, \sin\theta) &= \cos\theta \cdot \alpha(\theta) - \frac{1}{2}\sin\theta \cdot \beta(\theta), \\ \hat{b}_{11}(\cos\theta, \sin\theta) &= \sin^2\theta \cdot \beta(\theta), \quad \hat{b}_{22}(\cos\theta, \sin\theta) = \cos^2\theta \cdot \beta(\theta), \end{aligned}$$

and

$$\hat{b}_{12}(\cos\theta, \sin\theta) = -\cos\theta \sin\theta \cdot \beta(\theta).$$

In addition,

$$\alpha(\theta) = \hat{a}_2(\cos\theta, \sin\theta)\cos\theta - \hat{a}_1(\cos\theta, \sin\theta)\sin\theta$$

and

$$\beta(\theta) = \hat{b}_{11}(\cos\theta, \sin\theta) + \hat{b}_{22}(\cos\theta, \sin\theta).$$

Denote by $\theta_\varepsilon\bigl(t/\varepsilon\bigr)$ the stochastic process on R for which

$$\Psi_1\left(x_\varepsilon\left(\frac{t}{\varepsilon}\right), y_\varepsilon\left(\frac{t}{\varepsilon}\right)\right) = \cos\theta_\varepsilon\left(\frac{t}{\varepsilon}\right)$$

and

$$\Psi_2\left(x_\varepsilon\left(\frac{t}{\varepsilon}\right), y_\varepsilon\left(\frac{t}{\varepsilon}\right)\right) = \sin\theta_\varepsilon\left(\frac{t}{\varepsilon}\right).$$

That is, $\theta_\varepsilon\,(t/\varepsilon) = 2\pi\Psi_0\,(x_\varepsilon\,(t/\varepsilon)\,, y_\varepsilon\,(t/\varepsilon))$.

Corollary 1 The stochastic process $\theta_\varepsilon\bigl(t/\varepsilon\bigr)$ converges weakly to the process $\theta(t)$ that was constructed in Remark 7.
Set

$$\Theta_\varepsilon(t,x,y) = 2\pi\,\Psi_0(X_\varepsilon(t,x,y), Y_\varepsilon(t,x,y)),$$

where the functions X_ε, Y_ε were introduced in the beginning of Section 2.2.1. Introduce the stopping time

$$T^\varepsilon = \inf\{t : |\Theta_\varepsilon(t,0,y_0) - \Theta_\varepsilon(0,0,y_0)| \ge 2\pi\}.$$

Then for $t < T^\varepsilon$ the following inequality holds:

$$\Phi_{X_\varepsilon(t,0,y_0)}(y_0) - 1 < Y_\varepsilon(t,0,y_0) < \Phi_{X_\varepsilon(t,0,y_0)}(y_0) + 1,$$

and

$$Y_\varepsilon(T^\varepsilon,0,y_0) = \Phi_{X_\varepsilon(T^\varepsilon,0,y_0)}(y_0) + \frac{1}{2\pi}\bigl(\Theta_\varepsilon(T^\varepsilon,0,y_0) - \Theta_\varepsilon(0,0,y_0)\bigr).$$

This means that at the instant T^ε the coordinate y of the perturbed system has one additional rotation with respect to the averaged system in the positive direction if $\lambda = +1$ or in the negative direction if $\lambda = -1$, where

$$\lambda = (2\pi)^{-1}\bigl(\Theta_\varepsilon(T^\varepsilon,0,y_0) - \Theta_\varepsilon(0,0,y_0)\bigr).$$

Therefore, $[(2\pi)^{-1}\bigl(\Theta_\varepsilon(T^\varepsilon,0,y_0) - \Theta_\varepsilon(0,0,y_0)\bigr)]$ is the number of additional rotations of the perturbed system with respect to the averaged system at time t. (Recall that $[x]$ is the integer part of x.)

Corollary 2 The stochastic process

$$\nu_\varepsilon(t) = \left[\frac{\theta_\varepsilon(t) - \theta_\varepsilon(0)}{2\pi}\right]$$

converges weakly as $\varepsilon \to 0$ to the stochastic process

$$\nu(t) = \left[\frac{\theta(t) - \theta(0)}{2\pi}\right],$$

where $\theta(t)$ is the diffusion process on R that was constructed in Remark 7.

We next consider ergodic properties of the process $\zeta(t)$. Suppose that the coefficients $\alpha(\theta)$ and $\beta(\theta)$ for the diffusion process $\theta(t)$ constructed in Remark 7 are periodic smooth functions.

Lemma 5 *Assume that $\beta(\theta) > 0$. Then $\zeta(t)$ is an ergodic process on the circle $\mathbb{C}_1$,and its ergodic distribution is absolutely continuous with respect to Lebesgue measure on $\mathbb{C}_1$ with the density $g(z)$ for which the function $g_1(\theta) = g(e^{i\theta})$ is a 2π-periodic function satisfying the differential equation*

$$(\alpha(\theta)g_1(\theta))' - \frac{1}{2}(\beta(\theta)g_1(\theta))'' = 0. \tag{9.36}$$

Proof It is known that a diffusion process on the real line with continuous bounded coefficients $\alpha(\theta)$ and $\beta(\theta)$ and with $\inf_\theta \beta(\theta) > 0$ has its transition probability function absolutely continuous with respect to Lebesgue measure. Therefore, the transition probability of the process $\zeta(t)$ is also absolutely continuous with respect to Lebesgue measure on $\mathbb{C}_1$. It is easy to see that for any $t > 0$, any interval $\Delta \subset \mathbb{C}_1$, and for $z \in \mathbb{C}_1$ we have

$$P\{\zeta(t) \in \Delta/\zeta(0) = z_1\} > 0.$$

So the Markov process $\zeta(t)$ has no invariant proper subsets. That implies the existence of a unique invariant measure that is absolutely continuous with respect to Lebesgue measure because of the properties of the transition probabilities. So the existence of $g(z)$ is proved.
Let $h(\theta)$ be a 2π-periodic function with continuous derivatives $h'(\theta)$ and $h''(\theta)$. Then

$$\begin{aligned}\lim_{t\to\infty} \frac{1}{t}\int_0^t h(\theta(s))\,ds &= \lim_{t\to\infty} \frac{1}{t}\int_0^t \hat{h}(e^{i\theta(s)})\,ds \\ &= \int_{\mathbb{C}_1} \hat{h}(z)g(z)\,dz = \int_0^{2\pi} h(\theta)g_1(\theta)\,d\theta,\end{aligned}$$

where $\hat{h}(e^{i\theta}) = h(\theta)$, for $\theta \in R$.
Note that the stochastic process $\theta(t)$ can be considered the solution to the stochastic differential equation

$$d\theta(t) = \alpha(\theta(t))\,dt + \sqrt{\beta(\theta(t))}\,dw(t). \tag{9.37}$$

If there is a periodic function $f(\theta)$ for which

$$h(\theta) = \alpha(\theta)f'(\theta) + \frac{1}{2}\beta(\theta)f''(\theta), \tag{9.38}$$

then

$$df(\theta(t)) = \Big(\alpha(\theta(t))f'(\theta(t)) + \frac{1}{2}\beta(\theta(t))f''(\theta(t))\Big)dt + f'(\theta(t))\sqrt{\beta(\theta(t))}\,dw(t)$$

and

$$\frac{1}{t}[f(\theta(t)) - f(\theta(0))] = \frac{1}{t}\int_0^t h(\theta(s))\,ds + \frac{1}{t}\int_0^t f'(\theta(s))\sqrt{\beta(\theta(s))}\,dw(s).$$

Since

$$E\left(\frac{1}{t}\int_0^t f'(\theta(s))\sqrt{\beta(\theta(s))}\,dw(s)\right)^2 = O\left(\frac{1}{t}\right),$$

we have

$$\lim_{t\to\infty}\frac{1}{t}\int_0^t h(\theta(s))\,ds = 0.$$

So it is proved that

$$\int_0^{2\pi} h(\theta)g_1(\theta)\,d\theta = 0 \tag{9.39}$$

for $h(\theta)$ given by formula (9.38) with an arbitrary periodic function $f \in C^{(2)}$. This implies equation (9.36), and the proof of Lemma 5 is complete. □

Remark 8 We can construct a positive periodic solution $g_1(\theta)$ to equation (9.36) in the following way. Set

$$\gamma = \int_0^{2\pi} 2\alpha(\theta)\beta^{-1}(\theta)\,d\theta.$$

Then

$$\int_0^u 2\alpha(\theta)\beta^{-1}(\theta)\,d\theta = \gamma u + z(u),$$

where $z(u)$ is a 2π-periodic function.
Equation (9.36) can be written in the form

$$\big(g_1(u)\beta(u)\exp\{-\gamma u - z(u)\}\big)' = c\exp\{-\gamma - z(u)\}, \tag{9.40}$$

where c is a constant. Then the function

$$Z_1(u) = e^{-z(u)}$$

is a 2π-periodic function, and

$$\int_0^u e^{-\gamma\theta}Z_1(\theta)d\theta = e^{-\gamma u}Z_2(u) - Z_2(0) \tag{9.41}$$

is also 2π-periodic. Relations (9.40) and (9.41) imply that

$$g_1(\theta)\beta(\theta)Z_1(\theta)e^{-\gamma\theta} = c_1 + ce^{-\gamma\theta}Z_2(\theta),$$

where c_1 is another constant. If $\gamma = 0$, we set $c = 0$,

$$g_1(\theta) = \frac{c_1}{\beta(\theta)Z_1(\theta)}, \quad c_1 > 0,$$

$$g_1(\theta) = \frac{cZ_2(\theta)}{\beta(\theta)Z_1(\theta)}, \quad c > 0.$$

Theorem 10 *Let $h(\theta)$ be a 2π-periodic continuous function. Then*

$$\lim_{\varepsilon\to 0} \limsup_{t\to\infty} \left| \frac{1}{t} \int_0^t h\big(\theta_\varepsilon\big(\tfrac{s}{\varepsilon}\big)\big)\, ds - \int_0^{2\pi} h(\theta) g_1(\theta)\, d\theta \right| = 0$$

with probability 1.

Proof It follows from Theorem 9 that for any $\delta > 0$, $t_0 > 0$ we can find $\varepsilon_0 > 0$ such that for $\varepsilon < \varepsilon_0$,

$$E \left| \frac{1}{t_0} \int_0^{t_0} h\big(\theta_\varepsilon\big(\tfrac{s}{\varepsilon}\big)\big)\, ds - \int_0^{2\pi} h(\theta) g_1(\theta) d\theta \right| < \delta. \tag{9.42}$$

Then

$$\begin{aligned}
\frac{1}{nt_0} \int_0^{nt_0} h\left(\theta_\varepsilon\left(\frac{s}{\varepsilon}\right)\right) ds &= \frac{1}{nt_0} \sum_{k=0}^{n-1} \int_{kt_0}^{(k+1)t_0} h\left(\theta_\varepsilon\left(\frac{s}{\varepsilon}\right)\right) ds \\
&= \frac{1}{nt_0} \sum_{k=0}^{n-1} \left(\int h\left(\theta_\varepsilon\left(\frac{s}{\varepsilon}\right)\right) ds - E\left(\int h\left(\theta_\varepsilon\left(\frac{s}{\varepsilon}\right)\right) ds \Big/ \mathcal{F}^\varepsilon_{\frac{kt_0}{\varepsilon}} \right) \right) \\
&+ \frac{1}{nt_0} \sum_{k=0}^{n-1} E\left(\int h\left(\theta_\varepsilon\left(\frac{s}{\varepsilon}\right)\right) ds / \mathcal{F}^\varepsilon_{\frac{kt_0}{\varepsilon}} \right).
\end{aligned}$$

where the integrals are over $[kt_0, (k+1)t_0]$. It follows from Lemma 3 that

$$\limsup_{n\to\infty} \left| \frac{1}{nt_0} \sum_{k=0}^{n-1} \left(\int h\big(\theta_\varepsilon\big(\tfrac{s}{\varepsilon}\big)\big)\, ds - E\left(\int h\big(\theta_\varepsilon\big(\tfrac{s}{\varepsilon}\big)\big)\, ds \Big/ \mathcal{F}^\varepsilon_{\frac{kt_0}{\varepsilon}} \right) \right) \right|$$

is zero with probability 1. And (9.42) implies the inequality

$$\left| \frac{1}{nt_0} \sum_{k=0}^{n-1} \left(\int_{kt_0}^{(k+1)t_0} h\big(\theta_\varepsilon\big(\tfrac{s}{\varepsilon}\big)\big)\, ds \Big/ \mathcal{F}^\varepsilon_{\frac{kt_0}{\varepsilon}} \right) - \int_0^{2\pi} \alpha(\theta) g_1(\theta)\, d\theta \right| < \delta.$$

This completes the proof of the theorem. □

Remark 9

$$P\left\{ \lim_{t\to\infty} \frac{\theta(t)}{t} = \int_0^{2\pi} \alpha(u) g_1(u)\, du \right\} = 1.$$

This is a consequence of the relation

$$\frac{\theta(t)}{t} = \frac{\theta(0)}{t} + \frac{1}{t} \int_0^t \alpha(\theta(s))\, ds + \frac{1}{t} \int_0^t \sqrt{\beta(\theta(s))}\, dw(s),$$

which follows from equation (9.37).

Corollary 3

$$\limsup_{t\to\infty}\left|\frac{\nu_\varepsilon(t)}{t} - \varepsilon\int_0^{2\pi}\alpha(u)g_1(u)\,du\right| = o(\varepsilon).$$

This relation is a consequence of the relation

$$\lim_{\varepsilon\to 0}\limsup_{t\to\infty}\left|\frac{\nu_\varepsilon(t)}{t} - \varepsilon\int_0^{2\pi}\alpha(u)g_1(u)\,du\right| = 0.$$

Corollary 4

$$y_\varepsilon(t) = \theta\,x_\varepsilon(t) + \varepsilon t\int_0^{2\pi}\alpha(u)g_1(u)\,du + t\,o(\varepsilon).$$

9.2.4 Ergodic Systems

Assume that the averaged system (9.15) is ergodic; equivalently, the rotation number θ for it is an irrational number. In addition, we assume that the function $\bar{c}(x,y) = \bar{b}(x,y)/\bar{a}(x,y)$ has a continuous derivative $\bar{c}_y(x,y)$ that is of bounded variation.

We will write $\Phi_x(y)$ for the solution to the equation

$$\frac{d\Phi_x}{dx}(y) = \bar{c}(x,\Phi_x(y))$$

that satisfies the initial condition $\Phi_0(y) = y$. As before, we write $\Psi_0(x,y) = \Phi_x^{-1}(y)$, where $\Phi_x^{-1}(y)$ is the inverse function for y.

Introduce the sequence of stopping times $\{\tau_k^\varepsilon,\ k = 0,1,\dots\}$, where

$$\tau_k^\varepsilon = \inf\{t : x_\varepsilon(t) = x_0 + k\},\quad x_0 = x_\varepsilon(0).$$

Denote by $\hat{X}_\varepsilon(t,x_0,y_0,s)$, $\hat{Y}_\varepsilon(t,x_0,y_0,s)$ the solution to system (9.14) on the interval $[s,\infty)$ satisfying the initial conditions

$$\hat{X}_\varepsilon(s,x_0,y_0,s) = x_0,\quad \hat{Y}_\varepsilon(s,x_0,y_0,s) = y_0.$$

We consider the sequence of random mappings

$$S_k(\varepsilon,y) = \hat{Y}_\varepsilon(\tau_{k+1}^\varepsilon, x_0 + k, y, \tau_k^\varepsilon).$$

Then $y_\varepsilon(\tau_{k+1}^\varepsilon) = S_k(\varepsilon, y_\varepsilon(\tau_k^\varepsilon))$, and we see that by successive back-substitutions we get

$$y_\varepsilon(\tau_n^\varepsilon) = S_{n-1}\big(\varepsilon, S_{n-2}(\varepsilon,\dots,S_0(\varepsilon,y_0)\dots\big). \tag{9.43}$$

We will investigate the asymptotic behavior of the ratio

$$y_\varepsilon(\tau_n^\varepsilon)/n$$

as $n\to\infty$ and $\varepsilon\to 0$. It follows from Theorem 7 that

$$\lim_{\varepsilon\to 0}\limsup_{n\to\infty}|y_\varepsilon(\tau_n^\varepsilon)/n - \theta| = 0,$$

but we can find a more precise asymptotic representation for the ratio that is based on a diffusion approximation.

Introduce the sequence of random variables

$$u_n^\varepsilon = \Psi_0(n + x_0, y_\varepsilon(\tau_n^\varepsilon)), \tag{9.44}$$

which is equivalent to

$$y_\varepsilon(\tau_n^\varepsilon) = \Phi_{x_0+n}(u_n^\varepsilon). \tag{9.45}$$

Using Theorem 6 we can write $y_\varepsilon(\tau_n^\varepsilon)$ as

$$y_\varepsilon(\tau_n^\varepsilon) = g(u_n^\varepsilon) + \theta(x_0 + n) + \chi(x_0 + n, g(u_n^\varepsilon) + \theta(x_0 + n)). \tag{9.46}$$

Since $\chi(x, y) = O(1)$, we have $g(y) = y + O(1)$, and we can write

$$\frac{1}{n} y_\varepsilon(\tau_n^\varepsilon) = \frac{1}{n} u_n^\varepsilon + \theta + O\left(\frac{1}{n}\right). \tag{9.47}$$

We will investigate the asymptotic behavior of $\frac{1}{n} u_n^\varepsilon$ as $n \to \infty$ and $\varepsilon \to 0$. Set

$$u_\varepsilon(t) = \Psi_0(x_\varepsilon(t), y_\varepsilon(t)). \tag{9.48}$$

Then

$$u_\varepsilon(\tau_n^\varepsilon) = u_n^\varepsilon.$$

The following theorem describes the asymptotic behavior of the stochastic process $u_\varepsilon(t)$ as $t \to \infty$.

Theorem 11 *The stochastic process $u_\varepsilon(t/\varepsilon)$ converges weakly as $\varepsilon \to 0$ to a diffusion Markov process on R having generator $\hat{L}$ defined for $F \in C^{(2)}(R)$ by*

$$\hat{L}F(u) = AF'(u) + \frac{1}{2} BF''(u). \tag{9.49}$$

The constants A and B are defined by

$$A = \iiiint L(x, y, z)[L(x, y, z')\Psi_0(x, y)]R(z, dz')\,\rho(dz)\,m(dx, dy), \tag{9.50}$$

and $B/2 =$

$$\iiiint [L(x, y, z)\Psi_0(x, y)][L(x, y, z')\Psi_0(x, y)]R(z, dz')\,\rho(dz)\,m(dx, dy), \tag{9.51}$$

where $L(x, y, z)$ is the differential operator for which

$$L(x, y, z)G(x, y) = a(x, y, z)G_x(x, y) + b(x, y, z)G_y(x, y).$$

Here $m(dx, dy)$ is the ergodic measure for system (9.15) that is determined in Theorem 5, namely,

$$\iint f(x, y)m(dx, dy) = \lim_{t\to\infty} \frac{1}{t} \int_0^t f(\bar{x}(s), \bar{y}(s))\,ds$$

for any continuous function for which $f(x, y) \equiv f(x+1, y) \equiv f(x, y+1)$.

The proof of the theorem is based on several auxiliary lemmas.

Lemma 6 *Set*

$$G(x, y, z) = L(x, y, z)\Psi_0(x, y). \tag{9.52}$$

Then

$$\dot{u}_\varepsilon(t) = G\left(x_\varepsilon(t), y_\varepsilon(t), z\left(\frac{t}{\varepsilon}\right)\right) \tag{9.53}$$

and

$$\int G(x, y, z)\rho(dz) = 0. \tag{9.54}$$

Proof Equation (9.53) is a consequence of (9.48) and (9.52). In addition,

$$\int G(x, y, z)\rho(dz) = \bar{a}(x, y)\frac{\partial \Psi_0}{\partial x}(x, y) + \bar{b}(x, y)\frac{\partial \Psi_0}{\partial y}(x, y). \tag{9.55}$$

The relation

$$y = \Psi_0(x, \Phi_x(y))$$

implies that

$$\begin{aligned} 0 &= \frac{\partial \Psi_0}{\partial x}(x, \Phi_x(y)) + \frac{\partial \Psi_0}{\partial y}(x, \Phi_x(y))\frac{\partial \Phi_0}{\partial x}(y) \\ &= \frac{\partial \Psi_0}{\partial x}(x, \Phi_x(y)) + \frac{\partial \Psi_0}{\partial y}(x, \Phi_x(y))\frac{\bar{b}(x, \Phi_x(y))}{\bar{a}(x, \Phi_x(y))}, \end{aligned}$$

so

$$0 = \bar{a}(x, \Phi_x(y))\frac{\partial \Psi_0}{\partial x}(x, \Phi_x(y)) + \bar{b}(x, \Phi_x(y))\Psi_0(x, \Phi_x(y)).$$

Since for any x and y' we can find y for which $\Phi_x(y) = y'$, (9.54) is proved.

□

Lemma 7 *Let a function* $\Phi : R^2 \times Z \to R$ *satisfy the following conditions:*

(i) There exist measurable and bounded derivatives $\Phi_x(x, y, z)$, $\Phi_y(x, y, z)$.

(ii) $\Phi(x, y, z) \equiv \Phi(x+1, y, z) \equiv \Phi(x, y+1, z)$.

(iii) $\int \Phi(x, y, z)\rho(dz) = 0$.

Then for $0 \le t_1 < t_2$ *we have*

$$E\Big(\int_{t_1}^{t_2} \Phi\big(x_\varepsilon(s), y_\varepsilon(s), z\big(\frac{s}{\varepsilon}\big)\big)\, ds \Big/ \mathcal{F}_{t_1}^\varepsilon\Big)$$
$$= \varepsilon E\Big(\int_{t_1}^{t_2} \iint L(x_\varepsilon(s), y_\varepsilon(s), z)\Phi(x_\varepsilon(s), y_\varepsilon(s), z')\rho(dz)R(z, dz')\, ds \Big/ \mathcal{F}_{t_1}^\varepsilon\Big)$$
$$+ O(\varepsilon), \tag{9.56}$$

where the expression $L(\cdot)\Phi(\cdot)$ *means*

$$L(x, y, z)\Phi(x, y, z')\Big|_{x=x_\varepsilon(s),\, y=y_\varepsilon(s)}.$$

The proof follows from formula (5.7).

Lemma 8 *Let* $F(u) \in C^{(2)}(R)$. *Then for* $0 < t_1 < t_2$,

$$E\big(F(u_\varepsilon(t_2)) - F(u_\varepsilon(t_1))/\mathcal{F}_{t_1}^\varepsilon\big) = O(\varepsilon)$$
$$+ \varepsilon E\Big(\int_{t_1}^{t_2} \iint L(x_\varepsilon(s), y_\varepsilon(s), z)\big[F'(\Psi_0(x_\varepsilon(s), y_\varepsilon(s))) \tag{9.57}$$
$$\times L(x_\varepsilon(s), y_\varepsilon(s), z')\Psi_0(x_\varepsilon(s), y_\varepsilon(s))\big]\big)R(z, dz')\rho(dz)\, ds \Big/ \mathcal{F}_{t_1}^\varepsilon\Big).$$

Proof This result follows from the relation

$$F(u_\varepsilon(t_2)) - F(u_\varepsilon(t_1))$$
$$= \int_{t_1}^{t_2} F'(u_\varepsilon(s))L\left(x_\varepsilon(s), y_\varepsilon(s), z\left(\frac{s}{\varepsilon}\right)\right)\Psi_0(x_\varepsilon(s), y_\varepsilon(s))\, ds$$

and Lemma 6.

□

Proof of Theorem 11

We will use relation (9.57) with $t_1 = (t + kh)/\varepsilon$ and $t_2 = (t + (k+1)h)/\varepsilon$. Denote by $\hat{F}(x_\varepsilon(s), y_\varepsilon(s))$ the expression under the integral $\int_{t_1}^{t_2}$ in relation (9.57). Then

$$E\left(F\left(u_\varepsilon\left(\frac{t_0 + (k+1)h}{\varepsilon}\right)\right) - F\left(u_\varepsilon\left(\frac{t_0 + kh}{\varepsilon}\right)\right) \Big/ \mathcal{F}_{\frac{t_0+kh}{\varepsilon}}^\varepsilon\right)$$
$$= O(\varepsilon) + \varepsilon E\left(\int_{(t_0+kh)\varepsilon^{-1}}^{(t_0+(k+1)h)\varepsilon^{-1}} \hat{F}(\tilde{x}_\varepsilon(s), \tilde{y}_\varepsilon(s))\, ds \Big/ \mathcal{F}_{\frac{t_0+kh}{\varepsilon}}^\varepsilon\right), \tag{9.58}$$

where $\tilde{x}_\varepsilon(s)$, $\tilde{y}_\varepsilon(s)$ is the solution of system (9.15) on the interval

$$[(t_0 + kh)\varepsilon^{-1}, (t_0 + (k+1)h)\varepsilon^{-1}]$$

with initial values

$$\tilde{x}_\varepsilon\left(\frac{t_0+kh}{\varepsilon}\right)=x_\varepsilon\left(\frac{t_0+kh}{\varepsilon}\right), \quad \tilde{y}_\varepsilon\left(\frac{t_0+kh}{\varepsilon}\right)=y_\varepsilon\left(\frac{t_0+kh}{\varepsilon}\right).$$

It is easy to see that

$$E\left(\int_{(t_0+kh)\varepsilon^{-1}}^{(t_0+(k+1)h)\varepsilon^{-1}} \hat{F}(\tilde{x}_\varepsilon(s),\tilde{y}_\varepsilon(s))\,ds \Big/ \mathcal{F}^\varepsilon_{\frac{t_0+kh}{\varepsilon}}\right)$$
$$=\hat{\Phi}\left(\frac{h}{\varepsilon},\tilde{x}_\varepsilon\left(\frac{t_0+kh}{\varepsilon}\right),\tilde{y}_\varepsilon\left(\frac{t_0+kh}{\varepsilon}\right),z\left(\frac{t_0+kh}{\varepsilon^2}\right)\right),$$

where the function $\hat{\Phi}$ does not depend on t_0 or k, since the Markov process $\left(x_\varepsilon(t),y_\varepsilon(t),z\left(\frac{t}{\varepsilon}\right)\right)$ is homogeneous. So

$$\hat{\Phi}\left(\frac{h}{\varepsilon},x_0,y_0,z_0\right)$$
$$=E\left(\int_0^{h/\varepsilon}\hat{F}(x_\varepsilon(s),y_\varepsilon(s))\,ds \Big/ x_\varepsilon(0)=x_0,\ y_\varepsilon(0)=y_0,\ z_\varepsilon(0)=z_0\right).$$

Let $h=h_\varepsilon$, where $h_\varepsilon\to 0$, $h_\varepsilon/\varepsilon\to\infty$, and

$$P\left\{\lim_{\varepsilon\to 0}\sup_{s\le h_\varepsilon/\varepsilon}\left(|x_\varepsilon(s)-\bar{x}(s)|+|y_\varepsilon(s)-\bar{y}(s)|\right)=0\right\}=1.$$

We can find such a function h_ε satisfying these conditions because of statement I.
Therefore,

$$\hat{F}(\bar{x}(s),\bar{y}(s))=F'\left(\Psi_0(\bar{x}(s),\bar{y}(s))\right)A(\bar{x}(s),\bar{y}(s))$$
$$+\ \frac{1}{2}F''\left(\Psi_0(\bar{x}(s),\bar{y}(s))\right)B(\bar{x}(s),\bar{y}(s)),$$

where

$$A(x,y)=\iint L(x,y,z)[L(x,y,z')\Psi_0(x,y)]\rho(dz)R(z,dz')$$

and

$$B(x,y)=\iint[L(x,y,z)\Psi_0(x,y)][L(x,y,z')\Psi_0(x,y)]\rho(dz)R(z,dz').$$

The function $\Psi_0(\bar{x}(s),\bar{y}(s))$ is a constant in s, since

$$\frac{d\Psi_0}{ds}(\bar{x}(s),\bar{y}(s))=\bar{a}(\bar{x}(s),\bar{y}(s))\frac{\partial\Psi_0}{\partial x}(\bar{x}(s),\bar{y}(s))$$
$$+\bar{b}(\bar{x}(s),\bar{y}(s))\frac{\partial\Psi_0}{\partial y}(\bar{x}(s),\bar{y}(s)),$$

and relations (9.54) and (9.55) hold. So

$$\hat{F}(\bar{x}(s),\bar{y}(s)) = F'\big(\Psi_0(x(0),y(0))\big)A(\bar{x}(s),\bar{y}(s)) + \frac{1}{2}F''\big(\Psi_0(x(0),y(0))\big)B(\bar{x}(s),\bar{y}(s)).$$

Using the fact that $h_\varepsilon/\varepsilon \to \infty$, we can write

$$\varepsilon\int_0^{h_\varepsilon/\varepsilon} A(\bar{x}(s),\bar{y}(s))\,ds = h_\varepsilon\left(\frac{\varepsilon}{h_\varepsilon}\int_0^{h_\varepsilon/\varepsilon} A(\bar{x}(s),\bar{y}(s))\,ds\right)$$

$$= h_\varepsilon\left(\iint A(x,y)m(dx,dy) + o(1)\right) = h_\varepsilon A + o(h_\varepsilon)$$

and

$$\varepsilon\int_0^{h_\varepsilon/\varepsilon} B(\bar{x}(s),\bar{y}(s))\,ds = h_\varepsilon A + o(h_\varepsilon).$$

So

$$\varepsilon\hat{\Phi}\left(\frac{h_\varepsilon}{\varepsilon},x_0,y_0,z_0\right) = h_\varepsilon\left[AF'(\Psi_0(x_0,y_0)) + \frac{1}{2}BF''(\Psi_0(x_0,y_0))\right] + o(h_\varepsilon).$$

Note that $\varepsilon = o(h_\varepsilon)$ and $\Psi_0(x_0,y_0) = u_\varepsilon(0)$.
Therefore, we can rewrite relation (9.58) in the form

$$E\big(F\big(u_\varepsilon((t_0+(k+1)h_\varepsilon)\varepsilon^{-1})\big) - F\big(u_\varepsilon((t_0+kh_\varepsilon)\varepsilon^{-1})\big)/\mathcal{F}^\varepsilon_{(t_0+kh_\varepsilon)\varepsilon^{-1}}\big)$$

$$-h_\varepsilon\left[AF'\big(u_\varepsilon(\frac{t_0+kh_\varepsilon}{\varepsilon})\big) + \frac{1}{2}BF''\big(u_\varepsilon(\frac{t_0+kh_\varepsilon}{\varepsilon})\big)\right] = o(h_\varepsilon). \tag{9.59}$$

Let $n_\varepsilon h_\varepsilon \to t_2 - t_1$. Then taking the sum in k from 0 to $n_\varepsilon - 1$ on the left–hand side of relation (9.59), we obtain the following relation:

$$\lim_{\varepsilon\to 0} E\left|E\left(F\left(u_\varepsilon\left(\frac{t_2}{\varepsilon}\right)\right) - F\left(u_\varepsilon\left(\frac{t_1}{\varepsilon}\right)\right) - \int_{t_1}^{t_2}\hat{L}F\left(u_\varepsilon\left(\frac{s}{\varepsilon}\right)\right)ds/\mathcal{F}^\varepsilon_{t_1/\varepsilon}\right)\right| = 0. \tag{9.60}$$

Now the theorem follows from Theorem 2 of Chapter 2.

□

Remark 10 Let $\hat{u}(t)$ be the diffusion process having the generator $\hat{L}$ given by formula (9.49). Then

$$\hat{u}(t) = \hat{u}(0) + tA + \sqrt{B}\,w(t),$$

where $w(t)$ is a Wiener process. It follows that

$$P\left\{\lim_{t\to\infty}\frac{\hat{u}(t)}{t} = A\right\} = 1.$$

Remark 11

$$P\left\{\lim_{\varepsilon\to 0}\limsup_{n\to\infty}\frac{1}{\varepsilon}|u_n^\varepsilon - \varepsilon A| = 0\right\} = 1.$$

Corollary 5

$$y_\varepsilon(t) = \theta x_\varepsilon(t) + \varepsilon t A + o(\varepsilon)t.$$

9.2.5 *Periodic Systems*

We consider now a randomly perturbed system of type (9.14) for which the averaged system defined by equations (9.15) is a periodic one. Let $\theta = k/l$ be the rotation number for the averaged system, where $k, l \in \mathbf{Z}$ and the largest common divisor of k and l equals 1. Then the set Π of those $y \in [0,1]$ for which

$$\Phi_l(y) = y + k$$

is not empty. Note that Π is the set of those y for which the solution of system (9.15) with the initial data $x(0) = 0$, $y(0) = y$ is a periodic function on the torus $T^{(2)}$. Without loss of generality we can assume that $0 \in \Pi$. Define

$$\varphi(y) = \Phi_l(y) - k, \quad \varphi_0(y) = \varphi(y) - y.$$

In Section 9.2.3 we considered the case $\Pi = [0,1]$, then $\varphi(y) = y$, $\varphi_0(y) = 0$. Now we will consider the case where $[0,1]\setminus\Pi$ is an open nonempty set. Note that we can consider functions $\varphi(y)$, $\varphi_0(y)$ for all $y \in R$, where $\varphi(y)$ satisfies the circle mapping condition $\varphi(y+1) \equiv \varphi(y) + 1$, $\varphi_0(y)$ is a 1–periodic function, $\varphi(y)$ is an increasing function, $\varphi(0) = 0$, and $\varphi(1) = 1$.
We will investigate the number of additional y-rotations, or cycle slips, for a perturbed system that are caused by the random perturbation. Introduce the sequence of stopping times

$$\hat{\tau}_n^\varepsilon = \inf\{t : x_\varepsilon(t) = nl\}, \quad n = 0, 1, 2, \ldots.$$

The number $[y_\varepsilon(\hat{\tau}_n^\varepsilon)]$ is the number of y-rotations during the time that the system has nl x-rotations. The averaged system has for this time kn y-rotations. So the number of additional y-rotations is

$$[y_\varepsilon(\hat{\tau}_n^\varepsilon)] - kn = [y_\varepsilon(\hat{\tau}_n^\varepsilon) - kn].$$

Therefore, determining the asymptotic behavior of the number of additional y-rotations is equivalent to determining $y_\varepsilon(\hat{\tau}_n^\varepsilon) - kn$ as $n \to \infty$ and $\varepsilon \to 0$.

Lemma 9 *There exist continuous 1-periodic functions $\gamma(y)$ and $\sigma(y) > 0$ for which*

$$\begin{aligned} y_\varepsilon(\hat\tau^\varepsilon_{n+1}) = \varphi(y_\varepsilon(\hat\tau^\varepsilon_n)) + k + \varepsilon\gamma(y_\varepsilon(\hat\tau^\varepsilon_n)) + \sqrt{\varepsilon}\sigma(y_\varepsilon(\hat\tau^\varepsilon_n))\eta^\varepsilon_{n+1} \\ + \varepsilon\delta\big(y_\varepsilon(\hat\tau^\varepsilon_n), z\big(\frac{1}{\varepsilon}\hat\tau^\varepsilon_n\big)\big) + o(\varepsilon), \end{aligned} \tag{9.61}$$

where η^ε_n satisfies the conditions

$$E(\eta^\varepsilon_n/\mathcal{F}^\varepsilon_{\hat\tau^\varepsilon_n}) = 0, \quad E\big((\eta^\varepsilon_n)^2/\mathcal{F}^\varepsilon_{\hat\tau^\varepsilon_n}\big) = 1.$$

Moreover, the conditional distribution of η^ε_{n+1} with respect to the σ-algebra $\mathcal{F}^\varepsilon_{\hat\tau^\varepsilon_n}$ depends only on $y_\varepsilon(\hat\tau^\varepsilon_n)$ and $z\big(\frac{1}{\varepsilon}\hat\tau^\varepsilon_n\big)$, and it converges to the standard Gaussian distribution uniformly in these random variables. In this formula δ is given by

$$\delta(y,z) = \int b(0,y,z')R(z,dz'). \tag{9.62}$$

Proof Note that the sequence $\big\{\big(y_\varepsilon(\hat\tau^\varepsilon_n), z\big(\frac{1}{\varepsilon}\hat\tau^\varepsilon_n\big)\big),\ n = 0,1,\dots\big\}$ is a homogeneous Markov chain, so we have to consider only the case $n = 0$. It follows from statement II that $\hat\tau^\varepsilon_1 \to t_1(y)$ in probability, where

$$t_1(y) = \inf\{s : X(s,0,y) = l\}.$$

So we consider

$$Y_\varepsilon(t_1(y),0,y) - \Phi_l(y) = Y_\varepsilon(t_1(y),0,y) - Y(t_1(y),0,y).$$

The remainder of the proof from this point follows from statement II and Theorem 4 of Chapter 2.

□

Remark 12 It follows from formula (9.62) that

$$\int \delta(y,z)\rho(dz) = 0,$$

so the uniform ergodicity of $z(t)$ implies the relation

$$E\left(\delta\left(y_\varepsilon(\hat\tau^\varepsilon_n), z\left(\frac{1}{\varepsilon}\hat\tau^\varepsilon_n\right)\right)/\mathcal{F}^\varepsilon_{\hat\tau^\varepsilon_n}\right) = o(1)$$

as $\varepsilon \to 0$.

Remark 13 For any n, the joint distribution of the random variables $\eta^\varepsilon_1,\dots,\eta^\varepsilon_n$ converges to the joint distribution of the random variables $\eta_1,\dots,\eta_n$, which are independent standard Gaussian random variables.

Lemma 10 *Introduce the Markov chain $\{\hat u_\varepsilon(n),\ n = 0,1,2,\dots\}$ for which*

$$\hat u_\varepsilon(n+1) = \varphi(\hat u_\varepsilon(n)) + \sqrt{\varepsilon}\sigma(\hat u_\varepsilon(n))\eta_{n+1}, \tag{9.63}$$

where $\{\eta_1, \ldots, \eta_n\}$ *is a sequence of independent standard Gaussian random variables. Let*

$$u_\varepsilon(n) = y_\varepsilon(\hat{\tau}_n^\varepsilon) - kn.$$

Assume that $u_\varepsilon(0) = \hat{u}_\varepsilon(0) = y_0 \in (\alpha, \beta)$, *where* $(\alpha, \beta) \subset [0,1] \setminus \Pi$, *but* $\alpha, \beta \in \Pi$. *Set*

$$T_\varepsilon = \inf[k : \ u_\varepsilon(k) \notin (\alpha, \beta)],$$

$$\hat{T}_\varepsilon = \inf[k : \ \hat{u}_\varepsilon(k) \notin (\alpha, \beta)].$$

Then

$$\lim_{\varepsilon \to 0} \left(E 1_{\{u_\varepsilon(T_\varepsilon) \geq \beta\}} 1_{\{T_\varepsilon < \lambda_\varepsilon\}} - E 1_{\{\hat{u}_\varepsilon(\hat{T}_\varepsilon) \geq \beta\}} 1_{\{\hat{T}_\varepsilon < \lambda_\varepsilon\}} \right) = 0,$$

$$\lim_{\varepsilon \to 0} \left(E 1_{\{u_\varepsilon(T_\varepsilon) \leq \alpha\}} 1_{\{T_\varepsilon < \lambda_\varepsilon\}} - E 1_{\{\hat{u}_\varepsilon(\hat{T}_\varepsilon) \leq \alpha\}} 1_{\{\hat{T}_\varepsilon < \lambda_\varepsilon\}} \right) = 0,$$

for any $\lambda_\varepsilon \to \infty$ *as* $\varepsilon \to 0$ *and*

$$\lim_{\varepsilon \to 0} E(T_\varepsilon / E\hat{T}_\varepsilon) = 1.$$

The proof of this statement can be obtained from Lemma 9 and Remark 13.

Next, we investigate the asymptotic behavior of $\hat{T}_\varepsilon$ and $\hat{u}_\varepsilon(T_\varepsilon)$ as $\varepsilon \to 0$, $y_0 \to \alpha$ and as $y_0 \to \beta$ under the condition of Lemma 9.

Lemma 11 *Assume that* α, y_0, β *are the same as in Lemma 9 and* $\varphi_0(y) > 0$ *for* $y \in (\alpha, \beta)$,

$$\varphi_0(y) = \delta(y - \alpha)^\gamma + o((y - \alpha)^\gamma) \quad \text{if } y \to \alpha,$$

and

$$\varphi_0(y) = \delta(\beta - y)^\gamma + o((\beta - y)^\gamma) \quad \text{if } y \to \beta,$$

where $\gamma > 1$.
Define stochastic processes

$$y_\varepsilon^-(t) = \sum_n (\hat{u}_\varepsilon(n) - \alpha) \varepsilon^{-\frac{1}{1+\gamma}} 1_{\{n\varepsilon^\kappa \leq t < (n+1)\varepsilon^\kappa\}}, \tag{9.64}$$

$$y_\varepsilon^+(t) = \sum_n (\beta - \hat{u}_\varepsilon(n)) \varepsilon^{-\frac{1}{1+\gamma}} 1_{\{n\varepsilon^\kappa \leq t < (n+1)\varepsilon^\kappa\}}, \tag{9.65}$$

where $\kappa = (\gamma - 1)/(\gamma + 1)$.
Let $y^-(t)$ *and* $y^+(t)$ *be the solutions to the stochastic differential equations*

$$dy^-(t) = \delta |y^-(t)|^\gamma dt + \sigma(\alpha)\, dw(t), \tag{9.66}$$

$$dy^+(t) = -\delta |y^+(t)|^\gamma dt + \sigma(\beta)\, dw(t), \tag{9.67}$$

where $w(t)$ is the standard Wiener process.
For $h > 0$ define the stopping times

$$\tau_\varepsilon^-(h) = \inf\{t : y_\varepsilon^-(t) \notin (0, h)\},$$
$$\tau_\varepsilon^+(h) = \inf\{t : y_\varepsilon^+(t) \notin (0, h)\},$$
$$\tau^-(h) = \inf\{t : y^-(t) \notin (0, h)\},$$
$$\tau^+(h) = \inf\{t : y^+(t) \notin (0, h)\}.$$

Then the distribution of $\left(\tau_\varepsilon^\pm(h), y_\varepsilon^\pm(\tau_\varepsilon^\pm(h))\right)$ converges weakly to the distribution of $\left(\tau^\pm(h), y^\pm(\tau^\pm(h))\right)$.

Proof It follows from relation (9.64) that $y_\varepsilon^-(t)$ satisfies the difference equation

$$y_\varepsilon^-(t + \varepsilon^\kappa) - y_\varepsilon^-(t) = \delta\big(y_\varepsilon^-(t)\big)^{\gamma+1}\varepsilon^\kappa + \sigma(\alpha)[w_\varepsilon(t + \varepsilon^\kappa) - w_\varepsilon(t)] + o(\varepsilon),$$

where

$$w_\varepsilon(t) = \sum_{n\varepsilon^\kappa \le t} \varepsilon^{\kappa/2}\eta_n. \tag{9.68}$$

The solution of this difference equation on the interval $[0, \tau_\varepsilon^-(h))$ converges in distribution to the solution of equation (9.66) on the interval $[0, \tau^-(h)]$ because of the weak convergence of the stochastic process $w_\varepsilon(t)$ to $w(t)$. The proof for the other case is similar.

□

Remark 14 Set

$$U^-(x) = \int_0^x \exp\left\{-\frac{\delta v^{\gamma+1}}{2(\gamma+1)\sigma^2(\alpha)}\right\} dv$$

and

$$V^-(x) = \int_0^x 2\exp\left\{-\frac{\delta v^{\gamma+1}}{2(\gamma+1)\sigma^2(\alpha)}\right\} \int_0^v \exp\left\{-\frac{\delta u^{\gamma+1}}{2(\gamma+1)\sigma^2(\alpha)}\right\} \frac{du\, dv}{\sigma^2(\alpha)}.$$

Then for $0 < y < h$,

$$P\{y^-(\tau^-(h)) = h/y^-(0) = y\} = \frac{U^-(y)}{U^-(h)},$$

$$E(\tau^-(h)/y^-(0) = y) = \frac{V^-(y)}{U^-(h)}U^-(y) - V^-(y).$$

Analogously, for $0 < y < h$,

$$P\{y^+(\tau^+(h)) = h/y^+(0) = y\} = \frac{U^+(y)}{U^+(h)}$$

and

$$E(\tau^+(h)/y^+(0) = y) = \frac{V^+(y)}{U^+(h)}U^+(y) - V^+(y),$$

where

$$U^+(x) = \int_0^x \exp\left\{\frac{\delta v^{\gamma+1}}{2(\gamma+1)\sigma^2(\beta)}\right\} dv$$

and

$$V^+(x) = \int_0^x \frac{2}{\sigma^2(\beta)} \exp\left\{\frac{\delta v^{\gamma+1}}{2(\gamma+1)\sigma^2(\beta)}\right\} \int_0^v \exp\left\{-\frac{\delta u^{\gamma+1}}{2(\gamma+1)\sigma^2(\beta)}\right\} du\, dv.$$

The proof of this is presented in [40].

Corollary 6 Assume that $h \to 0$, $y \in [0, \lambda h]$, $0 < \lambda < 1$. Then

$$\begin{aligned}
P\{y^-(\tau^-(h))/y^-(0) = y\} &= 1 + O\left(\exp\left\{-\frac{\delta h^{\gamma+1}}{2(\gamma+1)\sigma^2(\alpha)}\right\}\right), \\
P\{y^+(\tau^-(h))/y^+(0) = y\} &= O\left(\exp\left\{-\frac{\delta h^{\gamma+1}}{2(\gamma+1)\sigma^2(\beta)}\right\}\right), \\
E(\tau^-(h)/y^-(0) = y) &= O(1), \\
E(\tau^+(h)/y^+(0) = y) &= O(1).
\end{aligned}$$

Consider the behavior of the sequence $\{\hat{u}_\varepsilon(n)\}$ in a neighborhood of an isolated point $\alpha \in \Pi$.

Lemma 12 *Assume that $\varphi_0(y)$ satisfies the relation*

$$\varphi_0(y) = \delta(\text{sign}(y-\alpha))|y-\alpha|^\gamma + o(y-\alpha),$$

where $\delta(-1)$, $\delta(+1) \in R$; $y_\varepsilon(t)$ is determined by the right-hand side of formula (9.64); and $\hat{y}(t)$ is the solution of the stochastic differential equation

$$d\hat{y}(t) = \delta(\text{sign}(y))|\hat{y}|^\gamma dt + \sigma(\alpha) dw(t). \tag{9.69}$$

For an $h > 0$, denote by $\tau_\varepsilon(h)$ and $\hat{\tau}(h)$ the stopping times

$$\tau_\varepsilon(h) = \inf\{t : y_\varepsilon(t) \notin (-h, h)\},$$

$$\hat{\tau}(h) = \inf\{t : \hat{y}(t) \notin (-h, h)\}.$$

Then the distribution of $(\tau_\varepsilon(h), y_\varepsilon(\tau_\varepsilon(h)))$ converges weakly to the distribution of $(\tau(h), y(\tau(h)))$.

The proof is the same as in Lemma 10.

Corollary 7 (1) Let $\delta(-1) > 0$, $\delta(+1) > 0$, $h \to \infty$, $0 < \lambda < 1$, and $|y| < \lambda h$. Then

$$P\{\hat{y}(\hat{\tau}(h)) = h/\hat{y}(0) = y\} = 1 + O\left(\exp\left\{-\frac{\delta(-1)h^{\gamma+1}}{2(\gamma+1)\sigma^2(\alpha)}\right\}\right)$$

and

$$E(\hat{\tau}(h)/\hat{y}(0) = y) = O(1). \tag{9.70}$$

If $\delta(-1) < 0$, $\delta(+1) < 0$, then (9.70) is true and

$$P\{\hat{y}(\hat{\tau}(h)) = -h/\hat{y}(0) = y\} = 1 + O\Big(\exp\Big\{-\frac{\delta(+1)h^{\gamma+1}}{2(\gamma+1)\sigma^2(\alpha)}\Big\}\Big). \tag{9.71}$$

(2) Let $\delta(-1) > 0$, $\delta(+1) < 0$ (this means that α is an attracting point for the mapping $\varphi: [0,1] \to [0,1]$), $h \to \infty$, $0 < \lambda < 1$, and $|y| < \lambda h$. Then

$$E(\hat{\tau}(h)/\hat{y}(0) = y) \geq c_1 \exp\{c - 2h^{\gamma+1}\} \tag{9.72}$$

for some $c_1 > 0$, $c_2 > 0$.

(3) Let $\delta(-1) < 0$, $\delta(+1) > 0$ (this means that α is an repulsing point for the mapping $\varphi: [0,1] \to [0,1]$), $h \to \infty$, $\lambda_1 h \leq y \leq h$, and $\lambda_1 > 0$. Then

$$E(\hat{\tau}(h)/\hat{y}(0) = y) = O(1),$$

$$P\Big\{\inf_{t \leq \hat{\tau}(h)} \operatorname{sign} y \; \hat{y}(t) > \frac{1}{2}\lambda_1 h\Big\} = 1 + O(\exp\{-c_2 h^{\gamma+1}),$$

where $c_2 > 0$.

Lemma 13 *Assume that Π contains a closed interval $[c,d]$ and $\varphi_0(y) \geq 0$, $\varphi_0(y) = \delta_1(c-y)^{\gamma_1} + o((c-y)^{\gamma_1})$ for $y < c$, and $\varphi_0(y) = \delta_2(y-d)^{\gamma_1} + o((y-d)^{\gamma_1})$ for $y > d$. Let*

$$\zeta_\varepsilon(d) = \inf[n : u_\varepsilon(n) \geq d].$$

Then

$$\lim_{\varepsilon \to 0} \big(\varepsilon E(\zeta_\varepsilon(d)/u_\varepsilon(0) = c)\big) = E\hat{\zeta}(d),$$

where $\hat{\zeta}(d) = \inf[t : \theta^*(t) = d]$, and $\theta^*(t)$ is the diffusion process on $[c, \infty)$ with the same generator as the diffusion process $\theta(t)$ constructed in Remark 2, but with reflecting boundary at the point c.

This statement is a consequence of Theorem 9 and Lemma 11.

Lemma 14 *Assume that for $c \leq d$, $\varphi_0(y)$ satisfies the conditions of Lemma 13 with $\delta_1 < 0$ and $\delta_2 > 0$. Then*

$$\lim_{\varepsilon \to 0} \varepsilon \, ET_\varepsilon = +\infty. \tag{9.73}$$

Proof Let $y \in (\alpha, c+1)$, and let $h > 0$ be small enough. Then it follows from Corollaries 6 and 7 that

$$P\{u_\varepsilon(\tau_\varepsilon(d+h, c+1)) = d+h/y(0) = y\} = o(\varepsilon) \quad \text{for } y \in (d+h, c+1)$$

and

$$P\{u_\varepsilon(\tau_\varepsilon(d, c+1-h)) = c+1-h/y(0) = y\} = o(\varepsilon) \quad \text{for } y \in (d, c+1-h),$$

where

$$\tau_\varepsilon(\alpha, \beta) = \inf\{n : u_\varepsilon(n) \notin (\alpha, \beta)\}.$$

So before the sequence $\{u_\varepsilon(n)\}$ walks out of the interval $(d, c+1)$ it has to cross the interval $(d+h, c+1-h)$ approximately $1/o(\varepsilon)$ times. Because $T_\varepsilon > \tau_\varepsilon(d, c+1)$, this implies relation (9.73). In the same way we can consider the case $y \in (d-1, c)$.

□

The main result of this subsection is the following one.

Theorem 12 *Let the stochastic process $z(t)$ be uniformly ergodic and satisfy condition SMC II of Section 2.3. Assume that the function $\varphi_0(y)$ has a continuous first derivative. Then the following statements hold:*

(i) If $\varphi_0(y)$ is a function that changes sign, and there exist $c \le d$, $c, d \in \Pi$, and $\gamma > 1$ for which

$$\lim_{y \to c-} \varphi_0(y)/(c-y)^\gamma < 0, \qquad \lim_{y \to d+} \varphi_0(y)/(y-d)^\gamma > 0,$$

then with probability 1,

$$\limsup_{t\to\infty} |y_\varepsilon(t)/x_\varepsilon(t) - \theta| = o(\varepsilon). \tag{9.74}$$

(ii) If $\varphi_0(y) \ge 0$, $\varphi_0(y) \le H(r(y,\Pi))^\gamma$, where $r(y,\Pi)$ is the distance from the point y to the set Π, and $H > 0$ and $\gamma > 1$ are some constants, and if there exists an interval $[c, d]$ that belongs to Π, then there exists a positive constant A for which with probability 1,

$$\limsup_{t\to\infty} |y_\varepsilon(t)/x_\varepsilon(t) - \theta - A\varepsilon| = o(\varepsilon). \tag{9.75}$$

(iii) If $\varphi_0(y) \ge 0$, Π is a finite set, and for every $c \in \Pi$ there exist numbers $\delta_+(c) > 0$, $\delta_-(c) > 0$, $\gamma_+(c) > 1$, $\gamma_-(c) > 1$ for which

$$\varphi_0(y) = \delta_+(c)(y-c)^{\gamma_+(c)} + o\big((y-c)^{\gamma_+(c)}\big), \quad y > c,$$

$$\varphi_0(y) = \delta_-(c)(c-y)^{\gamma_-(c)} + o\big((c-y)^{\gamma_-(c)}\big), \quad y < c.$$

Then there exists a positive constant A for which with probability 1,

$$\limsup_{t\to\infty} |y_\varepsilon(t)/x_\varepsilon(t) - \theta - A\varepsilon^\kappa| = o(\varepsilon^\kappa), \tag{9.76}$$

where

$$\kappa = 1 - \frac{2}{\gamma+1}, \quad \gamma = \inf\{\gamma_-(c), \gamma_+(c), c \in \Pi\}.$$

Proof We introduce the sequence of stopping times: $T_\varepsilon^{(0)} = T_\varepsilon^{(0)}$ and for $k = 1, 2, \ldots,$

$$T_\varepsilon^{(k)} = \inf\{n > T_\varepsilon^{(k-1)} : |u_\varepsilon(n) - u_\varepsilon(T_\varepsilon^{(k-1)})| \ge 1\}.$$

Suppose that $u_\varepsilon(0) = y_0 \in \Pi$. We can prove that the conditional distribution of $T_\varepsilon^{(1)}$ with respect to a given y_0 and $z(0)$ does not depend on $y_0 \in \Pi$ and $z(0)$ as $\varepsilon \to 0$. So, $\{T_\varepsilon^{(k)} - T_\varepsilon^{(k-1)}, k = 1, 2, \ldots\}$ is asymptotically distributed like a sequence of independent random variables. This implies that

$$T_\varepsilon^{(k)}/ET_\varepsilon^{(k)} \sim T_\varepsilon^{(k)}/kET_\varepsilon.$$

Note that in case (i) the number of y-rotations for the time $T_\varepsilon^{(k)}$ is

$$\theta T_\varepsilon^{(k)} = O(k)$$

and the number of x-rotations is $T_\varepsilon^{(k)}$. So

$$\frac{y_\varepsilon\left(T_\varepsilon^{(k)}\right)}{x_\varepsilon\left(T_\varepsilon^{(k)}\right)} = \theta + \frac{O(k)}{kET_\varepsilon} = \theta + O\left(\frac{1}{T_\varepsilon}\right).$$

This and relation (9.73) imply relation (9.74). If $y \in (\alpha, \beta)$, where $\alpha, \beta \in \Pi$, then we can start the calculation of the rotation at the stopping time $\tau_\varepsilon(\alpha, \beta)$.

In the same way, we use Lemma 12 in case (ii), and we use Lemma 10 in case (iii).

□

10
Phase–Locked Loops

In this chapter we describe an important electronic circuit, the phase–locked loop (PLL). We first investigate the circuit's dynamics without and with random perturbations, but in the absence of external forcing. Then we analyze the response of the circuit to noisy external signals, and we describe a method for extracting signals from the result.

The phase–locked loop (PLL) is a standard electronic synchronous control circuit. The circuit is described in Figure 10.1.

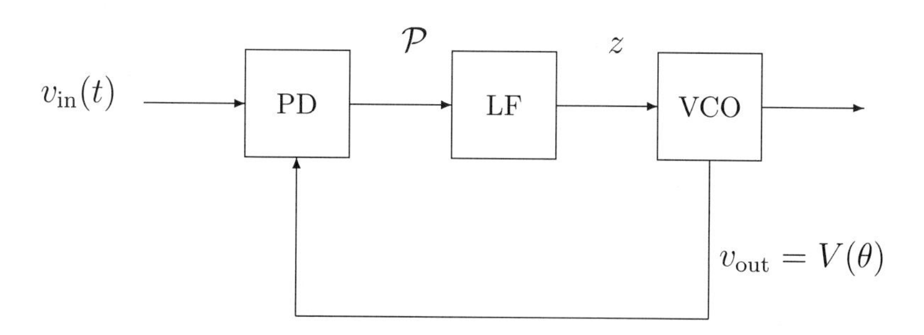

Figure 10.1. Phase–locked loop circuit. Included are a phase detector (PD) having output $\mathcal{P}$, a loop filter (LF) having output z, and a voltage–controlled oscillator (VCO) having output $V(\theta)$.

Here $v_{\text{in}}(t)$ is an external signal that is put into the PLL, and $V(\theta(t))$ is the voltage put out by the voltage–controlled oscillator. (V is a fixed waveform like a sinusoid, a saw-tooth, or a square wave.) The phase detector (PD) combines these two signals, usually by simply multiplying the two signals v_{in} and V and putting out a voltage $\mathcal{P}$ that is proportional to the product. This in turn passes into the loop filter (LF), and its output (z) is used as the controlling voltage of the VCO.

Part of the brilliance of this circuit is that it is described in terms of frequency–domain variables (or angles), making possible the use of powerful Fourier methods to analyze it and to design more complicated circuits in which it is used.

The VCO's output is $v_{\text{out}} = V(\theta(t))$ (volts), but its dynamics are described by a linear problem,

$$\dot{\theta} = \omega + z,$$

where ω is a constant, called the center frequency, that characterizes the VCO. The variable $\theta(t)$ is the VCO's phase, and it is the key variable in the model. In a PLL, the VCO output feeds back to the phase detector for comparison with the input and for generating the control signal $\mathcal{P}$.

In the case where the loop filter is a low–pass filter, the circuit is described in mathematical terms by the following set of differential equations

$$\begin{aligned} \tau\dot{z}(t) + z(t) &= \mathcal{P}(\theta(t), v_{\text{in}}), \\ \dot{\theta}(t) &= \omega + z(t), \end{aligned}$$

where:

1. τ is the time constant of the low–pass filter;
2. ω is the VCO's center frequency;
3. $v_{\text{in}}(t)$ is the input voltage;
4. $\mathcal{P} = \mathcal{P}(\theta(t), v_{\text{in}})$ is the output of the phase detector, which will be described later.

Usually filters are described as being linear time–invariant system of the form

$$z(t) = \int_0^t h(t - t')v(t')\,dt',$$

where h is called the impulse–response function, and its Laplace transform

$$H(s) = \int_0^\infty e^{-st}h(t)\,dt,$$

is called the system's transfer function.

For example, a low–pass filter is described by a differential equation of the form

$$\tau\dot{z} + z = f(t),$$

where τ is the filter time constant and $f(t)$ is an external signal. The solution is

$$z(t) = e^{-t/\tau} z(0) + \frac{1}{\tau} \int_0^t e^{(t'-t)/\tau} f(t')\, dt'.$$

where the first term on the right-hand side is a transient term and the second is a LTI with input $f(t)$. Note that in this case $H(s) = 1/(\tau s + 1)$. If f is purely oscillatory, say $f(t) = e^{i\omega t}$, then

$$z(t) = \frac{e^{i\omega t}}{(1 + i\omega\tau)} + O(e^{-t/\tau}) \approx \frac{e^{i\omega t}(1 + i\omega\tau)}{(1 + (\omega\tau)^2)}.$$

This solution comprises a steady oscillation plus a decaying transient. We write the steady oscillation as $z(t) = A(\omega)\exp\left[i\phi(\omega) + i\omega\tau\right]$. In fact, the amplitude and the phase deviation of the response are given by

$$A^2(\omega\tau) = \frac{1}{1 + (\omega\tau)^2}, \quad \phi(\omega\tau) = \frac{\pi}{2} \arctan(\omega\tau).$$

The functions A and ϕ can be plotted as functions of $\omega\tau$. Such plots are called Bode plots. For example, $A(\omega\tau) \to 0$ as the dimensionless parameter $\omega\tau \to \infty$. In practice, one says that the filter allows passage of only those frequencies with $\omega < 1/\tau$; the larger is the time constant, the narrower the pass band of frequencies.

If the loop filter is a band–pass filter, the model becomes more complicated. The simplest case is a filter with transfer function

$$H(s) = \frac{a\, s}{s^2 + a\, s + \Omega^2},$$

in which case the equation for z becomes a second–order system,

$$\begin{aligned}
\dot{z}(t) &= -az(t) - \Omega u + a\mathcal{P}(\theta(t), v_{\text{in}}), \\
\dot{u}(t) &= \Omega z, \\
\dot{\theta}(t) &= \omega + z(t),
\end{aligned}$$

which effectively allows passage of frequencies that are within a units of $\pm\Omega$.

Filters cannot always be reduced to equivalent differential equations, in particular when we consider them to have noisy components.

With more general filters, the equation for $z(t)$ is typically a Volterra integral equation that cannot be reduced to a differential equation. In this case, we write the system as

$$\begin{aligned}
z(t) &= \phi(t) + \int_0^t h(t - t')\mathcal{P}(\theta(t'), t')\, dt', \\
\dot{\theta}(t) &= \omega + z(t).
\end{aligned}$$

It is possible to adjust all of these data from outside the circuit. In circuit design we have flexibility in selecting components of the circuit, and this is often done in a way that facilitates the design of more complicated circuits. On the other hand, in modeling physical or biological systems, we are informed about the nature of the pass bands and the phase detector, and we must deal with them as given.

The outputs of the circuit, z, $\mathcal{P}$, and $V(\theta)$, are all useful in various applications. For example, z is used to demodulate the phase in FM radio, giving the part of the signal that we listen to, and $\mathcal{P}$ gives an error signal detecting phase shifts between the input and PLL output that is used in radar systems.

Both the free problem (no external forcing signal) and the forced problem, where an external signal is brought into the circuit through the phase detector, are of interest in applications. The terms in the circuit are usually taken with the following meanings:

1. $\mathcal{P}$ combines external signals with the VCO output, usually by a simple multiplication of the two. This output is used to obtain the frequency and the phase modulations in the signal.

2. The total controlling voltage is passed through the loop filter to reduce noise and unwanted interference.

3. z is the output of the loop filter, and it controls the VCO frequency.

4. V is a fixed waveform, namely a periodic function, say having period 2π.

For the remainder of this chapter, we consider only problems with a low–pass filter, sometimes even with $\tau = 0$, to illustrate our methods. There are obvious, but nontrivial, extensions of this work to systems having noisy filters represented by convolution integrals based on our earlier work on Volterra integral equations.

10.1 The Free System

When there is no external signal to the circuit, we say that the problem is free. In this case, the model is described by the system of differential equations

$$\begin{aligned} \tau \dot{z} &= -z + V(\theta), \\ \dot{\theta} &= \omega + z. \end{aligned} \tag{10.1}$$

The unperturbed system has τ and ω as constant parameters, and $V(\theta)$ is a 2π-periodic, nonrandom function describing the output waveform. The *randomly perturbed system* has τ, ω, $V(\theta)$ as stochastic processes depending on t/ε (as usual in this book), and system (10.1) with nonrandom parameters constitutes the averaged equations for the perturbed system. We consider the case

$$V(\theta) = \cos\theta,$$

but our results can be applied to $V(\theta)$ of general form.

10.1.1 The Nonrandom Free System

We consider system (10.1) with $V(\theta) = \cos\theta$. Introduce an auxiliary scalar differential equation

$$\tau \frac{dZ(\theta)}{d\theta} = \frac{-Z(\theta) + \cos\theta}{Z(\theta) + \omega}, \tag{10.2}$$

which is an implicit form of (10.1). We will investigate the solutions of system (10.1) and equation (10.2) for fixed $\tau > 0$ and various $\omega > 0$. Note that the point $Z = -\omega$ is a singularity for equation (10.2). For $Z \neq -\omega$ denote by $T(z, \omega) > 0$ the value for which equation (10.2) has a solution $Z(\theta, z)$ satisfying the initial condition $Z(0, z) = z$, having a maximal interval of existence $[0, T(z, \omega))$ and satisfying

$$\lim_{\theta \to T(z,\omega)} |Z(\theta, z) + \omega| = 0.$$

It is easy to see that if $-\omega < z_1 < z_2$, then $T(\omega, z_1) < T(\omega, z_2)$ and $Z(\theta, z_1) < Z(\theta, z_2)$ for $\theta \leq T(\omega, z_1)$. If $z < -\omega$, then $T(\omega, z) = +\infty$. So we can define the following functions:

$$Z_+(\theta, -\omega) = \lim_{z \to (-\omega)+} Z(\theta, z),$$

$$Z_-(\theta, -\omega) = \lim_{z \to (-\omega)-} Z(\theta, z),$$

$$T(-\omega, -\omega) = \lim_{z \to (-\omega)+} T(\omega, z).$$

Lemma 1 *Let $\theta(t)$ be the solution of the equation*

$$\dot{\theta}(t) = \omega + Z(\theta(t), z)$$

with the initial condition $\theta(0) = 0$ and $\theta(t) < T(\omega, z)$ for $t \in [0, t_1]$. Then the functions

$$\theta(t) \quad \textit{and} \quad z(t) = Z(\theta(t), z)$$

satisfy the system

$$\begin{aligned} \tau \dot{z}(t) &= -z(t) + \cos\theta(t), \\ \dot{\theta}(t) &= \omega + z(t), \end{aligned} \tag{10.3}$$

on the interval $[0, t_1]$ and satisfy the initial condition: $\theta(0) = 0$, $z(0) = z$.

The proof of this lemma is accomplished by direct calculation, which is left to the reader.
Define the mapping $\Phi_\omega(z) = Z(2\pi, z)$ when $T(\omega, z) \geq 2\pi$. It is easy to see that $\Phi_\omega(z)$ satisfies the following conditions:

(i) For fixed ω the function $\Phi_\omega(z)$ is defined on the interval (z^*_ω, ∞) where z^*_ω satisfies the relation

$$\lim_{\theta \to \theta_2(\omega)} Z(\theta, z^*_\omega) = -\omega,$$

where

$$\cos\theta_2(\omega) = -\omega, \quad \theta_2(\omega) \in [\pi, 2\pi].$$

(ii) $\Phi_\omega(z_1) < \Phi_\omega(z_2)$ for $z_1 < z_2$.

(iii) $\Phi_\omega(z) < z$ if $z \geq 1$.

(iv) If $\Phi_\omega(z^*_\omega+) \leq z^*_\omega$, then $\Phi_\omega(z) < z$ for all $z \in (z^*_\omega, \infty)$; if $\Phi_\omega(z^{*+}_\omega) > z^*_\omega$, then there exists the unique point $\bar{z}_\omega \in (z^*_\omega, \infty)$ for which

$$\Phi_\omega(\bar{z}_\omega) = \bar{z}_\omega. \tag{10.4}$$

Theorem 1 *Suppose that* $\Phi_\omega(z^*_\omega+) > z^*_\omega$. *Then*

(1) The function $Z_\omega(\theta, \bar{z}_\omega)$ *is a* 2π*-periodic function.*

(2) The solution of system (10.3) *with the initial condition* $\theta(0) = 0$, $z(0) = \bar{z}_\omega$ *has the form*

$$\begin{aligned} \bar{\theta}(t) &= t/c_\omega + p_\omega(t), \\ \bar{z}(t) &= Z_\omega(\bar{\theta}(t), \bar{z}_\omega), \end{aligned} \tag{10.5}$$

where

$$c_\omega = \frac{1}{2\pi}\int_0^{2\pi} \frac{du}{\omega + Z_\omega(u, \bar{z}_\omega)}$$

and $p_\omega(t)$ *is* $2\pi c_\omega$*-periodic.*

(3) For all $z \in (z^*_\omega, \infty)$,

$$\lim_{\theta \to \infty} |Z_\omega(\theta, \bar{z}_\omega) - Z_\omega(\theta, z)| = 0.$$

(4) If $(\theta(t), z(t))$ *is the solution to system* (10.3) *with the initial condition* $\theta(0) = 0$, $z(0) = z$, *and* $z \in (z^*_\omega, \infty)$, *then*

$$\lim_{t \to \infty} \left(|\theta(t) - \bar{\theta}(t)| + |z(t) - \bar{z}(t)|\right) = 0. \tag{10.6}$$

Proof Statement (1) is a consequence of relation (10.4), and (2) follows from Lemma 1. To prove (3) we consider $z > \bar{z}_\omega$. Then

$$z > \Phi_\omega(z) > \Phi_\omega(\Phi_\omega(z)) > \cdots > \Phi^n_\omega(z) \geq \bar{z}_\omega,$$

where Φ_ω^n is the nth iterate of the mapping Φ_ω. Define $\hat{z} = \inf_n \Phi_\omega^n(z)$. It is easy to see that $\Phi_\omega(\hat{z}) = \hat{z}$, so $\hat{z} = \bar{z}_\omega$. Since $\Phi_\omega^n(z) = Z_\omega(2n\pi, z)$, we have that

$$\lim_{n\to\infty} Z_\omega(2n\pi, z) = \bar{z}_\omega$$

for $z > \bar{z}_\omega$. In the same way we consider $z \in (z_\omega^*, \bar{z}_\omega)$. This implies (3). Statement (4) is a consequence of Lemma 1 and statement (3).

□

Remark 1 Let $\omega > 1$. Then $T(\omega, z) = +\infty$ for all $z \geq -\omega$. So $z_\omega^* = -\omega$. If $z < -\omega$, then $Z_\omega(\theta, z) \to -\infty$ as $\theta \to +\infty$ and

$$\lim_{\theta\to+\infty} \frac{1}{\theta} Z_\omega(\theta, z) = -\frac{1}{\tau}.$$

The last follows from the relation $Z_\omega(\theta, z) \to -\infty$ as $\theta \to +\infty$ in (10.2). We consider the solution of system (10.1) with the initial condition $z(0) = z_0 < -\omega$, $\theta(0) = 0$. Then $\dot{\theta}(t) < 0$ and $\dot{z}(t) > 0$ for $t < t_1$, where $t_1 = \sup\{s : z(s) < -\omega\}$ and $z(t_1) = -\omega$. So the solution of system (10.1) satisfies relation (10.6) for any initial condition $(z(0), \theta(0))$.

Remark 2 For $\omega = 1$ system (10.1) has singular points at $(-1, (2n+1)\pi)$ in the (z, θ)-plane for $n = 0, \pm1, \pm2, \ldots$. These points are unstable nodes, and for all $n \in \mathbf{Z}$ equation (10.2) has the solutions $Z_n^{(1)}(\theta)$, $\theta \in [(2n + 1)\pi, \infty)$ for which

$$Z_n^{(1)}(\theta) = -\frac{1}{\tau}(\theta - (2n+1)\pi) + O\big((\theta - (2n+1)\pi)^2\big).$$

Note that

$$Z_n^{(1)}(\theta) = Z_0^{(1)}(\theta - (2n+1)\pi).$$

Assume that the initial conditions $z(0) = z_0$, $\theta(0) = \theta_0$, for system (10.1) satisfy the relation

$$z_0 = Z_0^{(1)}(\theta_0 - (2n+1)\pi) \quad \text{for some } n. \tag{10.7}$$

Then

$$\lim_{t\to\infty} z(t) = -1, \quad \lim_{t\to\infty} \theta(t) = (2n+1)\pi,$$

where $(z(t), \theta(t))$ is the solution to system (10.1) with this initial condition. If $z_0 < -1$ and (z_0, θ_0) does not satisfy relation (10.7) for any n, then there exists t_1 for which $z(t_1) = -1$. For the solution of system (10.1) with such initial conditions, relation (10.6) is fulfilled.

Remark 3 Let $\omega \in (0, 1)$. Denote by $\theta_1^*(\omega)$ and $\theta_2^*(\omega)$ the solutions of the equation

$$\cos\theta = -\omega,$$

where $\theta_1^*(\omega) \in [0, \pi]$ and $\theta_2^*(\omega) \in [\pi, 2\pi]$. The points $(-\omega, \theta_i^*(\omega) + 2n\pi)$, $i = 1, 2$, $n \in \mathbf{Z}$, are singular points for system (10.1), and they are stable nodes for $i = 1$ and unstable nodes for $i = 2$. There exists a solution of equation (10.2),

$$Z^*(\omega, \theta), \quad \theta \in [0, \infty),$$

satisfying the following conditions:

(a) $Z^*(\omega, \theta) < -\omega$, $\theta > \theta_2^*(\omega)$;

(b) $Z^*(\omega, \theta_2^*(\omega)) = -\omega$, $Z^*(\omega, 0) = z_\omega^*$.

Consider the solution $(z(t), \theta(t))$ to system (10.1) satisfying the initial conditions $z(0) = z_0 < -\omega$, $\theta(0) = \theta_0$. The following statements hold:

(1) If for some $n \in \mathbf{Z}$

$$Z^*(\omega, \theta_0 + 2n\pi) = z_0, \tag{10.8}$$

then $z(t) < -\omega$ for all t and

$$\lim_{t\to\infty} z(t) = -\omega, \quad \lim_{t\to\infty} \theta(t) = \theta_2^*(\omega) + 2n\pi.$$

(2) If there exists no n for which relation (10.8) holds, then there exists $t_1 > 0$ for which $z(t_1) = -\omega$; in this case set

$$t_2 = \inf\{s \geq t_1 : \theta(s) = 2n\pi \text{ for some } n \in \mathbf{Z}\}.$$

Then $z(t_2) < \Phi_\omega(z_\omega^* +)$ and either relation (10.6) is fulfilled if $z(t_2) > z_\omega^*$ or

$$\lim_{t\to\infty} z(t) = -\omega, \quad \lim_{t\to\infty} \theta(t) = 2n\pi + \theta_1^*(\omega)$$

if $z(t_2) < z_\omega^*$.

Remark 4 Consider the solution to system (10.1) satisfying the initial condition $z(0) = z_\omega^*$, $\theta(0) = 0$. Then $z(t) + \omega > 0$ for all $t > 0$ and

$$\lim_{t\to\infty} z(t) = -\omega, \quad \lim_{t\to\infty} \theta(t) = \theta_1^*(\omega).$$

In the same way, if condition (2) of Remark 3 is fulfilled and $z(t_2) = z_\omega^*$, then

$$\lim_{t\to\infty} z(t) = -\omega, \quad \lim_{t\to\infty} \theta(t) = 2n\pi + \theta_2^*(\omega).$$

Introduce a number ω^* that is determined by the relation

$$\omega^* = \inf\{\omega : \Phi_\omega(z_\omega^*) > z_\omega^*\}.$$

So for $\omega > \omega^*$ the conditions of Theorem 1 hold, and the behavior of the solution to the system is described by Theorem 1 and Remarks 1–4. Now we consider the case $\omega \leq \omega^*$.

Theorem 2 *Suppose that $\omega \leq \omega^*$. Denote by $Z_+(\omega, \theta)$ the function satisfying the following conditions:*

(1) $Z_+(\omega, \theta)$ satisfies equation (10.2) for all

$$\theta \in \cup_n (\theta_2^*(\omega) + 2\pi n, \theta_2^*(\omega) + 2\pi(n+1)).$$

(2) $Z_+(\omega, 2\pi n + \theta_2^*(\omega)) = -\omega$ for all n.

Let $z(t)$, $\theta(t)$ be the solution of system (10.1) with initial conditions $z(0) = z_0$, $\theta(0) = \theta_0$. The following statements hold:

I. If $z_0 > Z_+(\omega, \theta_0)$, then $\theta(t) \to \infty$ as $t \to \infty$ and $z(t) - Z_+(\omega, \theta(t)) \to 0$ as $t \to \infty$.

II. If $z_0 = Z_+(\omega, \theta_0)$ and $\theta_2^*(\omega) + 2n\pi < \theta_0 < \theta_2^* + 2(n+1)\pi$, then $\theta(t) \to \theta_2^*(\omega) + 2(n+1)\pi$, $z(t) \to -\omega$ as $t \to \infty$.

III. If $z_0 = -\omega$ and $\theta_0 = \theta_2^*(\omega) + 2n\pi$, then $z(t) = z_0$, $\theta(t) = \theta_0$ for all t.

IV. If $-\omega \leq z_0 < Z_+(\omega, \theta_0)$, $\theta_2^*(\omega) + 2n\pi < \theta_0 < \theta_2^* + 2(n+1)\pi$, then $\theta(t) \to \theta_1^*(\omega) + 2n\pi$, $z(t) \to -\omega$ as $t \to \infty$.

V. If $z_0 < -\omega$ and for some n relation (10.8) is true, then

$$\lim_{t\to\infty} z(t) = -\omega, \quad \lim_{t\to\infty} \theta(t) = \theta_2^*(\omega) + 2n\pi.$$

VI. If $z_0 < -\omega$ and relation (10.8) holds for no n, then there exists n_1 for which

$$\lim_{t\to\infty} \theta(t) = \theta_1^*(\omega) + 2\pi n_1, \quad \lim_{t\to\infty} z(t) = -\omega.$$

Remark 5 It is easy to check that

$$Z_+(\omega, \theta) = Z^*(\omega, \theta) \quad \text{for } \theta \in [0, \theta_2^*(\omega))$$

and that the solution to equation (10.2) on the interval $(\theta_2^*(\omega) - 2\pi, 0]$ satisfying the initial condition $Z(0) = z_\omega^*$ has the property

$$\lim_{\theta \to \theta_2^*(\omega) - 2\pi} z(\theta) = -\omega.$$

Since this is true for all $\omega < 1$, we can define $Z_+(\omega, \theta)$ for all $\omega < 1$. If $\omega^* < \omega$, there exists the running periodic solution $z(t)$ and for all θ_0 and z_0 satisfying the relation $Z_+(\omega, \theta_0) > z_0$ the relation (10.6) holds.

Remark 6 The static state $(\theta_1(\omega), -\omega)$ is:

(i) a stable node if

$$\Delta = 4\tau(1-\omega)^2 - 1 < 0;$$

(ii) a stable star if $\Delta = 0$;

(iii) a stable focus if $\Delta > 0$.

Remark 7 We can consider the solution of system (10.1) on the cylinder $C = R \times [0, 2\pi)$; the points of the cylinder have coordinates $z \in R$, $\theta \in [0, 2\pi)$. The behavior of the system on the cylinder C can be described in the following way. Introduce on C the following curves:

(α) $z = Z_\omega(\theta, \bar{z}_\omega)$, $\theta \in [0, 2\pi)$, which is defined for $\omega > \omega^*$;

(β) $z = Z_+(\omega, \theta)$, $\theta \in [0, 2\pi)$, which is defined for $0 < \omega < 1$;

(γ) $\theta = \Theta_-(z)$, $z \in (-\infty, -\omega]$, $\omega \leq 1$, where $\Theta_-(z)$ is a continuous function for which for every $z < -\omega$ there exists $n \in \mathbf{Z}$ such that

$$Z^*(\omega, \Theta_-(z) + 2n\pi) = z$$

and $\Theta_-(-\omega) = \theta_2^*(\omega)$.

The curves (α) and (β) are closed on C, and each of them divides the cylinder into two parts.
The curves (α), (β), and (γ) are invariant manifolds for system (10.1), and the behavior of the system on these curves is described in Theorems 1 and 2. Introduce the sets

$$C_+ = \{(z, \theta) : z > Z_+(\omega, \theta)\},$$
$$C_- = \{(z, \theta) : z < Z_+(\omega, \theta)\} \setminus \{(z, \theta) : \theta = \Theta_-(z)\}.$$

Let $\omega > \omega^*$. Then for initial condition $(z_0, \theta_0) \in C_+$ the system tends to the curve (α) as $t \to \infty$, and for $(z_0, \theta_0) \in C_-$ the system tends to the stable static state $(-\omega, \theta_1^*(\omega))$.
Let $\omega \leq \omega^*$. Then for $(z_0, \theta_0) \in C_+$ the system tends to the curve (β), and for $(z_0, \theta_0) \in C_-$ the system tends to the stable state $(-\omega, \theta_1^*(\omega))$.

10.1.2 Example: Simulation of the Free PLL System

A computer simulation illustrates the primary behaviors of the free system. Consider

$$\tau\ddot{\theta} + \dot{\theta} = \omega + \cos\theta. \tag{10.9}$$

Simulations of two solutions for each of three choices of τ, ω are shown in Figure 10.2.

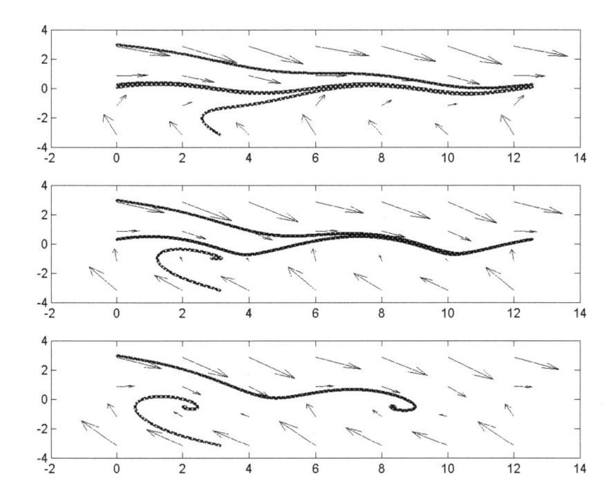

Figure 10.2. Three cases of the free phase–locked loop model (10.9): Top $\tau = 1.5$, $\omega = 2.0$. Middle: $\tau = 1.5$, $\omega = 0.963$. Bottom: $\tau = 1.5$, $\omega = 0.5$. Two solutions are shown in each case. The top one starts at $(\theta(0), \dot{\theta}(0)) = (0.0, 3.0)$ and the other starts at $(\theta(0), \dot{\theta}(0)) = (\pi, -\pi)$. In the top case, all solutions approach the running periodic solution. In the middle case, the top solution approaches a running periodic solution and the bottom one approaches a stable node. This illustrates the case of co–existence of the two stable solutions. In the bottom case, all solutions eventually approach the stable node.

10.1.3 Random Perturbations of the Free System

We consider a randomly perturbed system that is defined by the system of differential equations

$$\begin{aligned} \tau\left(y\left(\frac{t}{\varepsilon}\right)\right)\dot{z}_\varepsilon(t) &= -z_\varepsilon(t) + v\left(y\left(\frac{t}{\varepsilon}\right), \theta_\varepsilon(t)\right), \\ \dot{\theta}_\varepsilon(t) &= \omega\left(y\left(\frac{t}{\varepsilon}\right)\right) + z_\varepsilon(t), \end{aligned} \tag{10.10}$$

where $y(t)$ is a homogeneous Markov process in a noise space $(Y, \mathcal{C})$ satisfying condition SMC II of Section 2.3, $\tau(y)$ and $\omega(y)$ are nonnegative measurable functions on $(Y, \mathcal{C})$, $v(y, \theta) : Y \times R \to R$ is measurable in $y \in Y$ and sufficiently smooth in θ, and $\varepsilon > 0$ is a small parameter.
Let $\rho(dy)$ be the ergodic distribution for the Markov process $y(t)$. We assume that

$$\int \tau^{-1}(y)\rho(dy) < \infty,$$

$$\int |v(y,\theta)|\tau^{-1}(y)\rho(dy) < \infty,$$

$$\int v(y,\theta)\tau^{-1}(y)\rho(dy) = \cos\theta \cdot \int \tau^{-1}(y)\rho(dy),$$

and

$$\int \omega(y)\rho(dy) < \infty.$$

Let

$$\mathcal{T} = \left(\int \mathcal{T}^{-1}(y)\rho(dy)\right)^{-1},$$
$$\omega = \int \omega(y)\rho(dy).$$

Then system (10.1) is the averaged system for system (10.10).

The next statement follows from Theorem 6 of Chapter 3:

I. Let $(z(t), \theta(t))$ be the solution to system (10.1) satisfying the initial conditions $z(0) = z_\varepsilon(0)$ and $\theta(0) = \theta_\varepsilon(0)$, where $z_\varepsilon(0)$ and $\theta_\varepsilon(0)$ do not depend on $\varepsilon > 0$. Then for any $t_0 > 0$,

$$P\left\{\lim_{\varepsilon\to 0}\sup_{t\le t_0}(|z_\varepsilon(t) - z(t)| + |\theta_\varepsilon(t) - \theta(t)|) = 0\right\} = 1. \tag{10.11}$$

Next, suppose that

$$\frac{1}{\mathcal{T}(y)}\frac{\partial v}{\partial \theta}(y,\theta)$$

is a bounded function that is continuous in θ. Then the conditions of Theorem 2 of Chapter 4 are fulfilled, and so we have the following

II. With the initial condition $z(0) = z_\varepsilon(0)$, $\theta(0) = \theta_\varepsilon(0)$, the two-dimensional stochastic process $(\hat{z}_\varepsilon(t), \hat{\theta}_\varepsilon(t))$, where

$$\hat{z}_\varepsilon(t) = \varepsilon^{-1/2}(z_\varepsilon(t) - z(t)),$$
$$\hat{\theta}_\varepsilon(t) = \varepsilon^{-1/2}(\theta_\varepsilon(t) - \theta(t)),$$

converges weakly in C to the stochastic process $(\hat{z}(t), \hat{\theta}(t))$ defined by the system of integral equations

$$\hat{z}(t) = -\int_0^t \frac{1}{\mathcal{T}}\hat{z}(s)ds + \int_0^t \frac{1}{\mathcal{T}}\sin\theta(s)\cdot\hat{\theta}(s)ds + v_1(t) + v_2(t),$$
$$\hat{\theta}(t) = \int_0^t \hat{z}(s)ds + v_3(t),$$

where $(v_1(t), v_2(t), v_3(t))$ is a three-dimensional Gaussian stochastic process with independent increments for which

$$Ev_k(t) = 0, \quad k = 1,2,3,$$
$$E\big(\alpha_1 v_1(t) + \alpha_2 v_2(t) + \alpha_3 v_3(t)\big)^2$$
$$= 2\int_0^t \iint \big(-\alpha_1\mathcal{T}^{-1}(y) + \alpha_2\mathcal{T}^{-1}(y)v(y,\theta(s)) + \alpha_3\omega(y)\big)$$
$$\times\big(-\alpha_1\mathcal{T}^{-1}(y') + \alpha_2\mathcal{T}^{-1}(y')v(y',\theta(s)) + \alpha_3\omega(y')\big)R(y,dy')\rho(dy)ds.$$

It follows from statement I that there exists a function $t(\varepsilon) : R_+ \to R_+$ for which $\lim_{\varepsilon\to 0} t(\varepsilon) = +\infty$ and relation (10.11) is true if t_ε replaces t_0. So we can describe the behavior of the perturbed system using the results of Theorems 1 and 2 for the averaged system.
A more precise description of the perturbed system in neighborhoods of the running periodic solution and of the stable static state follows from the results on normal deviations.

10.1.4 Behavior of the Perturbed System near the Running Periodic Solution

Suppose that the averaged system (10.1) has $\omega > \omega^*$, so there exists a running periodic solution, say denoted by $(\bar{z}(t), \bar{\theta}(t))$. Let $(z_\varepsilon(t), \theta_\varepsilon(t))$ be the solution to system (10.10) satisfying the initial conditions

$$z_\varepsilon(0) = z_0, \quad \theta_\varepsilon(0) = 0.$$

Theorem 3 *Assume that the relation*

$$\Phi_\omega(z_\omega^* +) > z_\omega^*$$

holds for equation (10.2). *Let $T(\varepsilon) : R_+ \to R_+$ be a function such that $T(\varepsilon) \to +\infty$ and $\varepsilon T(\varepsilon) \to 0$ as $\varepsilon \to 0$. Then for any $z_0 > z_\omega^*$ there exist constants $C > 0$ and $\alpha > 0$ and a function $\delta(\varepsilon) : R_+ \to R_+$ such that $\delta(\varepsilon) \to 0$ as $\varepsilon \to 0$ for which*

$$\lim_{\varepsilon\to 0} P\left\{ \sup_{t\le T(\varepsilon)} \frac{|\bar{z}(t) - z_\varepsilon(t)| + |\bar{\theta}(t) - \theta_\varepsilon(t)|}{Ce^{-\alpha t} + \delta(\varepsilon)} \le 1 \right\} = 1. \tag{10.12}$$

Proof Define a sequence of stopping times $\{\tau_k^\varepsilon,\ k = 0, 1, 2, \dots\}$ by the relations $\tau_0^\varepsilon = 0$ and

$$\tau_{k+1}^\varepsilon = \inf\{s > \tau_k^\varepsilon : \theta_\varepsilon(s) = 2(k+1)\pi\}, \quad k \ge 0. \tag{10.13}$$

Set

$$\Phi_\varepsilon^{(k)}(z) = z_\varepsilon^*({\tau_{k+1}^\varepsilon}^*),$$

where $z_\varepsilon^*(t)$, $\theta_\varepsilon^*(t)$ is the solution to system (10.10) on the interval $[\tau_k^\varepsilon, \infty)$ with the initial conditions

$$z_\varepsilon^*(\tau_k^\varepsilon) = z, \quad \theta_\varepsilon^*(\tau_k^\varepsilon) = 2k\pi,$$

and ${\tau_{k+1}^\varepsilon}^*$ is determined by formula (10.13) in which $\theta_\varepsilon(s)$ is changed to $\theta_\varepsilon^*(s)$. It is easy to see that

$$z_\varepsilon(\tau_{k+1}^\varepsilon) = \Phi_\varepsilon^{(k)}(\Phi_\varepsilon^{(k-1)}(\cdots(\Phi_\varepsilon^{(0)}(z_0)\cdots))).$$

It follows from statement II that

$$\Phi_\varepsilon^{(k)}(z) = \Phi_\omega(z) + \sqrt{\varepsilon}\,\sigma(z)\,\eta_k^\varepsilon, \tag{10.14}$$

where $\sigma(z)$ is a bounded function and the distribution of η_k^ε converges to the standard Gaussian distribution. The function $\Phi_\omega(z)$ was introduced in Section 10.1.1 (see (i)–(iii)). Let $\bar{z}_\omega$ be determined by relation (10.4). To prove the theorem it suffices to prove that

$$\lim_{\varepsilon\to 0} P\left\{\sup_{t\le T(\varepsilon)} \frac{|\bar{z}_\omega - z_\varepsilon(\tau_k^\varepsilon)|}{Cr^k + \delta(\varepsilon)} \le 1\right\} = 1, \tag{10.15}$$

where $C > 0$ and $r < 1$ are some constants.
It can be proved that

$$\frac{d\Phi_\omega}{dz}(z)\bigg|_{z=\bar{z}_\omega} = \lambda, \quad 0 < \lambda < 1.$$

So

$$z_\varepsilon(\tau_{k+1}^\varepsilon) - \bar{z}_\omega = \Phi_\varepsilon^{(k)}(z_\varepsilon(\tau_k^\varepsilon)) - \bar{z}_\omega = \Phi_\omega(z_\varepsilon(\tau_k^\varepsilon)) + \sqrt{\varepsilon}\,\sigma(z_\varepsilon(\tau_k^\varepsilon))\,\eta_k^\varepsilon - \bar{z}_\omega$$

$$= \Phi_\omega(z_\varepsilon(\tau_k^\varepsilon)) - \Phi_\omega(\bar{z}_\omega) + \sqrt{\varepsilon}\,\sigma(z_\varepsilon(\tau_k^\varepsilon))\,\eta_k^\varepsilon.$$

There exist constants $\delta_1 > 0$ and $\lambda_1 < 1$ for which

$$|\Phi_\omega(z) - \Phi_\omega(\bar{z}_\omega)| < \lambda_1|z - \bar{z}_\omega|$$

if $|z - \bar{z}_\omega| \le \delta_1$. Note that the inequality $|z_\varepsilon(\tau_k^\varepsilon) - \bar{z}_\omega| \le \delta_1$ implies the relation

$$|z_\varepsilon(\tau_{k+1}^\varepsilon) - \bar{z}_\omega| \le \lambda_1\delta_1 + C_1\sqrt{\varepsilon}\,|\eta_k^\varepsilon|,$$

where $C_1 = \sup|\sigma(z)|$.
Suppose that

$$\sup_{i\le T(\varepsilon)} |\eta_i^\varepsilon| \le \frac{(1-\lambda_1)\delta_1}{C_1\sqrt{\varepsilon}}. \tag{10.16}$$

Then $|z_\varepsilon(\tau_n^\varepsilon) - \bar{z}_\omega| \le \delta_1$ for all $k \le n \le T(\varepsilon)$ if $|z_\varepsilon(\tau_k^\varepsilon) - \bar{z}_\omega| \le \delta_1$, and

$$|z_\varepsilon(\tau_{n+1}^\varepsilon) - \bar{z}_\omega| \le \lambda_1|z_\varepsilon(\tau_n^\varepsilon) - \bar{z}_\omega| + C_1\sqrt{\varepsilon}\,|\eta_n^\varepsilon| \tag{10.17}$$

for all $n \in [k, T(\varepsilon)]$. It follows from (10.17) that

$$|z_\varepsilon(\tau_{n+1}^\varepsilon) - \bar{z}_\omega| \le C_1\sqrt{\varepsilon}\sum_{l=0}^{n-k}\lambda_1^l|\eta_{n-l}^\varepsilon| + \lambda_1^{n-k}|z_\varepsilon(\tau_k^\varepsilon) - \bar{z}_\omega|$$

$$\le \delta\lambda_1^{n-k} + \frac{C_1}{1-\lambda_1}\sqrt{\varepsilon}\sup_{m\le T(\varepsilon)}|\eta_m^\varepsilon|.$$

Assume that

$$\sup_{m\le T(\varepsilon)} |\eta_m^\varepsilon| \le \frac{1-\lambda_1}{C_1\sqrt{\varepsilon}}\delta(\varepsilon) \tag{10.18}$$

and that $\delta(\varepsilon) < \delta_1$. The last relation implies relation (10.16).

Note that

$$P\Big\{\sup_{m\le T(\varepsilon)}|\eta_m^\varepsilon|>\frac{1-\lambda_1}{C_1\sqrt{\varepsilon}}\delta(\varepsilon)\Big\}\le T(\varepsilon)\Big(\frac{1-\lambda_1}{C_1\sqrt{\varepsilon}}\delta(\varepsilon)\Big)^{-2}E|\eta_1^\varepsilon|^2$$

$$=\frac{T(\varepsilon)\varepsilon}{[\delta(\varepsilon)]^2}O(1).$$

Let $\delta(\varepsilon)$ satisfy the relations $\delta(\varepsilon)\to 0$ and

$$\lim_{\varepsilon\to 0}T(\varepsilon)\varepsilon\delta^{-2}(\varepsilon)=0.$$

Then the probability that relation (10.18) holds tends to 1 as $\varepsilon\to 0$.
If (10.18) is satisfied, then

$$|z_\varepsilon(\tau_n^\varepsilon)-\bar{z}|\le\delta\lambda_1^{n-k}+\delta(\varepsilon) \tag{10.19}$$

for all $n\in[k,T(\varepsilon)]$.
Let $(z(t),\theta(t))$ be the solution of the averaged system with the initial condition $z(0)=z_0$, $\theta(0)=0$. Then $(|z(t)-\bar{z}(t)|+|\theta(t)-\bar{\theta}(t)|)\to 0$ as $t\to\infty$. It follows from statement II that if $|z(2k\pi)-\bar{z}_\omega|<\delta_1/2$, then

$$\lim_{\varepsilon\to 0}P\big\{|z_\varepsilon(\tau_k^\varepsilon)-\bar{z}_\omega|<\delta_1\big\}=1. \tag{10.20}$$

This completes the proof of the theorem.

□

Remark 8 Assume that for some $\gamma>2$,

$$\limsup_{\varepsilon\to 0}E|\eta_1^\varepsilon|^\gamma<\infty. \tag{10.21}$$

Then $T(\varepsilon)\to\infty$ can be chosen such that

$$\lim_{\varepsilon\to 0}T(\varepsilon)\,\varepsilon^{\gamma/2}=0.$$

If relation (10.21) is true for all $\gamma>0$, then we can choose such a function $T(\varepsilon)$ for which

$$\lim_{\varepsilon\to 0}T(\varepsilon)\,\varepsilon^{\gamma}=+\infty$$

for all $\gamma>0$.

10.1.5 Behavior of the Perturbed System near the Stable Static State

We assume here that $\omega<1$. Then $(-\omega,\theta_1^*(\omega))$ is a static stable state for the averaged system (10.1). Set

$$\tilde{z}_\varepsilon(t)=\omega+z_\varepsilon(t),$$
$$\tilde{\theta}_\varepsilon(t)=\theta_\varepsilon(t)-\theta_1^*(\omega).$$

Theorem 4 *Let $z_\varepsilon(0) = z_0$ and $\theta_\varepsilon(0) = \theta_0$. Assume that $z^*(\omega, \theta_0) > z_0$ if $\omega > \omega^*$. Let a function $T(\varepsilon) : R_+ \to R_+$ be such that $T(\varepsilon) \to \infty$, $\varepsilon T(\varepsilon) \to 0$ as $\varepsilon \to 0$.*
Then there exist constants $C > 0$ and $\alpha > 0$ and a function $\delta(\varepsilon) : R_+ \to R_+$, $\delta(\varepsilon) \to 0$ as $\varepsilon \to 0$, for which the following relation is fulfilled:

$$\lim_{\varepsilon\to 0} P\left\{ \sup_{t\le T(\varepsilon)} \frac{|\tilde{z}_\varepsilon(t)| + |\tilde{\theta}_\varepsilon(t)|}{Ce^{-\alpha t} + \delta(\varepsilon)} \le 1 \right\} = 1. \tag{10.22}$$

Proof Introduce the stochastic processes

$$v_\varepsilon^{(1)}(t) = \frac{1}{\sqrt{\varepsilon}} \int_0^t \left(\frac{1}{\tau(y(\frac{s}{\varepsilon}))} - \frac{1}{\tau} \right) ds,$$

$$v_\varepsilon^{(2)}(t) = \frac{1}{\sqrt{\varepsilon}} \int_0^t \tau^{-1}\left(y\left(\frac{s}{\varepsilon}\right)\right) \left[v\left(y\left(\frac{s}{\varepsilon}\right), \theta_1^*\right) + \omega \right] ds,$$

$$v_\varepsilon^{(3)}(t) = \frac{1}{\sqrt{\varepsilon}} \int_0^t \left(\tau^{-1}\left(y\left(\frac{s}{\varepsilon}\right)\right) v'\left(y\left(\frac{s}{\varepsilon}\right), \theta_1^*\right) - \frac{1}{\tau}\sqrt{1-\omega^2} \right) ds,$$

$$v_\varepsilon^{(4)}(t) = \frac{1}{\sqrt{\varepsilon}} \int_0^t \left[\omega\left(y\left(\frac{s}{\varepsilon}\right)\right) - \omega \right] ds.$$

It follows from system (10.10) that the functions $\tilde{z}_\varepsilon(t)$, $\tilde{\theta}_\varepsilon(t)$ satisfy the system of integral equations

$$\begin{aligned} \tilde{z}_\varepsilon(t) &= \tilde{z}_\varepsilon(0) - \frac{1}{\tau}\int_0^t \tilde{z}_\varepsilon(s)ds + \frac{1}{\tau}\sqrt{1-\omega^2}\int_0^t \tilde{\theta}_\varepsilon(s)ds, \\ &- \sqrt{\varepsilon}\int_0^t \tilde{z}_\varepsilon(s)dv_\varepsilon^{(1)}(s) + \sqrt{\varepsilon}\int_0^t \tilde{\theta}_\varepsilon(s)dv_\varepsilon^{(3)}(s) + \sqrt{\varepsilon}v_\varepsilon^{(2)}(t) + \int_0^t \delta_\varepsilon(s)ds, \end{aligned} \tag{10.23}$$

and

$$\tilde{\theta}_\varepsilon(t) = \int_0^t \tilde{z}_\varepsilon(s)ds + \sqrt{\varepsilon}v_\varepsilon^{(4)}(t), \tag{10.24}$$

where

$$\delta_\varepsilon(t) = v\left(y\left(\frac{s}{\varepsilon}\right), \theta_1^* + \tilde{\theta}_\varepsilon(t)\right) - v\left(y\left(\frac{s}{\varepsilon}\right), \theta_1^*\right) - v'\left(y\left(\frac{s}{\varepsilon}\right), \theta_1^*\right)\tilde{\theta}_\varepsilon(t)$$

and

$$v'(y,\theta) = \frac{\partial v}{\partial \theta}(y,\theta).$$

Let $A(t)$ be the 2×2 matrix

$$A(t) = \exp\{tB\},$$

where

$$B = \frac{1}{\tau}\begin{pmatrix} -1 & \sqrt{1-\omega^2} \\ \tau & 0 \end{pmatrix}.$$

Denote the elements of the matrix $A(t)$ by $a_{11}(t)$, $a_{12}(t)$, $a_{21}(t)$, and $a_{22}(t)$. Then the system of equations (10.23), (10.24) can be rewritten in the form

$$\begin{aligned}\tilde{z}_\varepsilon(t) &= \tilde{z}_\varepsilon(0)a_{11}(t) + \tilde{\theta}_\varepsilon(0)a_{12}(t) \\ &+ \sqrt{\varepsilon}\int_0^t a_{11}(t-s)\big(-\tilde{z}_\varepsilon(s)\,dv_\varepsilon^{(1)}(s) + \tilde{\theta}_\varepsilon(s)\,dv_\varepsilon^{(3)}(s) + dv_\varepsilon^{(2)}(s)\big) \\ &+ \sqrt{\varepsilon}\int_0^t a_{12}(t-s)\,dv_\varepsilon^{(4)}(s) + \int_0^t a_{11}(t-s)\delta_\varepsilon(s)\,ds,\end{aligned} \tag{10.25}$$

$$\begin{aligned}\tilde{\theta}_\varepsilon(t) &= \tilde{z}_\varepsilon(0)a_{21}(t) + \tilde{\theta}_\varepsilon(0)a_{22}(t) \\ &+ \sqrt{\varepsilon}\int_0^t a_{21}(t-s)\big(-\tilde{z}_\varepsilon(s)\,dv_\varepsilon^{(1)}(s) + \tilde{\theta}_\varepsilon(s)\,dv_\varepsilon^{(3)}(s) + dv_\varepsilon^{(2)}(s)\big) \\ &+ \int_0^t a_{21}(t-s)\Delta_\varepsilon(s)\,ds + \sqrt{\varepsilon}\int_0^t a_{22}(t-s)\,dv_\varepsilon^{(4)}(s).\end{aligned} \tag{10.26}$$

Note that

$$|a_{ij}(t)| + |\dot{a}_{ij}(t)| = O(e^{-\delta t}),$$

where $\delta > 0$;

$$\delta_\varepsilon(t) \le k\,(\tilde{\theta}_\varepsilon(t))^2,$$

where k is a constant, and

$$|\dot{\tilde{z}}_\varepsilon(t)| + |\dot{\tilde{\theta}}_\varepsilon(t)| = O(1 + |\tilde{z}_\varepsilon(t)|).$$

Using integration by parts for integrals with respect to $dv_\varepsilon^{(i)}(t)$ we can obtain from (10.25) and (10.26) the inequality

$$\begin{aligned}&|\tilde{z}_\varepsilon(t)| + |\tilde{\theta}_\varepsilon(t)| \le \big(|\tilde{z}_\varepsilon(0)| + |\tilde{\theta}_\varepsilon(0)|\big)e^{-\delta t} \\ &+ c_1\bigg[\int_0^t e^{-\delta(t-s)}|\tilde{\theta}_\varepsilon(s)|^2\,ds \\ &+ \sqrt{\varepsilon}\int_0^t e^{-\delta(t-s)}(1 + |\tilde{z}_\varepsilon(s)| + |\tilde{\theta}_\varepsilon(s)|)\sum_{i=1}^4 |v_\varepsilon^{(i)}(s)|\,ds \\ &+ \sqrt{\varepsilon}(1 + |\tilde{z}_\varepsilon(t)| + |\tilde{\theta}_\varepsilon(t)|)\sum_{i=1}^4 |v_\varepsilon^{(i)}(t)|\bigg].\end{aligned} \tag{10.27}$$

Assume that $|\tilde{z}_\varepsilon(0)| + |\tilde{\theta}_\varepsilon(0)| \le r$ and

$$\tau_\varepsilon = \inf\{s : |\tilde{z}_\varepsilon(s)| + |\tilde{\theta}_\varepsilon(s)| > 2r\}.$$

Then for $t \le \tau_\varepsilon$ the following inequality is fulfilled:

$$|\tilde{z}_\varepsilon(t)| + |\tilde{\theta}_\varepsilon(t)| \le re^{-\delta t} + \frac{c_1}{\delta}r^2 + \sqrt{\varepsilon}(1+2r)\left(\frac{c_1}{\delta}+1\right)\sup_{s\le t}\sum_{i=1}^{4}|v_\varepsilon^{(i)}(s)|. \tag{10.28}$$

Assume that

$$\sup_{t\le T(\varepsilon)}\sum_{i=1}^{4}|v_\varepsilon^{(i)}(t)| \le \frac{\lambda r}{\sqrt{\varepsilon}}. \tag{10.29}$$

Then

$$|\tilde{z}_\varepsilon(t)| + |\tilde{\theta}_\varepsilon(t)| \le r\biggl(1 + \frac{c_1}{\delta}r + \lambda(1+2r)\left(\frac{c_1}{\delta}+1\right)\biggr),$$

and if λ and r are sufficiently small, then $\tau_\varepsilon > T(\varepsilon)$. Using the convergence of the processes $v_\varepsilon^{(i)}(t)$ to Wiener processes, we can write

$$P\biggl\{\sup_{t\le T(\varepsilon)}\sum_{i=1}^{4}|v_\varepsilon^{(i)}(t)| > \frac{\lambda r}{\sqrt{\varepsilon}}\biggr\} = O\biggl(\frac{T(\varepsilon)\varepsilon}{\lambda^2 r^2}\biggr).$$

This implies the proof of the theorem, since the averaged system tends to a static state.

□

Remark 9 As in Remark 8, we can prove the existence of a function $T(\varepsilon)$ satisfying relation (10.22) for which

$$\lim_{\varepsilon\to 0} T(\varepsilon)\varepsilon^\gamma = +\infty$$

for all $\gamma > 0$.

10.1.6 Ergodic Properties of the Perturbed System

Consider the stochastic processes $\bigl(z_\varepsilon(t), \theta_\varepsilon^f(t), y(t/\varepsilon)\bigr)$ in the space $C \times Y$, where $C = R \times [0, 2\pi)$ is the cylinder that was introduced in Remark 7, and where

$$\theta_\varepsilon^f(t) = 2\pi\left\{\frac{1}{2\pi}\theta_\varepsilon(t)\right\},$$

where $\{x\} = x - [x]$, $x \in R$; $[x]$ is the integer part of a number x. The stochastic process $\bigl(z_\varepsilon(t), \theta_\varepsilon^f(t), y(t/\varepsilon)\bigr)$ is a homogeneous Markov process.

Lemma 2 *Assume that* $0 < \tau(y) < \tau_0$ *and* $|v(y,\theta)| \le v_0$. *Then*

$$|z_\varepsilon(t)| \le \frac{1}{2}\tau_0\, v_0$$

for all $t \in R$ *if* $|z_\varepsilon(0)| \le \frac{1}{2}\tau_0\, v_0$.

Proof It follows from equations (10.10) that

$$\begin{aligned}\frac{d}{dt}(z_\varepsilon(t))^2 &\leq -\frac{2}{\tau_0}(z_\varepsilon(t))^2 + z_\varepsilon(t)v\left(y\left(\frac{t}{\varepsilon}\right), \theta_\varepsilon(t)\right)\\ &\leq -\frac{2}{\tau_0}(z_\varepsilon(t))^2 + \frac{1}{\tau_0}(z_\varepsilon(t))^2 + \frac{1}{4}\tau_0 v^2\left(y\left(\frac{t}{\varepsilon}\right), \theta_\varepsilon(t)\right)\\ &< -\frac{1}{\tau_0}(z_\varepsilon(t))^2 + \frac{1}{4}\tau_0 v_0^2.\end{aligned}$$

This completes the proof.

□

So we can consider the Markov process on the set $[-a_0, a_0] \times [0, 2\pi] \times Y$, where $a_0 = \frac{1}{2}\tau_0\, v_0$.

Lemma 3 *Suppose that the three-dimensional Gaussian stochastic process that was introduced in statement* II *is nondegenerate, i.e.,*

$$E(\alpha_1 v_1(t) + \alpha_2 v_2(t) + \alpha_3 v_3(t))^2 > 0$$

if $\alpha_1^2 + \alpha_2^2 + \alpha_3^2 > 0$, $\alpha_i \in R$, $i = 1, 2, 3$. *Then there exists an open set* $G \subset [-a_0, a_0] \times [0, 2\pi)$ *that contains the curves* (α) *and* (β) *introduced in Remark 7 and the static state* $(-\omega, \theta_1(\omega))$ *satisfies the property that for all open sets* $G_1 \subset G$ *and sets* $C \in \mathcal{C}$ *with* $\rho(C) > 0$ *the following inequality holds for all sufficiently small* $\varepsilon > 0$*:*

$$P\left\{(z_\varepsilon(t), \theta_\varepsilon(t)) \in G_1, y\left(\frac{t}{\varepsilon}\right) \in C / z_\varepsilon(0) = z_0,\, \theta_\varepsilon(0) = \theta_0,\, y(0) = y_0\right\} > 0$$

for all $t > 0$ *and* $(z_0, \theta_0, y_0) \in [-a_0, a_0] \times [0, 2\pi) \times Y$.

The proof of this statement follows directly from statement II.

Lemma 4 *Assume that for all* $\varepsilon > 0$ *small enough the distribution of the pair of random variables* $(z_\varepsilon(t), \theta_\varepsilon(t))$ *has the joint density that is absolutely continuous with respect to Lebesgue measure on the plane. If the conditions of Lemmas 2 and 3 are fulfilled, then the Markov process* $(z_\varepsilon(t), \theta_\varepsilon(t), y(t/\varepsilon))$ *is ergodic, and its ergodic distribution has a density with respect to the product of Lebesgue measure on* $[-a_0, a_0] \times [0, 2\pi)$ *and the ergodic measure* $\rho(dy)$ *on* Y.

Proof We can verify that the Markov process is Harris recurrent with respect to the measure described in the statement of the lemma. (See Section 1.2).

□

Remark 10 It follows from Lemma 3 that the density of the ergodic distribution, denoted by $\varphi_\varepsilon(z, \theta, y)$, satisfies the property

$$\int \varphi_\varepsilon(z, \theta, y)\rho(dy) > 0 \quad \text{if } (z, \theta) \in G.$$

Therefore, for $\varepsilon > 0$,

$$P\left\{\limsup_{t\to\infty}(|z_\varepsilon(t)+\omega| + |\theta_\varepsilon(t)-\theta_1^*(\omega)|)=0\right\}=0$$

and

$$P\left\{\limsup_{t\to\infty}(|z_\varepsilon(t)-\bar{z}(t)| + |\theta_\varepsilon(t)-\bar{\theta}(t)|)=0\right\}=0$$

for all initial values z_0, θ_0, y_0.

Remark 11 Set

$$\varphi_\varepsilon(z,\theta)=\int \varphi_\varepsilon(z,\theta,y)\rho(dy).$$

This function satisfies the relation

$$\lim_{T\to\infty}\frac{1}{T}\int_0^T G\big(z_\varepsilon(t),\theta_\varepsilon(t)\big)dt=\int_{-a_0}^{a_0}\int_0^{2\pi} G(z,\theta)\varphi_\varepsilon(z,\theta)\,dz\,d\theta$$

for any bounded measurable function $G:\,[-a_0,a_0]\times[0,2\pi]\to R$.
It follows from Theorems 3 and 4 that if there exists the limit of the distribution

$$m_\varepsilon(A)=\int_A \varphi_\varepsilon(z,\theta)\,dz\,d\theta,\quad A\in\mathcal{B}([-a_0,a_0]\times[0,2\pi]),$$

say $m(A)$, then the support of the measure m is the static state $(-\omega,\theta_1^*(\omega))$ and the curve $\{(z,\theta):\, z=Z_\omega(\theta,\bar{z}_\omega)\}$, and

$$\int G(z,\theta)m(dz,d\theta)=\alpha G(-\omega,\theta_1^*(\omega))+\beta\frac{1}{2\pi c_\omega}\int_0^{2\pi}\frac{G(Z_\omega(\theta,\bar{z}_\omega))}{\omega+Z_\omega(\theta,\bar{z}_\omega)}\,d\theta$$

(for notation see Theorem 1), $\alpha\ge 0$, $\beta\ge 0$, $\alpha+\beta=1$.

10.2 The Forced Problem

The form of the phase detector characteristic $\mathcal{P}$ (see the beginning of the chapter) varies with various applications. In most applications it is simple multiplication of the external signal and the VCO output:

$$\mathcal{P}(\theta,t)=\alpha V(\theta)\cos(\mu t+\varphi),$$

where μ and φ are the frequency and phase deviation of the input signal, respectively. In the case $V(\theta)=\beta\cos\theta$ and if the loop filter is a low-pass filter with time constant τ, then the model is described by the system of differential equations

$$\begin{aligned}\tau\dot{z}&=-z+A\cos\theta\cos(\mu t+\varphi),\\ \dot{\theta}&=\omega+z.\end{aligned}\tag{10.30}$$

10.2.1 Systems Without a Loop Filter (Nonrandom).

In the case $\tau = 0$, system (10.30) is represented by one equation of the form

$$\dot{\theta} = F(\theta, \mu t + \varphi),$$

where the function $F(\theta, \eta)$ from R^2 to R is sufficiently smooth and satisfies the condition

$$F(\theta, \eta) = F(\theta + 2\pi, \eta) = F(\theta, \eta + 2\pi)$$

for all θ and η. Let

$$\eta(t) = \mu t + \varphi.$$

Then the pair of functions $(\eta(t), \theta(t))$ satisfies the system of the differential equations

$$\begin{aligned} \dot{\eta}(t) &= \mu, \\ \dot{\theta}(t) &= F\big(\theta(t), \eta(t)\big), \end{aligned} \tag{10.31}$$

which can be considered a system on the torus $C^2_{2\pi}$, where $C_{2\pi}$ is the circle of radius 1.

We can use the results of Section 9.1 to investigate the asymptotic behavior of the average value of the form

$$\lim_{T\to\infty} \bar{H}_T(\eta_1, \theta_1, \eta_2, \theta_2) \equiv \lim_{T\to\infty} \frac{1}{T} \int_0^T H(\eta_1(s), \theta_1(s), \eta_2(s), \theta_2(s))\, ds, \tag{10.32}$$

where $H(x_1, x_2, x_3, x_4)$ is a continuous real-valued function on R^4 that is 2π-periodic in all variables:

$$H(x_1, x_2, x_3, x_4) \equiv H(x_1 + 2k_1\pi, \ldots, x_4 + 2k_4\pi)$$

for all $k_1, \ldots, k_4 \in \mathbf{Z}$, and $(\eta_1(t), \theta_1(t))$, $(\eta_2(t), \theta_2(t))$ are two solutions to system (10.31) with initial values (η_1^0, θ_1^0) and (η_2^0, θ_2^0), respectively.

Let ρ be the rotation number for the system. If ρ is a rational number, denote by t_0 the period of the system, and let Π denote the set of those θ_0 for which the solution to the equation

$$\frac{d\Theta}{d\eta}(\theta_0, \eta) = \frac{1}{\mu} F\big(\Theta(\theta_0, \eta), \eta\big) \tag{10.33}$$

is a periodic function. For every θ there exists $\bar{\theta} \in \Pi$ for which

$$\lim_{\eta\to+\infty} |\Theta(\theta, \eta) - \Theta(\bar{\theta}, \eta)| = 0.$$

Set $\bar{\theta} = \pi(\theta)$. Then $\pi(\theta) = \theta$ if $\theta \in \Pi$ and $\pi(\theta) = \alpha$ if $\Theta(\theta, 2\pi) < \theta$ and $\theta \in (\alpha, \beta)$, $\alpha, \beta \in \Pi$, $(\alpha, \beta) \cap \Pi = \varnothing$, $\pi(\theta) = \beta$ if $\Theta(\theta, 2\pi) > \theta$ and θ satisfies the previous condition.

If ρ is an irrational number and system (10.31) is ergodic, then there exists a continuous function $\chi(\kappa_1, \kappa_2)$ for which

$$\chi(\kappa_1, \kappa_2) \equiv \chi(\kappa_1 + 2\pi, \kappa_2) \equiv \chi(\kappa_1, \kappa_2 + 2\pi),$$

and the solution to equation (10.33) is represented in the form

$$\Theta(\theta_0, \eta) = g(\theta_0) + \rho\eta + \chi(\eta, g(\theta_0) + \rho\eta), \tag{10.34}$$

where $g(y)$ is a continuous and strictly increasing function.
All of these results were presented in Section 9.1, but we require some additional notation here. Denote by $\Theta(\eta_0, \theta_0, \eta)$ the solution of equation (10.33) with the initial condition

$$\Theta(\eta_0, \theta_0, \eta_0) = \theta_0.$$

We first consider the arguments of the function $\Theta(\eta_0, \theta_0, \eta)$ in R. The function $\Theta(\eta_0, \theta_0, \eta)$ can be represented by $\Theta(\theta_0, \eta)$ in the following way:

$$\Theta(\eta_0, \theta_0, \eta) = \Theta(\Theta(\eta_0, \theta_0, -\eta_0), \eta_0 + \eta).$$

Theorem 5 *Let $H(x_1, x_2, x_3, x_4)$ be a continuous R-valued function that is 2π-periodic in all of its arguments and $\bar{H}_T$ is defined by formula (10.32).* *(1) If system (10.31) is periodic with period t_0 and if the set Π and the function $\pi(\theta)$ are defined as above, then*

$$\lim_{T\to\infty} \bar{H}_T(\eta_1, \theta_1, \eta_2, \theta_2) = \frac{1}{\mu t_0} \int_0^{\mu t_0} H\big(s, \Theta(\tilde{\theta}_1^0, s), \eta_2^0 - \eta_1^0 + s, \Theta(\tilde{\theta}_2^0, s)\big)\, ds,$$

where

$$\tilde{\theta}_i^0 = \pi\big(\Theta(\eta_i^0, \theta_i^0, -\eta_i^0)\big), \quad i = 1, 2.$$

(2) If system (10.31) is ergodic with rotation number ρ, then

$$\lim_{T\to\infty} \bar{H}_T(\eta_1, \theta_1, \eta_2, \theta_2)$$
$$= \frac{1}{4\pi^2} \int_0^{2\pi} \int_0^{2\pi} H(u, v + \chi(u, v), a + u, b + v + \chi(a + u, b + v))\, du\, dv,$$

where $a = \eta_2^0 - \eta_1^0$ and

$$b = \rho a + g\big(\Theta(\eta_2^0, \theta_2^0, -\eta_2^0)\big) - g\big(\Theta(\eta_1^0, \theta_1^0, -\eta_1^0)\big).$$

Proof Let

$$\hat{\theta}_i^0 = \Theta(\eta_i^0, \theta_i^0, -\eta_i^0).$$

Then

$$\theta_i(t) = \Theta(\hat{\theta}_i^0, \eta_i(t)), \quad i = 1, 2,$$

and

$$\eta_2(t) = \eta_1(t) + a,$$

$$\bar{H}_T(\eta_1, \theta_1, \eta_2, \theta_2)$$
$$= \frac{1}{T} \int_0^T H\big(\eta_1(s), \Theta(\hat{\theta}_1^0, \eta_1(s)), \eta_1(s) + a, \Theta(\hat{\theta}_2^0, \eta_1(s) + a)\big)\, ds.$$

Under condition (1) we have

$$\lim_{t\to\infty} \big(\Theta(\hat{\theta}_i^0, \eta_i(t)) - \Theta(\tilde{\theta}_i^0, \eta_i(t))\big) = 0,$$

and we can write

$$\lim_{T\to\infty} \bar{H}_T(\eta_1, \theta_1, \eta_2, \theta_2)$$
$$= \lim_{T\to\infty} \frac{1}{T} \int_0^T H\big(\eta_1(s), \Theta(\tilde{\theta}_1^0, \eta_1(s)), \eta_1(s) + a, \Theta(\tilde{\theta}_2^0, \eta_1(s) + a)\big)\, ds. \tag{10.35}$$

The function

$$H\big(\eta_1^0 + \mu s, \Theta(\tilde{\theta}_1^0, \eta_1^0 + \mu s), \eta_2^0 + \mu s, \Theta(\tilde{\theta}_2^0, \eta_2^0 + \mu s)\big)$$

is a periodic function with period t_0. So the limit on the right-hand side of equation (10.35) equals

$$\frac{1}{t_0} \int_0^{t_0} H\big(\eta_1^0 + \mu s, \Theta(\tilde{\theta}_1^0, \eta_1^0 + \mu s), \eta_2^0 + \mu s, \Theta(\tilde{\theta}_2^0, \eta_2^0 + \mu s)\big)\, ds.$$

This proves statement (1) of the theorem.
In case (2) we use the representation (10.34) and write

$$\bar{H}_T(\eta_1, \theta_1, \eta_2, \theta_2)$$
$$= \frac{1}{T} \int_0^T H\big(\eta_1(s), g(\hat{\theta}_1^0) + \rho\eta, \eta_1(s) + a,$$
$$b + g_1(\hat{\theta}_1^0) + \rho\eta_1(s) + \chi(a + \eta_1(s), b + \rho\eta_1(s))\big)\, ds$$
$$= \frac{1}{\mu T} \int_{\eta_1^0}^{\eta_1^0 + \mu T} H\big(u, g(\hat{\theta}_1^0) + \rho u + \chi(u, g(\hat{\theta}_1^0) + \rho u), u + a,$$
$$b + g_1(\hat{\theta}_1^0) + \rho u + \chi(u + a, b + g_1(\hat{\theta}_1^0) + \rho u)\big)\, du.$$

The remainder of the proof follows from the following lemma.

Lemma 5 *Let $\Phi(x, y)$ be a continuous 2π-periodic function in x and y. Then for any irrational number ρ,*

$$\lim_{T\to\infty} \frac{1}{T} \int_0^T \Phi(t, \rho t) dt = \frac{1}{4\pi^2} \int_0^{2\pi} \int_0^{2\pi} \Phi(u, v)\, du\, dv. \tag{10.36}$$

Proof It follows from the theorem on approximation of continuous functions by trigonometric polynomial functions that we have to prove (10.36) only for $\Phi(x, y) = e^{i2\pi lx + i2\pi ky}$. In this case the right-hand side of (10.36) is zero if $|l| + |k| > 0$. It is easy to check that the left–hand side of (10.36) is equal to zero if $|l| + |k| > 0$. For the function $\Phi = 1$ relation (10.36) holds. □

Now we return to the proof of the theorem. The function

$$\Phi(x, y) = H\big(x, \alpha + y + \chi(x, \alpha + y), x + a, b + \alpha + y, \chi(x + a, b + \alpha + y)\big)$$

satisfies the conditions of Lemma 5. So

$$\lim_{T\to\infty} \bar{H}_T(\eta_1, \theta_1, \eta_2, \theta_2) = \frac{1}{4\pi^2}\int_0^{2\pi}\int_0^{2\pi} H\big(x, \alpha + y + \chi(x, \alpha + y), x + a, b + \alpha + y, \chi(x + a, b + \alpha + y)\big)\, dx\, dy$$

$$= \frac{1}{4\pi^2}\int_0^{2\pi}\int_0^{2\pi} H\big(x, y + \chi(x, y), x + a, b + y, \chi(x + a, b + y)\big)\, dx\, dy.$$

Here $\alpha = g(\hat{\theta}_1^0)$.

□

Corollary 1 Let $H_1(x, y)$ be a continuous 2π-periodic function in x and y. Set

$$\zeta(t) = H_1\big(\eta(t), \theta(t)\big),$$

where $(\eta(t), \theta(t))$ is the solution to system (10.31) with initial conditions $\eta(0) = \eta_0$, $\theta(0) = \theta_0$. It follows from Theorem 5 that the autocovariance function of $\zeta(t)$ is defined:

$$R_\zeta(s) = \lim_{T\to\infty} \frac{1}{T}\int_0^T \zeta(t)\zeta(t+s)dt - \left(\lim_{T\to\infty}\frac{1}{T}\int_0^T \zeta(t)dt\right)^2. \qquad (10.37)$$

If $(\eta(t), \theta(t))$ is a periodic solution with period t_0, then

$$R_\zeta(s) = \frac{1}{\mu t_0}\int_0^{\mu t_0} H_1(u, \Theta(\hat{\theta}_0, u))H_1(u + \mu s, \Theta(\hat{\theta}_0, u + \mu s))\, du - \left(\frac{1}{\mu t_0}\int_0^{\mu t_0} H_1(u, \Theta(\hat{\theta}_0, u))\, du\right)^2, \qquad (10.38)$$

where $\hat{\theta}_0 = \Theta(\eta_0, \theta_0, -\eta_0)$.
If system (10.31) is ergodic with rotation number ρ, then

$$R_\zeta(s) = \frac{1}{4\pi^2}\int_0^{2\pi}\int_0^{2\pi} H_1(u, v + \chi(u, v))H_1(u + \mu s, v + \rho\mu s + \chi(u + \mu s, v + \rho\mu s)\, du\, dv - \frac{1}{4\pi^2}\left(\int_0^{2\pi}\int_0^{2\pi} H_1(u, v + \chi(u, v))\, du\, dv\right)^2. \qquad (10.39)$$

10.2.2 Randomly Perturbed Systems Without a Loop Filter

Here we consider randomly perturbed systems of the form represented by equations (10.31). Assume that

$$\big(\mu(t), F(t, \theta, \eta)\big) \tag{10.40}$$

is an ergodic stationary process in the space

$$R \times C_{p,[0,2\pi]^2},$$

where $C_{p,[0,2\pi]^2}$ is the space of continuous functions $f(x,y)$ on $[0,2\pi]^2$ for which $f(0,y) \equiv f(2\pi, y)$ and $f(x,0) \equiv f(x,2\pi)$.
We consider the stochastic process $(\eta_\varepsilon(t), \theta_\varepsilon(t))$ defined by the system of differential equations

$$\begin{aligned} \dot{\eta}_\varepsilon(t) &= \mu\left(\frac{t}{\varepsilon}\right), \\ \dot{\theta}_\varepsilon(t) &= F\left(\frac{t}{\varepsilon}, \theta_\varepsilon(t), \eta_\varepsilon(t)\right). \end{aligned} \tag{10.41}$$

We assume that the stationary process $F(t, \theta, \eta)$ satisfies the condition that the derivative $F_\theta(t, \theta, \eta)$ exists and there is a constant $C > 0$ for which $|F_\theta(t, \theta, \eta)| \le C$ with probability 1. Then the averaging theorem can be applied to system (10.41) (see Theorem 5 of Chapter 5). Let

$$\mu = E\mu(t), \quad F(\theta, \eta) = EF(t, \theta, \eta).$$

If $\eta_\varepsilon(0) = \eta_0$, $\theta_\varepsilon(0) = \theta_0$, then

$$P\left\{\lim_{\varepsilon\to 0} \sup_{t \le T}\big(|\eta_\varepsilon(t) - \eta(t)| + |\theta_\varepsilon(t) - \theta(t)|\big) = 0\right\} = 1 \tag{10.42}$$

for all $T > 0$, where $(\eta(t), \theta(t))$ is the solution to system (10.31) with the initial condition $\eta(0) = \eta_0$, $\theta(0) = \theta_0$.

Remark 12 Assume that there exists $\varepsilon_0 > 0$ for which the function

$$\zeta_\varepsilon(t) = H_1(\eta_\varepsilon(t), \theta_\varepsilon(t)), \tag{10.43}$$

where the function $H_1(x,y)$ is the same as in Corollary 1, has autocovariance function, for $\varepsilon < \varepsilon_0$,

$$R_{\zeta_\varepsilon}(s) = \lim_{T\to\infty} \frac{1}{T}\int_0^T \zeta_\varepsilon(t)\zeta_\varepsilon(t+s)\,dt - \left(\lim_{T\to\infty}\frac{1}{T}\int_0^T \zeta_\varepsilon(t)\,dt\right)^2, \tag{10.44}$$

and the limits in relation (10.44) exist uniformly in ε as $\varepsilon \to 0$.
Then

$$\lim_{\varepsilon\to 0} R_{\zeta_\varepsilon}(s) = R_\zeta(s),$$

where $R_\zeta(s)$ is represented by formula (10.38) if the averaged system is periodic and by formula (10.39) if it is ergodic.

We will next investigate the existence and asymptotic behavior of the autocovariance for the function $\zeta_\varepsilon(t)$ given by relation (10.43) for systems with Markov perturbation.

Let $y(t)$ be a homogeneous Markov process in a measurable space $(Y, \mathcal{C})$ with a transition probability $P(t, y, C)$ for $C \in \mathcal{C}$. We assume that the Markov process $y(t)$ is uniformly ergodic with an ergodic distribution $\rho(dy)$. Assume that $(\eta_\varepsilon(t), \theta_\varepsilon(t))$ is the solution to the system of differential equations

$$\begin{aligned} \dot{\eta}_\varepsilon(t) &= \mu\left(y\left(\frac{t}{\varepsilon}\right)\right), \\ \dot{\theta}_\varepsilon(t) &= F\left(\theta_\varepsilon(t), \eta_\varepsilon(t), y\left(\frac{t}{\varepsilon}\right)\right), \end{aligned} \tag{10.45}$$

where $\mu(y)$ is a nonnegative bounded measurable function for which

$$\mu = \int \mu(y)\rho(dy) < \infty, \quad \mu > 0,$$

and $F(\theta, \eta, y) : R^2 \times Y \to R$ is a bounded measurable function with respect to $\mathcal{B}(R^2) \otimes \mathcal{C}$ satisfying the following properties:

(1) F is continuous in θ, η uniformly with respect to y.

(2) $F(\theta, \eta, y) = F(\theta + 2\pi, \eta, y) = F(\theta, \eta + 2\pi, y)$ for all θ, η, y.

(3) There exist continuous derivatives $F_\theta(\theta, \eta, y)$, $F_\eta(\theta, \eta, y)$ in θ, η that are bounded in all arguments.

Let

$$F(\theta, \eta) = \int F(\theta, \eta, y)\rho(dy).$$

Then system (10.31) is the averaged system for (10.45), and relation (10.42) holds.

Remark 13 Denote by $(\eta_\varepsilon(\eta_0, \theta_0, t), \theta_\varepsilon(\eta_0, \theta_0, t))$ the solution to system (10.41) with the initial conditions

$$\eta_\varepsilon(\eta_0, \theta_0, 0) = \eta_0, \quad \theta_\varepsilon(\eta_0, \theta_0, 0) = \theta_0.$$

It follows from condition (3) that $\eta_\varepsilon(\eta_0, \theta_0, t)$ and $\theta_\varepsilon(\eta_0, \theta_0, t)$ are continuous in θ_0, η_0, uniformly with respect to $\varepsilon > 0$ and $t \le t_0$. So if $(\eta(\eta_0, \theta_0, t), \theta(\eta_0, \theta_0, t))$ is the solution to system (10.31) with the same initial condition, then

$$\begin{aligned} P\Big\{ \lim_{\varepsilon \to 0} \sup_{t \le t_0} \sup_{\eta_0, \theta_0} \big(|\eta_\varepsilon(\eta_0, \theta_0, t) - \eta(\eta_0, \theta_0, t)| \\ + |\theta_\varepsilon(\eta_0, \theta_0, t) - \theta(\eta_0, \theta_0, t)|\big) = 0 \Big\} = 1. \end{aligned} \tag{10.46}$$

Theorem 6 *Let $H_1(x, y)$ satisfy the conditions of Corollary 1, and let the averaged system be ergodic. Set*

$$R_{\zeta_\varepsilon}(T, s) = \frac{1}{T} \int_0^T \zeta_\varepsilon(t)\zeta_\varepsilon(t+s)\, ds - \left(\frac{1}{T} \int_0^T \zeta_\varepsilon(t)\, dt \right)^2,$$

where $\zeta_\varepsilon(t)$ is defined by relation (10.43).
Then for any $s_0 > 0$ and $\delta > 0$ the following relation holds:

$$\lim_{\varepsilon \to 0} P \left\{ \limsup_{T \to \infty} \sup_{|s| \le s_0} |R_{\zeta_\varepsilon}(T, s) - R_\zeta(s)| > \delta \right\} = 0, \tag{10.47}$$

where $R_\zeta(s)$ is given by formula (10.39).

Proof Set

$$A_{\zeta_\varepsilon}(T) = \frac{1}{T} \int_0^T \zeta_\varepsilon(t)\, dt,$$
$$B_{\zeta_\varepsilon}(T, s) = \frac{1}{T} \int_0^T \zeta_\varepsilon(t)\zeta_\varepsilon(t+s)\, dt,$$
$$A_\zeta = \lim_{T \to \infty} \frac{1}{T} \int_0^T \zeta(t)\, dt,$$
$$B_\zeta(s) = \frac{1}{T} \int_0^T \zeta(t)\zeta(t+s)\, dt.$$

To prove the theorem we will show that

$$\lim_{\varepsilon \to 0} P \left\{ \limsup_{T \to \infty} \sup_{|s| \le s_0} |B_{\zeta_\varepsilon}(T, s) - B_\zeta(s)| > \delta \right\} = 0 \tag{10.48}$$

for all $\delta > 0$, $s > 0$, and

$$\lim_{\varepsilon \to 0} P \left\{ \limsup_{T \to \infty} |A_{\zeta_\varepsilon}(T) - A_\zeta| > \delta \right\} = 0. \tag{10.49}$$

It follows from the boundedness of μ, F, and $\partial F / \partial \theta$ that $\zeta_\varepsilon(t)$ is uniformly continuous in t with respect to $\varepsilon > 0$. So $B_{\zeta_\varepsilon}(T, s)$ is uniformly continuous in s with respect to $\varepsilon > 0$ and $T > 0$. Therefore, relation (10.47) is fulfilled if

$$\lim_{\varepsilon \to 0} P \left\{ \limsup_{T \to \infty} |B_{\zeta_\varepsilon}(T, s) - B_\zeta(s)| > \delta \right\} = 0 \tag{10.50}$$

for any s. We prove this relation and note that relation (10.48) can be proved in the same way.
Set $T_k = k\,(\hat{T} + s_0)$, $k = 0, 1$, where $s_0 > |s|$ and $\hat{T}$ is chosen such that

$$\sup_{\eta_0, \theta_0} \left| \frac{1}{\hat{T}} \int_0^{\hat{T}} \zeta(t)\zeta(t+s)\, ds - B_\zeta(s) \right| + \frac{2C_1 s_0}{\hat{T}} < \delta_1,$$

where δ_1 is a given number and $C_1 = (\sup_{x,y} H(x,y))^2$.
Denote by $\mathcal{F}_t^\varepsilon$ the σ-algebra generated by $\{\eta_\varepsilon(u), \theta_\varepsilon(u), y(\frac{u}{\varepsilon}),\ u \le t\}$.
Let λ_k, $k = 0,1,2,\ldots$, be some random variables for which $|\lambda_k| \le 1$. Then we can write

$$B_{\zeta_\varepsilon}(T_n, s) - B_\zeta(s) = \frac{1}{T_n}\int_0^{T_n}\int_0^{T_n} \zeta_\varepsilon(t)\zeta_\varepsilon(t+s)\,ds - B_\zeta(s)$$

$$= \frac{1}{n(\hat{T}+s_0)}\sum_{k=1}^n\Big(\int_{T_k}^{T_k+\hat{T}} \zeta_\varepsilon(t)\zeta_\varepsilon(t+s)\,dt + \lambda_k C_1 s_0\Big) - B_\zeta(s)$$

$$= \frac{1}{n(\hat{T}+s_0)}\Big(\sum_{k=1}^n E\Big(\int_{T_k}^{T_k+\hat{T}} \zeta_\varepsilon(t)\zeta_\varepsilon(t+s)\,dt \Big/ \mathcal{F}_{T_k}^\varepsilon\Big) - B_\zeta(s)(\hat{T}+s_0)$$

$$+ \lambda_k C_1 s_0 + \sum_{k=1}^n\Big[\int_{T_k}^{T_k+\hat{T}} \zeta_\varepsilon(t)\zeta_\varepsilon(t+s)\,ds - E\int_{T_k}^{T_k+\hat{T}} \zeta_\varepsilon(t)\zeta_\varepsilon(t+s)\,dt\Big]\Big).$$

It follows from Lemma 3 of Chapter 9 that the expression on the last line tends to zero with probability 1 as $n \to \infty$. Set

$$\hat{Z}_{\hat{T}}(\eta_0, \theta_0, y(0), s) = E\Big(\int_0^{\hat{T}} \zeta_\varepsilon(t)\zeta_\varepsilon(t+s)\,dt \Big/ \mathcal{F}_0^\varepsilon\Big).$$

Then

$$E\Big(\int_{T_k}^{T_k+\hat{T}} \zeta_\varepsilon(t)\zeta_\varepsilon(t+s)\,dt \Big/ \mathcal{F}_{T_k}^\varepsilon\Big) = \hat{Z}_{\hat{T}}(\eta_\varepsilon(T_k), \theta_\varepsilon(T_k), y(T_k/\varepsilon), s).$$

Remark 13 implies that

$$\lim_{\varepsilon\to 0} E\Big|\hat{Z}_{\hat{T}}(\eta_\varepsilon(T_k), \theta_\varepsilon(T_k), y(T_k/\varepsilon), s)$$
$$- \int_0^{\hat{T}} H_1(\eta(\eta_\varepsilon(T_k), \theta_\varepsilon(T_k), t), \theta(\eta_\varepsilon(T_k), \theta_\varepsilon(T_k), t))$$
$$\times H_1(\eta(\eta_\varepsilon(T_k), \theta_\varepsilon(T_k), t+s), \theta(\eta_\varepsilon(T_k), \theta_\varepsilon(T_k), t+s))\,dt\Big| = 0 \tag{10.51}$$

uniformly in k.
So we write the difference

$$B_{\zeta_\varepsilon}(T_n, s) - B_\zeta(s) = \alpha_n(\varepsilon) + \frac{1}{n}\sum_{k=0}^{n-1}\Delta_k(\varepsilon)$$
$$+ \frac{1}{n}\sum_{k=1}^n\Big[\frac{1}{\hat{T}+s_0}\int_0^{\hat{T}} \zeta^{(k)}(t)\zeta^{(k)}(t+s)\,dt - B_\zeta(s) + \lambda_k C_1 s_0\Big],$$

where $\alpha_n(\varepsilon) \to 0$ with probability 1 as $n \to \infty$, $E|\Delta_k(\varepsilon)| \to 0$ as $\varepsilon \to 0$ uniformly in k, Δ_k is the expression under the absolute value on the left–

hand side of relation (10.51), and $\zeta^{(k)}(t)$ is calculated for the solution of the averaged system (10.31) with initial conditions

$$\eta(0) = \eta_\varepsilon(T_k), \quad \theta(0) = \theta_\varepsilon(T_k).$$

This implies the relation

$$\limsup_{\varepsilon \to 0} P\left\{\limsup_n |B_{\zeta_\varepsilon}(T_n, s) - B_\zeta(s)| > \delta_1\right\} = 0.$$

Note that for $T_n < T < T_{n+1}$ we have

$$|B_{\zeta_\varepsilon}(T) - B_{\zeta_\varepsilon}(T_n)| = O\left(\frac{1}{n}\right).$$

Thus is relation (10.49) proved.

□

Remark 14 Assume that the averaged system (10.31) for system (10.45) is pure periodic. Denote by $R_\zeta(s, \hat{\theta}_0)$ the expression on the right-hand side of (10.38). It follows from Theorems 9 and 10 of Chapter 9 that the function

$$\hat{\theta}_\varepsilon(t) = \Theta\left(\frac{t}{\varepsilon}, \theta_\varepsilon\left(\frac{t}{\varepsilon}\right), -\frac{t}{\varepsilon}\right)$$

satisfies the condition

$$\lim_{\varepsilon \to 0} \limsup_{t \to \infty} \left|\frac{1}{t}\int_0^t h\big(\hat{\theta}_\varepsilon\big(\frac{s}{\varepsilon}\big)\big)\, ds - \int_0^{2\pi} h(\theta) g_1(\theta)\, d\theta\right| = 0$$

with probability 1, where $h(\theta)$ is any 2π-periodic continuous function and $g_1(\theta)$ is defined in Remark 3 of Chapter 9.
This implies the following relation: For all $s \in R$,

$$\lim_{\varepsilon \to 0} \limsup_{T \to \infty} \left| R_{\zeta_\varepsilon}(T, s) - \int_0^{2\pi} R_\zeta(s, \theta) g_1(\theta)\, d\theta\right| = 0,$$

where $R_{\zeta_\varepsilon}(T, s)$ is defined in Theorem 6.

10.2.3 Forced Systems with Noisy Input Signals

Finally, we consider a system of the form

$$\begin{aligned} \tau \dot{z}_\varepsilon(t) &= -z_\varepsilon(t) + A\cos\theta_\varepsilon(t)\cos(\mu t + \varphi + n_\varepsilon(t)), \\ \dot{\theta}_\varepsilon(t) &= \omega + z_\varepsilon(t), \end{aligned} \tag{10.52}$$

where $n_\varepsilon(t)$ is a fast random noise process that can be expressed by either a stationary stochastic process $\xi(t/\varepsilon)$ or a Markov process $y(t/\varepsilon)$. We assume that the noise process satisfies the following conditions:

(i) The pair of stochastic processes

$$w_\varepsilon^1(t) = \frac{1}{\sqrt{\varepsilon}} \int_0^t \cos n_\varepsilon(s)\, ds, \quad w_\varepsilon^2(t) = \frac{1}{\sqrt{\varepsilon}} \int_0^t \sin n_\varepsilon(s)\, ds,$$

converges weakly to a two-dimensional process $\big(w^1(t), w^2(t)\big)$, which is Gaussian with independent increments.

(ii) Denote by $\mathcal{F}_t^\varepsilon$ the σ-algebra generated by $n_\varepsilon(s)$, $s \le t$. There exists a constant c_1 such that for any family of functions $f_\varepsilon(t)$ that are $\mathcal{F}_t^\varepsilon$-measurable and

$$\sup_t \big[|\dot{f}_\varepsilon(t)| + |f_\varepsilon(t)|\big] \le C$$

for some nonrandom $C > 0$ the following inequality holds:

$$\lim_{\varepsilon\to 0} \sup_t E\bigg(\bigg(\int_t^{t+T} f_\varepsilon(s) dw_\varepsilon^k(s)\bigg)^2 \Big/ \mathcal{F}_t^\varepsilon\bigg) \le c_1 C^2 T.$$

Let $z_\varepsilon(0) = z_0$.
We can investigate the behavior of the stochastic process $\theta_\varepsilon(t)$ as $\varepsilon \to 0$ and $t \to \infty$. In fact, it follows from the first equation of system (10.52) that

$$z_\varepsilon(t) = z_0 + \frac{A\sqrt{\varepsilon}}{\tau} \int_0^t e^{-\frac{(t-s)}{\tau}} \cos\theta_\varepsilon(t) \big[\cos(\mu s+\varphi) dw_\varepsilon^1(s) + \sin(\mu s+\varphi) dw_\varepsilon^2(s)\big]$$

Substituting this expression in the second equation of system (10.52) and integrating, we obtain the relation

$$\begin{aligned} \theta_\varepsilon(t) = \theta_0 + t(\omega + z_0) + A\sqrt{\varepsilon} \int_0^t \big(1 - e^{-\frac{1}{\tau}(t-s)}\big) \\ \times \cos\theta_\varepsilon(t) \big[\cos(\mu s + \varphi)\, dw_\varepsilon^1(s) - \sin(\mu s + \varphi)\, dw_\varepsilon^2(s)\big], \end{aligned} \tag{10.53}$$

where $\theta_0 = \theta_\varepsilon(0)$. It follows from condition (ii) that

$$\sup_t E\bigg(\frac{\theta_\varepsilon(t) - \theta_0 - t(\omega + z_0)}{\sqrt{\varepsilon}}\bigg)^2 < \infty.$$

Set

$$\nu_\varepsilon(t) = \varepsilon^{-1/2}(\theta_\varepsilon(t) - \theta_0 - t(\omega + z_0)).$$

Then

$$\begin{aligned} \nu_\varepsilon(t) = \int_0^t \big(1 - e^{-\frac{1}{\tau}(t-s)}\big) \cos(\theta_0 + t(\omega + z_0) + \sqrt{\varepsilon}\nu_\varepsilon(t)) \\ \times \big[\cos(\mu s + \varphi) dw_\varepsilon^1(s) - \sin(\mu s + \varphi) dw_\varepsilon^2(s)\big]. \end{aligned}$$

Let

$$\begin{aligned} \tilde{\nu}_\varepsilon(t) = \int_0^t \big(1 - e^{-\frac{1}{\tau}(t-s)}\big) \cos(\theta_0 + s(\omega + z_0)) \\ \times \big[\cos(\mu s + \varphi) dw_\varepsilon^1(s) - \sin(\mu s + \varphi) dw_\varepsilon^2(s)\big]. \end{aligned}$$

Then $E|\nu_\varepsilon(t) - \tilde{\nu}_\varepsilon(t)|^2 = O(\varepsilon)$. So

$$\theta_\varepsilon(t) = \theta_0 + t(\omega + z_0) + \sqrt{\varepsilon}\,\tilde{\nu}(s) + O(\varepsilon).$$

(That is,

$$E\big(\theta_\varepsilon(t) - \theta_0 - t(\omega + z_0) - \sqrt{\varepsilon}\,\tilde{\nu}(s)\big)^2 = O(\varepsilon^2)$$

uniformly in t.)

Lemma 6 *Let $H(\theta)$ be a 2π-periodic function with continuous derivatives $H'(\theta)$, $H''(\theta)$. Then*

$$\limsup_{T\to\infty} \left| \frac{1}{T}\int_0^T H(\theta_\varepsilon(s))\,ds - \frac{1}{T}\int_0^T H(\hat{\theta}(s))\,ds \right| = O(\varepsilon),$$

and

$$\limsup_{T\to\infty} \left| \frac{1}{T}\int_0^T H(\theta_\varepsilon(s))H(\theta_\varepsilon(s+u))\,ds - \frac{1}{T}\int_0^T H(\hat{\theta}(s))H(\hat{\theta}(s+u))\,ds \right| = O(\varepsilon),$$

for all $u \in R$, where

$$\hat{\theta}(t) = \theta_0 + t(\omega + z_0).$$

Proof We use the representations

$$H(\theta_\varepsilon(s)) = H(\hat{\theta}(s)) + H'(\hat{\theta}(s))\,\sqrt{\varepsilon}\,\tilde{\nu}(s) + O(\varepsilon)$$

and

$$\begin{aligned} H(\theta_\varepsilon(s))H(\theta_\varepsilon(s+u) &= H(\hat{\theta}(s))H(\hat{\theta}(s+u)) + H'(\hat{\theta}(s))H(\hat{\theta}(s+u))\,\sqrt{\varepsilon}\,\tilde{\nu}(s) \\ &\quad + H(\hat{\theta}(s))H'(\hat{\theta}(s+u))\,\sqrt{\varepsilon}\,\tilde{\nu}(s+u) + O(\varepsilon). \end{aligned}$$

The remainder of the proof is a consequence of the following fact: For any bounded measurable function $g(t) : R_+ \to R$

$$\begin{aligned} &E\left(\frac{1}{T}\int_0^T g(s)\tilde{\nu}_\varepsilon(s)\,ds\right)^2 \\ &= E\left(\frac{1}{T}\int_0^T g(t)\int_0^t \left(1 - e^{-\frac{1}{\tau}(t-s)}\right)\left[\lambda_1(s)dw_\varepsilon^1(s) + \lambda_2(s)dw_\varepsilon^2(s)\right]\,dt\right)^2 \\ &= E\left(\frac{1}{T}\int_0^T \int_s^T \left(1 - e^{-\frac{1}{\tau}(t-s)}\right) g(t)\,dt \left(1 - e^{-\frac{1}{\tau}(t-s)}\right)\right) \\ &\le \frac{2c-1}{T^2}\int_0^T \left(\int_s^T \left(1 - e^{-\frac{1}{\tau}(t-s)}\right) g(t)\,dt\right)^2 [\lambda_1^2(s) + \lambda_2^2(s)]\,ds \end{aligned}$$

for all T and for $\varepsilon > 0$ small enough, where $\lambda_1(s) = \cos\hat{\eta}(s)\cos(\mu s + \varphi)$ and $\lambda_2(s) = -\cos\hat{\eta}(s)\sin(\mu s + \varphi)$. Note that

$$\int_s^T (1 - e^{-\frac{1}{\tau}(t-s)})g(t)\,dt$$

is a bounded function. So

$$\lim_{T\to\infty} T\left(\frac{1}{T}\int_0^T g(s)\tilde{\nu}_\varepsilon(s)\,ds\right)^2 = 0$$

if $\varepsilon > 0$ is small enough.

□

Remark 15 Note that the function $\hat{\theta}(s)$ does not contain any information about the signal $\mu t+\varphi$. So the presence of noise in systems with filters does not provide much information about the signal.

10.2.4 Simulation of a Phase–Locked Loop with Noisy Input

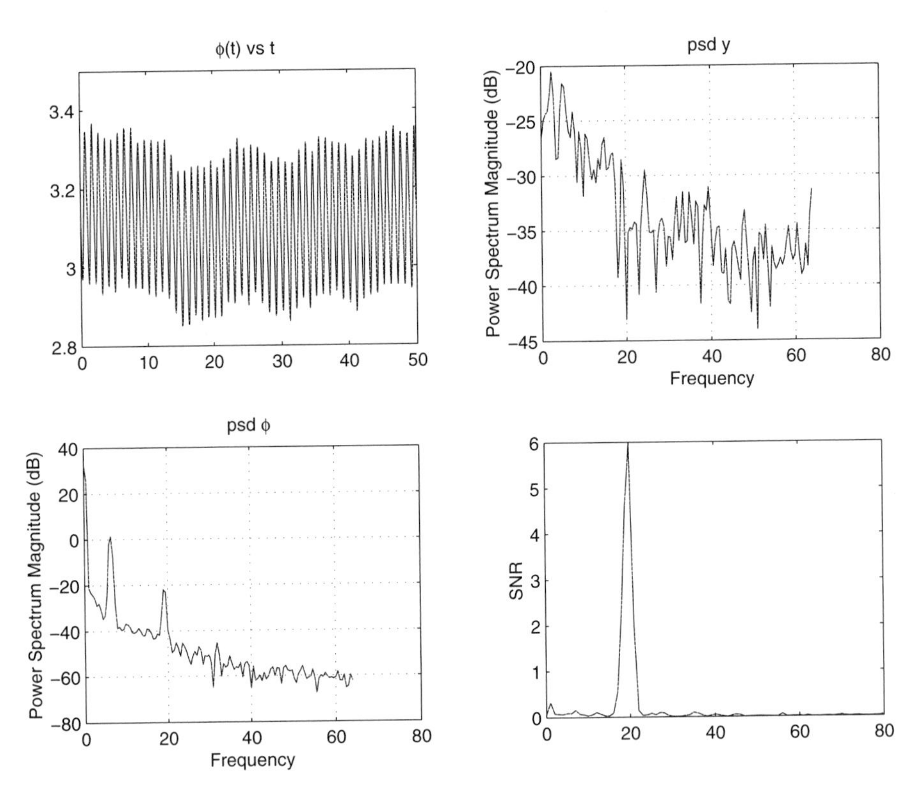

Figure 10.3. Upper left: The output voltage phase $\phi(t)$. Upper right: Power spectral density of the noise N. Lower left: Power spectral density of the output $\phi(t)$. Lower right: Signal–to–noise Ratio of the output to the noise evaluated at each frequency.

This computer simulation illustrates some of the results derived in this section. Consider a first–order phase–locked loop (i.e., one without a loop filter) of the form

$$\dot{\theta}_\varepsilon = \omega + \sin(\theta_\varepsilon(t) + N(t/\varepsilon) - S(t)),$$

where ω is the carrier frequency, $\theta_\varepsilon(t)$ is the loop's phase, $S(t)$ is the phase of the external signal, and N accounts for noise in the incoming signal. We suppose that the input is perfectly tuned, so $S(t) = \omega t + \psi(\omega_0 t)$, where $\omega_0 \ll \omega$ is the frequency of the input phase deviation. We rewrite this model in terms of the output phase variable $\phi(t) = \theta_\varepsilon(t) - \omega t$:

$$\dot{\phi} = \sin(\phi(t) + N(t/\varepsilon) - \psi(t)).$$

For the simulation, we take the input signal $\psi(t) = 2\cos 20t$ and the noise as a jump process with ergodic measure uniformly distributed on $[-0.1, 0.1]$, as in our earlier simulations. The results of this simulation are shown in Figure 10.3. (Recall that the power spectral density of a function is the Fourier transform of its autocorrelation function.)

11
Models in Population Biology

In this chapter we consider random perturbations of ecological systems, of epidemic disease processes, and of demographic models. In the next chapter we consider several problems from genetics. In each case, we describe classical models for these phenomena and some useful results concerning their asymptotic behavior for large time. Then we describe how random noise perturbations can be included in these models, and we investigate the asymptotic behavior of the resulting systems using the methods in this book. In each case, we compare the behavior of perturbed and unperturbed systems.

11.1 Mathematical Ecology

The Lotka–Volterra model is the paradigm for describing prey–predator systems involving interactions of two populations where one (the predator population) dies out in the absence of the other (the prey population), while the prey grow without bound in the absence of predation. We consider this model in detail first. Then we consider other ecological systems for two populations competing for the same resource and for a food chain of three interacting populations.

11.1.1 Lotka–Voltera Model

Consider two interacting species, one prey and the other its predator. Denote by $x(t)$ the number of prey and by $y(t)$ the number of predators at time t. The Lotka–Voltera model describes x and y by the system of ordinary differential equations

$$\begin{aligned} \dot{x} &= \alpha x - \beta xy, \\ \dot{y} &= -\gamma y + \delta xy, \end{aligned} \tag{11.1}$$

where $\alpha, \beta, \gamma, \delta$ are positive constants. Note that the growth rate for the prey, $\dot{x}/x$, is $\alpha - \beta y$, which decreases as the number of predators increases and becomes negative if $y > b = \alpha/\beta$. The growth rate of the predators, $\dot{y}/y$, is $\delta x - \gamma$, which increases as the number of prey increases, but is negative if $x < a = \gamma/\delta$ and positive if $x > a$. The population distribution (a, b) is an equilibrium of coexistence for system (11.1), and $(0, 0)$ is the equilibrium corresponding to extinction.
The first quadrant $(x \geq 0, y \geq 0)$ is relevant to the biological system. In it, we will see that orbits beginning with $x(0) > 0$ and $y(0) > 0$ revolve around the coexistence equilibrium point. Obviously, if $x(0) = 0$, then $y(t) = y(0)e^{-\gamma t}$, so $y(t) \to 0$ as $t \to \infty$, and in the absence of prey, the predators die out. On the other hand, if $y(0) = 0$, then $x(t) = x(0)e^{\alpha t}$ and $x(t) \to \infty$ as $t \to \infty$, and in the absence of predation the prey grow without bound.
The behavior of system (11.1) is described by the following theorem.

Theorem 1 *(i) System (11.1) has a first integral*

$$\Phi(x, y) = x^{\gamma} y^{\alpha} \exp\{-\delta x - \beta y\}.$$

(ii) For $c \in (0, \Phi(a, b)]$ *the orbits of the system with*

$$\Phi(x, y) = c$$

are closed curves.
(iii) If $\Phi(x_0, y_0) \in (0, \Phi(a, b))$, *then the solution of system (11.1) is a periodic function, and its period can be determined in the following way: Denote by* $y_1(c) < y_2(c)$ *the solutions of the equation*

$$\Phi(a, y_k) = c, c \in (0, \Phi(a, b)) \quad \textit{for} \quad k = 1, 2$$

and by $\phi_1(c, y) < \phi_2(c, y)$ *the solutions for* $k = 1, 2$ *of the equations*

$$\Phi(\phi_k(c, y), y) = c, \ y \in (y_1(c), y_2(c)).$$

Set

$$T(c) = \int_{y_1(c)}^{y_2(c)} \frac{\phi_2(c, y) - \phi_1(c, y)}{(\phi_2(c, y) - a)(a - \phi_1(c, y))} \frac{dy}{y}. \tag{11.2}$$

Then the solution of system (11.1) is periodic with period $T(\Phi(x(0), y(0)))$.

(iv) For any continuous function $g(x, y) : R^2 \to R$ there exists the limit

$$\lim_{t\to\infty} \frac{1}{t} \int_0^t g(x(t), y(t))\, dt = \frac{1}{T(c)} \int_{y_1(c)}^{y_2(c)} \left(\frac{g(\phi_1(c,y),y)}{a - \phi_1(c,y)} + \frac{g(\phi_2(c,y),y)}{\phi_2(c,y) - a} \right) \frac{dy}{y}. \tag{11.3}$$

Proof (i) It follows from system (11.1) that

$$\frac{-\gamma + \delta x}{x}\, dx = \frac{\alpha - \beta y}{y}\, dy.$$

Integrating this from $(x(0), y(0))$ to (x, y) gives

$$\Phi(x, y) = \Phi(x(0), y(0)),$$

where

$$\Phi(x, y) = \exp\big(\gamma \log x - \delta x + \alpha \log y - \beta y\big) = x^\gamma y^\alpha \exp\big(-\delta x - \beta y\big).$$

Then if $(x(t), y(t))$ is a solution of (11.1),

$$\frac{d\Phi}{dt}(x(t), y(t)) = 0.$$

(ii) The function $\log \Phi(x, y)$ is a convex function in the first quadrant $x > 0, y > 0$, and the point $x = a, y = b$ is its unique maximum. As a result, the contours of this surface $\Phi = c$ for $c < \Phi(a, b)$ are closed curves containing no equilibria of the system.
(iii) Assume that $x(0) = a, y(0) = y_1(c)$ as defined in the statement of the theorem. Denote by t_1 the first time for which $y(t) = y_2(c)$. For $t \in [0, t_1]$ we have $x(t) = \phi_2(c, y(t))$ and for $t \in (t_1, T(c))$, $x(t) = \phi_1(c, y(t))$. Then

$$dy(t) = [-\gamma y(t) + \delta\, y(t)\, \phi_2(c, y(t))]\, dt$$

for $t \in (0, t_1)$ and

$$dy(t) = [-\gamma y(t) + \delta y(t) \phi_1(c, y(t))]\, dt$$

for $t \in (t_1, T(c))$. Integrating these formulas gives

$$t_1 = \int_{y(0)}^{y(t_1)} \frac{dy}{-\gamma y + \delta y \phi_2(c, y)},$$

$$T(c) - t_1 = \int_{y(t_1)}^{y(T(c))} \frac{dy}{-\gamma y + \delta y \phi_1(c, y)}.$$

Combining these equations gives formula (11.2).
Formula (11.3) is a consequence of the fact that the temporal average of a periodic function is determined by the integral over one period and the

relation

$$\int_0^{T(c)} g(x(t), y(t))dt = \int_{y(0)}^{y(t_1)} \frac{g(\phi_2(c,y))\,dy}{-\gamma y + \delta y \phi_2(c,y)} + \int_{y(t_1)}^{y(T(c))} \frac{g(\phi_2(c,y))\,dy}{-\gamma y + \delta y \phi_2(c,y)}.$$

This completes the proof of the theorem.

□

11.1.2 Random Perturbations of the Lotka–Volterra Model

Consider a homogeneous Markov process $z(t)$ in a measurable space $(\mathcal{Z}, \mathcal{C})$ that represents random perturbations of (11.1). The system we consider is described by the differential equations

$$\begin{aligned} \dot{x}_\varepsilon &= \alpha(z_\varepsilon(t))\, x_\varepsilon - \beta(z_\varepsilon(t))\, x_\varepsilon y_\varepsilon, \\ \dot{y}_\varepsilon &= -\gamma(z_\varepsilon(t))\, y_\varepsilon + \delta(z_\varepsilon(t))\, x_\varepsilon y_\varepsilon, \end{aligned} \tag{11.4}$$

where $z_\varepsilon(t) = z(t/\varepsilon)$. Here $x_\varepsilon(t)$ and $y_\varepsilon(t)$ are numbers of prey and of predators, respectively, at time t, and $\alpha(z), \beta(z), \gamma(z)$, and $\delta(z)$ are positive, bounded measurable functions from $\mathcal{Z}$ into R. So the randomly perturbed system for (11.1) has coefficients that are randomly varying on a fast time scale as described by z_ε.

We assume that z is an ergodic Markov process with ergodic distribution $\rho(dz)$, and we write the averages of the data as

$$\begin{aligned} \alpha &= \textstyle\int \alpha(z)\rho(dz), \quad & \beta &= \textstyle\int \beta(z)\rho(dz), \\ \gamma &= \textstyle\int \gamma(z)\rho(dz), \quad & \delta &= \textstyle\int \delta(z)\rho(dz). \end{aligned}$$

Then the averaged equation for system (11.4) is given by (11.1). The next lemma follows from averaging theorems derived earlier (see Chapter 3, Theorem 6).

Lemma 1 *On any finite interval* $[0, T]$,

$$P\left\{ \lim_{\varepsilon \to 0} \sup_{t \le T} \left(|x_\varepsilon(t) - x(t)| + |y_\varepsilon(t) - y(t)| \right) = 0 \right\} = 1$$

if the initial data for the two systems are the same, $x_\varepsilon(0) = x(0), y_\varepsilon(0) = y(0)$, *where* $(x_\varepsilon(t), y_\varepsilon(t))$ *is the solution to the system (11.4) and* $(x(t), y(t))$ *is the solution to the system (11.1).*

The next lemma is a consequence of the theorem of normal deviations, Chapter 4, Theorem 2.

Lemma 2 *Assume that the Markov process* $z(t)$ *satisfies condition SMC II of Section 2.3, and let* $R(z, C) = \int_0^\infty (P(t, z, C) - \rho(C))\, dt$, *where* $P(t, z, C)$ *is the transition probability for* $z(t)$. *Set*

$A(x, y) =$

$$\iint \begin{pmatrix} \alpha(z)x - \beta(z)xy \\ -\gamma(z)y + \delta(z)xy \end{pmatrix} \begin{pmatrix} \alpha(z')x - \beta(z')xy \\ -\gamma(z')y + \delta(z')xy \end{pmatrix}^T R(z, dz')\rho(dz).$$

(Note that the integrand is a product of a column vector with a row vector and so is a 2×2 matrix). Further, we define

$$A(t) = A(x(t), y(t)) \equiv \begin{pmatrix} A_{11}(t) & A_{12}(t) \\ A_{21}(t) & A_{22}(t) \end{pmatrix}, \tag{11.5}$$

and the stochastic processes

$$\tilde{x_\varepsilon} = \varepsilon^{-1/2}(x_\varepsilon(t) - x(t)), \quad \tilde{y_\varepsilon} = \varepsilon^{-1/2}(y_\varepsilon(t) - y(t)),$$

where $(x_\varepsilon(t), y_\varepsilon(t))$ and $(x(t), y(t))$ are the same as in Lemma 1.
Then the two-dimensional stochastic process $(\tilde{x}_\varepsilon(t), \tilde{y}_\varepsilon(t))$ converges weakly in C as $\varepsilon \to 0$ to a two–dimensional stochastic process $(\tilde{x}(t), \tilde{y}(t))$ that satisfies the following system of stochastic integral equations:

$$\tilde{x}(t) = \int_0^t [(\alpha - \beta y(s))\tilde{x}(s) - \beta x(s)\tilde{y}(s)]\, ds + u_1(t),$$

$$\tilde{y}(t) = \int_0^t [\delta y(s)\tilde{x}(s) + (-\gamma + \delta x(s))\tilde{y}(s)]\, ds + u_2(t),$$

where $(u_1(t), u_2(t))$ is a two-dimensional Gaussian process with independent increments, mean values 0, and correlation matrix $R(t)$ given by the formula

$$R(t) = \int_0^t A(s)ds. \tag{11.6}$$

(The matrix $A(t)$ is defined in the formula (11.5).)

Lemmas 1 and 2 describe the behavior of the stochastic process $(x_\varepsilon(t), y_\varepsilon(t))$ on any finite interval $[t_0, t_0 + T]$. Namely, the system is moving close to the trajectory

$$\{(X(t_0, x_\varepsilon(t_0), y_\varepsilon(t_0), t), Y(t_0, x_\varepsilon(t_0), y_\varepsilon(t_0), t)) : t_0 \le t \le t_0 + T\}, \tag{11.7}$$

where $(X(t_0, x, y, t), Y(t_0, x, y, t))$ is the solution to system (11.1) on the interval (t_0, ∞) satisfying the initial conditions

$$X(t_0, x, y, t_0) = x, \quad Y(t_0, x, y, t_0) = y.$$

This solution is periodic with period $T(\Phi(x_\varepsilon(t_0), y_\varepsilon(t_0)))$ (this follows from Lemma 1). It follows from Lemma 2 that the deviation of the perturbed trajectory from the trajectory (11.7) is of order $\sqrt{\varepsilon}$. As a result, the trajectory of the perturbed system is close to another orbit of the average system. This orbit is determined by values $\Phi(x_\varepsilon(t), y_\varepsilon(t))$. So we can obtain a complete description of the perturbed system (as $\varepsilon \to 0$) by investigating the asymptotic behavior of the process $\phi_\varepsilon(t) = \Phi(x_\varepsilon(t), y_\varepsilon(t))$. To do this we use the diffusion approximation for first integrals that we derived in Chapter 5, Theorem 5.

Lemma 3 *Set $u_\varepsilon(t) = \log \phi_\varepsilon(t/\varepsilon)$ and define the functions*

$$\begin{aligned}
B(x,y) &= \iint [(\gamma - \delta x)(\alpha(z) - \beta(z)y) + (\alpha - \beta y)(-\gamma(z) + \delta(z)x)] \\
&\times [(\gamma - \delta x)(\alpha(z') - \beta(z')y) + (\alpha - \beta y)(-\gamma(z') + \delta(z')x)]\, R(z, dz')\rho(dz), \\
A(x,y) &= \iint [(\gamma - \delta x)(\alpha(z) - \beta(z)y)(\alpha(z') - \beta(z')y) \\
&- (\gamma - \delta x)\beta(z')(-\gamma(z) + \delta(z)x)y + (\alpha - \beta y)\delta(z')(\alpha(z') - \beta(z')y)x \\
&+ (\alpha - \beta y)(-\gamma(z) + \delta(z)x)(-\gamma(z') + \delta(z')x) \\
&- \gamma(\alpha(z) - \beta(z)y)(\alpha(z') - \beta(z')y) \\
&- \alpha(-\gamma(z) + \delta(z)x)(-\gamma(z') + \delta(z')x)]R(z, dz')\rho(dz),
\end{aligned}$$

$$\hat{b}(\theta) = \int_{y_1(c)}^{y_2(c)} \left(\frac{B(\phi_1(c,y), y)}{a - \phi_1(c,y)} + \frac{B(\phi_2(c,y), y)}{\phi_2(c,y) - a} \right) \frac{dy}{\delta y}, \tag{11.8}$$

and

$$\hat{a}(\theta) = \int_{y_1(c)}^{y_2(c)} \left(\frac{A(\phi_1(c,y), y)}{a - \phi_1(c,y)} + \frac{A(\phi_2(c,y), y)}{\phi_2(c,y) - a} \right) \frac{dy}{\delta y}, \tag{11.9}$$

where $\theta = \log c$.
Then the stochastic process $u_\varepsilon(t)$ converges weakly in C to the diffusion process $\theta(t)$ on the interval $(-\infty, \log \Phi(a,b))$ whose generator L^θ is

$$L^\theta f(u) = \hat{a}(u)f'(u) + \frac{1}{2}\hat{b}(u)f''(u). \tag{11.10}$$

Remark 1 This lemma implies that for any closed interval $[c,d] \subset (-\infty, \log \Phi(a,b))$ the process $u_\varepsilon(\tau_\varepsilon \wedge t)$ converges weakly in C to the process $\theta(\tau \wedge t)$, where

$$\begin{aligned}
\tau_\varepsilon &= \inf\{t : (u_\varepsilon(t) - c)(u_\varepsilon(t) - d) > 0\}, \\
\tau &= \inf\{t : (\theta(t) - c)(\theta(t) - d) > 0\}.
\end{aligned}$$

Next we investigate the behavior of the processes $\theta(t)$ and $u_\varepsilon(t)$ near the boundary of the interval $(-\infty, \log \Phi(a,b))$.

First, consider the behavior of the averaged system near the point (a,b).

Lemma 4 *Assume that the initial values $(x(0), y(0))$ satisfy the condition that $\Phi(x(0), y(0)) = (1-h)\Phi(a,b)$, where $h > 0$ is sufficiently small. Then* *the solution to the system (11.1) can be written in the form*

$$\begin{aligned} x(t) &= a + a\sqrt{h/\gamma}\cos(\lambda t + \psi_0) + O(h), \\ y(t) &= b + b\sqrt{h/\alpha}\sin(\lambda t + \psi_0) + O(h), \end{aligned} \tag{11.11}$$

where ψ_0 *is determined by the relations*

$$x(0) = a + a\sqrt{h/\gamma}\,\cos\psi_0, \quad y(0) = b + b\sqrt{h/\alpha}\,\sin\psi_0,$$

and

$$\lambda = \sqrt{\alpha\gamma}. \tag{11.12}$$

Proof Define new variables

$$\begin{aligned} x^*(t) &= (x(t) - a)/(a\sqrt{\gamma/h}), \\ y^*(t) &= (y(t) - b)/(b\sqrt{\alpha/h}). \end{aligned} \tag{11.13}$$

Then $x^*(0) = \cos\psi_0$, $y^*(0) = \sin\psi_0$, and $(x^*(t), y^*(t))$ satisfies the differential equations

$$\begin{aligned} \dot{x}^* &= -\lambda\, y^* + O(\sqrt{h}), \\ \dot{y}^* &= \lambda\, x^* + O(\sqrt{h}). \end{aligned}$$

Let $(x_1(t), y_1(t))$ be the solution to the system of differential equations

$$\begin{aligned} \dot{x}_1 &= -\lambda\, y_1, \\ \dot{y}_1 &= \lambda\, x_1, \end{aligned}$$

for which $x_1(0) = \cos\psi_0$ and $y_1(0) = \sin\psi_0$. Then

$$|x^*(t) - x_1(t)| + |y^*(t) - y_1(t)| = O\left(\sqrt{h}\right) \tag{11.14}$$

uniformly for $t \in [0, T]$ and for any fixed $T > 0$. It is easy to see that

$$\begin{aligned} x_1(t) &= \cos(\lambda t + \psi_0), \\ y_1(t) &= \sin(\lambda t + \psi_0). \end{aligned} \tag{11.15}$$

The proof of the lemma follows from relations (11.13), (11.14), (11.15). □

Corollary 1 (i) Let $g(x, y) : R^2 \to R$ be a continuous function with continuous derivatives of the first and second order. Then for $c = \Phi(a, b)(1-h)$ and $\Phi(x(0), y(0)) = c$ we have

$$\begin{aligned} \frac{1}{T(c)} \int_o^{T(c)} g(x(t), y(t))dt = g(a, b)\frac{h}{4} + \left(g_{xx}(a, b)\frac{a^2}{\gamma^2} + g_{yy}(a, b)\frac{b^2}{\alpha^2}\right) \\ + (g_x(a, b) + g_y(a, b))O(h) + o(h) \end{aligned} \tag{11.16}$$

and

$$\lim_{c \to \Phi(a,b)} T(c) = 2\pi(\alpha\gamma)^{-1/2}.$$

Lemma 5 *Let $\theta = \log \Phi(a,b) - h$ and let $h > 0$ be sufficiently small. Then the functions $\hat{a}(\theta)$ and $\hat{b}(\theta)$ have the following representations*

$$\begin{aligned}\hat{a}(\theta) = &-\iint [\gamma\,(\alpha(z) - b\beta(z))(\alpha(z') - b\beta(z')) \\ &+ \alpha\,(-\gamma(z) + a\delta(z))(-\gamma(z') + a\delta(z'))]R(z,dz')\rho(dz) + O(h),\\ \hat{b}(\theta) = &\frac{h}{2}\iint [(\alpha(z) - b\beta(z))(\alpha(z') - b\beta(z')) \\ &+ (-\gamma(z) + a\delta(z))(-\gamma(z') + a\delta(z'))]R(z,dz')\rho(dz) + o(h).\end{aligned}$$

The proof of the lemma follows from formula (11.16) and the representation of the functions $\hat{a}(\theta)$ and $\hat{b}(\theta)$ given in Lemma 3.

Remark 2 Set

$$\begin{aligned}\Delta_1(z) &= \det\begin{pmatrix}\alpha & \beta\\ \alpha(z) & \beta(z)\end{pmatrix}, \quad \Delta_2(z) = \det\begin{pmatrix}\gamma & \delta\\ \gamma(z) & \delta(z)\end{pmatrix},\\ \bar{\Delta}_k &= \iint \Delta_k(z)\Delta_k(z')R(z,dz')\rho(dz), \quad k = 1,2.\end{aligned} \tag{11.17}$$

Then

$$\begin{aligned}\hat{a}(\theta) &= -\frac{\gamma}{\beta^2}\bar{\Delta}_1 - \frac{\alpha}{\delta^2}\bar{\Delta}_2 + O(h),\\ \frac{2\hat{b}(\theta)}{h} &= \frac{1}{\beta^2}\bar{\Delta}_1 + \frac{1}{\delta^2}\bar{\Delta}_2 + o(1).\end{aligned}$$

Corollary 2 If $\bar{\Delta}_1 + \bar{\Delta}_2 > 0$, then the point $\log \Phi(a,b)$ is a natural boundary point for the diffusion process $\theta(t)$ defined in Lemma 3 (see Remark 4 of Chapter 5).

Next we consider the behavior of the diffusion coefficients $\hat{a}(\theta)$ and $\hat{b}(\theta)$ as $\theta \to -\infty$ using their representations given in formulas (11.8) and (11.9). For this purpose we determine asymptotic formulas for functions $y_k(c), \phi_k(c), k = 1,2$, and the integrals on the right-hand side of formula (11.3) as $c \to 0$.

Lemma 6 *As $c \to 0$,*

(i) $\log y_1(c) \sim \frac{1}{\alpha}\log c$ *and* $y_2(c) \sim -\frac{1}{\beta}\log c,$.

(ii) $T(c) \sim -\frac{1}{\alpha a}\log c$.

(iii) If in addition $y \to \infty$ and $y_2(c) - y \to \infty$, then

$$\phi_1(c,y) \to 0, \qquad \phi_2(c,y) \sim \frac{\beta}{\delta}(y_2(c) - y).$$

(iv) Denote by $I(c,g)$ the expression on the right-hand side of relation (11.3).

Then for $g_1(x,y) = x^m, m > 0,$

$$I(c, g_1) \sim \frac{\alpha a}{\delta^{m-1}} \left(\log \frac{1}{c} \right)^{m-1},$$

and for $g_2(x,y) = x^m y^n,$

$$I(c, g_2) \sim \frac{\alpha a}{\delta^{m-1} \beta^{n-1}} \left(\log \frac{1}{c} \right)^{n+m-1} \frac{(m-1)!(n-1)!}{(m+n-2)!}.$$

Proof (i) follows from the relations

$$\begin{aligned} \log c &\sim \alpha \log y_k(c) - \beta y_k(c), \quad k = 1, 2, \\ y_1(c) &= o(1), \quad \log y_2(c) = o(y_2(c)). \end{aligned}$$

(iii) follows, since $\phi_k(c, y),\ k = 1, 2,$ satisfies the relation

$$\log c = \alpha \log y + \gamma \log \phi_k(c, y) - \beta y - \delta \phi_k(c, y),$$

and, in addition,

$$\log y = o(y), \quad \log \phi_2(c, y) = o(\phi_2(c, y)), \quad \phi_1(c, y) = o(\log(\phi_1(c, y)).$$

(ii) Using formula (11.2) we can write

$$T(c) \sim \int_{y_1(c)}^{y_2(c)} \frac{1}{a - \phi_1(c, y)} \frac{dy}{y} \sim \frac{1}{a} \int_{y_1(c)}^{y_2(c)} \frac{dy}{y} \sim \frac{1}{a} \log \frac{y_2(c)}{y_1(c)}.$$

(iv) Using formula (11.3) and taking into account that

$$|\phi_k(c, y) - a| > h|y - y_i(c)|^{1/2} \text{ if } |y - y_i(c)| \le 1, \quad k = 1, 2,\ i = 1, 2,$$

we can write for $g = g_j,\ j = 1, 2$

$$I(c, g) \sim \frac{1}{T(c)} \int_{y_1(c)}^{y_2(c)} \frac{g(\phi_2(c, y), y)}{\phi_2(c, y) - a} \frac{dy}{y} \sim \frac{1}{T(c)} \int_{y_1(c)}^{y_2(c)} \frac{g(\phi_2(c, y), y)}{\phi_2(c, y)} \frac{dy}{y}.$$

So

$$\begin{aligned} I(c, g_1) &\sim \frac{1}{T(c)} \int_{y_1(c)}^{y_2(c)} (\phi_2(c, y))^{m-1} \frac{dy}{y} \\ &\sim \frac{1}{T(c)} \left(\frac{\beta}{\delta} \right)^{m-1} \int_{y_1(c)}^{y_2(c)} (y_2(c) - y)^{m-1} \frac{dy}{y} \\ &= \frac{1}{T(c)} \left(\frac{\beta}{\delta} \right)^{m-1} (y_2(c))^{m-1} \int_{\frac{y_1(c)}{y_2(c)}}^{1} (1 - y)^{m-1} \frac{dy}{y} \\ &\sim \frac{1}{T(c)} \left(\frac{\beta}{\delta} \right)^{m-1} (y_2(c))^{m-1} \int_{\frac{y_1(c)}{y_2(c)}}^{1} \frac{dy}{y} \\ &= \frac{1}{T(c)} \left(\frac{\beta}{\delta} \right)^{m-1} \left(\frac{1}{\beta} \log \frac{1}{c} \right)^{m-1} \log \frac{y_2(c)}{y_1(c)}. \end{aligned}$$

In the same way,

$$\begin{aligned} I(c, g_2) &\sim \frac{1}{T(c)}\left(\frac{\beta}{\delta}\right)^{m-1}\int_{y_1(c)}^{y_2(c)}(y_2(c)-y)^{m-1}y^{n-1}\frac{dy}{y} \\ &\sim \frac{1}{T(c)}\left(\frac{\beta}{\delta}\right)^{m-1} y_2(c)^{m+n-1}\int_0^1 (1-y)^{m-1}y^{n-1}dy. \end{aligned}$$

□

Lemma 7 *Assume that*

$$\hat{c} = \iint (\beta\delta(z) + \delta\beta(z))(\beta\delta(z') + \delta\beta(z'))R(z, dz')\rho(dz) > 0.$$

Then the relations

$$\begin{aligned} I(c, A) &\sim -\frac{\alpha a \hat{c}}{2\delta\beta^2}\log\frac{1}{c}, \\ I(c, B) &\sim -\frac{\alpha a \hat{c}}{6\delta\beta^2}\left(\log\frac{1}{c}\right)^2, \end{aligned}$$

hold as $c \to 0$*, where the functions* $A(x,y)$ *and* $B(x,y)$ *were introduced in Lemma 3.*

The proof follows from the formulas for $A(x,y)$, $B(x,y)$, and Lemma 3.

Corollary 3 If $\hat{c} > 0$, then the point $-\infty$ is an absorbing boundary for the diffusion $\theta(t)$; this means that with positive probability the stopping time

$$\tau_\infty = \inf\left\{t : \inf_{s\le t}\theta(s) > -\infty\right\}$$

is finite and

$$\lim_{\theta_0\to-\infty} P\{\tau_\infty < t \,|\, \theta(0) = \theta_0\} = 1$$

for all $t > 0$.

Note that if $\theta \to -\infty$, then

$$\begin{aligned} \hat{a}(\theta) &= I(e^\theta, A) \sim -\frac{1}{2}a_0|\theta|, \\ \hat{b}(\theta) &= 2I(e^\theta, A) \sim -\frac{1}{3}a_0\theta^2, \end{aligned}$$

where

$$a_0 = \frac{\alpha a \hat{c}}{\delta\beta^2}.$$

So

$$\frac{2\hat{a}(\theta)}{\hat{b}(\theta)} \sim -3|\theta|^{-1} \quad \text{as } \theta \to -\infty,$$

$$\exp\left\{-\int_1^z \frac{2\theta}{\hat{b}(\theta)} d\theta\right\} \sim c_1 \exp\left\{-3\int_{-1}^z \frac{d\theta}{|\theta|}\right\} \sim \frac{c_1}{|z|^3} \qquad \text{as } z \to -\infty.$$

The point $-\infty$ is a regular boundary for the diffusion process $\theta(t)$ because

$$\left|\int_{-1}^{-\infty} |z| \exp\left\{-\int_1^z \frac{2\theta}{\hat{b}(\theta)} d\theta\right\} dz\right| < \infty$$

(see Remark 4 of Chapter 5).

We summarize the results on the asymptotic behavior of the perturbed system in the following statement.

Theorem 2 *Let $(x_\varepsilon(t), y_\varepsilon(t))$ be the solution to system (11.4) with initial conditions $x_\varepsilon(0) = x^0$, $y_\varepsilon(0) = y^0$. Set $u_\varepsilon(t) = \log \Phi(x_\varepsilon(t/\varepsilon), y_\varepsilon(t/\varepsilon))$, and $\tau_\varepsilon(h) = \inf\{s : u_\varepsilon(s) < \log h\}$. Let $(X(t_0, x, y, t), Y(t_0, x, y, t))$ be the solution to the system (11.1) on the interval $[t_0, \infty)$ with the initial condition $X(t_0, x, y, t_0) = x$, $Y(t_0, x, y, t_0) = y$; let $\theta(t)$ be a diffusion process on the interval $(-\infty, \log \Phi(a, b))$ with the diffusion coefficients $\hat{a}(\theta)$, $\hat{b}(\theta)$ defined in Lemma 3; and let $\tau(h) = \inf\{s : \theta(s) < \log h\}$.*
Then:

(i) For any $h > 0$, the stochastic process $u_\varepsilon(t \wedge \tau_\varepsilon(h))$ converges weakly in C to the stochastic process $\theta(t)$, as $\varepsilon \to 0$.

(ii)

$$\limsup_{\varepsilon\to 0}\sup_{t_0} E\left[\sup_{t_0\le t\le T(\Phi(x_\varepsilon(t_0), y_\varepsilon(t_0)))} (|x_\varepsilon(t) - X(t_0, x_\varepsilon(t_0), y_\varepsilon(t_0), t)| + |y_\varepsilon(t) - Y(t_0, x_\varepsilon(t_0), y_\varepsilon(t_0), t)|)\right] = 0.$$

(iii) Assume in addition that

$$\iint [(\alpha\beta(z) + \beta\alpha(z))(\alpha\beta(z') + \beta\alpha(z')) + (\gamma\delta(z) + \delta\gamma(z))(\gamma\delta(z') + \delta\gamma(z'))]R(z, dz')\rho(dz) > 0$$

and

$$\iint (\beta\delta(z) + \delta\beta(z))(\beta\delta(z') + \delta\beta(z'))R(z, dz')\rho(dz) > 0.$$

Then the point $\log \Phi(a, b)$ is a natural boundary and the point $-\infty$ is a regular boundary for the diffusion process $\theta(t)$, and $P\{\tau(0) < \infty\} > 0$, where $\tau(0)$ is the stopping time for which $\theta(\tau(0)-) = -\infty$.

(iv) $\Phi(x_\varepsilon(t), y_\varepsilon(t)) \in (0, \Phi(a,b))$ *for all* t *with probability* 1 *if* $\Phi(x_0, y_0) < \Phi(a,b)$.

Proof *(i)* and *(ii)* are proved in Lemma 3, and *(iii)* is proved in Corollaries 2 and 3. Concerning *(iv)*, it follows from statement *(i)* and Lemma 2 that $P\{\Phi(x_\varepsilon(t), y_\varepsilon(t)) \in (0, \Phi(a,b))\} = 1$ for all t and for $\varepsilon > 0$ small enough. Using system (11.4) we can represent $x_\varepsilon(t)$, $y_\varepsilon(t)$ in the form

$$x_\varepsilon(t) = x_0 \exp\left\{\int_0^t \left[\alpha\left(z\left(\frac{s}{\varepsilon}\right)\right) - \beta\left(z\left(\frac{s}{\varepsilon}\right)\right) y_\varepsilon(s)\right] ds\right\},$$

$$y_\varepsilon(t) = y_0 \exp\left\{\int_0^t \left[\gamma\left(z\left(\frac{s}{\varepsilon}\right)\right) - \delta\left(z\left(\frac{s}{\varepsilon}\right)\right) x_\varepsilon(s)\right] ds\right\},$$

so $x_\varepsilon(t) > 0$, $y_\varepsilon(t) > 0$ for all t.

□

Remark 3 In the theorem we considered $h > 0$ as a small but fixed number. We can also consider $h = h(\varepsilon)$, where $h(\varepsilon) \to 0$ as $\varepsilon \to 0$. Note that it follows that the deviation of the perturbed system from its averaged one for one period is of order $\sqrt{\varepsilon}$. The proof of the theorem on diffusion approximation (Chapter 5, Theorem 5) is based on the assumption that this deviation is small with respect to the other variables; in our case it has to be small with respect to h. Therefore, we can consider $h(\varepsilon)$ for which $\varepsilon = o(h^2(\varepsilon))$.

11.1.3 Simulation of the Lotka–Volterra Model

Figure 14 in the introduction shows results of the simulation described in this section, but for a (relatively) large value of $\varepsilon \approx 0.5$. The averaged system exhibits a regular oscillation, but the system with noisy data exhibits an interesting behavior that might not be expected. Namely, the noise in the system can drive it toward extinction, but in such a way that there are increasingly rare emergences of prey. Figure 11.1 shows the population dynamics in this case. The model is of limited realism in extreme cases: For example, it is not accurate when either population is very small, since either can be eliminated by other causes of death, such as accidents or disease; and of course, the prey population cannot grow without bound. When the prey population becomes very small, it will probably die out.
Figure 11.2 shows the same calculation, but with $\varepsilon \approx 0.01$. The theorems of this section show that the perturbed trajectory should lie near the unperturbed one, to order $\sqrt{\varepsilon}$ for a reasonable length of time. This figure corroborates this.

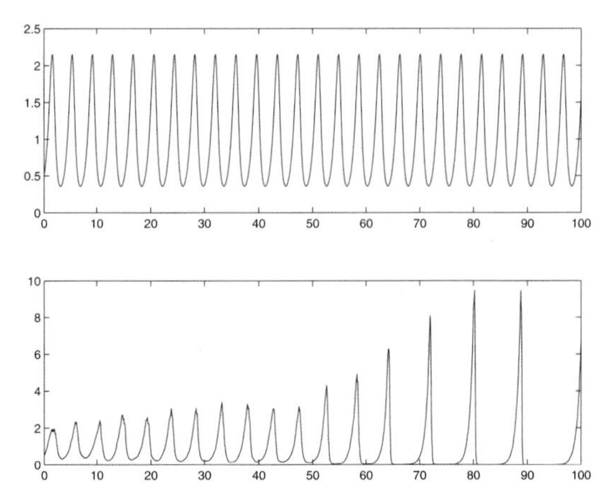

Figure 11.1. Top: The predator population for the averaged system. Bottom: The predator population for the system with noise as in Figure 11.2. Note that as noise drives Φ smaller, the system remains longer near extinction, but there is a pulsatile behavior to the system. If this were to continue, the period of the population cycle would grow without bound.

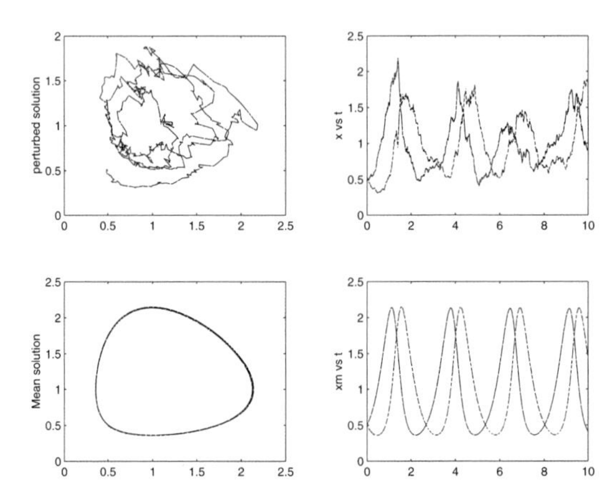

Figure 11.2. Bottom left: The averaged Lotka–Volterra model. Note that there is a limit cycle. Bottom right: The prey cycles followed by the predator cycles. Top left: The trajectory of the Lotka–Volterra model with noise beginning at the same point as above. Top right: The prey and predator cycles.

11.1.4 Two Species Competing for a Limited Resource

Consider two species that are competing for a common substrate, say with populations at time t being $x(t)$ and $y(t)$, respectively. We suppose that both require the same limiting substrate. Assume that $(x(t), y(t))$ satisfies the following system of differential equations:

$$\begin{aligned}\dot{x} &= \lambda x(1 - (\alpha x + \beta y)/K),\\ \dot{y} &= \mu y(1 - (\alpha x + \beta y)/K),\end{aligned} \tag{11.18}$$

where $\lambda, \mu, \alpha, \beta, K$ are positive constants characterizing speed of reproduction (λ for x and μ for y), the intensity of utilizing of the substrate (α for x and β for y), and the carrying capacity of the substrate K. The phase space of the system is the set $\{(x, y) : x \geq 0, y \geq 0, \alpha x + \beta y \leq K\}$. So we consider only those solutions of system (11.18) that satisfy the condition

$$\alpha x(0) + \beta y(0) \leq K.$$

These equations are a simple generalization of the logistic equation for a single species that has limiting carrying capacity:

$$\dot{x} = \lambda x(1 - x/K). \tag{11.19}$$

Note that under the condition $\lambda = \mu$, $\alpha = \beta$ the system (11.18) can be reduced to the logistic equation (11.19).
It is easy to see that system (11.18) has a first integral of the form

$$\Phi(x, y) = yx^{-\gamma}, \qquad \text{where } \gamma = \mu/\lambda. \tag{11.20}$$

The asymptotic behavior of the solution to system (11.18) is described in the following lemma.

Lemma 8 *For any initial condition $x(0) = x^0$, $y(0) = y^0$ satisfying the inequality $\alpha x^0 + \beta y^0 < K$, the solution $(x(t), y(t))$ to the system (11.18) has an ω-limit*

$$\lim_{t\to\infty} x(t) = \phi_1(x^0, y^0), \quad \lim_{t\to\infty} y(t) = \phi_2(x^0, y^0),$$

where the functions $\phi_1(x^0, y^0)$ and $\phi_2(x^0, y^0)$ are defined by the system of algebraic equations

$$\begin{aligned}\alpha\phi_1(x^0, y^0) + \beta\phi_2(x^0, y^0) &= K,\\ \phi_2(x^0, y^0) &= C\big(\phi_1(x^0, y^0)\big)^{\gamma}.\end{aligned}$$

Remark 4 It is easy to see that since the functions $\phi_1(x^0, y^0)$, $\phi_2(x^0, y^0)$ depend only on $y^0(x^0)^{-\gamma}$, there exist functions $\hat{\phi}_1 : R \to (0, K/\alpha)$, $\hat{\phi}_2 : R \to (0, K/\beta)$ for which

$$\phi_k(x^0, y^0) = \hat{\phi}_k(\log y^0 - \gamma \log x^0).$$

Consider the randomly perturbed system

$$\begin{aligned}\dot{x}_\varepsilon(t) &= \lambda(z_\varepsilon(t))x_\varepsilon(t)(1 - (\alpha(z_\varepsilon(t))x_\varepsilon(t) + \beta(z_\varepsilon(t))y_\varepsilon(t)/K)),\\ \dot{y}_\varepsilon(t) &= \mu(z_\varepsilon(t))y_\varepsilon(t)(1 - (\alpha(z_\varepsilon(t))x_\varepsilon(t) + \beta(z_\varepsilon(t))y_\varepsilon(t)/K)),\end{aligned} \tag{11.21}$$

where $\lambda(z)$, $\mu(z)$, $\alpha(z)$, and $\beta(z)$ are bounded nonnegative measurable functions and $z_\varepsilon(t) = z(t/\varepsilon)$ is the same as in Section 11.1.2. Set

$$\int \lambda(z)\rho(dz) = \lambda, \quad \int \mu(z)\rho(dz) = \mu,$$
$$\int \alpha(z)\rho(dz) = \alpha, \quad \int \beta(z)\rho(dz) = \beta,$$

so system (11.18) is the averaged system for system (11.21).
We will consider the solution to this system, which has the same initial condition as in Lemma 8. Then we can formulate statements on averaging and normal deviation as was done in Lemmas 1 and 2. We consider here only the diffusion approximation for the function

$$u_\varepsilon(t) = \log \Phi\left(x_\varepsilon\left(\frac{t}{\varepsilon}\right), y_\varepsilon\left(\frac{t}{\varepsilon}\right)\right), \tag{11.22}$$

where $\Phi(x, y)$ is given by formula (11.20).

Theorem 3 *We define the functions*

$$B(x, y, z, z') = (-\gamma\lambda(z) + \mu(z))\left(1 - \frac{\alpha(z)x + \beta(z)y}{K}\right) \times (-\gamma\lambda(z') + \mu(z'))\left(1 - \frac{\alpha(z')x + \beta(z')y}{K}\right),$$

$$A_1(x, y, z, z') = \left(1 - \frac{\alpha(z)x + \beta(z)y}{K}\right)\left(1 - \frac{\alpha(z')x + \beta(z')y}{K}\right) \times (\gamma\lambda(z)\lambda(z') - \mu(z)\mu(z')),$$

$$\begin{aligned} A_2(x, y, z, z') &= -\gamma\lambda(z)\lambda(z')\left(1 - \frac{\alpha(z)x + \beta(z)y}{K}\right)\left(1 - \frac{2\alpha(z')x + \beta(z')y}{K}\right) \\ &- \frac{\lambda(z)\mu(z')\alpha(z')x}{K}\left(1 - \frac{\alpha(z)x + \beta(z)y}{K}\right) \\ &+ \gamma\frac{\lambda(z)\beta(z')\mu(z')y}{K}\left(1 - \frac{\alpha(z)x + \beta(z)y}{K}\right) \\ &+ \mu(z)\mu(z')\left(1 - \frac{\alpha(z')x + 2\beta(z')y}{K}\right)\left(1 - \frac{\alpha(z)x + \beta(z)y}{K}\right), \end{aligned}$$

$$B(x, y) = \iint B(x, y, z, z')R(z, dz')\rho(dz),$$
$$A(x, y) = \iint (A_1(x, y, z, z') + A_2(x, y, z, z'))R(z, dz')\rho(dz),$$
$$a(\theta) = A(\hat{\phi}_1(\theta), \hat{\phi}_2(\theta)), \quad b(\theta) = 2B(\hat{\phi}_1(\theta), \hat{\phi}_2(\theta)),$$

where the functions $\hat{\phi}_k(\theta)$ *were defined in Remark 4.*
Denote by $\theta(t)$ *the diffusion process in* R *with the generator*

$$L^\theta f(u) = a(u)f'(u) + \frac{1}{2}b(u)f''(u), \tag{11.23}$$

and define the stopping times

$$\begin{aligned}\tau_\varepsilon(p,q) &= \inf\{t:\ (q-u_\varepsilon(t))(u_e(t)-p)\le 0\},\\ \tau(p,q) &= \inf\{t:\ (q-\theta(t))(\theta(t)-p)\le 0\}.\end{aligned}$$

Then the stochastic process $u_\varepsilon(t\wedge\tau_\varepsilon(p,q))$ *converges weakly in* C *to the stochastic process* $\theta(t\wedge\tau(p,q))$ *for all* $p<0<q$.

The proof of the theorem is a consequence of the general theorem on diffusion approximation (see Chapter 5, Theorem 6).
To investigate the asymptotic behavior of the diffusion process $\theta(t)$ as $t\to\infty$ we consider the asymptotic behavior of the functions $a(\theta)$, $b(\theta)$ as $\theta\to\infty$.

Lemma 9 *The following limits exist:*

$$\begin{aligned}\lim_{\theta\to-\infty} a(\theta) &= a(-\infty), \quad \lim_{\theta\to+\infty} a(\theta) = a(\infty),\\ \lim_{\theta\to-\infty} b(\theta) &= b(-\infty), \quad \lim_{\theta\to+\infty} b(\theta) = b(\infty),\end{aligned}$$

where

$$\begin{aligned}a(-\infty) &= \frac{1}{\beta^2}\iint(\beta-\beta(z))\beta(z')\mu(z)(\gamma\lambda(z')-\mu(z'))R(z,dz')\rho(z'),\\ a(+\infty) &= \frac{1}{\alpha^2}\iint(\alpha-\alpha(z))\alpha(z')\lambda(z)(\mu(z')-\gamma\lambda(z'))R(z,dz')\rho(z'),\\ b(-\infty) &= \frac{2}{\beta^2}\iint(-\gamma\lambda(z)+\mu(z))(\beta-\beta(z))\\ &\qquad\times(-\gamma\lambda(z')+\mu(z'))(\beta-\beta(z'))R(z,dz')\rho(z'),\\ b(+\infty) &= \frac{2}{\alpha^2}\iint(-\gamma\lambda(z)+\mu(z))(\alpha-\alpha(z))\\ &\qquad\times(-\gamma\lambda(z')+\mu(z'))(\alpha-\alpha(z'))R(z,dz')\rho(z').\end{aligned}$$

Proof It follows from Theorem 3 that

$$\begin{aligned}B(x,y) &= b_{00}+b_{10}x+b_{01}y+b_{20}x^2+b_{11}xy+b_{02}y^2,\\ A(x,y) &= a_{00}+a_{10}x+a_{01}y+a_{20}x^2+a_{11}xy+a_{02}y^2,\end{aligned}$$

where a_{ik}, b_{ik} are some constants. It is easy to see that

$$\begin{aligned}\lim_{\theta\to-\infty}\hat{\phi}_2(\theta) &= 0, \quad \lim_{\theta\to-\infty}\hat{\phi}_1(\theta) = \frac{K}{\alpha},\\ \lim_{\theta\to+\infty}\hat{\phi}_2(\theta) &= \frac{K}{\beta}, \quad \lim_{\theta\to+\infty}\hat{\phi}_1(\theta) = 0.\end{aligned}$$

This implies that

$$\lim_{\theta\to+\infty} a(\theta) = a_{00} + a_{10}\frac{K}{\beta} + a_{20}\left(\frac{K}{\beta}\right)^2,$$

$$\lim_{\theta\to-\infty} a(\theta) = a_{00} + a_{10}\frac{K}{\alpha} + a_{20}\left(\frac{K}{\alpha}\right)^2,$$

$$\lim_{\theta\to+\infty} b(\theta) = b_{00} + b_{10}\frac{K}{\beta} + b_{20}\left(\frac{K}{\beta}\right)^2,$$

$$\lim_{\theta\to-\infty} b(\theta) = b_{00} + b_{10}\frac{K}{\alpha} + b_{20}\left(\frac{K}{\alpha}\right)^2.$$

Calculating the coefficients a_{ij}, b_{ij} using the representation for $B(x, y, z, z')$, $A_k(x, y, z, z')$, $k = 1, 2$, we complete the proof of the lemma.

□

Corollary 4 Assume that $b(-\infty) > 0$, $b(+\infty) > 0$, and $b(\theta) > 0$, $\theta \in (-\infty, +\infty)$. Then

(i) If $a(-\infty) > 0$ and $a(\infty) < 0$, then the process $\theta(t)$ is an ergodic diffusion process and its ergodic distribution has density $g(\theta)$, which satisfies the differential equation

$$\frac{1}{2}(b(\theta)g(\theta))'' - (a(\theta)g(\theta))' = 0.$$

(ii) If $a(-\infty) > 0$, $a(\infty) > 0$, then $-\infty$ is a repelling boundary and $+\infty$ is an attracting boundary for $\theta(t)$.

(iii) If $a(-\infty) < 0$, $a(\infty) < 0$, then $-\infty$ is an attracting boundary and $+\infty$ a repelling boundary for $\theta(t)$.

(iv) If $a(-\infty) < 0$, $a(\infty) > 0$, then $-\infty$ and $+\infty$ are both attracting boundaries. In this case there exists a function

$$\pi(\theta) = P\left\{\lim_{t\to\infty} \theta(t) = +\infty/\theta(0) = \theta\right\}.$$

This function satisfies the differential equation

$$\frac{1}{2}b(\theta)\pi''(\theta) + a(\theta)\pi'(\theta) = 0,$$

and boundary conditions

$$\lim_{\theta\to-\infty} \pi(\theta) = 0, \quad \lim_{\theta\to+\infty} \pi(\theta) = 1$$

All these statements follow from known properties of one-dimensional diffusion process (see Remark 4 in Chapter 5).

Note that the competition between the two species is won by X in case *(ii)*, by Y in case *(iii)*, and by either of the two with probability depending on the initial sizes of their populations in case *(iv)*. There is no winner in case *(i)*.

Lemma 10 *Assume that* $a(+\infty) = 0$. *Then*

$$a(\theta) = O\left(e^{-\frac{1}{\gamma}\theta}\right) \quad \text{as} \quad \theta \to +\infty.$$

Assume that $a(-\infty) = 0$. *Then*

$$a(\theta) = O\left(e^{\theta}\right) \quad \text{as} \quad \theta \to -\infty.$$

Proof It follows from Lemma 8 and Remark 4 that the functions $\hat{\phi}_1(\theta)$ and $\hat{\phi}_2(\theta)$ satisfy the relations

$$\alpha\hat{\phi}_1(\theta) + \beta\hat{\phi}_2(\theta) = K,$$
$$\hat{\phi}_2(\theta) = e^{\theta}\left(\hat{\phi}_1(\theta)\right)^{\gamma}.$$

This implies that

$$\hat{\phi}_2(\theta) = O(e^{\theta}), \quad \frac{K}{\alpha} - \hat{\phi}_1(\theta) = O(e^{\theta}) \text{ as } \theta \to -\infty,$$

and

$$\hat{\phi}_1(\theta) = O(e^{-\theta/\gamma}), \quad \frac{K}{\beta} - \hat{\phi}_2(\theta) = O(e^{-\theta/\gamma}) \text{ as } \theta \to +\infty.$$

There exists a constant $C > 0$ for which

$$|A(x, y) - A(x_1, y_1)| \leq C(|x - x_1| + |y - y_1|),$$

so

$$\begin{aligned} |a(-\infty) - a(\theta)| &= \left|A(K/\alpha, 0) - A(\hat{\phi}_1(\theta), \hat{\phi}_2(\theta))\right| \\ &\leq C\left(\hat{\phi}_2(\theta) + \left(\frac{K}{\alpha} - \hat{\phi}_1\right)\right), \\ |a(+\infty) - a(\theta)| &= \left|A(0, K/\beta) - A(\hat{\phi}_1(\theta), \hat{\phi}_2(\theta))\right|. \end{aligned}$$

□

Corollary 5 Assume that $a(-\infty) = 0,\ b(\theta) > 0,\ b(-\infty) > 0$. Then

$$\int_{-\infty}^{0} \frac{2a(\theta)}{b(\theta)} d\theta < \infty,$$

and $\mathcal{U}(-\infty) > 0$, where $\mathcal{U}$ is the function introduced in Remark 4 of Chapter 5. This means that $-\infty$ is a natural boundary for the diffusion process $\theta(t)$; i.e., it is neither a repelling, an attracting, nor an absorbing boundary. The same is true for $+\infty$ if $a(+\infty) = 0,\ b(\theta) > 0,\ b(+\infty) > 0$, because of the fact that

$$\int_0^{+\infty} \frac{2a(\theta)}{b(\theta)}\,d\theta < \infty.$$

Remark 5 Assume that the point $-\infty$ is a natural boundary. Then species x wins the competition if the point $+\infty$ is an attracting boundary, but there is no winner if it is a repelling or natural boundary. An analogous statement is fulfilled if the point $+\infty$ is a natural boundary.

11.1.5 A Food Chain with Three Species

Now consider three species, say S_1, S_2, and S_3, where S_1 is prey for S_2, and S_2 is prey for S_3. Denote the size of the population of species S_k at time t by $x_k(t)$ for $k = 1, 2, 3$. The model is described by the system of differential equations

$$\begin{aligned} \dot{x}_1 &= x_1(\alpha_1 - \alpha_2 x_2), \\ \dot{x}_2 &= x_2(-\beta_1 + \beta_2 x_1 - \beta_3 x_3), \\ \dot{x}_3 &= x_3(-\gamma_1 + \gamma_2 x_2), \end{aligned} \tag{11.24}$$

where $\alpha_1, \alpha_2, \beta_1, \beta_2, \beta_3, \gamma_1, \gamma_2$ are some positive constants. We consider here only the special case where

$$\frac{\gamma_1}{\alpha_1} = \frac{\gamma_2}{\alpha_2} = \nu. \tag{11.25}$$

In this case, the behavior of the dynamical system (11.24) is similar to the behavior of the Lotka–Volterra system (11.1). The full description of solutions to system (11.24) is described in the next result.

Theorem 4 *Assume that relation (11.25) holds. Then:*

(i) System (11.24) has two independent first integrals:

$$\Phi_1(x_1, x_3) = x_1^\nu x_3, \tag{11.26}$$

and

$$\Phi_2(x_1, x_2, x_3) = x_1^{\beta_1} x_2^{\alpha_1} \exp\left\{-\beta_2 x_1 - \alpha_2 x_2 - \frac{\beta_3}{\nu} x_3\right\}. \tag{11.27}$$

(ii) If the initial condition for system (11.24) satisfies $[x_1(0)]^\nu x_3(0) = c_1$, then the functions $x_1(t)$ and $x_2(t)$ satisfy the system of differential equations

$$\begin{aligned} \dot{x}_1 &= x_1(\alpha_1 - \alpha_2 x_2) \\ \dot{x}_2 &= x_2(-\beta_1 + \beta_2 x_1 - \beta_3 c_1 x_1^{-\nu}). \end{aligned} \tag{11.28}$$

(iii) *The function* $\Phi_{c_1}(x_1, x_2) = \Phi_2(x_1, x_2, c_1 x_1^{-\nu})$ *is a first integral for (11.28). Set* $b = \alpha_1/\alpha_2$ *and denote by* $a(c_1)$ *the unique solution to the equation*

$$\beta_1 x_1^\nu + c_1\beta_3 = \beta_2 x_1^{\nu+1}.$$

Then $\Phi(x_1, x_2)$ *has a unique maximum in the region* $x_1 \geq 0, x_2 \geq 0$ *at the point* $x_1 = a(c_1)$, $x_2 = b$, *and for any* $c_2 \in (0, \Phi(a(c_1), b))$ *the curve*

$$\{(x_1, x_2) : \Phi_{c_1}(x_1, x_2) = c_2\} \tag{11.29}$$

is an orbit for system (11.28). It is a closed convex curve that is described perimetrically by the pair of functions

$$x_2 = \phi_k(x_1, c_1, c_2), \quad x_1 \in [a_1(c_1, c_2), a_2(c_1, c_2)],$$

for $k = 1, 2$, *where* $a_1(c_1, c_2) \leq a_2(c_1, c_2)$ *are the solutions of the equation*

$$\Phi_{c_1}(x_1, b) = c_2,$$

and $\phi_1(x_1, c_1, c_2) \leq \phi_2(x_1, c_1, c_2)$ *satisfy the equation*

$$\Phi_{c_1}(x_1, \phi_k(x_1, c_1, c_2)) = c_2, \quad k = 1, 2.$$

(iv) *The functions* $(x_1(t), x_2(t), x_3(t))$ *are periodic with period*

$$T(c_1, c_2) = T_1(c_1, c_2) + T_2(c_1, c_2), \tag{11.30}$$

where

$$T_k(c_1, c_2) = \int_{a_1(c_1,c_2)}^{a_2(c_1,c_2)} \frac{(-1)^{k-1} du}{u(\alpha_1 - \alpha_2\phi_k(u, c_1, c_2))}, \quad k = 1, 2, \tag{11.31}$$

where the constants c_1, c_2 *are determined by the initial conditions*

$$c_1 = \Phi_1(x_1(0), x_3(0)), \quad c_2 = \Phi_2(x_1(0), x_2(0), x_3(0)).$$

(v) *If* $g(x_1, x_2, x_3)$ *is a continuous function in* R^3_+, *then*

$$\lim_{t\to\infty} \frac{1}{t}\int_0^t g(x_1(s), x_2(s), x_3(s))ds$$
$$= \frac{1}{T(c_1, c_2)} \int_{a_1(c_1,c_2)}^{a_2(c_1,c_2)} \sum_{k=1}^{2} (-1)^{k-1} \frac{g(\phi_k(u, c_1, c_2), c_1 u^{-\nu})}{u(\alpha_1 - \alpha_2\phi_k(u, c_1, c_2))} du. \tag{11.32}$$

The proof of the theorem is left to the reader.

Next, we consider random perturbations of this system. Let $(Z, \mathcal{C})$ be a measurable space, and let $\alpha_k(z), k = 1, 2,\ \beta_i(z), i = 1, 2, 3,\ \gamma_j(z), j = 1, 2,$ be measurable bounded positive functions from Z into R. Furthermore,

$z(t)$ is the same Markov process that was considered earlier in this section, and

$$\alpha_k = \int \alpha_k(z)\rho(dz), \quad \beta_i = \int \beta_i(z)\rho(dz), \quad \gamma_j = \int \gamma_j(z)\rho(dz)$$

for $k = 1, 2$, $i = 1, 2, 3$, and $j = 1, 2$.
The perturbed system is

$$\begin{aligned} \dot{x}_1^\varepsilon &= x_1^\varepsilon\,[\alpha_1(z_\varepsilon(t)) - \alpha_2(z_\varepsilon(t))x_2^\varepsilon], \\ \dot{x}_2^\varepsilon &= x_2^\varepsilon\,[-\beta_1(z_\varepsilon(t)) + \beta_2(z_\varepsilon(t))x_1^\varepsilon - \beta_3(z_\varepsilon(t))x_3^\varepsilon], \\ \dot{x}_3^\varepsilon &= x_3^\varepsilon\,[-\gamma_1(z_\varepsilon(t)) + \gamma_2(z_\varepsilon(t))x_2^\varepsilon], \end{aligned} \tag{11.33}$$

where $z_\varepsilon(t) = z(t/\varepsilon)$.
System (11.24) is the averaged system for this one. So the statements of the averaging theorem and the theorem on normal deviations can be reformulated for system (11.33) directly. Here we will consider in detail only the asymptotic behavior of the two–dimensional stochastic process

$$\begin{aligned} u_\varepsilon(t) &= \log \Phi_1\left(x_1^\varepsilon\left(\frac{t}{\varepsilon}\right), x_2^\varepsilon\left(\frac{t}{\varepsilon}\right)\right), \\ v_\varepsilon(t) &= \log \Phi_2\left(x_1^\varepsilon\left(\frac{t}{\varepsilon}\right), x_2^\varepsilon\left(\frac{t}{\varepsilon}\right), x_3^\varepsilon\left(\frac{t}{\varepsilon}\right)\right), \end{aligned}$$

using the theorem on the diffusion approximation for first integrals.
Define the function

$$V(u) = \beta_1 \log a(e^u) - \beta_2 a(e^u) - \frac{\beta_3}{\nu} e^u (a(e^u))^{-\nu} + \alpha_1 \log b - \alpha_2 b, \quad u \in R,$$

where b and $a(c)$ were defined in Theorem 4 *(iii)*. It is easy to check that if

$$\Phi_1(x_1, x_3) = e^u,$$

then

$$\Phi_2(x_1, x_2, x_3) \le V(u).$$

So the region $\Delta = \{(u, v) : \ v \le V(u)\}$ in the (u, v)-plane coincides with the region

$$\{(u, v) : \ u = \log \Phi_1(x_1, x_3), \ v = \log \Phi_2(x_1, x_2, x_3), x_1 > 0, \ x_2 > 0, \ x_3 > 0\}.$$

To describe the limit diffusion process, we introduce further notation. Set

$$\begin{aligned}
A_1(x_1, x_2, x_3, z, z') &= x_2(-\beta_1(z) + \beta_2(z)x_1 - \beta_3(z)x_3)(\gamma_2(z') + \nu\alpha_2(z')),\\
A_2(x_1, x_2, x_3, z, z') &= -\beta_2 x_2(\alpha_1(z) - \alpha_2(z)x_2)(\alpha_1(z') - \alpha_2(z')x_2)\\
&\quad - (\beta_1 - \beta_2 x_1)(-\beta_1(z) + \beta_2(z)x_1 - \beta_3(z)x_3)\alpha_2(z')x_2\\
&\quad + (\alpha_1 - \alpha_2 x_2)(\alpha_1(z) - \alpha_2(z)x_2)\beta_2(z')x_1\\
&\quad - \alpha_2 x_2(-\beta_1(z) + \beta_2(z)x_1 - \beta_3(z)x_3)\\
&\quad \times (-\beta_1(z') + \beta_2(z')x_1 - \beta_3(z')x_3)\\
&\quad - (\alpha_1 - \alpha_2 x_2)(-\gamma_1(z) + \gamma_2(z)x_2)\beta_3(z')x_3\\
&\quad - \frac{\beta_3 x_3}{\nu}(-\beta_1(z) + \beta_2(z)x_1 - \beta_3(z)x_3)\gamma_2(z')x_2\\
&\quad - \frac{\beta_3 x_3}{\nu}(-\gamma_1(z) + \gamma_2(z)x_2)(-\gamma_1(z') + \gamma_2(z')x_2),
\end{aligned}$$

$$\begin{aligned}
A_{11}(x_1, x_2, x_3, z, z') &= 2(\nu\alpha_1(z) - \gamma_1(z) + \gamma_2(z)x_2 - \nu\alpha_2(z)x_2)\\
&\quad \times (\nu\alpha_1(z') - \gamma_1(z') + \gamma_2(z')x_2 - \nu\alpha_2(z')x_2),\\
A_{12}(x_1, x_2, x_3, z, z') &= (\nu\alpha_1(z) - \gamma_1(z) + \gamma_2(z)x_2 - \nu\alpha_2(z)x_2)\\
&\quad \times [(\beta_1 - \beta_2 x_1)(\alpha_1(z') - \alpha_2(z')x_2)\\
&\quad + (\alpha_1 - \alpha_2 x_2)(-\beta_1(z') + \beta_2(z')x_1 - \beta_3(z')x_3)\\
&\quad - \frac{\beta_3 x_3}{\nu}(-\gamma_1(z') + \gamma_2(z')x_2)]\\
&\quad + (\nu\alpha_1(z') - \gamma_1(z') + \gamma_2(z')x_2 - \nu\alpha_2(z')x_2)\\
&\quad \times [(\beta_1 - \beta_2 x_1)(\alpha_1(z) - \alpha_2(z)x_2)\\
&\quad + (\alpha_1 - \alpha_2 x_2)(-\beta_1(z) + \beta_2(z)x_1 - \beta_3(z)x_3)\\
&\quad - (\beta_3 x_3/\nu)(-\gamma_1(z) + \gamma_2(z)x_2)],
\end{aligned}$$

$$\begin{aligned}
A_{22}(x_1, x_2, x_3, z, z') &= 2[(\beta_1 - \beta_2 x_1)(\alpha_1(z) - \alpha_2(z)x_2)\\
&\quad + (\alpha_1 - \alpha_2 x_2)(-\beta_1(z) + \beta_2(z)x_1 - \beta_3(z)x_3)\\
&\quad - (\beta_3 x_3/\nu)(-\gamma_1(z) + \gamma_2(z)x_2)]\\
&\quad \times [(\beta_1 - \beta_2 x_1)(\alpha_1(z') - \alpha_2(z')x_2)\\
&\quad + (\alpha_1 - \alpha_2 x_2)(-\beta_1(z') + \beta_2(z')x_1 - \beta_3(z')x_3)\\
&\quad - (\beta_3 x_3/\nu)(-\gamma_1(z') + \gamma_2(z')x_2)].
\end{aligned}$$

Theorem 5 *Define the averages*

$$\hat{A}_\star(u, v) = \iiint A_\star(x_1, x_2, x_3, z, z')R(z, dz')\rho(dz)m_{u,v}(dx),$$

where the symbol $\star$ *means to take one of the values* $1, 2, 11, 12, 22$, *and the measure* $m_{u,v}(dx)$ *in* R_+^3 *is determined by the following relation: For any bounded function* $g(x_1, x_2, x_3)$ *on* R_+^3, $\int g(x_1, x_2, x_3)m_{u,v}(dx)$ *is equal to the expression on the right-hand side of formula (11.32), where* $c_1 =$

$e^u, c_2 = e^v$. *Let* $(\hat{u}(t), \hat{v}(t))$ *be the diffusion process in* Δ *with generator*

$$\hat{L}\Psi(u,v) = \hat{A}_1(u,v)\Psi_u(u,v) + \hat{A}_2(u,v)\Psi_v(u,v) + \frac{1}{2}\Big(\hat{A}_{11}(u,v)\Psi_{uu}(u,v) + 2\hat{A}_{12}(u,v)\Psi_{uv}(u,v)\hat{A}_{22}(u,v)\Psi_{vv}(u,v)\Big)$$

with absorption on the boundary of Δ.
For $\theta > 0$ *set*

$$\Delta_\theta = \left\{(u,v): \ |u| \le \frac{1}{\theta}, \frac{1}{\theta} \le v \le V(u) - \theta\right\},$$
$$\tau_\theta = \inf\{t: \ (\hat{u}(t), \hat{v}(t)) \notin \Delta_\theta\},$$
$$\tau_\theta^\varepsilon = \inf\{t: \ (u_\varepsilon(t), v_\varepsilon(t)) \notin \Delta_\theta\}.$$

Then the stochastic process $(u_\varepsilon(t \wedge \tau_\theta^\varepsilon), v_\varepsilon(t \wedge \tau_\theta^\varepsilon))$ *converges weakly in* C *to the diffusion process* $(\hat{u}(t \wedge \tau_\theta), \hat{v}(t \wedge \tau_\theta))$.

Proof To prove this theorem, we use Theorem 7 of Chapter 5. According to that theorem,

$$\hat{L}\Psi(u,v) = \iiint \left(\frac{\partial}{\partial x}\left([\Psi(\phi_1(x), \phi_2(x))]', a(x,z')\right), a(x,z)\right) \times R(z, dz')\rho(dz)m_{u,v}(dx),$$

where $x \in R_+^3$, $x = (x_1, x_2, x_3)$, $a(x,z) = (a_1(x,z), a_2(x,z), a_3(x,z))$, and

$$a_1(x,z) = x_1(\alpha_1(z) - \alpha_2(z)x_2),$$
$$a_2(x,z) = x_2(-\beta_1(z) + \beta_2(z)x_2 - \beta_3(z)x_3),$$
$$a_3(x,z) = x_3(-\gamma_1(z) + \gamma_2 x_2),$$

$\phi_1(x) = \nu \log x_1 + \log x_3$, $\phi_3(x) = \beta_1 \log x_1 - \beta_2 x_1 + \alpha_1 \log x_2 - \alpha_2 x_2 - \frac{\beta_3}{\nu} x_3$. The proof follows directly from these calculations.

□

11.1.6 *Simulation of a Food Chain*

Dynamics similar to the Lotka–Volterra case are observed in simulating system (11.33). Figure 11.3 shows the results.

11.2 Epidemics

A big problem in epidemiology is to predict whether or not an infectious disease will propagate in a population. In this section we study two cases: This first is for a one-pass situation in which once individuals have had the disease, they are permanently immune to it. This is depicted by the graph

$$S \to I \to R,$$

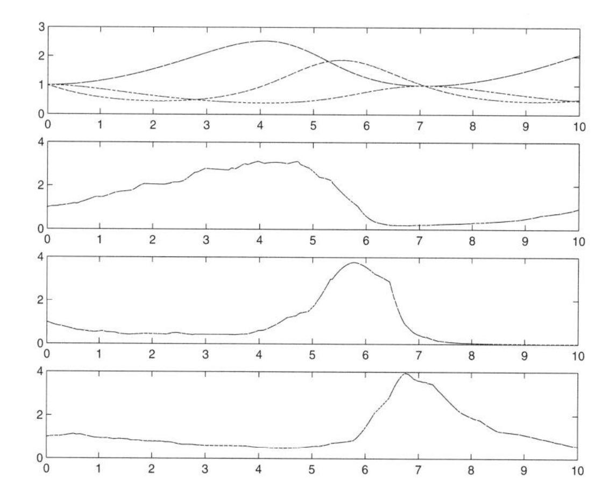

Figure 11.3. Computer simulation of the system in (11.33). Top: All three populations described by the averaged equation. The prey lead the way. Second: Prey population with noisy data. Third: Intermediate species. Fourth: Top of the food chain.

where S denotes the susceptible population, I the infectious population, and R those who are removed from the disease process. While this model is quite simple, it captures important aspects of disease processes. It was successfully modeled and analyzed by W. O. Kermack and A. G. McKendrick, and their work from the 1930s resulted in a threshold theorem on herd immunity that continues to be used as a measure of a susceptible population's vulnerability to an infectious disease. In particular, biweekly reports of estimates of susceptible population sizes and disease thresholds are published by the Communicable Disease Center for various diseases like measles, mumps, and influenza.

The second case pertains to diseases that do not impart permanent immunity, but rather individuals can pass from the infective population directly back into the susceptible population. Sexually transmitted diseases such as gonorrhea are of this type. These are depicted by the graph

$$S \to I \to S.$$

The interest in this case is how to design a control strategy to minimize the impact of such diseases.

In both cases, we first develop nonrandom theories, and then we investigate those models in the presence of random noise.

11.2.1 The Kermack–McKendrick Model

Diseases of the type $S \to I \to R$ are modeled as follows: Consider a population separated by a disease into three subpopulations: susceptibles, infectives, and removals. Let the population be of fixed size, and let $S(t)$

denote the proportion of the population that is susceptible at time t, while $I(t)$ is the proportion of infectives, and $R(t)$ is the proportion of removals. We treat these as continuous functions, and use the law of mass action to describe the infection process. Namely, we write

$$\begin{aligned} \dot{S} &= -rIS, \\ \dot{I} &= rIS - \lambda I, \\ \dot{R} &= \lambda I, \end{aligned} \tag{11.34}$$

where r is the infection rate and λ is the rate at which infectives are removed from the process (e.g., through cure, quarantine, death, or recovery from the illness.)

The main result in this case is called the *threshold theorem.* It is based (most simply) on the observation that if $S(0) > \lambda/r$, then $\dot{I}(0) > 0$, so the initial group of infectives in the population will more than replace themselves. In this model we can ignore the size of the R population, since it plays no role in the disease process other than to count how many have passed through it.

The severity of an epidemic, say measured by $S(\infty)$, can be predicted for this model by solving the equation

$$\frac{dI}{I} = \left(\frac{\lambda}{rS} - 1\right) dS. \tag{11.35}$$

Since $S(t) \geq 0$ and $\dot{S} \leq 0$, the limit $S(\infty)$ exists, and the next theorem shows how to compute it.

Theorem 6 *(i) System (11.35) has the first integral*

$$\Phi(S, I) = \alpha \log S - I - S, \tag{11.36}$$

where $\alpha = \lambda/r$.

(ii) Assume that $S_0 = S(0) > \alpha$. *Then there exists a time* $t_m > 0$ *at which* I *peaks, that is, for which* $\dot{I}(t) < 0$ *for all* $t > t_m$, $\dot{I}(t) > 0$ *for all* $t < t_m$,

$$I_m \equiv I(t_m) = \max_t I(t),$$

and

$$I_m = S_0 + I_0 - \alpha - \alpha \log \frac{S_0}{\alpha}, \tag{11.37}$$

where $I_0 = I(0)$, $S_0 = S(0)$.

(iii) $I(t) > 0$ *for all* $t > 0$, $\lim_{t\to\infty} I(t) = 0$, *and there exists the limit*

$$\hat{S} = \lim_{t\to\infty} S(t), \quad \hat{S} > 0,$$

where $\hat{S}$ *is the unique solution of the equation*

$$-\hat{S} - \alpha \log \frac{S_0}{\hat{S}} + I_0 + S_0 = 0 \tag{11.38}$$

satisfying the inequality $\hat{S} < \alpha$.
(iv) Set $\beta = \lambda - r\hat{S} = r(\alpha - \hat{S})$, $\beta > 0$. *Then there exist limits*

$$\lim_{t\to\infty} I(t)e^{\beta t} = \tilde{I}, \quad \tilde{I} \in (0,\infty),$$

$$\lim_{t\to\infty} (S(t) - \hat{S})e^{\beta t} = \tilde{S}, \quad \tilde{S} \in (0,\infty).$$

An interesting corollary to this theorem is that any such epidemics will leave some susceptibles uninfected. This phenomenon, which is referred to as *herd immunity*, is often observed in epidemics.

Proof First, we know that $\dot{S}(t) < 0$ for all t, so $S(t)$ is a decreasing function and there exist $\hat{S} = \lim_{t\to\infty} S(t)$ with $\hat{S} < S_0$. On the interval $(\hat{S}, S_0]$ there exists the inverse function $g : (\hat{S}, S_0] \to [0,\infty)$, $g(S(t)) \equiv t$, and the function $i(S) = I(g(S))$ satisfies the differential equation

$$\frac{di(S)}{dS} = -1 + \frac{\alpha}{S},$$

so

$$i(S) = -S + \alpha \log S + C.$$

This implies statement *(i)*.
It follows from the first equation of the Kermack–McKendrick system that

$$S(t) = S_0 \exp\left\{-r\int_0^t I(u)\,du\right\}. \tag{11.39}$$

Denote by t_m the solution of the equation $S(t_m) = \alpha$. Then for $t < t_m$, $I(t) > I(0)$. So

$$\alpha < S_0 \exp\{-rt_m I_0\}$$

and $t_m < (1/r)\log(S_0/\alpha)$. Note that

$$\alpha \log S_0 - I_0 - S_0 = \alpha \log S(t_m) - I_m - S(t_m)$$

and $S(t_m) = \alpha$. This implies statement *(ii)*.
The second equation of the Kermack–McKendrick system implies the following relation: If $0 \le t_0 < t$, then

$$I(t) = I(t_0) \exp\left\{\int_{t_0}^t [rS(u) - \lambda]\,du\right\}. \tag{11.40}$$

So $I(t) > 0$. Assume that $t_0 > t_m$. Then $rS(u) - \lambda - \delta = rS(t_0) - \lambda < 0$, where $\delta > 0$, and so

$$I(t) < I(t_0)\exp\{-\delta(t - t_0)\}.$$

This implies that

$$\lim_{t\to\infty} I(t) = 0, \quad \text{and} \quad \int_0^\infty I(u)\,du < \infty.$$

Since $\hat{S} = S_0 \exp\{-r \int_0^\infty I(u)\, du\}$, $\hat{S} > 0$. This implies *(iii)*.
To prove the last statement we first note that for $t > t_0 > t_m$,

$$S(t) - \hat{S} = r \int_t^\infty I(s)S(s)\, ds \le rS(t) \int_t^\infty O\left(e^{-\delta s}\right)\, ds = O\left(e^{-\delta t}\right).$$

It follows that

$$\int_0^\infty (S(s) - \hat{S})\, ds < \infty. \tag{11.41}$$

For $t > t_0$ we have

$$\begin{aligned} I(t)e^{\beta(t-t_0)} &= I(t_0) \exp\left\{\int_{t_0}^t [rS(u) - \lambda]\, du - [r\hat{S} - \lambda](t - t_0)\right\} \\ &= I(t_0) \exp\left\{r \int_{t_0}^t (S(u) - \hat{S})\, du\right\}, \\ \lim_{t\to\infty} I(t)e^{\beta t} &= I_0 \exp\left\{\int_0^\infty (S(u) - \hat{S}) du\right\}. \end{aligned}$$

The expression on the left–hand side is finite because of relation (11.41).
It follows from (11.39) that

$$S(t) - \hat{S} \sim r\hat{S} \int_t^\infty I(u)du,$$

$$[S(t) - \hat{S}]e^{\beta t} \sim \hat{S}[e^{\beta t} \int_t^\infty I(u)du],$$

and the relation

$$\lim_{t\to\infty} I(t)e^{\beta t} = c$$

implies

$$\lim_{t\to\infty} e^{\beta t} \int_t^\infty I(u)du = \frac{c}{\beta}.$$

This completes the proof of the theorem.

□

Primary characteristics of an epidemic are the numbers S_0, α, t_m, I_m, $\hat{S}$, and β. We can see that S_0, α, and β do not depend on I_0. It is reasonable to consider the case where I_0 is small with respect to S_0, to test the vulnerability of the population to the spread of infection. In the next lemma we consider the behavior of some of these characteristics as $I_0 \to 0$.

Lemma 11 *The following relations hold:*
(i)

$$\lim_{I_0\to 0} I_m = S_0 - \alpha - \alpha \log \frac{S_0}{\alpha}.$$

(ii)

$$\lim_{I_0 \to 0} \hat{S} = \hat{S}_0,$$

where $\hat{S}_0$ *is the solution of the equation*

$$S_0 - \hat{S}_0 - \alpha \log \frac{S_0}{\hat{S}_0} = 0$$

satisfying the inequality $\hat{S}_0 < \alpha$.
(iii) *Let* $I_0 \to 0$. *Then*

$$t_m \sim \frac{1}{r(S_0 - \alpha)} \log \frac{1}{I_0}.$$

Proof *(i)* and *(ii)* follow from Theorem 6. To prove statement *(iii)* we can use the representation

$$t_m = \frac{1}{r} \int_\alpha^{S_0} \frac{du}{u[\alpha \log \frac{u}{S_0} + S_0 - u + I_0]}$$
$$= \frac{1}{rS_0} \int_{\frac{\alpha}{S_0}}^1 \frac{du}{u[\frac{\alpha}{S_0} \log u + (1-u) + \frac{I_0}{S_0}]} \sim \frac{1}{rS_0} \int_{\frac{\alpha}{S_0}}^1 \frac{du}{\left(1 - \frac{\alpha}{S_0}\right)(1-u) + \frac{I_0}{S_0}}.$$

(iii) follows from direct evaluation of this integral.

□

11.2.2 Randomly Perturbed Epidemic Models

We consider the system of differential equations

$$\dot{S}_\varepsilon(t) = -r\left(\frac{t}{\varepsilon}\right) I_\varepsilon(t) S_e(t),$$
$$\dot{I}_\varepsilon(t) = r\left(\frac{t}{\varepsilon}\right) I_\varepsilon(t) S_e(t) - \lambda\left(\frac{t}{\varepsilon}\right) I_\varepsilon(t), \tag{11.42}$$

where $(r(t), \lambda(t))$ is a two-dimensional stationary process on the interval $(0, \infty)$ satisfying:

(i) $0 < r(t) < c$ and $0 < \lambda(t) < c$ where $c > 0$ is a nonrandom constant.

(ii) The process $(r(t), \lambda(t))$ is ergodic, and we write $Er(t) = r$ and $E\lambda(t) = \lambda$, so system (11.34) is the averaged system for (11.42).

It follows from the averaging theorem (see Chapter 3, Theorem 5) that for any $t_0 > 0$,

$$P\left\{\lim_{\varepsilon \to 0} \sup_{t \le t_0} (|I_\varepsilon(t) - I(t)| + |S_\varepsilon(t) - S(t)|) = 0\right\} = 1 \tag{11.43}$$

if $(I_\varepsilon(t), S_\varepsilon(t))$ is the solution to the system (11.42), $(I(t), S(t))$ is the solution to the system (11.34), and $I_\varepsilon(0) = I(0) = I_0$, $S_\varepsilon(0) = S(0) = S_0$. The behavior of the solution $(I_\varepsilon(t), S_\varepsilon(t))$ on the interval $[0, \infty)$ is described in the following theorem.

Theorem 7 *Let $(I_\varepsilon(t), S_\varepsilon(t))$ be the solution to the system (11.42) with the initial values $I_\varepsilon(0) = I_0$, $S_\varepsilon(0) = S_0$. Then:*

(i) For any $\varepsilon > 0$, with probability 1 there exists the limit

$$\hat{S}_\varepsilon = \lim_{t\to\infty} S_\varepsilon(t),$$

and $P\{\hat{S}_\varepsilon \le \alpha\} = 1$, where $\alpha = \lambda/r$.

(ii) On the set $\{\hat{S}_\varepsilon \le \alpha - \delta\}$ for any $\delta > 0$ (this is a measurable set in the probability space on which the stochastic process $(r(t), \lambda(t))$ is defined) the functions $I_\varepsilon(t)$ and $S_\varepsilon(t)$ have the following properties:

(a) $\hat{S}_\varepsilon > 0$.

(b) $I_\varepsilon(t) > 0$ and $\lim_{t\to\infty} I_\varepsilon(t) = 0$.

(c) If $\beta_\varepsilon = \lambda - r\hat{S}_\varepsilon$, then

$$\lim_{t\to\infty} I_\varepsilon(t)e^{\nu t} = 0, \qquad \lim_{t\to\infty} (S_\varepsilon(t) - \hat{S}_\varepsilon)e^{\nu t} = 0 \quad \text{if } \nu < \beta_\varepsilon,$$

$$\lim_{t\to\infty} I_\varepsilon(t)e^{\nu t} = +\infty, \quad \lim_{t\to\infty} (S_\varepsilon(t) - \hat{S}_\varepsilon)e^{\nu t} = +\infty \quad \text{if } \nu > \beta_\varepsilon,$$

for some real number ν.

(iii) Set $I_m(\varepsilon) = \sup_t I_\varepsilon(t)$. Then

$$\hat{S}_\varepsilon \to \hat{S}, \quad I_m(\varepsilon) \to I_m, \quad \beta_\varepsilon \to \beta \text{ in probability as } \varepsilon \to 0,$$

where $\hat{S}$, I_m, and β are defined in Theorem 6 for the solution $(I(t), S(t))$ to system (11.34) for which $I(0) = I_0$, $S(0) = S_0$.

(iv)

$$\sup_{t>0}(|I_\varepsilon(t) - I(t)| + |S_\varepsilon(t) - S(t)|) \to 0$$

in probability as $\varepsilon \to 0$.

Proof Using the representation

$$S_\varepsilon(t) = S_0 \exp\left\{-\int_0^t r\left(\frac{u}{\varepsilon}\right) I_\varepsilon(u)\, du\right\}, \tag{11.44}$$

$$I_\varepsilon(t) = I_0 \exp\left\{\int_0^t \left[r\left(\frac{u}{\varepsilon}\right) S_\varepsilon(u) - \lambda\left(\frac{u}{\varepsilon}\right)\right] du\right\}, \tag{11.45}$$

we can see that $I_\varepsilon(t) \ge 0$ and $S_\varepsilon(t)$ is a decreasing function of t. Thus there exists

$$\lim_{t\to\infty} S_\varepsilon(t) = \hat{S}_\varepsilon,$$

and $\hat{S}_\varepsilon \le S_\varepsilon(t)$. Let $\delta > 0$. Consider the set $\{\hat{S}_\varepsilon \ge \alpha + \delta\}$. On this set

$$I_\varepsilon(t) \ge I_0 \exp\left\{\int_0^t \left[(\alpha+\delta)\, r\left(\frac{u}{\varepsilon}\right) - \lambda\left(\frac{u}{\varepsilon}\right)\right] du\right\}.$$

It follows from the ergodic theorem that

$$\int_0^t \left[(\alpha+\delta) r\left(\frac{u}{\varepsilon}\right) - \lambda\left(\frac{u}{\varepsilon}\right)\right] du \sim \int_0^t \left[(\alpha+\delta) r - \lambda\right] du \sim t\, r\, \delta$$

with probability 1. So $I_\varepsilon(t) \to +\infty$ and $S_\varepsilon(t) \to 0$ because of (11.44). The last relation is impossible because $\hat{S}_\varepsilon \ge \alpha + \delta$. This implies that

$$P\{\hat{S}_\varepsilon \ge \alpha + \delta\} = 0$$

for all $\delta > 0$. Statement *(i)* is proved.
On the set $\{\hat{S}_\varepsilon \ge \alpha - \delta\}$ we have, using the ergodic theorem, that

$$\limsup_{t\to\infty} \frac{1}{t} \int_0^t \left[r\left(\frac{u}{\varepsilon}\right) S_\varepsilon - \lambda\left(\frac{u}{\varepsilon}\right)\right] du \le -r\delta, \tag{11.46}$$

and this implies that

$$I_\varepsilon(t) = O\left(\exp\left\{-\frac{1}{2} r\delta t\right\}\right) \quad \text{and} \quad \int_0^\infty I_\varepsilon(u)du < \infty,$$

and statements (a) and (b) follow. Statement (c) can be proved in the same way as *(iv)* in Theorem 6.
It follows from (11.42) that

$$\lim_{\varepsilon\to 0} P\{\hat{S}_\varepsilon > \hat{S} + \delta\} = 0 \quad \text{for all } \delta > 0. \tag{11.47}$$

Since $\hat{S} < \alpha$, we can find $\delta > 0$ for which $\hat{S} + \delta < \alpha - \delta$ and

$$P\{\hat{S}_\varepsilon < \alpha - \delta\} \to 1 \text{ as } \varepsilon \to 0.$$

Next we derive some estimates for $S_\varepsilon(t)$ and $I_\varepsilon(t)$ when $\varepsilon < \varepsilon_0$ and ε_0 is small enough. Let ε_1 be chosen in such a way that

$$P\{|I_\varepsilon(t_0) - I(t_0)| + |S_\varepsilon(t_0) - S(t_0)|\} < \rho,$$

where ρ is a positive number and t_0 is such that

$$S(t_0) - \hat{S} < \delta,$$

where $\delta > 0$ satisfies the inequality $S(t_0) + \delta < \alpha - \delta_0$ where $\delta_0 = (\alpha - \hat{S})/2$. Denote by $\mathcal{E}(\varepsilon)$ the event that $\{S_\varepsilon(t_0) < \alpha - \delta_0\}$. Then $P\{\mathcal{E}(\varepsilon)\} > 1 - \rho$ for all $\varepsilon < \varepsilon_1$. The event $\{\mathcal{E}(\varepsilon)\}$ implies that

$$\begin{aligned} I_\varepsilon(t) &\le I_\varepsilon(t_0) \exp\left\{\int_{t_0}^t (r(u/\varepsilon)(\alpha - \delta_0) - \lambda(u/\varepsilon))\, du\right\} \\ &= I_\varepsilon(t_0) \exp\left\{-\delta_0 r(t - t_0) + \varepsilon \int_0^{(t-t_0)\varepsilon^{-1}} \xi\left(\frac{t_0}{\varepsilon} + u\right) du\right\}, \end{aligned}$$

where

$$\xi(t) = (r(t) - r)(\alpha - \delta_0) - \lambda(t) + \lambda$$

is an ergodic stationary stochastic process. Set

$$\Delta(s, T) = \sup_{t \geq T} \left| \frac{1}{t} \int_0^t \xi(s+u)\, du \right|.$$

Then $\Delta(s, T)$ is a stationary stochastic process in s for any fixed $T > 0$, and

$$\lim_{T \to \infty} P(\Delta(0, T) > \gamma) = 0, \quad \text{for } \gamma > 0.$$

Let T_0 satisfy the relations

$$P\left(\Delta(0, T_0) > \frac{1}{2}\delta_0 r\right) < \rho$$

and

$$\mathcal{E}_1(\varepsilon) = \left\{\Delta\left(\frac{t_0}{\varepsilon}, T_0\right) \leq \frac{1}{2}\delta_0 r\right\}.$$

Since

$$I_\varepsilon(t) \leq I_\varepsilon(t_0) \exp\left\{-\delta_0 r(t - t_0) + (t - t_0)\Delta\left(\frac{t_0}{\varepsilon}, \frac{t - t_0}{\varepsilon}\right)\right\},$$

the event $\mathcal{E}(\varepsilon) \cap \mathcal{E}_1(\varepsilon)$ implies the inequality

$$I_\varepsilon(t) \leq I_\varepsilon(t_0) \exp\left\{-\frac{1}{2}\delta_0 r(t - t_0)\right\} \tag{11.48}$$

for $t > t_0 + \varepsilon T_0$. Note that $|\xi(t)| \leq c_0$, where c_0 is a constant, so

$$I_\varepsilon(t) \leq I_\varepsilon(t_0) \exp\{\varepsilon c_0 T_0\}$$

and

$$\int_{t_0}^{\infty} I_\varepsilon(t) dt \leq \frac{2}{\delta_0 r} I_\varepsilon(t_0) \exp\{\varepsilon c_0 T_0\} \tag{11.49}$$

if the event $\mathcal{E}(\varepsilon) \cap \mathcal{E}_1(\varepsilon)$ holds. The last inequality implies that

$$S_\varepsilon(t_0) - \hat{S}_\varepsilon \leq S_\varepsilon(t_0) \int_{t_0}^{\infty} I_\varepsilon(t) \lambda\left(\frac{t}{\varepsilon}\right) dt \leq c_1 S_\varepsilon(t_0) \int_{t_0}^{\infty} I_\varepsilon(t) dt, \tag{11.50}$$

where c_1 is a constant for which $P\{\lambda(t) \leq c_1\} = 1$.
We have proved that

$$P\left\{\hat{S}_\varepsilon < \hat{S} - \delta - c_1(S(t_0) + \delta)\frac{2}{\delta_0 r}(I(t_0) + \delta)\exp\{\varepsilon c_0 T_0\}\right\} < 2\rho$$

for all $\varepsilon, \varepsilon_0$. Since $\delta > 0$, $I(t_0)$, and ρ can all be chosen arbitrarily small, we have proved, from (11.47), that $\hat{S}_\varepsilon \to \hat{S}$ in probability. Since the probability

of the event that the inequalities (11.48), (11.49), (11.50) hold is greater than $1 - 2\rho$, statement *(iv)* follows from relation (11.43). Statement *(iv)* implies that $I_m(\varepsilon) \to I_m$ in probability.
This completes the proof of the theorem.

□

11.2.3 Recurrent Epidemics

Recovery from diseases like gonorrhea does not impart permanent immunity. As a result, the susceptible pool is constantly being replenished, and the population can reach an endemic state with respect to the disease. To investigate this, we consider the model

$$\begin{aligned} \dot{S} &= -rIS + \lambda I, \\ \dot{I} &= rIS - \lambda I, \end{aligned} \tag{11.51}$$

in which $S + I$ is constant, so that the whole population is separated into two subpopulations, susceptibles and infectives, and we suppose that the size of the total population does not change over time. The behavior of the solution to system (11.51) is described in the following statement.

Lemma 12 *Let* $(S(t), I(t))$ *be the solution to system* (11.51) *satisfying the initial conditions* $S(0) = S_0,\ I(0) = I_0$, *where* $S_0 > 0,\ I_0 > 0$. *We define* $\alpha = \lambda/r,\ c_0 = S_0 + I_0$, *and* $\gamma = |c_0 - \alpha|$.
(i) If $c_0 < \alpha$, *then*

$$\begin{aligned} I(t) &= \frac{\gamma I_0}{I_0 + \gamma}\left(1 - \frac{I_0}{I_0 + \gamma} e^{-r\gamma t}\right)^{-1} e^{-r\gamma t}, \\ S(t) &= c_0 - I(t). \end{aligned}$$

(ii) If $c_0 > \alpha$ *and* $I_0 \neq \gamma$, *then*

$$I(t) = \gamma\left(1 + \frac{\gamma - I_0}{I_0} e^{-r\gamma t}\right)^{-1}.$$

(iii) If $c_0 > \alpha$ *and* $I_0 = \gamma$, *then* $I(t) = I_0,\ S(t) = S_0$ *for all* $t \geq 0$.
(iv) If $c_0 = \alpha$, *then*

$$I(t) = \frac{I_0}{1 + rI_0 t}.$$

The proof follows directly from the equation for I:

$$\frac{dI}{dt}(t) = rI(t)(c_0 - I(t)) - \lambda I(t),$$

which is equivalent to a logistic equation.

Now we consider a corresponding randomly perturbed system:

$$\begin{aligned}\dot{S}_\varepsilon(t) &= -r\left(\frac{t}{\varepsilon}\right) I_\varepsilon(t) S_\varepsilon(t) + \lambda\left(\frac{t}{\varepsilon}\right) I_\varepsilon(t),\\ \dot{I}_\varepsilon(t) &= r\left(\frac{t}{\varepsilon}\right) I_\varepsilon(t) S_\varepsilon(t) - \lambda\left(\frac{t}{\varepsilon}\right) I_\varepsilon(t).\end{aligned} \tag{11.52}$$

The two–dimensional process $(r(t), \lambda(t))$ satisfies the conditions that were introduced in Section 11.2.2, so the averaged system for (11.52) is (11.51). We will investigate the asymptotic behavior of system (11.52) as $t \to \infty$.

Lemma 13 *If $c_0 < \alpha$, then for any $\nu < \gamma r$,*

$$P\left\{\lim_{t\to\infty} I_\varepsilon(t) e^{\nu t} = 0\right\} = 1,$$

and for any $\nu > \gamma r$,

$$P\left\{\lim_{t\to\infty} I_\varepsilon(t) e^{\nu t} = +\infty\right\} = 1$$

for all $\varepsilon > 0$.

Proof Using the relation $S_\varepsilon(t) = c_0 - I_\varepsilon(t)$ we can rewrite the second equation of system (11.52) in the form

$$\frac{dI_\varepsilon}{dt}(t) = -r\left(\frac{t}{\varepsilon}\right) I_\varepsilon^2(t) + \left[c_0 r\left(\frac{t}{\varepsilon}\right) - \lambda\left(\frac{t}{\varepsilon}\right)\right] I_\varepsilon(t). \tag{11.53}$$

Therefore,

$$\begin{aligned}I_\varepsilon(t) &\le I_\varepsilon(0)\exp\left\{\int_0^t [c_0 r(\frac{s}{\varepsilon}) - \lambda(\frac{s}{\varepsilon})]\, ds\right\},\\ e^{\nu t} I_\varepsilon(t) &\le I_\varepsilon(0)\exp\left\{\int_0^t [\nu + c_0 r(\frac{s}{\varepsilon}) - \lambda(\frac{s}{\varepsilon})]\, ds\right\}.\end{aligned}$$

The relation

$$\lim_{t\to\infty} \frac{1}{t}\int_0^t \left[\nu + c_0 r\left(\frac{s}{\varepsilon}\right) - \lambda\left(\frac{s}{\varepsilon}\right)\right] ds = \nu + c_0 r - \lambda = \nu - \gamma r \tag{11.54}$$

implies the first statement of the lemma. In addition, we see that

$$P\left\{\int_0^\infty r\left(\frac{t}{\varepsilon}\right) I_\varepsilon(t)\, dt < \infty\right\} = 1.$$

It follows from (11.53) that

$$e^{\nu t} I_\varepsilon(t) \ge I_0(t)\exp\left\{\int_0^\infty r\left(\frac{s}{\varepsilon}\right) I_\varepsilon(s)\, ds + \int_0^t \left[\nu + c_0 r\left(\frac{s}{\varepsilon}\right) - \lambda\left(\frac{s}{\varepsilon}\right)\right] ds\right\},$$

and the second statement of the lemma follows from (11.54).

□

Theorem 8 *Assume that $(Z_1^\varepsilon(t), Z_2^\varepsilon(t))$ is a two-dimensional stochastic process, where*

$$Z_1^\varepsilon(t) = \varepsilon^{-1/2} \int_0^t \left(r\left(\frac{s}{\varepsilon}\right) - r \right) ds, \quad Z_2^\varepsilon(t) = \varepsilon^{-1/2} \int_0^t \left(\lambda\left(\frac{s}{\varepsilon}\right) - \lambda \right) ds$$

converges weakly in C to a two-dimensional homogeneous Gaussian process with independent increments $(Z_1(t), Z_2(t))$ for which

$$\begin{aligned} EZ_k(t) &= 0, \quad k = 1, 2, \\ EZ_k(t)Z_i(t) &= b_{ki}, \quad k = 1, 2, \text{ and } i = 1, 2. \end{aligned}$$

Suppose $I_\varepsilon(0) = I_0$. Then (i) for $c_0 > \alpha$ the stochastic process

$$v_\varepsilon(t) = (I_\varepsilon^{-1}(T_\varepsilon + t) - \gamma^{-1})/\sqrt{\varepsilon}$$

converges weakly to the Gaussian stationary process $v(t)$ for $t \in R$, and $Ev(t)v(s) = a \exp\{-2\gamma r|t-s|\}$, where $a = b_{11}(c_1 - 1/\gamma)^2 + 2b_{12}(c_1 - 1/\gamma) + b_{22}$ as $\varepsilon \to 0$ and as $\log \varepsilon / T_\varepsilon \to 0$.
(ii) For $c_0 = \alpha$ the stochastic process

$$u_\varepsilon(t) = \varepsilon I_\varepsilon^{-1}\left(\frac{t}{\varepsilon}\right)$$

converges weakly as $\varepsilon \to 0$ to the process

$$u(t) = r \exp\{-Z(t)\} \int_0^t \exp\{Z(s)\}\, ds,$$

where $Z(s) = c_0 Z_1(t) - Z_2(t)$.

Proof It follows from the relation (11.53) that the function I_ε^{-1} satisfies the differential equation

$$\frac{dI_\varepsilon^{-1}}{dt}(t) = r\left(\frac{t}{\varepsilon}\right) - \left(c_0 r\left(\frac{t}{\varepsilon}\right) - \lambda\left(\frac{t}{\varepsilon}\right) \right) I_\varepsilon^{-1}(t), \tag{11.55}$$

from which we obtain the formula

$$I_\varepsilon^{-1}(t) = I_\varepsilon^{-1}(0) \exp\{-Z_\varepsilon(t)\} + \exp\{-Z_\varepsilon(t)\} \int_0^t r\left(\frac{t}{\varepsilon}\right) \exp\{Z_\varepsilon(s)\}\, ds, \tag{11.56}$$

where

$$Z_\varepsilon(t) = \int_0^t \left(c_0 r\left(\frac{t}{\varepsilon}\right) - \lambda\left(\frac{t}{\varepsilon}\right) \right) ds. \tag{11.57}$$

Note that

$$Z_\varepsilon(t) = r\gamma t + c_0\sqrt{\varepsilon}\,(Z_1^\varepsilon(t) - Z_2^\varepsilon(t)),$$

and for $\gamma > 0$,

$$\lim_{t\to\infty} \exp\{-Z_\varepsilon(t)\} = 0.$$

To prove *(i)* we define the stochastic processes

$$\hat{Z}_k^\varepsilon(t) = Z_k^\varepsilon(t + T_\varepsilon),\ k = 1, 2, \quad \hat{Z}^\varepsilon(t) = c_0 \hat{Z}_1^\varepsilon(t) - \hat{Z}_2^\varepsilon(t).$$

It follows that

$$\begin{aligned} v_\varepsilon(t) = \int_{-T_\varepsilon}^t & \exp\{-r\gamma(t-s) - \sqrt{\varepsilon}\,(\hat{Z}^\varepsilon(t) - \hat{Z}^\varepsilon(s))\}\, d\hat{Z}_1^e(s) \\ & - r \int_{-T_\varepsilon}^t \exp\{-r\gamma(t-s)\}(\hat{Z}^\varepsilon(t) - \hat{Z}^\varepsilon(s))\}\, ds + \eta_\varepsilon(t), \end{aligned}$$

where $\eta_\varepsilon(t) \to 0$ in probability as $\varepsilon \to 0$. It follows from the condition of the theorem that the distribution of $v_\varepsilon(t)$ coincides asymptotically with the distribution of the random variable

$$\begin{aligned} \hat{v}_\varepsilon(t) = \int_{-T_\varepsilon}^t & \exp\{-r\gamma(t-s)\} dZ_1^e(T_\varepsilon + s) \\ & - r \int_{-T_\varepsilon}^t \exp\{-r\gamma(t-s)\}(Z^\varepsilon(t+T_\varepsilon) - \hat{Z}^\varepsilon(s+T_\varepsilon))\}\, ds. \end{aligned}$$

It easy to check that $\hat{v}_\varepsilon(t)$ is the Gaussian process for which

$$E\hat{v}_\varepsilon(t) = 0 \quad \text{and} \quad \lim_{\varepsilon\to 0} E\hat{v}_\varepsilon(t)\hat{v}_\varepsilon(s) = ae^{-\gamma r|t-s|}.$$

Next we prove statement *(ii)*. In this case $\gamma = 0$, and the stochastic process $Z_\varepsilon(t/\varepsilon)$ converges weakly to the process $Z(t)$. Using formula (11.56) we write

$$\begin{aligned} u_\varepsilon(t) = \varepsilon I^{-1} & \exp\left\{-Z_\varepsilon\left(\frac{t}{\varepsilon}\right)\right\} \\ & + \exp\left\{-Z_\varepsilon\left(\frac{t}{\varepsilon}\right)\right\} \int_0^t r\left(\frac{2}{\varepsilon^2}\right) \exp\left\{Z_\varepsilon\left(\frac{s}{\varepsilon}\right)\right\} ds. \end{aligned} \tag{11.58}$$

It is easy to see that

$$\lim_{\varepsilon\to 0} \int_0^t r\left(\frac{s}{\varepsilon}\right) f(s)\, ds \to r \int_0^t f(s)\, ds$$

uniformly in $t \in [0, T]$ for $f \in K$, a compact set in $C_{[0,T]}$. So the proof of statement *(ii)* follows from weak convergence in C of the process $Z_\varepsilon(t/\varepsilon)$ to the process $Z(t)$.
This completes the proof of the theorem.

□

Remark 6 It follows from statement *(iv)* of Lemma 12 that for the unperturbed system we have

$$\lim_{\varepsilon\to 0} \varepsilon I^{-1}\left(\frac{t}{\varepsilon}\right) = r\,t.$$

For the perturbed system this limit, under the condition $\gamma = 0$, is a stochastic process $u(t)$ that satisfies a stochastic differential equation

$$du(t) = r dt - u(t) dZ(t),$$

so it is a diffusion process. If $Z(t) \equiv 0$, then $u(t) = r\,t$.

11.2.4 Diffusion Approximation to Recurrent Epidemics

Consider the randomly perturbed model defined by the system

$$\begin{aligned} \dot{S}_\varepsilon(t) &= -r_1\left(y\left(\frac{t}{\varepsilon}\right)\right) I_\varepsilon(t)\, S_\varepsilon(t) + \lambda_1\left(y\left(\frac{t}{\varepsilon}\right)\right) I_\varepsilon(t), \\ \dot{I}_\varepsilon(t) &= r_2\left(y\left(\frac{t}{\varepsilon}\right)\right) I_\varepsilon(t)\, S_\varepsilon(t) - \lambda_2\left(y\left(\frac{t}{\varepsilon}\right)\right) I_\varepsilon(t), \end{aligned} \tag{11.59}$$

where $y(t)$ is a Markov process in a measurable space $(Y, \mathcal{C})$ satisfying the usual conditions, and the data $r_1(y)$, $r_2(y)$, $\lambda_1(y)$, $\lambda_2(y)$ are nonnegative bounded measurable functions satisfying

$$r = \int r_1(y)\rho(dy) = \int r_2(y)\rho(dy), \quad \lambda = \int \lambda_1(y)\rho(dy) = \int \lambda_2(y)\rho(dy),$$

where $\rho(dy)$ is the ergodic distribution for $y(t)$. As a result, the averaged equation for system (11.59) is system (11.51). Although $S_\varepsilon(t) + I_\varepsilon(t)$ is not constant for system (11.59), $S(t) + I(t)$ is the first integral for system (11.51). Under the assumption that $y(t)$ satisfies condition SMC II of Section 2.3 we can describe the diffusion approximation for the stochastic process

$$s_\varepsilon(t) = S_\varepsilon\left(\frac{t}{\varepsilon}\right) + I_\varepsilon\left(\frac{t}{\varepsilon}\right).$$

Theorem 9 *Let $I_\varepsilon(0) = I_0$, $S_\varepsilon(0) = S_0$. Denote by τ_ε the stopping time*

$$\tau_\varepsilon = \inf\{t : \ s_\varepsilon(t) = \alpha\}.$$

Let $\hat{s}(t)$ be the diffusion process on the interval (α, ∞) with generator $L^{\hat{s}}$ of the form

$$L^{\hat{s}} f(x) = \hat{a}(x) f'(x) + \frac{1}{2}\hat{b}(x) f''(x)$$

for $f \in C^2[\alpha, \infty)$, where the coefficients $\hat{a}(x)$ and $\hat{b}(x)$ are given by the formulas

$$\begin{aligned} \hat{a}(x) &= a_2(x-\alpha)^2 + a_1(x-\alpha), \\ \hat{b}(x) &= b_2(x-\alpha)^2, \end{aligned}$$

and

$$a_2 = \iint (-\alpha r_1(y) + \lambda_1(y))(r_2(y') - r_1(y'))R(y, dy')\rho(dy),$$

$$a_1 = \iint (-\alpha r_2(y) + \lambda_2(y))$$
$$\times (\alpha r_2(y') - \alpha r_1(y') + \lambda_1(y') - \lambda_2(y'))R(y, dy')\rho(dy),$$

$$b_2 = \iint (\alpha r_2(y) - \alpha r_1(y) + \lambda_1(y) - \lambda_2(y))$$
$$\times (\alpha r_2(y') - \alpha r_1(y') + \lambda_1(y') - \lambda_2(y'))R(y, dy')\rho(dy).$$

Set

$$\tau = \sup\left\{t : \inf_{u \le t} \hat{s}(u) > \alpha\right\}.$$

Then the stochastic process $s_\varepsilon(t \wedge \tau_\varepsilon)$ converges weakly in C to the stochastic process $\hat{s}(t \wedge \tau_\varepsilon)$ with the initial value $\hat{s}(0) = S_0 + I_0$.

The proof is a consequence of Theorem 5 of Chapter 5 and is not discussed further here.

Remark 7 The behavior of the diffusion process $\hat{s}(t)$ at the boundary α is determined by the coefficients a_1, a_2, and b_2 in the following way:

(1) If $b_2 > 0$ and $a_1 \neq 0$, we define $\gamma = 2a_1/b_2$. Then for $\gamma > 1$ the boundary α is a natural boundary for the process, i.e., $P\{\tau = +\infty\} = 1$ and

$$P_x\left\{\lim_{t\to\infty} = \alpha\right\} = 0$$

for all $x > \alpha$. For $\gamma < 1$, the boundary α is absorbing:

$$\lim_{x\to\alpha} P_x\{\tau < \infty, \hat{s}(\tau-) = \alpha\} = 1.$$

(2) If $b_2 > 0$ and $a_1 = 0$, then the boundary is attracting; i.e.,

$$\lim_{x\to\alpha} P_x\left\{\tau = \infty, \lim_{t\to\infty} \hat{s}(t) = \alpha\right\} = 1.$$

These results follow from Remark 4 of Chapter 5.

Remark 8 It can be proved in the same way as in Lemma 13 that for $I_0 + S_0 < \alpha$,

$$P\left\{\lim_{t\to\infty} I_\varepsilon(t) = 0\right\} = 1.$$

With this result, the averaging theorem, and the first equation of system (11.59), we can prove that $S_\varepsilon(t)$ tends to $S_0 + I_0$ in probability as $t \to \infty$ and $\varepsilon \to 0$. So $s_\varepsilon(t)$ converges to the constant $I_0 + S_0 = s_\varepsilon(0)$. Thus, in this case, the infection (probably) dies out of the population.

11.3 Demographics

Let $x(t)$ denote a population's birth rate at time t. The general (nonlinear) population renewal theory involves the equation

$$x(t) = \phi(t) + \int_0^t K(t, s, x(s))\, ds, \tag{11.60}$$

where the kernel K describes the number of births at time t due to those born at time s. (K is referred to as the *maternity function.*) The function ϕ describes contributions to the birth rate at time t from those present at the start of our model ($t = 0$), and it has compact support (i.e., it is zero for large t). Our primary interest is to study the behavior of $x(t)$ as $t \to \infty$. This problem has been approached in various important settings using various mathematical methods [74]. Some of those cases are considered next.
The randomly perturbed renewal equation takes the form

$$x_\varepsilon(t) = \phi(t) + \int_0^t K\Big(t, s, x_\varepsilon(s), y\big(\frac{s}{\varepsilon}\big)\Big)\, ds, \tag{11.61}$$

where $y(s/\varepsilon)$ is a noise process satisfying the usual conditions. Note that it is reasonable to consider the perturbation of the maternity function at the time when the mother was born (the time s). Therefore, $K(t, s, \dots)$ does not depend on $y(t/\varepsilon)$. Otherwise, we would consider the equation (11.61) with the maternity function

$$\hat{K}\left(t, s, x, y_\varepsilon\left(\frac{s}{\varepsilon}\right)\right) = E\left(K\left(t, y\left(\frac{t}{\varepsilon}\right), s, x, y_\varepsilon\left(\frac{s}{\varepsilon}\right)\right) / y_\varepsilon\left(\frac{s}{\varepsilon}\right)\right).$$

Let $\rho(dy)$ be the ergodic distribution for the process $y(t)$. If

$$K(t, s, x) = \int K(t, s, x, y)\rho(dy),$$

then equation (11.60) is the averaged equation for (11.61). We will investigate the relations between solutions of the perturbed equation and those of its average.

11.3.1 Nonlinear Renewal Equations

We assume that $\phi(t)$ is a positive continuous function with compact support, that $K(t, s, x) = K(t - s, x)$, and that there exist $0 < a < b < \infty$ for which

$$K(t, x) = 0 \text{ if } t \le a, \text{ or } t \ge b.$$

That is, births take place only to those with ages between a and b. We assume that $K(t, x)$ satisfies additional conditions:

(1) $K(t, x)$ is a nonnegative, bounded continuous function on $R_+ \times R$.

(2) $K(t, 0) = 0$, and for some constant $l > 0$,

$$|K(t, x_1) - K(t, x_2)| \leq l|x_1 - x_2|, \quad t \in R_+, \ x_1, x_2 \in R.$$

Denote by $\mathbf{C}_a^+$ the metric space of nonnegative continuous functions on $[0, a]$ with the distance

$$\text{dist}(x_1(\cdot), x_2(\cdot)) = \sup_{t \in [0,a]} |x_1(t) - x_2(t)|.$$

We define the functions

$$A_k(x(\cdot), t) = \int_0^a K(t + ka - s, x(s))\, ds \qquad (11.62)$$

and

$$x_n(t) = x(t + na), \ \phi_n(t) = \phi(t + na), \quad t \in [na, (n+1)a],$$

where $x(t)$ and $\phi(t)$ satisfy equation (11.59). (Under conditions 1 and 2, equation (11.59) has a unique solution for any given continuous function $\phi(t)$.)

Lemma 14 *Equation (11.59) is equivalent to the recursive system of equations*

$$x_n(t) = \phi_n(t) + \sum_{k=1}^{n} A_k(x_{n-k}(\cdot), t). \qquad (11.63)$$

The proof is a consequence of the fact that $K(t - s, x) = 0$ if $t - s \leq a$.

Remark 9 $A_k(x(\cdot), t) = 0$ if $(k-1)a \geq b$, since $K(t, x) = 0$ if $t \leq a$. So there exists $r > 0$ for which

$$x_n(t) = \phi_n(t) + \sum_{k=1}^{r} A_k(x_{n-k}(\cdot), t). \qquad (11.64)$$

Since $x_0(t) = \phi_0(t)$ and $\phi_k(t)$ are given for all k, we can sequentially calculate $x_n(t)$ for all $n > 0$ using formula (11.64).

Remark 10 $\phi_n(t) = 0$ for all $n > n_0$ because ϕ has compact support.

Consider the space $(\mathbf{C}_a^+)^r$ and the function

$$\mathcal{A} : (\mathbf{C}_a^+)^r \longrightarrow (\mathbf{C}_a^+)^r$$

defined recursively by

$$\mathcal{A}(z_1(\cdot), \ldots, z_r(\cdot)) = (z_2(\cdot), \ldots, z_r(\cdot), \sum_{k=1}^{r} A_{r-k}(z_k(\cdot), \cdot)).$$

Lemma 15 *(i) The mapping $\mathcal{A} : (\mathbf{C}_a^+)^r \to (\mathbf{C}_a^+)^r$ is a compact mapping, i.e., $\mathcal{A}(B)$ is a compact set in $(\mathbf{C}_a^+)^r$ for any bounded closed set $B \subset (\mathbf{C}_a^+)^r$.*

(ii) For $(z_1(\cdot), \dots, z_r(\cdot)) \in (\mathbf{C}_a^+)^r$ *set*

$$N(z_1(\cdot), \dots, z_r(\cdot)) = \max_{k \le r} \max_{t \in [0,a]} |z_k(t)|,$$
$$\mathcal{A}^2(z_1(\cdot), \dots, z_r(\cdot)) = \mathcal{A}(\mathcal{A}(z_1(\cdot), \dots, z_r(\cdot))),$$
$$\mathcal{A}^n(z_1(\cdot), \dots, z_r(\cdot)) = \mathcal{A}(\mathcal{A}^{n-1}(z_1(\cdot), \dots, z_r(\cdot))), \quad n = 3, 4, \dots.$$

Then there exists the limit

$$\lim_{n \to \infty} \frac{1}{n} \log \sup_{N(z_1(\cdot), \dots, z_r(\cdot)) \le 1} N(\mathcal{A}^n(z_1(\cdot), \dots, z_r(\cdot))) = \alpha,$$

where α *is a constant,* $-\infty \le \alpha < \infty$.

Proof Statement *(i)* follows from properties 1 and 2. Define the function

$$\Psi(n) = \sup\{N(\mathcal{A}^n(z_1(\cdot), \dots, z_r(\cdot))) : N(z_1(\cdot), \dots, z_r(\cdot)) \le 1\}.$$

The reader can check that

$$\Psi(n + m) \le \Psi(n)\Psi(m),$$

so $\Psi(n) \le (\Psi(1))^n$ and $(1/n) \log \Psi(n) \le \log \Psi(1)$. Define

$$\alpha = \inf \frac{1}{n} \log \Psi(n).$$

The function $\log \Psi(n)$ is semiadditive, so

$$\lim_{n \to \infty} \frac{1}{n} \log \Psi(n) = \alpha.$$

This completes the proof of the lemma.

□

Remark 11 The constant α/a is called the intrinsic growth rate of the population, and it can be proved that

$$\lim_{t \to \infty} \frac{1}{t} \log x(t) = \frac{\alpha}{a}.$$

Next, consider a randomly perturbed equation of the form (11.61). Suppose that $y(t)$ is an ergodic stationary process in Y with ergodic distribution $\rho(dy)$. The function $K(t, s, x, y)$ satisfies the following conditions:

(3) $K(t, s, x, y) = K(t - s, x, y)$ and $K(t, x, y) = 0$ if $t \le a,\ t \ge b$, where $0 < a < b < \infty$.

(4) $K(t - s, x, y)$ is continuous in t uniformly in y.

(5) $|K(t, x_1, y) - K(t, x_2, y)| \le l(y)|x_1 - x_2|$, where $l(y)$ is a bounded measurable function and $K(t, 0, y) = 0$.

(6) $K(t, x, y)$ is nonnegative and measurable with respect to all variables.

Define the random mapping $\mathcal{A}_n^\varepsilon(\omega, z_1(\cdot), \ldots, z_r(\cdot)) : (\mathbf{C}_a^+)^r \to (\mathbf{C}_a^+)^r$ for which

$$\mathcal{A}_n^\varepsilon(\omega, z_1(\cdot), \ldots, z_r(\cdot)) = (z_2(\cdot), \ldots, z_r(\cdot), \sum_{k=1}^r \mathcal{A}_{r-k}^{n,\varepsilon}(z_k(\cdot), \cdot)),$$

where

$$\mathcal{A}_{r-k}^{n,\varepsilon}(z_k(\cdot), t) = \int_0^a K\left(t + (r-k)a - s, z_k(s), y\left(\frac{s + (n-r+k)a}{\varepsilon}\right)\right) ds.$$

Let

$$x_n^\varepsilon = x_\varepsilon(t + na), \quad t \in [0, a].$$

Then

$$x_n^\varepsilon = \phi_n(t) + \sum_{k=1}^r A_k^{n,\varepsilon}(x_{n-k}^\varepsilon(\cdot), t), \tag{11.65}$$

$$\mathcal{A}_n^\varepsilon(\omega, x_{n-r}^\varepsilon(\cdot), \ldots, x_{n-1}^\varepsilon(\cdot)) = (x_{n-r+1}^\varepsilon(\cdot), \ldots, x_n^\varepsilon(\cdot)).$$

Define

$$\mathcal{A}_{n,n+1}^\varepsilon(\omega, z_1(\cdot), \ldots, z_r(\cdot)) = \mathcal{A}_{n+1}^\varepsilon(\omega, \mathcal{A}_n^\varepsilon(z_1(\cdot), \ldots, z_r(\cdot))),$$
$$\mathcal{A}_{n,n+k}^\varepsilon(\omega, z_1(\cdot), \ldots, z_r(\cdot)) = \mathcal{A}_{n+k}^\varepsilon(\omega, \mathcal{A}_{n,n+k-1}^\varepsilon(z_1(\cdot), \ldots, z_r(\cdot))), \ k \geq 2.$$

Then for all $n \geq r, \ k > 0$,

$$\mathcal{A}_{n,n+k}^\varepsilon(\omega, x_{n-r}^\varepsilon(\cdot), \ldots, x_{n-1}^\varepsilon(\cdot)) = (x_{n+k-r+1}^\varepsilon(\cdot), \ldots, x_{n+k}^\varepsilon(\cdot)).$$

In particular,

$$\mathcal{A}_{r,n}^\varepsilon(\omega, x_0^\varepsilon(\cdot), \ldots, x_{r-1}^\varepsilon(\cdot)) = (x_{n-r+1}^\varepsilon(\cdot), \ldots, x_n^\varepsilon(\cdot)). \tag{11.66}$$

With these preliminaries, we can prove the next theorem.

Theorem 10 *For $n < m$ define*

$$\theta_{n,m}^\varepsilon = \log \sup\{N(\mathcal{A}_{n,m}^\varepsilon(\omega, z_1(\cdot), \ldots, z_r(\cdot))) : \ N(z_1(\cdot), \ldots, z_r(\cdot)) \leq 1\}.$$

Then for any $\varepsilon > 0$ there exists a number $\theta^\varepsilon \geq -\infty$ for which:

(1)

$$\lim_{m\to\infty} \frac{1}{m}\theta_{n,m}^\varepsilon = \theta^\varepsilon \tag{11.67}$$

in probability.

(2) If $\theta^\varepsilon > -\infty$,

$$\lim_{m\to\infty} E\left|\frac{1}{m}\theta_{n,m}^\varepsilon - \theta^\varepsilon\right| = 0.$$

(3) $\theta^\varepsilon \to \alpha$ as $\varepsilon \to 0$, where α is determined in Lemma 15.

Proof First, we prove the existence of θ^ε. Note that the function $\theta^\varepsilon_{n,m}$ has the following properties:

(a) $\theta^\varepsilon_{n+kl,m+(k+1)l}$ is a stationary process in $k = 0, 1, 2, \ldots$ for fixed n and l, and this process is ergodic because it is a specific function of the process $y(t)$.

(b) $\theta^\varepsilon_{n,m+l} \le \theta^\varepsilon_{n,m} + \theta^\varepsilon_{n+1,m+l}$ if $n < m < m + l$.

Using the inequalities

$$\theta^\varepsilon_{n,m} \le \sum_{k=n}^{m-1} \theta^\varepsilon_{k,k+1}$$

we have

$$\theta^\varepsilon_{n,m} \le \Theta < \infty,$$

where Θ is a constant that depends on a, b, and $\sup_y |l(y)|$. (The function $l(y)$ was defined in condition 4).)
We can prove that

$$\limsup_{m\to\infty} \frac{1}{m}\theta^\varepsilon_{n,m} \le E\theta^\varepsilon_{r,r+1},$$
$$\limsup_{m\to\infty} \frac{1}{ml}\theta^\varepsilon_{n,ml} \le \frac{1}{l}E\theta^\varepsilon_{r,r+l}.$$

Therefore,

$$\limsup_{m\to\infty} \frac{1}{m}\theta^\varepsilon_{n,m} \le \inf_{l>r} \frac{1}{l}E\theta^\varepsilon_{r,r+l}. \tag{11.68}$$

Note that

$$\inf_{l>r} \frac{1}{l}E\theta^\varepsilon_{r,r+l} = \lim_{l\to\infty} \frac{1}{l}E\theta^\varepsilon_{r,r+l}.$$

(for the same reason as in the proof of statement *(ii)* of Lemma 15). If the left-hand-side of (11.68) is $-\infty$, then $\theta^\varepsilon = -\infty$. Assume that

$$\inf_{l>r} \frac{1}{l}E\theta^\varepsilon_{r,r+l} = \theta^\varepsilon > -\infty.$$

Then

$$\begin{aligned} E & \left| \frac{1}{l(m+1)} \theta^\varepsilon_{n,n+l(m+1)} - \frac{1}{l(m+1)} E\theta^\varepsilon_{n,n+l(m+1)} \right| \\ & \le E \left| \frac{1}{l(m+1)} \theta^\varepsilon_{n,n+l(m+1)} - \frac{1}{l(m+1)} \sum_{k=1}^m \theta^\varepsilon_{n+kl,n+(k+1)l} \right| \\ & + E \left| \frac{1}{l(m+1)} \sum_{k=1}^m \theta^\varepsilon_{n+kl,n+(k+1)l} - \frac{1}{l} E\theta^\varepsilon_{n,n+l} \right| \\ & + E \left| \frac{1}{l} E\theta^\varepsilon_{n,n+l} - \frac{1}{l(m+1)} E\theta^\varepsilon_{n,n+l(m+1)} \right|, \end{aligned} \tag{11.69}$$

where $n, m, l \in \mathbf{Z}_+$. Note that

$$\frac{1}{l(m+1)} \left(\sum_{k=1}^m \theta^\varepsilon_{n+kl,n+(k+1)l} - \theta^\varepsilon_{n,n+l(m+1)} \right) \ge 0,$$

so the first term on the right–hand side of inequality (11.69) equals

$$\frac{1}{l} E\theta^\varepsilon_{n,n+l} - \frac{1}{l(m+1)} E\theta^\varepsilon_{n,n+l(m+1)}.$$

Similarly, for the third term,

$$\lim_{m\to\infty} E \left| \frac{1}{(m+1)} \sum_{k=1}^m \theta^\varepsilon_{n+kl,n+(k+1)l} - \frac{1}{l} E\theta^\varepsilon_{n,n+l} \right| = 0$$

because of the ergodic theorem. Therefore,

$$\lim_{m\to\infty} E \left| \frac{1}{l(m+1)} \theta^\varepsilon_{n,n+l(m+1)} - \theta^\varepsilon \right| \le \frac{1}{l} E\theta^\varepsilon_{n,n+l} - \theta^\varepsilon.$$

This implies the relation

$$\lim_{m\to\infty} E \left| \frac{1}{m} \theta^\varepsilon_{n,m} - \theta^\varepsilon \right| = 0.$$

To prove that $\theta^\varepsilon \to 0$ as $\varepsilon \to 0$ we have only to note that

$$\lim_{\varepsilon\to 0} \frac{1}{l} E\theta^\varepsilon_r = \frac{1}{l} \sup\{N(\mathcal{A}^{r+l}(z_1(\cdot), \dots, z_r(\cdot))) : \; N(z_1(\cdot), \dots, z_r(\cdot)) \le 1\}.$$

This completes the proof of the theorem.

□

Remark 12 It follows from (11.68) that with probability 1,

$$\limsup_{\varepsilon\to 0} \frac{1}{m} \theta^\varepsilon_{n,m} \le \theta^\varepsilon.$$

11.3.2 Linear Renewal Equations

The linear renewal equation is

$$x(t) = \phi(t) + \int_0^t M(t-s)x(s)\,ds, \tag{11.70}$$

where $M(s)$ is the probability that a newborn at time zero produces an offspring at time s. This accounts for the probability of survival to age s and the fertility of those of age s. The maternity function in this case is linear in x and homogeneous in time, so this is a linear time–invariant system.

A general approach, which was proposed separately by McKendrick and von Forster, is based on the partial differential equation

$$\frac{\partial u}{\partial t}(t,x) + \frac{\partial u}{\partial x}(t,x) = -\delta(x,t)u(t,x), \tag{11.71}$$

where $u(t,x)$ is the number of population members having age x at the time t and $\delta(t,x)$ is the age–specific death rate. This equation is considered for $x > 0,\ t > 0$ with given initial data $u(0,x)$ and boundary condition

$$u(0,t) = \int_0^\infty \beta(x,t)u(x,t)dx, \tag{11.72}$$

which describes the birth of new population members. Under some natural assumptions, this equation can be reduced to the renewal equation for $x(t) = u(0,t)$. The main results related to the asymptotic behavior of the solution to equation (11.70) as $t \to \infty$ are summarized in the next theorem.

Theorem 11 *Suppose that the functions $\phi(t)$ and $M(t)$ on R_+ are continuous nonnegative functions and $\phi(t)$ has compact support. Define by $\alpha \in R$ the number for which*

$$\int_0^\infty e^{\alpha t}M(t)dt = 1.$$

Assume that there exists $\alpha_1 > \alpha$ for which $\int_0^\infty e^{\alpha_1 t}M(t)dt < \infty$. Then there exists a constant $B_0 > 0$ for which

$$\lim_{t\to\infty} x(t)e^{\alpha t} = B_0,$$

where $x(t)$ is the solution to the equation (11.70) and

$$B_0 = \frac{\int_0^\infty \phi(t)e^{\alpha t}dt}{\int_0^\infty tM(t)e^{\alpha t}dt}.$$

The proof of the theorem follows from the theory of convolution integral equations (see [200, Chapter 2]).

The constant α is called the *intrinsic growth rate* of the population, and B_0 characterizes the stable age distribution's birth rate.

A randomly perturbed version of the renewal equation is of the form

$$x_\varepsilon(t) = \phi(t) + \int_0^t M\left(t - s, y\left(\frac{s}{\varepsilon}\right)\right) x_e(s)\, ds, \tag{11.73}$$

where the function $\phi(t)$ is the same as in equation (11.70); $M(s,y)$ is a function from $R_+ \times Y$ into R_+ that is measurable in both variables; and

$$M(s,y) \le M^*(s),$$

where $M^*(s)$ satisfies the condition

$$\int_0^\infty M^*(s) e^{\alpha^* s}\, ds < \infty \qquad \text{for some } \alpha^* \in R.$$

Here $y(t)$ is a Y-valued ergodic homogeneous Markov process with an ergodic distribution $\rho(dy)$ satisfying SMC II of Section 2.3.
Define

$$M(t) = \int M(t,y)\rho(dy).$$

Then equation (11.70) is the averaged equation for equation (11.73). The asymptotic behavior of the solution to equation (11.73) was considered in Section 6.3, Theorem 12. We reformulate that theorem here, taking into account the conditions on $\phi(t)$, $M(t)$, and $M(t,y)$. The results are summarized in the next theorem.

Theorem 12 *Let $\phi(t)$ and $M(t)$ satisfy the conditions of Theorem 11, and suppose that $\alpha^* > \alpha$, where α was defined in Theorem 11. Suppose also that the following conditions are satisfied:*

(1)

$$\lim_{h\to 0} \sup_t \sup_y \left| \frac{M(t+h,y) - M(t,y)}{M(t,y)} \right| = 0.$$

(2) There exists a measurable function $\theta : R \to R_+$ with $\int \theta(u)\, du < \infty$ for which

$$|\hat{\phi}(\lambda + iu)| + |\hat{M}(\lambda + iu, y)| \le \theta(u), \quad u \in R,\ y \in Y,\ \lambda \ge -\alpha^*,$$

where

$$\hat{\phi}(\lambda) = \int_0^\infty e^{-\lambda t}\phi(t)\, dt, \quad \hat{M}(\lambda, y) = \int_0^\infty e^{-\lambda t} M(t,y)\, dt,$$

are the Laplace transforms of these functions.

Then there exist constants $\varepsilon_0 > 0$ and $\gamma > 0$ such that

$$P\left\{ \limsup_{t\to\infty} \frac{1}{t} \log x_\varepsilon(t) \le -\alpha + \gamma\varepsilon \right\} = 1 \tag{11.74}$$

for $\varepsilon < \varepsilon_0$.

Remark 13 In Theorem 12 we estimate only an upper bound for the growth rate of $x_\varepsilon(t)$. Using the averaging theorems and some limit theorems for integral equations we can also prove that

$$\lim_{c\to\infty} \limsup_{\varepsilon\to 0} P\left\{ \sup_{c\le t\le T/\varepsilon} \left| \frac{1}{t}\log x_\varepsilon(t) + \alpha \right| > \delta \right\} = 0$$

for all $T > 0$ and $\delta > 0$. The proof of this statement is presented in an article by the authors (see [79, Theorem 4.1]).

11.3.3 Discrete–Time Renewal Theory

The age distribution of a population can be determined by census, at which the numbers in various age classes, say $\{\nu_1, \dots, \nu_m\}$, define a vector whose dynamics can usefully describe how the population's age structure changes. If a census is taken every five years, in which females are counted, and if data are kept through the end of reproductive ages, say age 55, then 11 age classes would be tracked.

The population dynamics in this case can be described by the system of equations

$$x_{n+1} = \Lambda x_n,$$

where x_n is the vector of age classes at the nth census and Λ is Leslie's matrix (see [74]) of the form

$$\Lambda = \begin{pmatrix} \alpha_1 & \alpha_2 & \dots & \alpha_{m-1} & \alpha_m \\ \lambda_1 & 0 & \dots & 0 & 0 \\ \vdots & \vdots & \ddots & \vdots & \vdots \\ 0 & 0 & \dots & \lambda_{m-1} & 0 \end{pmatrix}, \tag{11.75}$$

where the numbers α_i are the fertilities of the various age groups and the numbers $\lambda_i \in (0,1]$ are the survival probabilities of the various age groups to the next census.

The first problem we consider here is to determine the asymptotic behavior of $\Lambda^n x$ as $n \to \infty$.

Lemma 16 *Assume that $\Lambda + \Lambda^2 + \cdots + \Lambda^m$ is a matrix with positive terms. Then (1) there exists a unique positive eigenvalue for Λ, denote it by ρ^*, and a corresponding eigenvector $b \in R^m$ having positive coordinates; (2) if $a \in R^M$ is an eigenvector of the adjoint matrix Λ^* corresponding to the eigenvalue ρ^*, then the coordinates of a are also positive; and (3) for any $x \in R^m$,*

$$\Lambda^n x = \rho^{*n} \cdot (x, a) \cdot b + O(\theta^n), \tag{11.76}$$

where $0 < \theta < \rho$.

The proof is based on the theory of nonnegative matrices (see [54, p.65, Theorem 2]).

A randomly perturbed version of Leslie's model is described by the system of equations

$$x_{n+1}^{\varepsilon} = (\Lambda + \varepsilon\tilde{\Lambda}(y_{n+1}))x_n^{\varepsilon}, \quad x_0^{\varepsilon} = x_0, \tag{11.77}$$

where $\tilde{\Lambda}(y)$ is the matrix of form (11.75), where $\alpha_i(y)$, $i = 1, \dots, m$, and $\lambda_i(y)$, $i = 1, \dots, m-1$, are bounded measurable functions, Λ is given by (11.75) with nonnegative $\alpha_1, \dots, \alpha_m, \lambda_1, \dots, \lambda_{m-1}$, and $\varepsilon > 0$ is sufficiently small that $\Lambda + \varepsilon\tilde{\Lambda}$ is a nonnegative matrix for all $y \in Y$. Here $\{y_n\}$ is a homogeneous Markov process in the space $(Y, \mathcal{C})$, a discrete–time process that is assumed to be ergodic with an ergodic distribution $\rho(dy)$.

With these preliminaries we have the following theorem.

Theorem 13 *There exist $\varepsilon_0 > 0$ and $\gamma > 0$ such that for any $\varepsilon < \varepsilon_0$,*

$$P\left\{\limsup_{n\to\infty}\left|\frac{\log|x_n^{\varepsilon}|}{n} - \log\rho^*\right| \le \gamma\varepsilon\right\} = 1, \tag{11.78}$$

where ρ^* is the same as in the relation (11.76).

Proof It follows from equations (11.77) that

$$x_n^{\varepsilon} = \Lambda^n x_0 + \sum_{k=1}^{n} \varepsilon^k S_{n,k} x_0,$$

where

$$S_{n,k} = \sum_{1\le i_k < i_{k-1} < \dots < i_1 \le n} \Lambda^{n-i_1}\tilde{\Lambda}(y_{i_1})\Lambda^{i_1-i_2-1}\tilde{\Lambda}(y_{i_2})\cdots\tilde{\Lambda}(y_{i_k})\Lambda^{i_k-1}.$$

Define Q to be the $m \times m$ matrix for which

$$Qx = (x, a)b, \quad x \in R^m.$$

It follows from relation (11.76) that

$$\Lambda^l = \rho^l Q + O(\theta^l).$$

Using this formula we can obtain the following representation:

$$\begin{aligned} S_{n,k} = \rho^{n-k}\Bigg(&\sum_{1\le i_k < i_{k-1} < \dots < i_1 \le n} (1_{\{n=i_1\}}I + 1_{\{n>i_1\}}Q)\tilde{\Lambda}(y_{i_1}) \\ &\times (1_{\{i_1-i_2=1\}}I + 1_{\{i_1>i_2\}}Q)\tilde{\Lambda}(y_{i_2})\cdots\tilde{\Lambda}(y_{i_k})(1_{\{i_k=1\}}I + 1_{\{i_k>1\}}Q) \\ &+ O\left(\left(\frac{\theta}{\rho-\theta}\right)^k\right)\Bigg). \end{aligned}$$

We use here the notation that if X is a matrix, we write

$$X = O(c) \text{ if } ||X|| = O(c).$$

Therefore,

$$\|S_{n,k}\| \le c_1^k \rho^n \left(1 + \sum_{1 \le i_k < i_{k-1} < \cdots < i_1 \le n} \|\tilde{\Lambda}(y_{i_1})\| \cdots \|\tilde{\Lambda}(y_{i_k})\| \right)$$

$$\le c_1^k \rho^n \left(1 + \frac{1}{k!} \left(\sum_{i=1}^{n} \|\tilde{\Lambda}(y_i)\| \right)^k \right).$$

So

$$|x_n^\varepsilon - \rho^n (x_0, a) b| \le \rho^n \left(\frac{\varepsilon}{1 - c_1 \varepsilon} + \exp \left\{ \varepsilon \sum_{i=1}^{n} \|\tilde{\Lambda}(y_i)\| \right\} \right) \tag{11.79}$$

if $\varepsilon < 1/c_1$. Inequality (11.79) implies the proof of this theorem.

□

Remark 14 This result describes the intrinsic growth rate for the random discrete–time Leslie model. The theorem is also valid if $\int \|\tilde{\Lambda}(y)\| \rho(dy) < \infty$.

12
Genetics

Genetics continues to be one of the outstanding applications of mathematics in the life sciences. Mathematical research has benefited from this interaction by being motivated to create mathematical structures that describe randomness in ways that are used to interpret data and to predict experimental outcomes. Our understanding of random processes was greatly enhanced by this interaction. On the other hand, the development of theories for random processes and the derivation of statistical methods to link them to experimental observations have been pivotal in our understanding of how genetic information is transmitted from one generation to the next. These show how genetic structures can be engineered to create important products such as novel drugs, strains of bacteria that degrade toxic waste, and a variety of crops and animals used in agriculture. This fruitful interaction between biology and mathematics has benefited biology with greater understanding of genetics and benefited mathematics with the creation of new structures of random processes and of statistics. As with most mathematical structures, these have found numerous applications elsewhere.

In this chapter we describe some of the fundamental models of population genetics and show how random perturbations of them behave. First, we consider a model of diploid genetics in a fixed population that was derived and studied by three early workers in the field of mathematical genetics: R. Fisher, S. Wright, and J.B.S. Haldane. We consider the influence of random noise on this model using the methods we developed in this book. Second, we consider the stability of extrachromosomal DNA elements in bacterial cells. These elements are called plasmids, and their stability in the presence of random noise is investigated in Section 12.2. Interest in this grows out of problems in biotechnology where one must determine conditions under which a genetically engineered organism will persist in a production environment. Finally, we investigate how changes in an entire collection of genes, called the genome, can be modeled, and how changes respond to random perturbations.

12.1 Population Genetics: The Gene Pool

Organisms, such as farm animals and humans, whose chromosomes appear in matched pairs are called diploids. We study here a population of diploids whose reproduction is synchronized and whose population size is kept fixed, say N, by a fixed carrying capacity, by culling, or by some other removal mechanism. Consider a gene at one location on a chromosome pair that can occur in two possible variant forms, say A or B. So, all members of the population are of exactly one of the three types: type AA, type AB (we cannot distinguish between AB and BA), or type BB as defined by this gene. The population carries a pool of $2N$ genes at this location, and in this section we describe how this gene pool changes from one reproductive interval to the next.

Let the proportion of the gene pool that is of type A at the start of the nth generation be denoted by g_n. The proportion at the next reproduction time depends on the fitness of the A genes from the previous reproduction that survive [75]. The varying ability of different genetic types to survive and reproduce is described by selection coefficients. For example, the Fisher–Wright–Haldane (FWH) model for the proportions $\{g_n\}$ is

$$g_{n+1} = \frac{r_n g_n^2 + s_n g_n(1 - g_n)}{r_n g_n^2 + 2s_n g_n(1 - g_n) + t_n(1 - g_n)^2}, \tag{12.1}$$

where the numbers r_n, s_n, and t_n describe the selection coefficients of the genotypes AA, AB, and BB, respectively, in the nth generation. They account for the probability of survival to the next reproduction time and the fertility of those that do.

A problem of particular interest is that of *slow selection*, where there are only small differences between the selection coefficients r_n, s_n, and t_n, say

$$r_n = r + \varepsilon\rho_n, \quad s_n = r + \varepsilon\sigma_n, \quad t_n = r + \varepsilon\tau_n, \tag{12.2}$$

where $0 \leq \varepsilon \ll 1$ and the other data are bounded. This case arises when the pressures of natural selection act over a longer time scale than the time scale of reproduction. When $\varepsilon = 0$, there is no selection, and the model becomes

$$g_{n+1} = g_n. \tag{12.3}$$

This fact was observed by G. Hardy and W. Weinberg in 1907, and it is referred to in the genetics literature as the Hardy–Weinberg law. However, several interesting phenomena can occur when $\varepsilon > 0$: One gene can dominate the gene pool, both can be maintained in the population in comparable numbers, or the system can be bistable, where each gene is stable, but there is also an unstable intermediate distribution.

12.1.1 The FHW Model with Stationary Slow Selection

Assume that r_n, s_n, t_n in equation (12.2) are

$$r_n = r + \varepsilon\rho, \quad s_n = r + \varepsilon\sigma, \quad t_n = r + \varepsilon\tau. \tag{12.4}$$

Then equation (12.1) can be rewritten as a difference equation of the form

$$g_{n+1}^{\varepsilon} - g_n^{\varepsilon} = \frac{\varepsilon Q(g_n^{\varepsilon})}{1 + \varepsilon P(g_n^{\varepsilon})}, \tag{12.5}$$

where g_0^{ε} is given and

$$\begin{aligned} Q(x) &= x(1-x)(ax+b), \\ P(x) &= \rho x^2 + 2\sigma x(1-x) + \tau(1-x)^2, \\ a &= \rho + \tau - 2\sigma, \quad b = \rho - \tau. \end{aligned}$$

It follows from the theory of difference equations (see Chapter 3, Lemma 2) that for any $T > 0$,

$$\lim_{\varepsilon \to 0} \sup_{k\varepsilon \leq T} |g(\varepsilon k) - g_k^{\varepsilon}| = 0$$

if $g(t)$ is the solution to the differential equation

$$\frac{dg}{dt}(t) = Q(g(t)) \tag{12.6}$$

with the initial condition $g(0) = g_0^{\varepsilon}$. (We assume here and below that g_0^{ε} does not depend on ε.) The asymptotic behavior of the solution to equation (12.6) as $t \to \infty$ is described by the following lemma, which is proved in [75].

Lemma 1 *Assume* $g(0) \in (0,1)$. *Then*

(i) If $b \geq 0$, $a + b \geq 0$, *and* $|a| + |b| > 0$, *then A dominates the gene pool:*

$$\lim_{t \to \infty} g(t) = 1.$$

(ii) If $b \leq 0$, $a + b \leq 0$, *and* $|a| + |b| > 0$, *then B dominates the gene pool:*

$$\lim_{t \to \infty} g(t) = 0.$$

(iii) If $b > 0$ *and* $a + b < 0$, *then a genetic polymorphism occurs:*

$$\lim_{t \to \infty} g(t) = -b/a,$$

so both genes are maintained in the gene pool.

(iv) If $b < 0$ and $a + b > 0$, then disruptive selection occurs:

$$\lim_{t\to\infty} g(t) = 1 \quad \text{if } g(0) \in \left(-\frac{b}{a}, 1\right),$$

$$\lim_{t\to\infty} g(t) = 0 \quad \text{if } g(0) \in \left(0, -\frac{b}{a}\right),$$

and in this case the value $g = -b/a$ is an unstable equilibrium for the system.

12.1.2 Random Perturbation of the Stationary Slow Selection Model

Suppose that ρ_n, σ_n, and τ_n in formula (12.2) are of the form

$$\rho_n = \rho(y_n), \quad \sigma_n = \sigma(y_n), \quad \tau_n = \tau(y_n),$$

where $\{y_n\}$ is a discrete–time stochastic process on a measurable space $(Y, \mathcal{C})$, and $\rho(y)$, $\sigma(y)$, and $\tau(y)$ are bounded real-valued measurable functions on this space. We will use the following condition:
(**CP**) The process $\{y_n\}$ is either a stationary ergodic process or an ergodic homogeneous Markov process. The ergodic distribution of the process is $m(dy)$, and we define

$$\rho = \int \rho(y) m(dy), \quad \sigma = \int \sigma(y) m(dy), \quad \tau = \int \tau(y) m(dy). \tag{12.7}$$

Denote by $\{\tilde{g}_n^\varepsilon\}$ the sequence satisfying the difference equation

$$\tilde{g}_{n+1}^\varepsilon - \tilde{g}_n^\varepsilon = \frac{\varepsilon \tilde{Q}_n(\tilde{g}_n^\varepsilon)}{1 + \varepsilon \tilde{P}_n(\tilde{g}_n^\varepsilon)} \tag{12.8}$$

with the initial condition $\tilde{g}_0^\varepsilon = g_0$, where

$$\begin{aligned}
\tilde{Q}_n(x) &= x(1-x)(\tilde{a}_n x + \tilde{b}_n), \\
\tilde{P}_n(x) &= \rho(y_n)x^2 + 2\sigma(y_n)x(1-x) + \tau(y_n)(1-x)^2, \\
\tilde{a}_n &= \rho(y_n) + \tau(y_n) - 2\sigma(y_n), \quad \tilde{b}_n = \rho(y_n) - \tau(y_n).
\end{aligned}$$

It follows from the averaging theorem for difference equations (see Chapter 3, Theorem 8) that with probability 1 for any $T > 0$,

$$\lim_{\varepsilon\to 0} \sup_{k\varepsilon \le T} |\tilde{g}_k^\varepsilon - g(k\varepsilon)| = 0, \tag{12.9}$$

where $g(t)$ is the solution to equation (12.6) with the initial value $g(0) = g_0$. So the behavior of the sequence $\{\tilde{g}_n^\varepsilon : n\varepsilon < T\}$ for any T as $\varepsilon \to 0$ can be described by the solutions to equation (12.6). To investigate the asymptotic behavior of $\{g_n^\varepsilon\}$ as $n \to \infty$ we can use the results of Section 6.2. However, the form of the difference equation in (12.7) is not exactly the same as in the theorems in Section 6.2. So, we must first extend the earlier results, which we state next specifically for the FWH model.

Theorem 1 *Let condition* (**CP**) *be fulfilled and define the averages of* a *and* b *to be*

$$a = \rho + \tau - 2\sigma, \quad b = \rho - \tau,$$

where ρ, σ, τ *are determined by formula (12.7). Then the following statements hold:*
(i) If $b \geq 0, \ a + b > 0, \ g_0 > 0$, *then*

$$\lim_{\varepsilon \to 0} P\left\{ \lim_{n \to \infty} \tilde{g}_n^{\varepsilon} = 1 \right\} = 1.$$

(ii) If $b < 0, \ a + b \leq 0, \ g_0 < 1$, *then*

$$\lim_{\varepsilon \to 0} P\left\{ \lim_{n \to \infty} \tilde{g}_n^{\varepsilon} = 0 \right\} = 1.$$

(iii) If $b < 0, \ a + b < 0$, *then*

$$(\alpha) \quad \lim_{\varepsilon \to 0} P\left\{ \lim_{n \to \infty} \tilde{g}_n^{\varepsilon} = 1 \right\} = 1 \quad \text{if } g_0 \in \left(-\frac{b}{a}, 1 \right),$$

$$(\beta) \quad \lim_{\varepsilon \to 0} P\left\{ \lim_{n \to \infty} \tilde{g}_n^{\varepsilon} = 0 \right\} = 1 \quad \text{if } g_0 \in \left(0, -\frac{b}{a} \right).$$

Proof All of these statements can be proved in similar ways, so we present here only the proof of statement *(i)*. Set

$$1 - \tilde{g}_n^{\varepsilon} = \tilde{f}_n^{\varepsilon}.$$

Then $\{\tilde{f}_n^{\varepsilon}\}$ satisfies the difference equation

$$\tilde{f}_{n+1}^{\varepsilon} - \tilde{f}_n^{\varepsilon} = -\varepsilon \tilde{f}_n^{\varepsilon} U_n^{\varepsilon}(\tilde{f}_n^{\varepsilon}),$$

where

$$U_n^{\varepsilon}(x) = (1 - x)(\tilde{a}_n + \tilde{b}_n - \tilde{a}_n x) \frac{1}{1 + \varepsilon \tilde{P}_n(1 - x)}.$$

Using successive back-substitutions for $\tilde{f}_n^{\varepsilon}$ gives

$$\tilde{f}_n^{\varepsilon} = (1 - g_0) \exp\left\{ \sum_{k=1}^{n-1} \log(1 - \varepsilon U_n^{\varepsilon}(\tilde{f}_k^{\varepsilon})) \right\},$$

and for $m > n$,

$$\tilde{f}_m^{\varepsilon} = \tilde{f}_n^{\varepsilon} \exp\left\{ \sum_{k=n+1}^{m-1} \log(1 - \varepsilon U_n^{\varepsilon}(\tilde{f}_k^{\varepsilon})) \right\}.$$

It follows from relation (12.9) and Lemma 1 *(i)* that for any $\delta_1 > 0$ and $\delta_2 > 0$ we can find n_0 and ε_0 such that for all $\varepsilon < \varepsilon_0$,

$$P\{\tilde{f}^{\varepsilon}_{n_0} < \delta_1\} \geq 1 - \delta_2. \tag{12.10}$$

Note that there exist constants $c_1 > 0,\ c_2 > 0$ depending on ε_0 such that for all $\varepsilon < \varepsilon_0$ the following inequality is fulfilled:

$$\log(1 - \varepsilon U^{\varepsilon}_n(x)) \leq -\varepsilon c_1(\tilde{a}_n + \tilde{b}_n)1_{\{x \leq \frac{1}{2}\}} + \varepsilon c_2 1_{\{x > \frac{1}{2}\}}.$$

So

$$\tilde{f}^{\varepsilon}_m \leq \delta_1 \exp\left\{-c_1 \sum_{k=n_0+1}^{m-1} \varepsilon(\tilde{a}_n + \tilde{b}_n)1_{\{\tilde{f}^{\varepsilon}_k \leq \frac{1}{2}\}} + \varepsilon \sum_{k=n_0+1}^{m-1} 1_{\{\tilde{f}^{\varepsilon}_k > \frac{1}{2}\}}\right\} \tag{12.11}$$

if $\tilde{f}^{\varepsilon}_{n_0} \leq \delta_1$. It follows from the ergodic theorem that

$$\sum_{k=n_0+1}^{m-1} \varepsilon(\tilde{a}_k + \tilde{b}_k) \sim \varepsilon(m - n_0)(a + b) \tag{12.12}$$

and

$$\lim_{c \to \infty} P\left\{\sup_{m > n_0}\left[-\sum_{n_0+1}^{m-1}(\tilde{a}_k + \tilde{b}_k)\right] > c\right\} = 0.$$

We can find $c_0 > 0$ for which

$$P\left\{\sup_{m > n_0}\left[-c_1\varepsilon \sum_{n_0+1}^{m-1}(\tilde{a}_k + \tilde{b}_k)\right] > c_0\right\} < \delta_2.$$

If $c_0 = \frac{1}{c_1 \varepsilon_0} \log \frac{1}{2\delta_1}$, then for all $\varepsilon < \varepsilon_0$,

$$P\left\{\sup_{m > n_0} \tilde{f}^{\varepsilon}_{n_0} \leq \frac{1}{2}\right\} \geq 1 - 2\delta_2,$$

so

$$P\left\{\tilde{f}^{\varepsilon}_m \leq \delta_1 \exp\left\{-c_1\varepsilon \sum_{n_0+1}^{m-1}(\tilde{a}_k + \tilde{b}_k)\right\} \text{ for all m}\right\} \geq 1 - 2\delta_2 \tag{12.13}$$

and

$$P\left\{\lim_{m \to \infty} \tilde{f}^{\varepsilon}_m = 0\right\} \geq 1 - 2\delta_2$$

for all $\varepsilon < \varepsilon_0$. Thus, statement *(i)* of the theorem is proved.

□

The next theorem shows how the sequence $\{g^{\varepsilon}_n\}$ is related to a Gaussian process.

Theorem 2 *Suppose that $\{y_k\}$ is a Markov process that satisfies SMC II of Section 2.3, and suppose $b > 0$, $a + b < 0$, and $g_0 \in (0, 1)$ (the notation here is as in Theorem 1). For $N \in \mathbf{Z}_+$ and $n \in \mathbf{Z}$, $n \geq -N$, set*

$$z_n^{\varepsilon,N} = \varepsilon^{-\frac{1}{2}} \left(\tilde{g}_{N+n}^{\varepsilon} + \frac{b}{a} \right), \tag{12.14}$$

and let

$$z_N^{\varepsilon}(t) = \sum 1_{\{n \geq -N\}} 1_{\{\varepsilon n \leq t < \varepsilon(n+1)\}} z_n^{\varepsilon,N}, \quad t \in R. \tag{12.15}$$

Let $W(dt)$ be the Gaussian measure on the real line with independent values for which for any measurable set A we have $EW(A) = 0$ and

$$EW^2(A) = v\, l(A),$$

where l is Lebesgue measure on the real line and

$$\begin{aligned} v = \frac{b^2(a+b)^2}{a^6} \Bigg[& \int (ab(y) - ba(y))^2 \rho(dy) \\ & + 2 \iint (ab(y) - ba(y))(ab(y') - ba(y')) R(y, dy') \rho(dy) \Bigg]. \end{aligned} \tag{12.16}$$

Here

$$R(y, C) = \sum_{n=1}^{\infty} (P_n(y, C) - \rho(C)),$$

where $\{P_n\}$ are the transition probabilities for $\{y_n\}$. Define the stationary Gaussian stochastic process $\eta(t)$, $t \in R$, by the formula

$$\eta(t) = \int_{-\infty}^{t} \exp\{-\gamma(t - s)\} W(ds), \tag{12.17}$$

where $\gamma = b(a + b/a) > 0$. Assume that $\varepsilon N \to \infty$, $N = o\left(\frac{1}{\varepsilon} \log \frac{1}{\varepsilon}\right)$ and $|\tilde{g}_0 + \frac{b}{a}| = O\left(\sqrt{\varepsilon}\right)$ as $\varepsilon \to 0$.
Then the stochastic process $z_N^{\varepsilon}(t)$ converges weakly in $\mathbf{C}$ to the stochastic process $\eta(t)$ on any finite interval as $\varepsilon \to 0$.

Proof We modify Theorem 9 of Chapter 6 to begin the proof. Because of the form of the difference equation (12.8) we need to modify the proof of that theorem. Define

$$\tilde{g}_n^{\varepsilon} + \frac{b}{a} = h_n^{\varepsilon}.$$

It follows from equation (12.8) that

$$h_{n+1}^{\varepsilon} - h_n^{\varepsilon} = -\varepsilon r(y_n) h_n^{\varepsilon} + O\left(\varepsilon (h_n^{\varepsilon})^2\right) + O\left(\varepsilon^2\right) + \varepsilon q(y_n),$$

where

$$r(y) = \frac{b(a+b)}{a^2} a(y), \quad q(y) = \frac{b(a+b)}{a^3} (ab(y) - ba(y)).$$

It is easy to see that

$$\int r(y)\rho(dy) = \gamma, \quad \int q(y)\rho(dy) = 0.$$

So

$$\begin{aligned} h^\varepsilon_{n_0+m} &= h^\varepsilon_{n_0} \prod_{n=n_0+1}^{m-1} (1 - \varepsilon r(y_n) + O(\varepsilon h^\varepsilon_n)) + O(m\varepsilon^2) \\ &+ \varepsilon \sum_{n=n_0+1}^{m-1} q(y_n) \prod_{k=n_0+1}^{n-1} (1 - \varepsilon r(y_k) + O(\varepsilon h^\varepsilon_k)). \end{aligned} \tag{12.18}$$

The remainder of the proof is the same as in Theorem 9 of Chapter 6, and it is left to the reader.

□

Remark 1 Using (12.18) we can estimate the time τ_ε for which $|h^\varepsilon_{n_0+\tau_\varepsilon}| \leq c\sqrt{\varepsilon}$, where $c > 0$ is a constant: Assume that $|h^\varepsilon_{n_0}| < \delta$, where δ is sufficiently small. Then

$$\prod_{n=n_0+1}^{m-1} (1 - \varepsilon r(y_n) + O(\varepsilon h^\varepsilon_n)) \sim \exp\left\{-\varepsilon \sum_{n=n_0+1}^{m-1} r(y_n) + O(\varepsilon h_n)\right\}$$

is less than 1 for $m = O(1/\varepsilon)$. This implies that

$$E\left(\sum_{n=n_0+1}^{m-1} q(y_n) \prod_{k=n_0+1}^{n-1} (1 - \varepsilon r(y_k) + O(\varepsilon h^\varepsilon_k))\right)^2 = O(\varepsilon^2 m).$$

It follows from these estimates that $\tau_\varepsilon = O\left(\frac{1}{\varepsilon} \log \frac{1}{\varepsilon}\right)$.

12.2 Bacterial Genetics: Plasmid Stability

Bacteria have a single chromosome made up of DNA, but in addition, they can have extrachromosomal DNA elements, called *plasmids*. Each newborn cell contains exactly one chromosome, but it might have many copies, say N, of a particular plasmid. In this section we consider how the plasmids are distributed at each reproductive event, which in this case is cell division into two daughter cells. Suppose that each newborn cell in the population has exactly N of these plasmids, so a newborn cell having N plasmids will have $2N$ plasmids just prior to division, and each of its daughters will receive exactly N plasmids. While these are essentially the same plasmid, they can be of various types depending on what variants of information they carry. For example, some plasmids might code for resistance to the antibiotic ampycillin and some for resistance to tetracycline. We describe here the distribution of various plasmid types as cells grow and divide.

Suppose each newborn cell has exactly N plasmids. These might be of types $T_1, \ldots, T_r$, and denote by $(n_1, \ldots, n_r)$ the vector representing the distribution of plasmids in a cell, so n_1 is the number of plasmids in the cell having type T_1, n_2 is the number of plasmids in the cell having type T_2, n_3 is the number of plasmids in a cell having type T_3, etc. Since the number of plasmids per cell is fixed, we have that $\sum n_k = N$. Each newborn cell has such a vector associated with it.
There are

$$m = \binom{N + r - 1}{N}$$

possible different types of cells as described by their plasmid-type distribution. This distribution is defined by the vector $(n_1, \ldots, n_r)$, and m is the number of vectors with integer–valued nonnegative coordinates $\{n_i,\ i = 1, 2, \ldots, r\}$ with $\sum_{k=1}^{r} n_k = N$.
Denote by $\{C_1, \ldots, C_m\}$ the possible types of cells. A population of these bacteria can be described by the proportions of cells among the various plasmid types. We denote by

$$\vec{p} = (p^1, \ldots, p^m)$$

the vector of the proportions: p^i is the proportion of the population that is of type C_i. Clearly, $\sum_k p^k = 1$. This vector changes from generation to generation, and we write $\vec{p}_t$, where $\vec{p}_t$ is the vector of proportions at time t, and $t = 0, 1, 2, \ldots,$ is discrete time. We suppose that each cell of the population divides at these times, and $\vec{p}_t$ is the vector just after the tth division. We assume that $\{\vec{p}_t,\ t > 0\}$ satisfies the relation

$$\vec{p}_{t+1} = \vec{p}_t\, A(t), \quad t = 0, 1, \ldots, \tag{12.19}$$

where $A(t)$ is a stochastic matrix of size $m \times m$ that is defined by sampling without replacement.
During the synthesis phase of the mitotic cycle, each copy is duplicated, so that just before splitting the plasmid pool is of size $2N$ with $2n_i$ plasmids of type T_i for $i = 1, \ldots, r$. We suppose that this process is contaminated with noise, perhaps due to mutation or by replication errors, so that the pool becomes

$$2n_1, \ldots, 2n_r \longrightarrow k_1, \ldots, k_r, \quad \sum k_i = 2N,$$

with probability

$$\tilde{P}_{2\vec{n},\vec{k}}(t) = 1_{\{\vec{k}=2\vec{n}\}} + \varepsilon Q_{2\vec{n},\vec{k}}(t), \tag{12.20}$$

where $\varepsilon > 0$ is a small positive number, $\vec{n} = (n_1, \ldots, n_r)$ and $\vec{k} = (k_1, \ldots, k_r)$ are vectors of integers, and the functions in $\tilde{P}$ are bounded stochastic processes with $Q_{2\vec{n},\vec{k}}(t)$ such that

$$\sum_{\vec{k}} Q_{2\vec{n},\vec{k}}(t) = 0,$$

and $Q_{2\vec{n},\vec{k}}(t) \geq 0$ if $\vec{k} \neq 2\vec{n}$.
At cell division one daughter receives a distribution of plasmids that is characterized by a vector $\vec{l} = (l_1, \ldots, l_r)$ with probability

$$R_{\vec{k},\vec{l}} = \frac{\binom{k_1}{l_1} \cdots \binom{k_r}{l_r}}{\binom{2N}{N}}. \tag{12.21}$$

Thus

$$(n_1, \ldots, n_r) \to (l_1, \ldots, l_r)$$

with probability

$$\hat{P}_{\vec{n},\vec{l}}(t) = \sum_{\vec{k}} \tilde{P}_{2\vec{n},\vec{k}}(t) R_{\vec{k},\vec{l}}. \tag{12.22}$$

This implies relation (12.19) with the matrix $A(t)$ having components that are defined by equation (12.22).

12.2.1 Plasmid Dynamics

Denote by V_N the set of vectors $\vec{n}$ having nonnegative integer-valued coordinates $(n_1, \ldots, n_r)$ with $\sum n_i = N$. We consider a homogeneous Markov chain in V_N describing the system when $\varepsilon = 0$. In this case,

$$\hat{P}_{\vec{n},\vec{l}}(t) = R_{2\vec{n},\vec{l}}.$$

Thus, the matrix $A(t)$ is nonrandom, and it does not depend on t. Denote this matrix by A. The elements of the matrix A are given by the formula

$$a_{\vec{n},\vec{l}} = \frac{\binom{2n_1}{l_1} \cdots \binom{2n_r}{l_r}}{\binom{2N}{N}}. \tag{12.23}$$

The asymptotic behavior of

$$\vec{p}(t) = \vec{P}_0 \, A^t$$

as $t \to \infty$ is determined by the homogeneous finite Markov chain having transition probability matrix A.

Lemma 2 *Denote by $C(S)$, where $S \subset \{1, \ldots, r\}$, the set of vectors $\vec{n} \in V_N$ for which $\sum_{i \in S} n_i = N$.*

(1) $C(S)$, for any nonempty S, is a connected class of states for the Markov chain having transition probability matrix A, and any such class of states is of the form $C(S)$, where S is a nonempty subset of $\{1, \ldots, r\}$.

(2) *$C(S)$ is an essential connected class of states if and only if S contains exactly one element.*

(3) *Denote by $\vec{n}^{(i)}$ the vector for which $n_i^{(i)} = N$, $n_j^{(i)} = 0$, $j \neq i$. The states $\vec{n}^{(1)}, \ldots, \vec{n}^{(r)}$ are all essential states of the Markov chain; each of them forms a connected class. Let $P_{\vec{n},\vec{l}}(t)$ be the transition probability from the state $\vec{n}$ to the state $\vec{l}$ after t steps. Then*

$$\lim_{t\to\infty} P_{\vec{n},\vec{n}^{(i)}}(t) = \frac{n_i}{N},$$

where $\vec{n} = (n_1, \ldots, n_r)$.

(4)

$$\lim_{t\to\infty} \vec{p}(t) = \vec{p}(\infty) = \{p_{\vec{n}}(\infty), \vec{n} \in V_N\},$$

where

$$p_{\vec{n}}(\infty) = 0 \quad \textit{if} \quad \vec{n} \in V_N \setminus \bigcup_{i=1}^{r} \{\vec{n}^{(i)}\},$$

and otherwise,

$$p_{\vec{n}^{(i)}}(\infty) = \frac{1}{N} \sum_{\vec{n}\in V_N} n_i \, p_{\vec{n}}(0).$$

Proof (1) follows from formula (12.23).
(2) It is easy to see that $\vec{n}^{(i)}$ is an absorbing state for the Markov chain for $i \in \overline{1,r}$, and the state $\vec{n}$ is connected with $\vec{n}^{(i)}$ if $n_i > 0$.
(3) Define

$$q_i(t) = \sum_{\vec{n}\in V_N} n_i \, p_{\vec{n}}(t).$$

Since

$$\sum_{\vec{n}\in V_N} n_i \, a_{\vec{l},\vec{n}} = l_i$$

because of formula (12.23),

$$\begin{aligned} q_i(t) &= \sum_{\vec{n}\in V_N} n_i \sum_{\vec{l}\in V_N} p_{\vec{l}}(t-1) a_{\vec{l},\vec{n}} = \sum_{\vec{l}\in V_N} p_{\vec{l}}(t-1) \sum_{\vec{n}\in V_N} n_i \, a_{\vec{l},\vec{n}} \\ &= \sum_{\vec{l}\in V_N} p_{\vec{l}}(t-1) l_i = q_i(t-1). \end{aligned}$$

This implies statements (3) and (4).

□

12.2.2 Randomly Perturbed Plasmid Dynamics

We assume that the matrices $A(t)$ in relation (12.19) are of the form

$$A(t) = A_\varepsilon(t) = A + \varepsilon B(t), \quad t = 0, 1, 2, \ldots, \tag{12.24}$$

where A is a constant matrix with elements given by relation (12.23), and $\{B(t),\ t = 0, 1, 2, \ldots\}$ is a stationary Q_m-valued stochastic process, where Q_m is the set of $m \times m$ matrices B for which $\|B\|_1 \le 1,\ \vec{1}B = \vec{0}$, where $\vec{1} = (1, \ldots, 1),\ \vec{0} = (0, \ldots, 0)$. We assume additionally that $B(t)$ satisfies the following conditions:

I. Denote the elements of the matrix $B(t)$ by $b_{\vec{n},\vec{l}}(t)$, where $\vec{n}, \vec{l} \in V_N$. Then $b_{\vec{n},\vec{l}}(t) \ge 0$ if $a_{\vec{n},\vec{l}} = 0$.

II. Denote by $\mathcal{E}_n$ the σ-algebra generated by $B(0), \ldots, B(n)$, and let $\mathcal{E}^l$ be the σ-algebra generated by $B(l+1), B(l+2), \ldots$. There exists a sequence $\mu_l,\ l = 0, 1, 2, \ldots$, for which $\mu_l \to 0$ and

$$|E\xi_1\xi_2 - E\xi_1 E\xi_2| \le \mu_l$$

if ξ_1 is an $\mathcal{E}_n$-measurable random variable and ξ_2 is an $\mathcal{E}^{n+l}$-measurable random variable, and $|\xi_1| \le 1,\ |\xi_2| \le 1$.

Note that condition II is analogous to condition EMC of Section 7.2.

We will investigate the asymptotic behavior of the matrix

$$P_\varepsilon(t) = (A + \varepsilon B(0))(A + \varepsilon B(1)) \cdots (A + \varepsilon B(t-1)) \tag{12.25}$$

as $t \to \infty$ and $\varepsilon \to 0$ using the results of Section 7.2. Note that this Markov chain is aperiodic with r essential classes of states.

Define the average $\bar{B} = EB(t)$, and let

$$\bar{P}_\varepsilon(t) = (A + \varepsilon\bar{B})^t.$$

If $\alpha(A + \varepsilon\bar{B}) > 0$ for all $\varepsilon > 0$ small enough (the definition of α can be found in Section 7.1), then there exists the limit

$$\Pi_\varepsilon = \lim_{t\to\infty} \bar{P}_\varepsilon(t).$$

It follows from Theorem 5 of Chapter 7 that

$$\hat{\Pi}_0 = \lim_{\varepsilon\to 0} \Pi_\varepsilon = \lim_{t\to\infty} \Pi_0 \exp\{t\Pi_0\bar{B}\Pi_0\}, \tag{12.26}$$

where $\Pi_0 = \lim_{n\to\infty} A^n$. This limit exists because the matrix A is an aperiodic matrix.

Lemma 3 *Let $\alpha(A + \varepsilon\bar{B}) > 0$ for $\varepsilon > 0$ small enough. Denote by $\bar{b}_{\vec{n},\vec{l}}$ the elements of matrix $\bar{B}$ and set*

$$q_{ij} = \frac{1}{N} \sum_{\vec{n}} \bar{b}_{\vec{n}^{(i)},\vec{n}} n_j, \quad i, j \in \overline{1, r}.$$

The following statements hold:

(i) *The* $r \times r$ *matrix* $Q = (q_{ij})_{i,j\in\overline{1,r}}$ *is a stochastic matrix for which* $\alpha(Q) > 0$.

(ii) *If* $\vec{q} = (q_1, \dots, q_r) \in D_r$ *is the unique vector satisfying the relation* $\vec{q}Q = \vec{q}$, *then* $\hat{\Pi}_0$ *is an* $m \times m$ *matrix with its rows equal to each other and coinciding with the vector*

$$\vec{d}^0 = (d^0_{\vec{n}} : \ \vec{n} \in V_N),$$

where

$$d^0_{\vec{n}^{(i)}} = q_i, \ i = 1, \dots, r, \qquad d^0_{\vec{n}} = 0 \ \textit{if} \ \vec{n} \notin \{\vec{n}^{(i)}, \dots, \vec{n}^{(r)}\}.$$

Proof If $\alpha(Q) = 0$, then $\alpha(A + \varepsilon\bar{B}) = 0$ for all $\varepsilon > 0$ small enough. This implies statement *(i)*. Statement *(ii)* follows from Theorem 5 of Chapter 7.

□

Our main result on the behavior of the matrix-valued process $P_\varepsilon(t)$ is formulated as follows.

Theorem 3 *Let conditions I and II be fulfilled and* $E\alpha(A + \varepsilon B_1) > 0$ *for all* $\varepsilon > 0$ *small enough. Then:*

(i) *There exists a matrix* $\tilde{\Pi}_\varepsilon$ *with* $\|\tilde{\Pi}_\varepsilon\|_0 = 0$ *(this means that the matrix has identical rows) such that for* $\varepsilon > 0$ *small enough it satisfies the relation*

$$\lim_{t\to\infty} \frac{1}{t} \sum_{k=1}^{t} P_\varepsilon(t) = \tilde{\Pi}_\varepsilon.$$

(ii)

$$\lim_{\varepsilon\to 0} \tilde{\Pi}_\varepsilon = \hat{\Pi}_0.$$

(iii) *There exists a stationary* D_m*-valued process* $\vec{d}^\varepsilon_n$ *for which*

$$\lim_{n\to\infty} |\vec{d}^\varepsilon_n - \vec{a}P_\varepsilon(n)|_1 = 0$$

for any $\vec{a} \in D_m$ *with probability* 1.

(iv)

$$\lim_{n\to\infty} \frac{1}{n} \sum_{k=1}^{n} \vec{a}P_\varepsilon(k) = \vec{a}\,\tilde{\Pi}_\varepsilon, \quad \vec{a} \in D_m,$$

with probability 1.

The proof of the theorem follows from Theorems 4 and 5 of Chapter 7, and the details are not presented here.
The assumption in Theorem 3 concerning $\alpha(A + \varepsilon B_1)$ can be weakened. In order to do this, we need some further estimates.

Lemma 4 *Let conditions I and II be fulfilled and*

$$\alpha(A + \varepsilon \bar{B}) > 0. \tag{12.27}$$

Then there exist l_0 and ε_0 for which

$$E\alpha(P_\varepsilon(l)) > 0$$

if $\varepsilon < \varepsilon_0$, $l > l_0$, and εl is small enough.

Proof We have the relation

$$P_\varepsilon(l) = (A + \varepsilon B_0) \cdots (A + \varepsilon B_{l-1}) = A^l + \varepsilon \sum_{k=1}^{l} A^{k-1} B_{k-1} A^{l-k} + O(l^2 \varepsilon^2),$$

$$(A + \varepsilon \bar{B})^l = A^l + \varepsilon \sum_{k=1}^{l} A^{k-1} \bar{B} A^{l-k} + O(l^2 \varepsilon^2).$$

So

$$P_\varepsilon(l) - (A + \varepsilon \bar{B})^l = \varepsilon \sum_{k=1}^{l} A^{k-1} (B_{k-1} - \bar{B}) A^{l-k} + O(l^2 \varepsilon^2).$$

It follows from the ergodicity of $\{B_k\}$ that

$$\lim_{l \to \infty} \frac{1}{l} \sum_{k=1}^{l} A^{k-1} (B_{k-1} - \bar{B}) A^{l-k} = 0$$

with probability 1, because of the boundedness of $\|A^k\|_1$ and the existence of a limit

$$\Pi_0 = \lim_{n \to \infty} A^n.$$

So

$$\|P_\varepsilon(l) - (A + \varepsilon \bar{B})^l\|_1 = O(l^2 \varepsilon^2) + l \varepsilon \, o(1),$$

where $o(1) \to 0$ with probability 1 as $l \to \infty$. This implies the relation

$$E\alpha(P_\varepsilon(l)) \geq \alpha((A + \varepsilon \bar{B})^l) - O(l^2 \varepsilon^2) + l \varepsilon \beta_l, \tag{12.28}$$

where $\beta_l \to 0$ as $l \to \infty$. We have

$$\alpha((A + \varepsilon \bar{B})^l) \geq l \alpha(A + \varepsilon \bar{B}) \geq \theta l \varepsilon,$$

since (12.27) implies the inequality $\alpha(A + \varepsilon \bar{B}) \geq \theta \varepsilon$, where $\theta > 0$ for $\varepsilon > 0$ small enough. The proof is a consequence of (12.28).

□

Theorem 4 *Let conditions I and II be fulfilled and $\alpha(A + \varepsilon \bar{B}) > 0$ for $\varepsilon > 0$ small enough. Then all the statements of Theorem 3 hold.*

Proof We can find $\varepsilon_0 > 0$ and l such that

$$E\alpha(P_\varepsilon(l)) > 0$$

for $0 < \varepsilon < \varepsilon_0$. Then the statement of the theorem will be established if we consider the stationary process

$$P_\varepsilon^{(l)}(n) = P_\varepsilon(ln).$$

It is easy to check using Theorem 2 from Chapter 7 that the statement holds for $P_\varepsilon(n)$.

□

Next, we investigate the asymptotic behavior of the D_m-valued stochastic process

$$\vec{p}_\varepsilon(s) = \sum \vec{p}_0 \, P_\varepsilon(n) \, 1_{\{\varepsilon n \le s \le \varepsilon(n+1)\}}, \quad s \in R_+,$$

as $\varepsilon \to 0$, where $\vec{p}_0$ is a vector of the form

$$\vec{p}_0 = \sum_{i=1}^r q_i \vec{n}^{(i)}, \quad \sum_{i=1}^r q_i = \frac{1}{N}. \tag{12.29}$$

Define the matrix-valued function $Q(s)$, $s \in R_+$, by the relation

$$Q(s) = \exp\{-sI_r + sQ\}, \tag{12.30}$$

where I_r is the $r \times r$ identity matrix and Q was introduced in Lemma 3. We consider the Markov chain with m states and transition probability matrix for n steps

$$(A + \varepsilon \bar{B})^n.$$

Denote by ν_n^ε the state of the Markov chain at time n.

Lemma 5

$$\lim_{\varepsilon \to 0, \; \varepsilon n \to s} P\left\{\nu_n^\varepsilon = \vec{n} \,/\, \nu_0^\varepsilon = \vec{l}\right\} = \sum_{i=1}^r \frac{l_i}{N} \, q_{ij}(s) \, 1_{\{\vec{n} = \vec{n}^{(j)}\}}, \tag{12.31}$$

where $q_{ij}(s)$ are the elements of the matrix $Q(s)$.

Proof Note that

$$P\left\{\nu_n^\varepsilon = \vec{n} \mid \nu_0^\varepsilon = \vec{l}\right\} = \vec{\delta}_{\vec{l}}(A + \varepsilon \bar{B})^n,$$

where $\vec{\delta}_{\vec{l}}$ is the row vector in R^m with coordinates $1_{\{\vec{n} = \vec{l}\}}$. Using the same estimation as in Theorem 5 of Chapter 7 and Lemma 5 we can prove

$$\lim_{\varepsilon \to 0, \; \varepsilon n \to s} (A + \varepsilon \bar{B})^n = \Pi_0 \exp\{s \Pi_0 \bar{B} \Pi_0\}.$$

So

$$\lim_{\varepsilon \to 0, \; \varepsilon n \to s} P\{\nu_n^\varepsilon = \vec{n} \mid \nu_0^\varepsilon = \vec{l}\} = \vec{\delta}_{\vec{l}} \Pi_0 \exp\{s \Pi_0 \bar{B} \Pi_0\}. \tag{12.32}$$

Note that

$$\vec{\delta}_{\vec{l}}\Pi_0 = \sum_{i=1}^{r} \frac{l_i}{N}\vec{\delta}_{\vec{n}^{(i)}}. \tag{12.33}$$

This follows from statement 3) of Lemma 2. The definition of Q implies that

$$\vec{\delta}_{\vec{n}^{(i)}}\Pi_0\bar{B}\Pi_0 = \sum q_{ij}\vec{\delta}_{\vec{n}^{(j)}} - \vec{\delta}_{\vec{n}^{(i)}}.$$

Denote by J_r the subspace of R^m generated by the vectors $\{\vec{\delta}_{\vec{n}^{(i)}},\ i = \overline{1,r}\}$. It is easy to check that the matrix representation of the operator $\exp\{s\Pi_0\bar{B}\Pi_0\}$ on the subspace J_r with the basis $\{\vec{\delta}_{\vec{n}^{(1)}},\dots,\vec{\delta}_{\vec{n}^{(r)}}\}$ is $Q(s)$. Therefore, formula (12.31) is a consequence of formulas (12.32), (12.33).

□

Theorem 5 *Let the conditions of Theorem 4 be satisfied. Then for any $T > 0$,*

$$\lim_{\varepsilon\to 0}\int_0^T \left|\vec{p}_\varepsilon(s) - \sum_{i,j=1}^{r} q_i q_{ij}(s)\vec{n}^{(j)}\right| ds = 0$$

with probability 1.

The proof follows from statement *(iv)* of Theorem 3 and Lemma 5.

12.3 Evolutionary Genetics: The Genome

The chromosome of a bacterium is completely described by a vector having components taken from among the four symbols A, C, G, T. If the number of base pairs on the chromosome is N, then the DNA structure of the organism is described by a corresponding chromosome vector, say $s \in S = \{A, C, G, T\}^N$, where S denotes the set of all possible chromosomes and s lists the DNA sequence on one strand of the chromosome (s determines uniquely the complementary strand). Let M denote the number of elements in S. For example, a cell might have 3000 base pairs in its chromosome, so S has $M = 4^{3000}$ elements, a large number. While most of these are apparently not viable, we continue to work in this large space of sequences. We suppose that all cells in the population reproduce at the same time, so reproduction is synchronized. A population of these organisms can be described by a vector whose components give the population proportions of various types s for all $s \in S$. Our interest here is in how this vector of proportions changes from one generation to the next.

Denote by $\nu_s(t)$ the number of cells of type s in the tth generation, and denote by $\vec{\nu}(t)$ the vector whose components are these numbers indexed by elements of S. The vector of proportions is given by

$$\vec{p}(t) = \vec{\nu}(t)/|\vec{\nu}(t)|_1,$$

where

$$|\vec{\nu}(t)|_1 = \sum_{s \in S} \nu_s(t)$$

is the number of bacteria in the tth generation.

12.3.1 Branching Markov Process Model

At reproduction a cell will produce two daughter cells each having one chromosome that will generally be the same as the mother's except for possible mutations, recombinations, or transcription errors. We suppose that all changes of the type of a chromosome at the time of reproduction are of random character.

The evolution of the population can be described as a branching Markov process with the set of types S. (The interested reader can find all necessary information on this subject in the book [89].) Denote by $\pi(s, s_1, s_2)$ the probability that a cell of type s splits at the time of reproduction into two daughters, one of type s_1 and the other of type s_2. Then for all $s \in S$,

$$\sum_{s_1, s_2 \in S} \pi(s, s_1, s_2) = 1. \tag{12.34}$$

To describe the evolution of the population we introduce a collection of random variables H,

$$H = \{\eta_k(t, s, s_1, s_2), \quad k \in \mathbf{Z}_+, \ t \in \mathbf{Z}_+, \ s, s_1, s_2 \in S\},$$

satisfying the following conditions:

1. They are independent random variables for different values of $k, \ t, \ s$.
2. If $\eta \in H$, then $P\{\eta = 0\} + P\{\eta = 1\} = 1$.
3. For all $k \in \mathbf{Z}_+, \ t \in \mathbf{Z}_+, \ s \in S$,

$$\sum_{s_1, s_2 \in S} \eta_k(t, s, s_1, s_2) = 1,$$

$$P\{\eta_k(t, s, s_1, s_2) = 1\} = \pi(s, s_1, s_2).$$

We see next that the variables η_k are indicator variables related to the transition from s to (s_1, s_2) at time t.

The stochastic process $\vec{\nu}(t)$ can be expressed in terms of these counting variables as

$$\nu_s(t+1) = \sum_{s' \in S} \sum_{k=1}^{\nu_{s'}(t)} \left[2\eta_k(t, s', s, s) + \sum_{\substack{s_1 \in S \\ s \neq s_1}} \eta_k(t, s', s, s_1) \right] \tag{12.35}$$

for $t = 1, 2, \ldots$. The vector $\vec{\nu}(0)$ is assumed to be given, and using this and formula (12.35), we can (in principle) calculate $\vec{\nu}(t)$ for all $t > 0$.
It follows from formula (12.35) that $\eta_k(t, s, s_1, s_2)$ for $k \le \nu_s(t)$ is an indicator of the event that *a cell of the population at time t having type s and number k splits into two daughters of types s_1 and s_2*. We label the cells of type s by the numbers between 1 and $\nu_s(t)$.
Introduce an $M \times M$ matrix A with elements

$$A(s', s) = 2\pi(s', s, s) + \sum_{\substack{s_1 \in S \\ s \ne s_1}} \pi(s', s, s_1), \tag{12.36}$$

where M is the number of elements of S. Here $A(s', s)$ is the average number of daughters of a cell of type s' that will be of type s. It is easy to see that

$$E\vec{\nu}(t) = E\vec{\nu}(0)\, A^t. \tag{12.37}$$

Note that

$$\sum_s A(s', s) = 2,$$

since each mother has two daughters. So the matrix $\frac{1}{2}A$ is a stochastic matrix. Using this, we can describe the asymptotic behavior of $\vec{\nu}(t)$.

Lemma 6 *Let $\Pi = \frac{1}{2}A$. Let $S_1, \ldots, S_r$ be the subsets of S that are classes of communicating recurrent states for the Markov chain whose transition probability matrix is Π. So, $S_k \cap S_l = \emptyset$ for $k, l \in \overline{1, r}$ with $k \ne l$. Denote by $\vec{p}_k$ the vector of ergodic probabilities for the class S_k, $k = 1, 2, \ldots, r$, i.e.,*

$$\vec{p}_k = \{p_k(s),\ s \in S\} \quad \text{and} \quad \sum_{s \in S_k} p_k(s) = 1, \quad \sum_{s \in S \setminus S_k} p_k(s) = 0.$$

Denote by d_k the period for class S_k. Then the following statements hold:

(i) There exists $\delta > 0$ such that for all $s \in S \setminus \bigcup_{k=1}^r S_k$,

$$E\nu_s(t) = o\left((2 - \delta)^t\right);$$

that is, those not among the recurrent sets are transient states.

(ii) There exist constants $\{q_s(i),\ s \in S, i \in \overline{1, r}\}$ satisfying the conditions

$$(a)\ 0 \le q_s(i) \le 1, \quad (b) \sum_{i=1}^r q_s(i) = 1, \quad (c)\ q_s(i) = 1 \text{ if } s \in S_i$$

for which

$$\lim_{t \to \infty} \frac{1}{d_i} \sum_{j=0}^{d_i - 1} E\nu_s(t + j) 2^{-(t+j)} = \sum_{s' \in S} E\nu_{s'}(0) q_{s'}(i) p_i(s) \quad \text{if } s \in S_i. \tag{12.38}$$

The proof of the lemma follows from the theory of finite Markov chains.

Corollary 1 Suppose that the matrix Π is irreducible and aperiodic. Then there exists a unique ergodic distribution $\vec{p}$ and

$$\lim_{t\to\infty} 2^{-t} E\vec{\nu}(t) = E|\vec{\nu}(0)|_1 \vec{p}, \tag{12.39}$$

where

$$E|\vec{\nu}(0)|_1 = \sum_{s\in S} E\nu_s(0).$$

In this case we have only one class, $r = 1$, and $q_s(1) = 1$ for all $s \in S$, so formula (12.39) follows from (12.38).

Next, we consider a limit theorem for the stochastic processes

$$\vec{\rho}(t) = 2^{-t}\vec{\nu}(t). \tag{12.40}$$

First, we establish the following lemma.

Lemma 7 *There exists a number $d \in \mathbf{Z}_+$ for which for all $k \in \mathbf{Z}_+$ there exists the limit*

$$\lim_{n\to\infty} \Pi^{k+nd} = \Pi_k,$$

and Π_k satisfies

$$(a)\, \Pi_{k+d} = \Pi_k, \quad (b)\, \Pi_k = \Pi_0\Pi^k = \Pi^k\Pi_0 \ \textit{for}\ 0 \le k < d.$$

If d is the least common multiple of all periods d_i, $i = 1, \ldots, k$, introduced in Lemma 6, then properties (a), (b) are evident.

Remark 2 There exists $\theta \in (0, 1)$ for which

$$\|\Pi^{k+md} - \Pi_k\|_1 \le b\,\theta^{k+nd},$$

where b is a suitable constant.

Theorem 6 *With probability 1 there exists a limit*

$$\lim_{n\to\infty} \vec{\rho}(k + nd) = \rho_\infty(k) \quad \textit{for} \quad k \in \mathbf{Z}_+,$$

and the limit stochastic process $\rho_\infty(k)$ has the following properties: With probability 1,

$$(a)\ \vec{\rho}_\infty(k + d) = \vec{\rho}_\infty(k),$$

and

$$(b)\ \vec{\rho}_\infty(k) = \vec{\rho}_\infty(0)\Pi^k\Pi_0,\ 0 \le l < d.$$

Proof Denote by $\mathcal{F}_t$ the σ-algebra generated by $\{\vec{\nu}(k),\ k \le t\}$. Then

$$E(\vec{\nu}(t + 1) \mid \mathcal{F}_t) = \vec{\nu}(t)A.$$

Set

$$\vec{\mu}(t) = \vec{\nu}(t+1) - \vec{\nu}(t)A. \tag{12.41}$$

The components of the vector $\vec{\mu}(t)$ are represented by the formula

$$\begin{aligned}\mu_s(t) = \sum_{s' \in S} \sum_{k=1}^{\nu_{s'}(t)} \Big[& 2(\eta_k(t, s', s, s) - \pi(t, s', s, s)) \\ & + \sum_{\substack{s_1 \in S \\ s \neq s_1}} (\eta_k(t, s', s, s_1) - \pi(t, s', s, s_1)) \Big]\end{aligned} \tag{12.42}$$

(we used formula (12.35) here).
It follows from formula (12.42) that for some constant c_1,

$$E((\vec{\mu}(t), \vec{\mu}(t))/\mathcal{F}) \leq c_1 2^t. \tag{12.43}$$

Using formula (12.41), we obtain the following representation:

$$\vec{\nu}(t) = \vec{\nu}(0)A^t + \sum_{k=0}^{t-1} \vec{\mu}(k) A^{t-1-k}.$$

So

$$\vec{\rho}(t) = \vec{\rho}(0)\Pi^t + \sum_{k=0}^{t-1} \frac{1}{2^{k+1}} \vec{\mu}(k) \Pi^{t-1-k}. \tag{12.44}$$

Define the vector-valued random variables

$$\vec{\mu}^{(i)} = \sum_{l=0}^{\infty} \frac{1}{2^{ld+i}} \vec{\mu}(ld+i), \quad i = 0, 1, \ldots, d-1. \tag{12.45}$$

Note that

$$\begin{aligned} E\left(\sum_{l=l_1}^{\infty} \frac{1}{2^{ld+i}} \vec{\mu}(ld+i), \sum_{l=l_1}^{\infty} \frac{1}{2^{ld+i}} \vec{\mu}(ld+i) \right) \\ = \sum_{l=l_1}^{\infty} \frac{1}{2^{2ld+2i}} E(\vec{\mu}(ld+i), \vec{\mu}(ld+i)) \leq c_2 2^{-l_1 d} \end{aligned}$$

because of relation (12.43), where c_2 is a constant. So

$$\sum_{l_1} P\left\{ \left| \sum_{l=l_1}^{\infty} \frac{1}{2^{ld+i}} \vec{\mu}(ld+i) \right|_1 > \frac{1}{(l_1+1)^2} \right\} \leq \sum_{l_1} \frac{c_2 2^{-l_1 d}}{(l_1+1)^2} < \infty.$$

This implies that the series on the right-hand side of relation (12.45) converges with probability 1.

Using formula (12.44) we write

$$\vec{\rho}(k+nd) = \vec{\rho}(0)\,\Pi^{nd+k} + \sum_{j=0}^{nd+k-1} \frac{1}{2^{j+1}}\vec{\mu}(j)\,\Pi^{nd+k-1-j}$$

$$= \vec{\rho}(0)\,\Pi^{nd+k} + \sum_{i=0}^{d-1}\sum_{l} \frac{1}{2^{ld+i}}\vec{\mu}(ld+i)\,\Pi^{(n-l)d+k-1-i}1_{\{0\le ld+i<nd+k\}}$$

$$= \vec{\rho}(0)\,\Pi_k + \sum_{i=0}^{d-1}\vec{\mu}^{(i)}\,\Pi^{k-1-i+d}1_{\{k<i+1\}} + \vec{\rho}(0)(\Pi^{nd+k} - \Pi_k)$$

$$+ \sum_{i=0}^{d-1}\sum_{l} \frac{1}{2^{ld+i}}\vec{\mu}(ld+i)\big(\Pi^{(n-l)d+k-1-i}1_{\{0\le ld+i<nd+k\}}$$

$$- \Pi^{k-1-i+d}1_{\{k<i+1\}}\big).$$

From the convergence with probability 1 of the series of $\vec{\mu}^{(i)}$ and Remark 2, it follows that with probability 1,

$$\lim_{n\to\infty}\vec{\rho}(k+nd) = \vec{\rho}(0)\Pi_k + \sum_{i=1}^{d-1}\vec{\mu}^{(i)}\Pi_{k-1-i+d}1_{\{k<i+1\}}. \tag{12.46}$$

Let the expression on the right-hand-side of equation (12.46) be denoted by $\vec{\rho}_\infty(k)$. It is easy to check that $\vec{\rho}_\infty(k)$ satisfies properties (a) and (b). □

Corollary 2 If $d = 1$, then with probability 1,

$$\lim_{n\to\infty}\vec{\rho}(n) = \left(\vec{\rho}(0) + \sum_{l=0}^{\infty}\frac{1}{2^l}\vec{\mu}(l)\right)\Pi_0, \tag{12.47}$$

where $\Pi_0 = \lim_{n\to\infty}\Pi^n$. (This limit exists if $d = 1$.)

Corollary 3 Suppose that $d = 1$, $r = 1$ (meaning that Π is the transition probability matrix for an aperiodic ergodic Markov chain). Then with probability 1,

$$\lim_{n\to\infty}\vec{\rho}(n) = |\vec{\rho}(0)|_1\vec{p}, \tag{12.48}$$

where $\vec{p}$ is the vector of ergodic probabilities.

To prove formula (12.48) we first note that all rows of the matrix Π_0 coincide with the vector $\vec{p}$. Therefore, for any vector $\vec{a}$ we have the following relation:

$$\vec{a}\,\Pi_0 = (\vec{a},\vec{1})\vec{p},$$

where $\vec{1} = (1,1,\ldots,1)$. It follows that $(\vec{\mu}(l),\vec{1}) = 0$ for all $l \in \mathbf{Z}_+$. So formula (12.48) is a consequence of formula (12.47).

Corollary 4 If $r = 1,\ d > 1$. Then with probability 1,

$$\lim_{n\to\infty} \frac{1}{d} \sum_{k=0}^{d-1} \vec{\rho}(nd+k) = |\vec{\rho}(0)|_1 \vec{p}. \tag{12.49}$$

To see this, let

$$\hat{\Pi} = \frac{1}{d} \sum_{k=0}^{d-1} \Pi_k. \tag{12.50}$$

It follows from formula (12.46) that

$$\begin{aligned}\lim_{n\to\infty} \frac{1}{d} \sum_{k=0}^{d-1} \vec{\rho}(nd+k) &= \vec{\rho}(0)\,\hat{\Pi} + \frac{1}{d} \sum_{k=0}^{d-1}\sum_{i=0}^{d-1} \vec{\mu}^{(i)}\,\Pi_{k-1-i+d1_{\{k<i+1\}}} \\ &= \left(\vec{\rho}(0) + \sum_{i=0}^{d-1} \vec{\mu}^{(i)}\right)\hat{\Pi}.\end{aligned}$$

Then in the case $r = 1$, $\hat{\Pi} = \Pi_0$, which implies formula (12.49).

Remark 3 If $r = 1$, then with probability 1,

$$\lim_{n\to\infty} \frac{1}{n} \sum_{k=1}^{n} \vec{\rho}(k) = |\vec{\rho}(0)|_1 \vec{p}. \tag{12.51}$$

So, for nonrandom $\vec{\rho}(0)$, the sequence $\{\vec{\rho}(k)\}$ satisfies the strong law of large numbers. If $|\vec{\rho}(0)|_1$ (this is the number of cells at the initial time) is random, then the right–hand side of relation (12.51) is random, and the strong law of large numbers does not apply.

Next, consider the case $r > 1$. Let $\vec{p}_j$ be the vector of the ergodic probabilities for the class S_j, and let $\vec{1}_{S_j}$ be the vector indicator of the class S_j (i.e., the components $\{x_s,\ s \in S\}$ of this vector are $x_s = 1_{S_j}(s)$).

Theorem 7 *(i) For all $j \in \overline{1,r}$ there exists the limit*

$$\lim_{n\to\infty} (\vec{p}(n), \vec{1}_{S_j}) = \eta_j, \tag{12.52}$$

where $\eta_1, \ldots, \eta_r$ are nonnegative random variables, $\sum_{j=1}^{r} \eta_j = |\vec{\rho}(0)|_1$.
(ii) If the period of the Markov chain with transition matrix P is d, then

$$\lim_{n\to\infty} \frac{1}{d} \sum_{k=0}^{d-1} \vec{\rho}(n+k) = \sum_{j=1}^{r} \eta_j \vec{p}^{j}. \tag{12.53}$$

Proof Note that

$$(\vec{p}(n), \vec{1}_{S_j}) \le (\vec{p}(n+1), \vec{1}_{S_j})$$

because of formula (12.35) and the relation

$$P\{\eta_k(m, s', s, s_1) = 0\}$$

if $s' \in S_j$ and $s \notin S_j$ or $s_1 \notin S_j$. Also,

$$\sum_{j=1}^{r} \vec{p}(n) \le |\vec{p}(n)|_1 = |\vec{p}(0)|_1.$$

Suppose that

$$\sum_{j=1}^{r} (\vec{p}(0), \vec{1}_{S_j}) = |\vec{p}(0)|_1.$$

We can consider r different populations, such that the jth population is the population with genotypes $s \in S_j$, $j = 1, \ldots, r$. Each population develops independently of the others, and to each population we can apply Corollary 4, which implies the following statement:

$$\lim_{n\to\infty} \frac{1}{d} \sum_{k=0}^{d-1} Q_j \vec{\rho}(n+k) = |Q_j \vec{\rho}(0)|_1 \vec{p^j},$$

where Q_j is the projector matrix for which

$$Q_j \vec{x} = Q_j \{x_s : \; s \in S\} = \{1_{S_j} x_s\}.$$

So

$$\lim_{n\to\infty} \frac{1}{d} \sum_{k=0}^{d-1} \vec{\rho}(n+k) = \sum_{j=1}^{r} (\vec{\rho}(0), \vec{1}_{S_j}) \vec{p^j}.$$

It follows from this formula that for all $l \in \mathbf{Z}_+$,

$$\liminf_{n\to\infty} \frac{1}{d} \sum_{k=0}^{d-1} \vec{\rho}(n+k) = \liminf_{n\to\infty} \frac{1}{d} \sum_{k=0}^{d-1} \vec{\rho}(n+l+k) \ge \sum_{j=1}^{r} (\vec{\rho}(0), \vec{1}_{S_j}) \vec{p^j}$$

and

$$\liminf_{n\to\infty} \frac{1}{d} \sum_{k=0}^{d-1} \vec{\rho}(n+k) \ge \sum_{j=1}^{r} \eta_j \vec{p^j}.$$

Since

$$\lim_{n\to\infty} \frac{1}{d} \sum_{k=0}^{d-1} (\vec{\rho}(n+k), \vec{1}) = |\vec{\rho}(0)|_1 = \sum_{j=1}^{r} \eta_j = \left(\sum_{j=1}^{r} \eta_j \vec{p^j}, \vec{1} \right),$$

we conclude that

$$\lim_{n\to\infty} \frac{1}{d} \sum_{k=0}^{d-1} \vec{\rho}(n+k) = \eta_j \vec{p^j}.$$

□

Remark 4 If $r > 1$, then with probability 1,

$$\lim_{n\to\infty} \frac{1}{n} \sum_{k=0}^{n} \vec{\rho}(k) = \sum_{j=1}^{r} \eta_j \vec{p^j}. \tag{12.54}$$

The expression on the right–hand side of relation (12.54) is essentially random because it can be proved that var $\eta_j > 0$, if for the Markov chain with transition probability Π there exist transient states and a state s for which $q_s(i) \in (0,1)$ for some i. It follows that the strong law of large numbers typically does not apply in the case $r > 1$.

12.3.2 Evolution of the Genome in a Random Environment

Now we assume that the probabilities $\pi(s, s_1, s_2)$ that determine reproduction of the cells are randomly changing in time. So they depend on some factors that can be described as being the "environment" in which the population develops. This environment will be described here by a stationary stochastic process. We consider a system of random variables

$$E^* = \{\pi_n^*(s, s_1, s_2),\ n = 0, 1, 2, \ldots,\ s, s_1, s_2 \in S\}$$

satisfying the following conditions:

(E^*1) $0 \le \pi_n^*(s, s_1, s_2) \le 1$.

(E^*2) $\sum_{s_1, s_2 \in S} \pi_n^*(s, s_1, s_2) = 1$ for $n \in \mathbf{Z}_+,\ s \in S$.

(E^*3) $\{\pi_t^*(s, s_1, s_2),\ s, s_1, s_2 \in S\}$, $t = 0, 1, \ldots,$ is an ergodic stationary stochastic process in the space $R^{M'}$, where $M' = M^2(M+1)/2$.

We will describe the evolution of the population as a Markov branching process in a random environment E^*. We introduce the system of random variables

$$H = \{\eta_k(t, s, s_1, s_2),\quad k \in \mathbf{Z}_+, t \in \mathbf{Z}_+,\ s, s_1, s_2 \in S\}$$

satisfying the following conditions:

(1) They are conditionally independent for various values of k, t, s with respect to the system E^*.

(2) For $\eta \in H$,

$$P\{\eta = 0/\mathcal{E}\} + P\{\eta = 1/\mathcal{E}\} = 1,$$

where $\mathcal{E}$ is the σ-algebra generated by system E^*.

(3) For all $t \in \mathbf{Z}_+,\ k \in \mathbf{Z}_+,\ s \in S$,

$$\sum_{\{s_1, s_2\}} \eta_k(t, s, s_1, s_2) = 1.$$

(4) $P\{\eta_k(t, s, s_1, s_2) = 1/\mathcal{E}\} = \pi_t^*(s, s_1, s_2)$.

The evolution of the population is described by the stochastic process $\vec{\nu}(t) = \{\nu_s(t),\ s \in S\}$, where $\nu_s(t)$ is defined by formula (12.35). Denote by $\mathcal{E}_t$ the σ-algebra generated by

$$\{\pi_t^*(s, s_1, s_2),\ n \le t,\ s, s_1, s_2 \in s\}.$$

Introduce the matrix-valued stochastic process $\{A_n^*,\ n \in \mathbf{Z}_+\}$, the elements of the matrix A_n^* being given by the formula

$$A_n^*(s_1, s_2) = 2\pi_n^*(s_1, s_2) + 2\sum_{s \ne s_2} \pi_n^*(s_1, s_2, s). \tag{12.55}$$

Note that $A_n^*(s_1, s_2)$ is the conditional average number of daughters of type s_2 with respect to the σ-algebra $\mathcal{E}$ if the mother is of type s_1.
Denote by $\mathcal{F}_n$ the σ-algebra generated by $\{\vec{\nu}(t),\ t \le n\}$. Then

$$E(\vec{\nu}(t+1)/\mathcal{F}_t \vee \mathcal{E}) = \vec{\nu}(t)A_t^* \tag{12.56}$$

and

$$E(\vec{\nu}(t)/\mathcal{E}) = \vec{\nu}(0)A_0^*A_1^* \cdots A_{t-1}^*, \tag{12.57}$$

First, we consider the asymptotic behavior of the conditional expectation of the process $\vec{\nu}(t)$. Note that the matrices

$$\Pi_n^* = \frac{1}{2}A_n^* \tag{12.58}$$

are stochastic matrices and $\{\Pi_n^*,\ n \in \mathbf{Z}\}$ is an ergodic stationary environment (see Section 7.1).

Theorem 8 *Suppose that conditions (E^*1)–(E^*3) are satisfied, and that $E\alpha(\Pi_0^*) > 0$. Then:*

(i) There exists a D_m^1-valued stochastic stationary process $\vec{d}_n$ for which $\vec{d}_{n+1} = \vec{d}_n\Pi_n^*$, and for all $\vec{a} \in D_m^1$ with probability 1 we have that

$$\lim_{n\to\infty} |\vec{d}_n - \vec{a}\Pi_0^* \cdots \Pi_{n-1}^*|_1 = 0.$$

(ii) There exists a matrix Π_0 with identical rows for which

$$\lim_{n\to\infty} E\Pi_0^* \cdots \Pi_n^* = \Pi_0.$$

(iii) With probability 1,

$$\lim_{n\to\infty} \frac{1}{n}(\Pi_0^* + \Pi_0^*\Pi_1^* + \cdots + \Pi_0^*\Pi_1^* \cdots \Pi_n^*) = \Pi_0.$$

The statements of the theorem follow from Theorem 2 of Chapter 7.
Let $\vec{\rho}(t)$ be defined by formula (12.40) with $\vec{\nu}(t)$ as defined in this section. We consider next the asymptotic behavior of $\vec{\rho}(t)$ as $t \to \infty$.

Theorem 9 *Let the conditions of Theorem 8 be fulfilled. Then*

$$\lim_{t\to\infty} E|\vec{\rho}(t) - \vec{d}_t|_1 = 0.$$

Proof Set

$$\vec{\mu}(t) = \vec{\nu}(t+1) - \vec{\nu}(t)A_t^*, \quad t \in \mathbf{Z}_+. \tag{12.59}$$

One can check that the following relations hold:

$$E(\vec{\mu}(t)/\mathcal{F}_t \vee \mathcal{E}) = 0, \tag{12.60}$$

$$E((\vec{\mu}(t), \vec{\mu}(t))/\mathcal{F}_t \vee \mathcal{E}) \leq c2^t, \tag{12.61}$$

where $c > 0$ is a constant.

It follows from formula (12.59) that for $t \in \mathbf{Z}_+$, $\vec{\nu}(t)$ can be represented by the formula

$$\vec{\nu}(t) = \vec{\nu}(0)A_0^* \cdots A_{t-1}^* + \sum_{k<t-1} \vec{\mu}(k)A_{k+1}^* \cdots A_{t-1}^*.$$

So for $\vec{\rho}(t)$ we have

$$\vec{\rho}(t) = \vec{\rho}(0)\Pi_0^* \cdots \Pi_{t-1}^* + \sum_{k<t-1} \frac{1}{2^{k+1}} \vec{\mu}(k)\Pi_{k+1}^* \cdots \Pi_{t-1}^*.$$

It follows from statement *(i)* of Theorem 8 that

$$\lim_{t\to\infty} |\vec{d}_t - \vec{\rho}(0)\Pi_0^* \cdots \Pi_{n-1}^*|_1 = 0,$$

and for all $k \in \mathbf{Z}_+$,

$$\lim_{t\to\infty} E\left|\frac{1}{2^{k+1}}\vec{\mu}(k)\Pi_{k+1}^* \cdots \Pi_{t-1}^*\right|_1$$

$$= \lim_{t\to\infty} E\left|\frac{1}{2^{k+1}}\vec{\nu}(k+1)\Pi_{k+1}^* \cdots \Pi_{t-1}^* - \frac{1}{2^k}\vec{\nu}(k)\Pi_k^* \cdots \Pi_{t-1}^*\right|_1$$

$$= \lim_{t\to\infty} E\left|\left(\vec{\rho}(k+1)\Pi_{k+1}^* \cdots \Pi_{t-1}^* - \vec{d}_{t-1}\right) - \left(\vec{\rho}(k)\Pi_k^* \cdots \Pi_{t-1}^* - \vec{d}_{t-1}\right)\right|_1,$$

which is zero. Therefore, for any $n \in \mathbf{Z}_+$,

$$\limsup_{t\to\infty} E|\vec{\rho}(t) - \vec{d}_t|_1 \leq \limsup_{t\to\infty} E|\sum_{n\leq k} \vec{\mu}(k)\Pi_{k+1}^* \cdots \Pi_{t-1}^*|_1.$$

Formulas (12.60) and (12.61) imply that for $n < t-1$,

$$E\left(\sum_{n\leq k<t-1} \frac{1}{2^{k+1}}\vec{\mu}_k\Pi_{k+1}^* \cdots \Pi_{t-1}^*, \sum_{n\leq k<t-1} \frac{1}{2^{k+1}}\vec{\mu}_k\Pi_{k+1}^* \cdots \Pi_{t-1}^*\right)$$

$$= E\sum_{n\leq k<t-1} \frac{1}{2^{k+1}}\left(\vec{\mu}_k\Pi_{k+1}^* \cdots \Pi_{t-1}^*, \vec{\mu}_k\Pi_{k+1}^* \cdots \Pi_{t-1}^*\right)$$

$$\leq E\sum_{n\leq k<t-1} \frac{1}{2^{k+1}}\left(\vec{\mu}_k, \vec{\mu}_k\right) \leq \frac{c}{4}\sum_{k=n}^{\infty} 2^{-k} = c2^{-n+1}$$

(because $\|\Pi_k^*\| = 1$ in the Euclidean space R^m).

□

Remark 5 Let $\vec{p}$ be a row vector of the matrix Π_0 defined in statement *(ii)* of Theorem 8. Then

$$\frac{1}{n}\sum_{k=0}^{n-1}\vec{\rho}(k) \to \vec{p} \quad \text{in probability.}$$

This is a consequence of Theorem 9 and statement *(iii)* of Theorem 8.

12.3.3 Evolution in a Random Environment

Suppose that the probabilities $\{\pi_n^*(s, s_1, s_2),\ n \in \mathbf{Z}_+, s, s_1, s_2 \in S\}$ that define the random environment are of the form

$$\pi_n^*(s, s_1, s_2) = \pi(s, s_1, s_2) + \varepsilon\hat{\pi}_n(s, s_1, s_2), \tag{12.62}$$

where $\pi(s, s_1, s_2)$ is a nonrandom function of the type considered in Section 12.3.1, $\varepsilon > 0$ is a small number, and $\{\pi_n^*(s, s_1, s_2),\ s, s_1, s_2 \in S\}$, as a function of $n \in \mathbf{Z}_+$, is an ergodic stationary process. We suppose that $\{\pi_n^*(s, s_1, s_2),\ n \in \mathbf{Z}_+, s, s_1, s_2 \in S\}$ in formula (12.62) satisfies conditions (E^*1)—(E^*2) of Section 12.3.2 for all $\varepsilon \in (0, \varepsilon_0)$, where $\varepsilon_0 > 0$ is a fixed constant. This implies that

$$\sum_{s_1, s_2 \in S} \hat{\pi}_n(s, s_1, s_2) = 0.$$

We suppose also that the following mixing condition holds

(E^*4) Denote by $\mathcal{E}^l$ the σ-algebra generated by the set of random variables

$$\{\hat{\pi}_n(s, s_1, s_2),\ n \geq l, s, s_1, s_2 \in S\}.$$

There exists a sequence of nonnegative numbers $\{\alpha_l,\ l \in \mathbf{Z}_+\}$ for which $\alpha_l \to 0$ and

$$|E\xi_1\xi_2 - E\xi_1 E\xi_2| \leq \alpha_l$$

for all pairs of random variables $\xi_1,\ \xi_2$ for which ξ_1 is $\mathcal{E}^n$-measurable, ξ_2 is $\mathcal{E}^{l+n}$-measurable, and $|\xi_k| \leq 1,\ k = 1, 2$.

Recall that the last condition was used in our investigation of Markov chains in a random environment (see Chapter 7).
The vector-valued stochastic process in R^m describing the evolution of the population in the random environment determined by the probabilities (12.62) is denoted by $\vec{\nu}_\varepsilon(t)$. We will investigate the asymptotic behavior of this process as $t \to \infty$ and $\varepsilon \to 0$.
Set $\vec{\rho}_\varepsilon(t) = 2^{-t}\vec{\nu}(t)$. Let Π be the matrix whose elements are $\frac{1}{2}A(s_1, s_2)$, where $A(s_1, s_2)$ is defined by formula (12.36). Denote by $\hat{\Pi}_n$ the matrix

whose elements are

$$\hat{\pi}_n(s_1, s_2) + \sum_{s \neq s_2} \hat{\pi}_n(s_1, s_2, s).$$

Then the matrices Π_n^* considered in Section 12.3.2 are

$$\Pi_n^* = \Pi + \varepsilon \hat{\Pi}_n.$$

Theorem 10 *Let conditions (E^*1)–(E^*4) be satisfied, let the matrix Π be aperiodic, and suppose that $\alpha(\Pi + \varepsilon E\hat{\Pi}_0) > 0$ for $\varepsilon \in (0, \varepsilon_0)$. Then the following statements hold: (i) There exists the limit*

$$\lim_{n\to\infty} E(\Pi + \varepsilon\hat{\Pi}_0) \cdots (\Pi + \varepsilon\hat{\Pi}_n) = \Pi_\varepsilon,$$

where the matrix Π_ε has identical rows (so $\|\Pi_\varepsilon\|_0 = 0$).
(ii) There exists the limit

$$\lim_{\varepsilon\to 0} \Pi_\varepsilon = \Pi_0,$$

where

$$\Pi_0 = \lim_{t\to\infty} \Pi_\infty \exp\{t\Pi_\infty Q \Pi_\infty\}, \quad \Pi_\infty = \lim_{n\to\infty} \Pi^n, \quad Q = E\hat{\Pi}_0,$$

and Π_0 is a matrix with identical rows.
(iii) For all $\varepsilon \in (0, \varepsilon_0)$,

$$\lim_{\varepsilon\to 0} E \left| \vec{\rho}_\varepsilon(t) - \vec{\rho}_\varepsilon(0)(\Pi + \varepsilon\hat{\Pi}_0) \cdots (\Pi + \varepsilon\hat{\Pi}_{t-1}) \right|_1 = 0.$$

(iv) With probability 1,

$$\lim_{n\to\infty} \frac{1}{n} \sum_{k=1}^{n} (\Pi + \varepsilon\hat{\Pi}_0) \cdots (\Pi + \varepsilon\hat{\Pi}_{n-1}) = \Pi_\varepsilon.$$

(v) Denote by $\vec{p}_0$ any of the rows of the matrix Π_0. Then

$$\lim_{\varepsilon\to 0} \limsup_{n\to\infty} \left|\frac{1}{n} \sum_{k=1}^{n} \vec{\rho}_\varepsilon(k) - \vec{p}_0\right| = 0.$$

The proof of this theorem is a consequence of Theorems 2, 5 of Chapter 7 and Theorems 4, 8, 9.
The results in this chapter provide methods for studying very large dynamical systems in genetics which are perturbed by random noise. For the most part, they show that if reasonable assumptions are made about the noise, then with high probability the system's behavior can be described using familiar limit procedures.

Appendix A
Some Notions of Probability Theory

This appendix provides some background on notation and concepts that play central roles in this book. This material is widely available in standard texts in the theory and applications of probability.

Probability Space and Random Variables

Probability Space

A probability space is a triple $\{\Omega, \mathcal{F}, P\}$, where Ω is a set that is called the set of *outcomes* or *samples*; $\mathcal{F}$ is a collection of subsets of Ω that forms a σ-algebra, which is called the collection of *events*; and P is a *probability measure* defined for each set in $\mathcal{F}$.

Remark 1 A collection of sets $\mathcal{F}$ is called a σ**-algebra** if:

(i) $\Omega \in \mathcal{F}$.

(ii) If $A \in \mathcal{F}$ and $B \in \mathcal{F}$, then the set difference $A \setminus B$ is also in $\mathcal{F}$.

(iii) If $A_n \in \mathcal{F}$ for $n = 1, 2, \ldots$, then $\bigcup_n A_n \in \mathcal{F}$.

Remark 2 A function $m : \mathcal{F} \to R_+$ is called a measure if

$$m(A \cup B) = m(A) + m(B) \text{ when } A \cap B = \emptyset$$

and

$$m\left(\bigcup_n A_n\right) = \lim_{k\to\infty} m\left(\bigcup_{n\le k} A_n\right).$$

Random variables

A function $\xi : \Omega \to R$ is called a *random variable* if the set

$$\{\omega \in \Omega : \xi(\omega) \le x\}$$

is in $\mathcal{F}$ for any $x \in R$. $\xi(\omega)$ measures some attribute of the sample ω. The function $F_\xi : R \to [0,1]$ defined by

$$F_\xi(x) = P\{\omega : \xi(\omega) \le x\}$$

is called the *distribution function* of the random variable ξ.

Convergence of Random Variables

Let $\xi_n(\omega)$, for $n = 1, 2, \ldots$, be a sequence of random variables defined on the probability space $\{\Omega, \mathcal{F}, P\}$. Then we say that:

(i) ξ_n converges to a random variable ξ *in probability* if for every $\varepsilon > 0$,

$$\lim_{n\to\infty} P\{|\xi_n(\omega) - \xi(\omega)| > \varepsilon\} = 0. \tag{A.1}$$

(ii) ξ_n converges to ξ *with probability* 1 (almost surely or almost always, i.e., for all ω except those in a set of P measure zero) if

$$P\left(\bigcap_{k>0}\bigcup_{n>k}\{\omega : |\xi_n(\omega) - \xi(\omega)| > \varepsilon\}\right) = 0. \tag{A.2}$$

(iii) The distribution of ξ_n converges *weakly* to the distribution of ξ if

$$\lim_{n\to\infty} F_{\xi_n}(x) = F_\xi(x)$$

for almost all $x \in R$. Here $F_{\xi_n}(x) = P\{\xi_n(\omega) \le x\}$.

Expectation

Let ξ be a random variable on the probability space $\{\Omega, \mathcal{F}, P\}$. We say that ξ has expected value or *expectation* $E\xi$ if the function $\xi(\omega)$ is integrable with respect to the measure P, and then

$$E\xi = \int_\Omega \xi(\omega)P(d\omega).$$

Remark 3 A measurable function ξ is integrable with respect to the measure P if

$$\sum_{n=1}^{\infty} nP\{n-1 \le |\xi(\omega)| < n\} < \infty. \tag{A.3}$$

In this case

$$E\xi = \lim_{h\to 0} \sum_{h=0}^{\infty} nhP\{(n-1)h \le \xi(\omega) < nh\}. \tag{A.4}$$

The main property of the expectation is that if c_1, c_2 are constants and if $E\xi_1, E\xi_2$ exist, then

$$E(c_1\xi_1 + c_2\xi_2) = c_1 E\xi_1 + c_2 E\xi_2.$$

Remark 4 For a nonnegative random variable $\xi \ge 0$, we set $E\xi = \infty$ if $E\xi$ does not exist. We denote by $L_1(\Omega, P)$ the set of random variables $\xi(\omega)$ for which $E|\xi| < \infty$. If we introduce into $L_1(\Omega, P)$ the norm $\|\xi\|_{L_1} = E|\xi|$. Then L_1 becomes a Banach space.
Suppose that $\xi^2 \in L_1(\Omega, P)$. Then the number

$$\mathrm{Var}(\xi) = E\xi^2 - (E\xi)^2$$

is called the *variance* of ξ.

Independence
A sequence of random variables, say $\{\xi_n(\omega)\}$ for $n = 1, 2, \ldots$, is a sequence of *independent random variables* if for every n and all real numbers $x_1, \ldots, x_n$, the following relation holds:

$$P\left(\bigcap_{k=1}^{n} \{\omega : \xi_k(\omega) \le x_k\}\right) = \prod_{k=1}^{n} P\{\xi_k(\omega) \le x_k\}. \tag{A.5}$$

If all $\xi_k(\omega)$ have the same distribution function, then the sequence is said to comprise *independent and identically distributed random variables* (i.i.d.r.v.).

Remark 5 The function

$$F_n(x_1, \ldots, x_n) = P\Big(\bigcap_{k=1}^{n} \{\omega : \xi_k(\omega) \le x_k\}\Big) \tag{A.6}$$

is called the *joint distribution function* for the random variables $\xi_1, \ldots, \xi_n$. Consider the vector $\vec{\xi}(\omega) \in R^n$ with coordinates $(\xi_1(\omega), \ldots, \xi_n(\omega))$. This vector is called a *random vector* or an R^n-valued random variable. Note that $\vec{\xi}$ is a measurable function from Ω into R^n. That is,

$$\{\omega : \vec{\xi}(\omega) \in A\} \in \mathcal{F}$$

if A is a *Borel set* in R^n. The Borel sets are defined to be the minimal collection of events for which the distribution functions are defined. They are the σ-algebra generated by the "intervals" $\{x \in R^d : a_1 \le x_1 < b_1, \ldots, a_d \le x_d < b_d\}$ for some constants $-\infty \le a_j < b_j \le \infty$. The measure

$$m_{\vec{\xi}}(A) = P(\{\omega : \vec{\xi}(\omega) \in A\}) \tag{A.7}$$

is called the *distribution function* for the random vector $\vec{\xi}$.

The Law of Large Numbers
(Also known as Chebyshev's theorem.) Suppose that $\{\xi_n\}$ are independent random variables for which $E|\xi_n|^2 < \infty$ for $n = 1, 2, \dots$, and

$$\frac{1}{n^2} \sum_{k=1}^{n} \mathrm{Var}(\xi_k) \to 0$$

as $n \to \infty$. Then

$$\frac{1}{n} \sum_{k=1}^{n} \xi_k(\omega) - \frac{1}{n} \sum_{k=1}^{n} E\xi_k \to 0$$

in probability.

Strong Law of Large Numbers
Consider a sequence of i.i.d.r.v. $\{\xi_k\}$ for which $E\xi_1$ is defined. Then

$$\lim_{n\to\infty} \frac{1}{n} \sum_{k=1}^{n} \xi_k(\omega) = E\xi_1$$

almost surely.

The Central Limit Theorem
(Also known as Laplace's theorem.) Let $\{\xi_k\}$ be i.i.d.r.v., each having the same mean and variance: $E\xi_k = a$ and $\mathrm{Var}(\xi_k) = b$ for all k. Then

$$\lim_{n\to\infty} P\{\xi_1(\omega) + \cdots + \xi_n(\omega) < \sqrt{n}x + na\} = \Phi(x), \tag{A.8}$$

where

$$\Phi(x) = \frac{1}{\sqrt{2\pi}} \int_{-\infty}^{x} \exp\left(-u^2/2\right) du.$$

This function is called the *Gaussian* or *normal distribution* function.

Very Important Remark: The argument ω in random variables is, as a rule, omitted in formulas.
Let $\{A_k\}$, for $k = 1, 2, \dots$, be elements of $\mathcal{F}$. These are called independent events if for every number r and every sequence $n_1, n_2, \dots, n_r$ we have

$$P\left(\bigcap_{i=1}^{r} A_{n_i}\right) = \prod_{i=1}^{r} P(A_{n_i}).$$

The following statement is useful in various applications of probability theory.

Borel–Cantelli Lemma: Let $A_k \in \mathcal{F}$ for $k = 1, 2, \dots$.
(i) If

$$\sum_{k} P(A_k) < \infty, \tag{A.9}$$

then

$$P\left(\bigcap_{n=1}^{\infty}\bigcup_{k=n}^{\infty}A_k\right)=0. \tag{A.10}$$

(ii) If the sets $\{A_k\}$ are independent, then equation (A.10) implies that equation (A.9) holds.

Conditional Expectations

Let $\mathcal{G}$ be a sub–σ-algebra of $\mathcal{F}$ and let $\xi \in L_1(\Omega, P)$. Then there exists a random variable $\eta(\omega)$ satisfying the following conditions:

(i) $\eta(\omega)$ is $\mathcal{G}$-measurable.

(ii) For any $B \in \mathcal{G}$,

$$E\xi 1_B = E\eta 1_B, \tag{A.11}$$

where 1_B denotes the indicator function for the set B.

This random variable η is unique in the sense that if $\tilde{\eta}$ is another random variable satisfying (i), (ii), then

$$P\{\eta=\tilde{\eta}\}=1.$$

The random variable $\eta(\omega)$ is called the *conditional expectation of* ξ *with respect to the* σ*-algebra* $\mathcal{G}$, and it is denoted by

$$\eta = E(\xi/\mathcal{G}).$$

Remark 6 Consider random variables $\{\xi_\lambda, \lambda \in \Lambda\}$, where Λ is a set of parameters labeling this set of random variables. We denote by $\sigma\{\xi_\lambda, \lambda \in \Lambda\}$ the minimal σ-algebra that contains all sets of the form

$$\{\omega : \xi_\lambda(\omega) \le x\}$$

for $\lambda \in \Lambda$ and $x \in R$. We say that this σ-algebra is generated by the set of random variables $\{\xi_\lambda, \lambda \in \Lambda\}$. If $\Lambda = \{1, 2, \ldots, n\}$, then the conditional expectation of a random variable ξ with respect to $\sigma\{\xi_k, k = 1, \ldots, n\}$ is denoted by

$$E(\xi/\xi_1(\omega), \ldots, \xi_k(\omega)).$$

There is a measurable function $g(x_1, \ldots, x_n) : R^n \to R$ for which

$$E(\xi/\xi_1(\omega), \ldots, \xi_n(\omega)) = g(\xi_1(\omega), \ldots, \xi_n(\omega)).$$

Remark 7 Let $\mathcal{G} \subset \mathcal{F}$ be an arbitrary σ-algebra and let $\{\xi_1, \ldots, \xi_n\}$ be random variables. Then the conditional distribution of these random variables with respect to the σ-algebra $\mathcal{G}$ exist and define a probability measure, say $\mu(\omega, B)$, that depends (measurably) on ω, where B is any Borel set on R^m. The measure μ satisfies the relation

$$E(F(\xi_1, \ldots, \xi_n)/\mathcal{G})(\omega) = \int_{R^n} F(x)\mu(\omega, dx) \tag{A.12}$$

for any measurable bounded function F.

Remark 8 The conditional probability with respect to the σ-algebra $\mathcal{G}$ is defined to be

$$P(A/G) \equiv E(1_A/G).$$

The conditional probability has the following properties:

(1) $0 \leq P(A/\mathcal{G}) \leq 1$, almost surely.

(2) $P(\Omega/\mathcal{G}) = 1$, almost surely.

(3) If $A_n \in \mathcal{F}$ for $n = 1, 2, \ldots$, and $A_n \cap A_m = \emptyset$ if $n \neq m$, then

$$P\left(\bigcup_n A_n/\mathcal{G}\right) = \sum_n P(A_n/\mathcal{G})$$

almost surely.

Martingales

A sequence of random variables, say $\{\xi_k\}$ in $L_1(\Omega, P)$ for $k = 0, 1, \ldots$, is called a *martingale* if for all k,

$$E(\xi_{k+1}/\xi_0(\omega), \ldots, \xi_k(\omega)) = \xi_k(\omega)$$

almost surely. If "=" in the last equation is replaced by "$\geq$," then the sequence is called a *submartingale*, and $\{-\xi_k\}$ is called a supermartingale. These structures are useful for studying limit behaviors because of the following result:

Theorem *Let $\{\xi_n\}$ be a submartingale and suppose that $E|\xi_n| < \infty$. Then $\lim_{n\to\infty} \xi_n$ exists with probability* 1.

Markov Chain in a Measurable Space

Let $(Y, \mathcal{C})$ be a measurable space where Y is an arbitrary set and $\mathcal{C}$ is a σ-algebra of its subsets. A measurable mapping $y : \Omega \to Y$ is called a Y-valued random variable if the set $\{\omega : y(\omega) \in C\}$ is in $\mathcal{F}$ for all sets $C \in \mathcal{C}$. The sequence of Y-valued random variables $\{y_k(\omega)\}, n = 0, 1, 2, \ldots$, is called a *Markov chain* (or discrete–time Markov process) if there is a sequence of functions $Q_n(y, C)$ for $y \in Y$ and $C \in \mathcal{C}$ satisfying the following properties:

(1) Q_n is $\mathcal{C}$-measurable in y.

(2) Q_n is a probability measure in $\mathcal{C}$.

(3) $P(y_{n+1}(\omega) \in C/y_0(\omega), \ldots, y_n(\omega)) = Q_n(y_n(\omega), C)$.

The function Q_n is called the *transition probability* of the Markov chain at the nth step. A Markov chain is called *homogeneous* if Q_n does not depend on n.

In the case $Y = \{1, 2, \ldots, n\}$, the σ-algebra $\mathcal{C}$ is the σ-algebra of all subsets of Y, and the function Q_n is defined by the $m \times m$ matrix

$$Q_n = (q_{i,j}(n)),$$

where $q_{i,j}(n) = P(Y_{n+1}(\omega) = j/Y_n(\omega) = i)$.

Stochastic Processes

Let $(Y, \mathcal{C})$ be a measurable space and let $S \subset R$. A function $y(t, \omega)$ from $S \times \Omega$ into Y is called a Y-valued *stochastic process* if $y(t, \omega)$ is $\mathcal{C}$-measurable for each $t \in S$. The set S is called the (time) domain of the process. The main characteristic of a stochastic process is its finite–dimensional distributions, which are defined by the sequence of functions

$$F_n(t_1, \ldots, t_n, C_1, \ldots, C_n) = P\Big(\bigcap_{k=1}^{n} \{\omega : y(t_k, \omega) \in C_k\}\Big) \tag{A.13}$$

for $n = 1, 2, \ldots$, and for $k = 1, ..., n$, $C_k \in \mathcal{C}$, and $t_k \in S$. If Y is a separable and complete metric space, then the sequence in equation (A.13) satisfies the following conditions:

(i) F_n does not change under a permutation of the numbers t_k and the sets C_k above.

(ii) F_n is a measure in C_k for all k.

(iii) The iteration

$$F_n(t_1, \ldots, t_n, C_1, \ldots, C_{n-1}, Y) = F_{n-1}(t_1, \ldots, t_{n-1}, C_1, \ldots, C_{n-1})$$

determines a unique probability measure μ on the space $\{Y^S, \mathcal{F}(Y^S)\}$, where Y^S is the set of all functions from S to Y, and $\mathcal{F}(Y^S)$ is the minimal σ-algebra of subsets of Y^S containing all sets of the form

$$\{y(\cdot) \in Y^S : y(t) \in C\}$$

for all $t \in S$ and $C \in \mathcal{C}$.

Gaussian Stochastic Processes

A real-valued random variable ξ has a Gaussian distribution if its distribution function $F_\xi(x)$ is represented by the formula

$$F_\xi(x) = \int_{-\infty}^{x} \frac{1}{\sqrt{2\pi b}} \exp\left(-\frac{(u-a)^2}{2b}\right) du,$$

where $a = E\xi$ and $b = \mathrm{Var}(\xi)$. Remarkably, this is equivalent to the relation

$$E\exp\{iz\xi\} = \exp\left(iaz - bz^2/2\right). \tag{A.14}$$

(Here $i = \sqrt{-1}$.) The expectation on the left–hand side of equation (A.14) is called the *characteristic function* of the random variable ξ.

Let $\xi_1, \ldots, \xi_n$ be real-valued random variables for which $E\xi_k^2 < \infty$, for $k = 1, 2, \ldots, n$. We suppose that these variables have a joint Gaussian distribution; i.e., for all $z_1, \ldots, z_n$,

$$E\exp\left(i\sum_{k=1}^{n} z_k\xi_k\right) = \exp\left(i\sum_{k=1}^{n} a_k z_k \xi_k - \frac{1}{2}\sum_{k,m=1}^{n} b_{k,m} z_k z_m\right), \tag{A.15}$$

where $a_k = E\xi_k$ and $b_{k,m} = E\xi_k\xi_m - a_k a_m$. In this case, we say that the vector $\vec{\xi}$ is Gaussian, or equivalently, it has an *n–dimensional Gaussian distribution*, with mean value $\vec{a}$ and *covariance matrix*

$$B = (b_{k,m})_{k,m\in\overline{1,n}}.$$

A function $\xi(t,\omega)$ for $t \in S$ is a real-valued *Gaussian stochastic process* if for all $t_1, \ldots, t_n \in S$, the joint distribution of the random variables $\xi(t_1,\omega), \ldots, \xi(t_n,\omega)$ is Gaussian. Let us set

$$a(t) = E\xi(t), \quad b(s,t) = E\xi(s)\xi(t) - a(s)a(t),$$

which are the mean value and covariance function of the Gaussian stochastic process $\xi(t,\omega)$, respectively. These functions uniquely determine ξ's finite–dimensional distributions.

Let $\vec{\xi}$ be an R^d-valued stochastic process with a domain S. It is called a *Gaussian process* if for all $n \geq 1$, $t_1, \ldots, t_n \in S$, and $\vec{x}_1, \ldots, \vec{x}_n \in R^d$, the real–valued random variable

$$\sum_{k=1}^{n} (\vec{x}_k, \vec{\xi}(t_k,\omega))$$

has a Gaussian distribution. Here (u,v) denotes the scalar (dot) product in R^d. Denote by $\{\xi_k(t,\omega)\}$ the coordinates of the vector $\vec{\xi}$. Set $\vec{a}(t) = E\vec{\xi}$ and $B(s,t) = (b_{k,m}(s,t))_{k,m\in\overline{1,n}}$, where the elements of the matrix B are

$$b_{k,m}(s,t) = E\xi_k(s,\omega)\xi_m(t,\omega) - E\xi_k(s,\omega)E\xi_m(t,\omega).$$

The vector $\vec{a}(t)$ is called the mean value of the stochastic process, and $B(s,t)$ is its covariance matrix function. These functions determine the finite–dimensional distribution of the stochastic process.

Processes with Independent Increments

Suppose that $Y = R^d$ and $S = [0,\infty)$. Let $x(t)$ be an R^d-valued stochastic process. (Note that we now use the notation $x(t)$ for $\vec{\xi}(t,\omega)$.) We say that x has independent increments if for every choice of numbers $0 \leq t_0 < t_1 <$

$\cdots < t_n$, the random vectors

$$x(t_0),\, x(t_1) - x(t_0), \ldots,\, x(t_n) - x(t_{n-1})$$

are independent; i.e.,

$$P\left(\bigcap_{k=1}^{n} \{x(t_k) - x(t_{k-1}) \in A_k\} \cap \{x(t_0) \in A_0\}\right)$$
$$= P\{x(t_0) \in A_0\} \prod_{k=1}^{n} P\{x(t_k) - x(t_{k-1}) \in A_k\}.$$

An R^d-valued stochastic process $x(t)$ is called *continuous in probability* if $|x(s) - x(t)| \to 0$ in probability as $s \to t$. This means that for every sequence $\{s_n\}$ converging to t, $|x(s_n) - x(t)| \to 0$ as $n \to \infty$ in probability.
Suppose that the process x has independent increments and that it is continuous in probability. Set $f(t, z) = E\exp\{i(x(t) \cdot z)\}$ for each vector $z \in R^d$ (this is the characteristic function of x). This function can be written in the form

$$f(t,z) = f(0,z)\exp\left\{ i(a(t),z) - \frac{1}{2}(B(t)z,z) + \int \left(e^{i(z\cdot x)} - 1 - \frac{i(z,x)}{1+|x|^2}\right)\Lambda(t,dx)\right\}, \tag{A.16}$$

where $a(t)$ is a continuous function, $B(t)$ is an $L(R^d)$-valued continuous function whose values are nonnegative symmetric matrices, and the matrix $B(t_2) - B(t_1)$ is nonnegative for $t_1 < t_2$. Here $\Lambda(t, \cdot)$ is for each $t > 0$ a measure on $R^d \setminus \{0\}$ for which the integral

$$\int_{R^d} \frac{|x|^2}{1+|x|^2}\Lambda(t,dx)$$

is an increasing continuous function in t.
A process with independent increments is called *homogeneous* if the distribution of $x(t+h) - x(t)$ does not depend on t for any $h > 0$. In this case, we have in formula (A.16) that

$$a(t) = ta,\ B(t) = tB,\ \Lambda(t,A) = t\Lambda(A),$$

where $a \in R^d$, $B \in L(R)$ is a symmetric nonnegative matrix, and $\Lambda(dx)$ is a measure on R^d for which

$$\int \frac{|x|^2}{1+|x|^2}\Lambda(dx) < \infty.$$

Gaussian Process with Independent Increments

The characteristic function of a Gaussian random vector $x(\omega)$ in R^d is of the form

$$E\exp\{i(x(\omega)\cdot z)\} = \exp\{i(a,z) - (Bz,z)/2\}, \tag{A.17}$$

where $a \in R^d$ is the mean value of $x(\omega)$ and $B \in L(R^d)$ is its correlation matrix. So a Gaussian process with independent increments has a characteristic function as in (A.16) in which $\Lambda(t, dx) = 0$.
A homogeneous Gaussian process $w(t) \in R^d$ is called a *Wiener process* if its distribution is invariant with respect to orthogonal transformations of R^d. In this case, the characteristic function of $w(t)$ is the form

$$\exp\{-t\sigma^2(z,z)\},$$

and the distribution of w is given by the formula

$$P\{w(t) \in A\} = \frac{1}{\sqrt{2\pi\sigma^2 t}} \int_A \exp\{-x^2/(2\sigma^2 t)\}\, dx, \qquad \text{(A.18)}$$

where A is any Borel set in R^d.

Remark 9 Let $w^1(t), \ldots, w^d(t)$ be the coordinates of a vector $\vec{w}(t)$. Then $w^k(t)$ are one-dimensional Wiener processes, and they are independent and identically distributed. In this case, independence of these processes means that the random variables

$$F_1(w^1(\cdot)), \ldots, F_d(w^d(\cdot))$$

are independent for any functions $F_1, \ldots, F_d$ of the form $F(x(\cdot)) = \Phi(x(t_1), \ldots, x(t_k))$, where $\Phi : R^k \to R$ is a measurable function.

One-Dimensional Wiener Processes

A *Wiener process* is a Gaussian process having independent increments, and its characteristic function is $\exp\{-\sigma^2 t z^2/2\}$ for, in the one–dimensional case, $z \in R$. When $\sigma = 1$, the process is called a *standard Wiener process*, and in that case, $Ew(t) = 0, E(w(t)^2) = t$. The next theorem characterizes such Wiener processes:

Theorem *(P. Levy) The process $w(t)$ is continuous with probability 1. On the other hand, if a process with independent increments is continuous with probability 1 and has mean value 0, then it is a standard Wiener process.*

Remark 10 Let $x(t,\omega)$ be an R^d-valued stochastic process that is continuous in probability. It is called *continuous with probability* 1 if there is a countable dense subset $\Lambda \subset R_+$ for which

$$P\left\{\lim_{h\to 0} \sup\{|x(t_1,\omega) - x(t_2,\omega)| : t_1, t_2 \in \Lambda \cap [0,t], |t_1 - t_2| \le h\} = 0\right\} = 1 \qquad \text{(A.19)}$$

for all $t > 0$.

Remark 11 The set $C_{[0,T]}(R^d)$ comprises continuous R^d-valued functions $x(t)$ on the interval $0 \le t \le T$. This is a separable Banach space. Let $\mathcal{B}(C_{[0,T]}(R^d))$ be the σ-algebra generated by Borel sets in $C_{[0,T]}(R^d)$. For any R^d-valued stochastic process $\xi(t,\omega)$ that is continuous with probability 1 on the interval $[0,T]$, a probability measure on this σ-algebra exists that satisfies

$$m_\xi(C_{t_1,\dots,t_k}(A_1,\dots,A_k)) = P\left(\bigcap_{j=1}^{k}\{\omega : \xi(t_j,\omega) \in A_j\}\right)$$

for all $t_1,\dots,t_k \in [0,T]$ and sets $A_1,\dots,A_k \in \mathcal{B}(R^d)$. Here

$$C_{t_1,\dots,t_k}(A_1,\dots,A_k) = \{x(\cdot) \in C_{[0,T]}(R^d) : x(t_j) \in A_j, j = 1,\dots,k\}.$$

The measure m_ξ is called the *distribution of the stochastic process* $\xi(t,\omega)$ in $C_{[0,T]}(R^d)$.

The distribution of the standard Wiener process in $C_{[0,T]}(R^d)$ is called the *Wiener measure.*

Filtrations and Stopping Times
Consider a family of σ-algebras $\{\mathcal{F}_t, t \in R_+\}$ that has the following properties:

(i) $\mathcal{F}_{t_1} \subset \mathcal{F}_{t_2} \subset \mathcal{F}$ for $0 \le t_1 < t_2$.

(ii) $\mathcal{F}_t = \bigcap_{s>t} \mathcal{F}_s$.

This family is called a *filtration.* A stochastic process $x(t,\omega)$ is adapted to the filtration $\{\mathcal{F}_t, t \in R_+\}$ if $x(t,\omega)$ is $\mathcal{F}_t$-measurable for all $t \in R_+$.
A nonnegative random variable $T(\omega)$ (which can take the value of $+\infty$) is called a *stopping time* with respect to the filtration $\{\mathcal{F}_t\}$ if $\{\omega : T(\omega) \le t\} \in \mathcal{F}_t$ for all $t \in R_+$. For any stopping time T we introduce the σ-algebra $\mathcal{F}_T$, which consists of those sets $A \in \mathcal{F}_\infty$ for which $A \cap \{T \le t\} \in \mathcal{F}_t$. Stopping times and their corresponding σ-algebras satisfy the following properties:

(I.) If T_1, T_2 are stopping times and $T_1 \le T_2$, then $\mathcal{F}_{T_1} \subset \mathcal{F}_{T_2}$.

(II.) If T_n is a sequence of stopping times, then $\bigwedge_n T_n$ is a stopping time and

$$\mathcal{F}_{(\bigwedge_n T_n)} = \bigcap_n \mathcal{F}_{T_n}.$$

Progressive Measurability
Let $x(t,\omega)$ for $t \in R_+$ be an adapted R-valued stochastic process. It is called a progressive measurable process if for any t the function $x(s,\omega)$ defined on $[0,t] \times \Omega$ is measurable with respect to the σ-algebra $\mathcal{B}([0,t]) \times \mathcal{F}_t$.
We note two important properties of progressive measurable processes:

(1) For any stopping time $T(\omega)$, the function $x(T(\omega),\omega)$ is an $\mathcal{F}_t$-measurable random variable.

(2) $\int_0^t \phi(s, x(s, \omega))ds$ is an $\mathcal{F}_t$-measurable random variable for any measurable bounded function $\phi : R_+ \times R \to R$.

Martingales and Semimartingales

An adapted R-valued stochastic process $\mu(t)$ is called a *martingale* (more precisely, an $\{\mathcal{F}_t, t \in R_+\}$-martingale) if $E|\mu(t)| < \infty$ and $E(\mu(t)/\mathcal{F}_s) = \mu(s)$ almost surely for $s < t$.

A stochastic process $\mu^*(t)$ is a *modification* of the stochastic process $\mu(t)$ if $P(\mu(t) = \mu^*(t)) = 1$ for all $t \in R_+$. Any martingale $\mu(t)$ has a modification μ^* that is right continuous, and $\mu^*(t-)$ exists for every $t > 0$. This modification is referred to as the *cadlag* representation of μ. (It is right–continuous and has left–limits at each point.)

Finally, we define a family of random variables, say $\{\xi_\lambda, \lambda \in \Lambda\}$, to be uniformly integrable if $E|\xi_\lambda| < \infty$ for all $\lambda \in \Lambda$ and

$$\lim_{c \to +\infty} \sup_\lambda E|\xi_\lambda| 1_{|\xi_\lambda| > c} = 0.$$

With these definitions, we have the following theorem:

Theorem *Let $\mu(t)$ be a uniformly integrable martingale (i.e., the family $\{\mu(t), t \geq 0\}$ is uniformly integrable). Then:*

(i) The limit

$$\mu(+\infty) = \lim_{t \to \infty} \mu(t)$$

exists with probability 1.

(ii) For any two stopping times $T_1 \leq T_2$, the condition

$$E(\mu(T_2)/\mathcal{F}_{T_1}) = \mu(T_1)$$

holds almost surely.

(iii) For any stopping time T the stochastic process

$$\mu_T(t) = \mu(t \wedge T)$$

is a uniformly integrable martingale.

A stochastic process $\xi(t)$ adapted to a filtration $\{\mathcal{F}_t, t \in R_+\}$ is called a supermartingale if $E|\xi(t)| < \infty$ and

$$E(\xi(t)/\mathcal{F}_s) \leq \xi(s)$$

almost surely for $s < t$. If the process $\xi(t)$ is uniformly integrable on any finite interval, then it has a cadlag representation $\xi^*(t)$, and if $\sup_t E|\xi(t)| < \infty$, then there exists the limit $\lim_{t \to \infty} \xi(t)$. A stochastic process is a *submartingale* if its negative is a supermartingale.

Square–Integrable Martingales

A martingale $\mu(t)$ is square–integrable if $E(\mu(t)^2) < \infty$ for all t. If μ is a square–integrable martingale, then an increasing, right-continuous stochastic process $\langle \mu \rangle_t$ exists for which the following properties hold:

(a) $\langle \mu \rangle_T$ is $\mathcal{F}_T$ measurable for any stopping time T, where the σ-algebra $\mathcal{F}_T$ is generated by the sets $A_t \cap \{T > t\}$ for $t \in R_+$ and $A_t \in \mathcal{F}_t$.

(b) $(\mu(t))^2 - \langle \mu \rangle_t$ is a martingale.

The stochastic process $\langle \mu \rangle_t$ is called the *square characteristic* of the martingale $\mu(t)$.

Let $\mu_1(t)$ and $\mu_2(t)$ be two square–integrable martingales. Then there exists a right-continuous stochastic process of bounded variation, say $\langle \mu_1, \mu_2 \rangle_t$, for which

$$\mu_1(t) \cdot \mu_2(t) - \langle \mu_1, \mu_2 \rangle_t$$

is a martingale and $\langle \mu_1, \mu_2 \rangle_T$ is $\mathcal{F}_{T-}$–measurable for any stopping time T. The stochastic process $\langle \mu_1, \mu_2 \rangle_t$ is called the *mutual square characteristic* of the martingales $\mu_1(t)$ and $\mu_2(t)$. It is determined almost surely by $\mu_1(t)$ and $\mu_2(t)$.

Martingale Characteristic of a Wiener Process

A Wiener process $w(t)$ is *adapted* to the filtration $\{\mathcal{F}_t, t \in R_+\}$ if it is $\mathcal{F}_t$ measurable for all t and $w(t+h) - w(t)$ is independent of the σ-algebra $\mathcal{F}_t$ for all $t, h \in R_+$.

Theorem *A continuous adapted stochastic process $\xi(t)$ is a Wiener process adapted to the filtration $\{\mathcal{F}_t, t \in R_+\}$ if and only if it is a square integrable martingale and $\langle \xi \rangle_t = t$.*

Remark 12 An R^d–valued adapted process $\vec{\xi}(t)$ is a d-dimensional adapted Wiener process if and only if $(\vec{\xi}(t), z) \equiv \xi_z(t)$ is a square–integrable martingale and $\langle \xi_z \rangle_t = \sigma^2(z, z)$ for every $z \in R^d$ and for some number $\sigma > 0$.

Markov Processes

Let $y(t, \omega)$ for $t \in R_+$ be a Y-valued stochastic process where $(Y, \mathcal{C})$ is a measurable space. We denote by $\mathcal{Y}_t$ the σ-algebra generated by $\{y(s, \omega), s \leq t\}$ and

$$\hat{\mathcal{Y}}_t = \bigcap_{s>t} \mathcal{Y}_s.$$

The function y is said to be a *Markov process* if there is a function $Q(s, y, t, C)$ defined for $0 \leq s < t, y \in Y, C \in \mathcal{C}$ such that

(1) $Q(s, y, t, \cdot)$ is a probability measure on $\mathcal{C}$.

(2) $Q(s, \cdot, t, C)$ is a $\mathcal{C}$-measurable function for fixed s, t, C.

(3) Q satisfies the Chapman–Kolmogorov equation

$$Q(s, y, u, C) = \int_Y Q(s, y, t, d\tilde{y})\, Q(t, y, u, C)$$

for all $y \in Y, C \in \mathcal{C}$, and $0 \le s < t < u$.

(4) $P\{y(t,\omega) \in C/\hat{\mathcal{Y}}_s\} = Q(s, y(s,\omega), t, C)$ almost surely.

The function Q is called the *transition probability function* for the Markov process $y(t, \omega)$.

Remark 12 The finite–dimensional distributions of the Markov process ξ are represented by the following formula: For $t_0 = 0 < t_1 < \cdots < t_n$,

$$P\{y(t_0,\omega) \in C_0, \ldots, y(t_n,\omega) \in C_n\}$$
$$= \int_{C_0} \int_{C_1} \cdots \int_{C_n} \pi(dy_0) \prod_{k=1}^{n} Q(t_{k-1}, y_{k-1}, t_k, dy_k),$$

where $\pi(C_0) = P\{y(t_0,\omega) \in C_0\}$ is the initial distribution of the process.

Jump Markov Processes

A Markov process $y(t, \omega)$ is called a *jump Markov process* if for every $T > 0$ there is a constant A_T for which

$$1 - Q(s, y, t, \{y\}) \le A_T(t - s)$$

for $0 \le s < t < T$ and $y \in Y$. Here $\{y\}$ is the set containing only the element y, which we suppose to be in $\mathcal{C}$. In this case, the transition probability function satisfies the Kolmogorov equations

$$\frac{\partial}{\partial t} Q(s, y, t, C) = \int Q(s, y, t, dz) q(t, z, C), \tag{A.20}$$

and

$$\frac{\partial}{\partial s} Q(s, y, t, C) = -\int q(s, y, dz) Q(s, z, t, C), \tag{A.21}$$

where the function $q(t, y, C)$ is a signed measure on $\mathcal{C}$ and $\mathcal{C}$-measurable in y. Moreover, $-q(t, y, C \setminus \{y\})/q(t, y, \{y\})$ is a probability measure.
A Markov process $y(t, \omega)$ is called *homogeneous* if its transition probability function satisfies the relation

$$Q(s, y, t, C) = Q(0, y, t - s, C) = Q(t - s, y, C),$$

and then $Q(t, y, C)$ is called the *transition probability function* for the homogeneous Markov process. For a homogeneous jump Markov process, the function q in equations (A.20) and (A.21) does not depend on t, so $q(t, y, C) = q(y, C)$.

The operator

$$G_t f(y) = \int f(z) q(t, y, dz) = \lim_{h \downarrow 0} \frac{1}{h} \left[\int Q(t, y, t+h, dz) f(z) - f(y) \right], \tag{A.22}$$

which is defined for all bounded measurable functions $f : Y \to R$, is called the *generator* of the Markov process $y(t, \omega)$. The next theorem describes a jump process through its generator:

Theorem *Let $y(t, \omega)$ be a jump Markov process whose generator is given by the formula (A.22). Define*

$$\lambda(t, y) = -q(t, y, \{y\}) \quad \textit{and} \quad \pi(t, y, C) = (\lambda(t, y))^{-1} q(t, y, C \setminus \{y\}).$$

Then there exists a sequence of stopping times $\{\tau_k, k = 0, 1, 2, \ldots\}$ for which

$$y(t, \omega) = y(\tau_k, \omega) \equiv \eta_k(\omega)$$

for $t \in [\tau_k, \tau_{k+1}), k = 0, 1, \ldots,$ where $\tau_0 = 0$, and $\{(\tau_k, \eta_k), k = 0, 1, 2, \ldots\}$ is a homogeneous Markov chain on $R_+ \times Y$ having transition probability

$$Q^*((s, y), B) = \int_s^\infty \exp\{-\int_s^u \lambda(v, y) dv\} \pi(u, y, B_u) \lambda(u, y)\, du,$$

where $B_u = \{y : (u, y) \in B\}$.

Diffusion Processes

A continuous Markov process $x(t, \omega)$ in R^d is called a *diffusion process* if its transition probability function $Q(s, y, t, B)$ satisfies the condition

$$\lim_{h \downarrow 0} \frac{1}{h} \int [Q(t, y, t+h, dx) f(x) - f(y)] = (f'(y) \cdot a(t, y)) + \frac{1}{2} \mathrm{Tr} f''(y) B(t, y), \tag{A.23}$$

for all $f \in C^{(2)}(R^d)$, where $a(t, y) : R_+ \times R^d \to R^d$ and $B(t, y) : R_+ \times R^d \to L_+(R^d)$ are continuous functions, $L_+(R^d)$ is the set of symmetric nonnegative matrices from $L(R^d)$, and Tr denotes the trace operator. Note that f'' is the Jacobian matrix of second derivatives of f. We denote the differential operator on the right–hand side of this relation by L_t. It is called the *generator of the diffusion process.*

Remark 13 Relation (A.23) is equivalent to the following statements for all $f \in C^{(2)}(R^d)$. The stochastic process

$$f(x(t, \omega)) - \int_0^t L_s(f(x(s, \omega)) ds$$

is a martingale with respect to the filtration $\{\mathcal{F}_t, t \in R_+\}$ generated by the stochastic process $x(t, \omega)$. This statement represents the martingale characteristic of a diffusion process. If this holds for a continuous process $x(t, \omega)$, then it is a diffusion process.

Stochastic Differential Equations

Diffusion processes can be constructed as solutions of stochastic differential equations. First, we introduce Ito's stochastic integral. We consider a filtration $\{\mathcal{F}_t, t \in R_+\}$ and an adapted Wiener process $w(t)$. Ito's integral

$$I_g(t) = \int_0^t g(s, \omega)\, dw(s)$$

is defined on progressive measurable functions $g(s, \omega) : R_+ \times \Omega \to R$ for which

$$E \int_0^t g^2(s, \omega)\, dw(s) < \infty$$

in such a way that $I_g(t)$ is a square–integrable martingale having square characteristic

$$\langle I_g \rangle_t = \int_0^t g^2(s, \omega)\, ds.$$

If $g(s, \omega) = 1_{\tau_1 \leq t < \tau_2}(s)$, where $\tau_1 \leq \tau_2$ are stopping times, then

$$I_g(t) = w(t \wedge \tau_2) - w(t \wedge \tau_1).$$

Let $a(t, x)$ and $C(t, x)$ be continuous functions with values in R^d and in $L(R^d)$, respectively. Let $\vec{w}(t)$ be an R^d-valued adapted Wiener process. We consider the differential equation

$$dx(t, \omega) = a(t, x(t, \omega))\, dt + C(t, x(t, \omega))\, d\vec{w}(t), \tag{A.24}$$

which (by definition) is equivalent to the integral equation

$$x(t, \omega) = x(0, \omega) + \int_0^t a(s, x(s, \omega))\, ds + \int_0^t C(s, x(t, \omega))\, d\vec{w}(t). \tag{A.25}$$

With these ideas, we have the following theorem.

Theorem *Let $a(t, x), C(t, x)$ satisfy the following conditions:*

(i) They are continuous.

(ii) For any $T > 0$, there is a constant l_T for which

$$|a(t, x_1) - a(t, x_2)| + ||C(t, x_1) - C(t, x_2)|| \leq l_T |x_1 - x_2|.$$

Then:

(i) If $x(0, \omega)$ is an $\mathcal{F}_0$-measurable R^d-valued random variable, then equation (A.25) has a unique solution that is a diffusion process with the generator of the form (A.23), where

$$B(t, x) = C(t, x) C^*(t, x).$$

(*ii*) *The transition probability function* $Q(s, x, t, A)$ *of this diffusion process is defined by the relation*

$$Q(s, x, t, A) = P(\xi_{x,s}(t) \in A), \tag{A.26}$$

where $\xi_{x,s}(t)$ *is the solution of equation (A.24) on the interval* (s, ∞) *and satisfies the initial condition*

$$\xi_{x,s}(s) = x.$$

Appendix B
Commentary

Randomly perturbed dynamical systems were first studied by N.N. Bogoliubov and K.M. Krylov [11] using methods from ergodic theory and Markov processes, for example deriving a Fokker–Planck equation for the transition probabilities of trajectories. These investigations were continued by I.I. Gikhman [56], a student of Bogoliubov's, who introduced ideas of stochastic differential equations and applied them to study randomly perturbed dynamical systems. A large literature has emerged from these fundamental works. In this commentary we mention only some of the works that are related to our study in this book.

Problems of randomly perturbed dynamical systems have been studied in several books. We consider here those that were written in a mathematical style. First, we acknowledge the book of R. Khashminski [97], which is devoted to the stability of randomly perturbed dynamical systems. We used some results of that book in Chapter 6 here. M.I. Friedlin and A.D. Wentzell [47] considered randomly perturbed dynamical systems that are described by Ito's stochastic differential equations with small diffusion. The main tools of that investigation were large deviation theorems that were obtained by those authors for such systems. They investigated the behavior of the system in a neighborhood of a stable static state and the asymptotic behavior of the system with finitely many stable static states, deriving both averaging and normal deviation approximations. In the book [130], H. Kushner studied some general results concerning systems that are described by stochastic differential equations. In particular, he considered phase-locked loop models randomly perturbed by small Wiener processes. These results are related to those in Chapter 10 of this book.

The book [179] by A.V. Skorokhod contains some results on ergodic properties for diffusion Markov processes, averaging theorems, normal deviations, and diffusion approximations for dynamical systems perturbed by fast ergodic Markov processes. There were considered some results concerning asymptotic behavior of nonergodic Markov processes. The book [98] is devoted to discrete time random dynamical systems determined by iteration of independent random transformations. In this book we did not consider such systems.

Chapter 1: The ergodic theorem was proved by G.D. Birkhoff [8], and in a different form by A.N. Kolmogorov [101]. The ergodic theory used here appears in the books by E. Hopf [72] and by P. Halmos [69]. Ergodic theory for Markov chains was developed by A.N. Kolmogorov [100] for countable phase spaces, and further for general phase spaces by W. Doeblin [26], T.E. Harris [70], S. Orey [142], D. Revuz [168], and V.M. Shurenkov [174].

Chapter 2: General limit theorems for stochastic processes were considered by M. Donsker [28], Yu.V. Prokhov [165], and A.V. Skorokhod [175]. Central limit theorems for Markov and stationary processes were studied by I.A. Ibragimov and Yu.V. Linik [85], I.A. Ibragimov [84], M. Rosenblatt [171], and results of M.I. Gordin, which are presented in the book of P. Billlingsley [7].

Chapter 3: Averaging theorems were first considered by N.M. Krylov and N.N. Bogoliubov [11] for differential equations, and these results are presented in the book of Yu.A. Mitropolski [137]. Averaging theorems for stochastic differential equations were considered by I.I. Gikhman [62] and R.Z. Khasminski [93]. The authors considered averaging theorems for perturbed Volterra integral equations and perturbed (discrete–time processes) difference equations in [79, 80].

Chapter 4: Normal deviations for differential equations were first studied by Khasminski [93]. Normal deviations were first considered for stochastic Volterra integral equations in [79] and for difference equations in [82].

Chapter 5: Diffusion approximations were considered by R.Z. Khasminkii [96] for some stochastic differential equations. A general theory for randomly perturbed differential equations was developed by G. Papanicolaou, D. Stroock, and S. R. S. Varadhan [150]. V. Sarafian and A. Skorokhod [172] obtained diffusion approximations for dynamical systems perturbed by fast jump Markov processes, and this problem was considered further in the book by A.V. Skorokhod [179]. Diffusion approximations for difference equations were considered by the authors in [82].

Chapter 6: The main results on stability of the solution to stochastic differential equations and randomly perturbed differential equations were obtained by R.Z. Kasminski [97]. Some further results on the subject were considered in the book by A.V. Skorokhod [175]. The stability of randomly perturbed difference equations was studied by the authors [82]. The asymptotic behavior of the solutions to randomly perturbed Volterra integral equations was also studied by the authors [79].

Chapter 7: Markov chains in random environments were introduced by K. Nawrotzki in 1976 [138] and by R. Cogburn [17]. The ergodic theory of Markov chains was studied by K. Nawrotzki [139], [140] and by R. Cogburn [18], [19]. The randomly perturbed Markov chains with small perturbations were considered by the authors in [81]. Chapter 7 is based mostly on these results. Nonrandom perturbations of Markov chains were considered by V.S. Koroluk and A.I. Turbin in 1978 [193] and by F. Hoppensteadt and W. Miranker in 1976 [73].

Chapter 8: Diffusion approximations for two-dimensional Hamiltonian systems were considered by M.I. Friedlin and A.D. Wentzell in [47]. Diffusion processes on graphs were introduced by the same authors in [50]. Mechanical systems perturbed by fast noise were considered by A.V. Skorokhod in 2000 [183].

Chapter 9: The authors used the book by E.A. Coddington and N. Levinson [16] for descriptions of nonrandom systems on a torus. The results presented here for random perturbations of torus flows are new.

Chapter 10: Phase–locked loops have been extensively studied for their responses to external signals by Levi [132]. Several uses of them in modern circuits are described by Horowitz and Hill [83]. These circuits also play a fundamental role in mathematical neuroscience, as shown by Hoppensteadt and Izhikevich [78] and Hoppensteadt [76]. Many engineering books have considered the response of such circuits to noisy signals, for example Papulous [153]. A more detailed mathematical investigation was conducted by Kushner [130]. The results presented here investigate system noise in such circuits and are apparently new.

Chapter 11: Population biology comprises studies of ecology, epidemics, and genetics, among many other phenomena. The definition of stochastic processes by McKendrick [136], Feller [41], R. Fisher [44], S. Wright, and J.B.S. Haldane have represented some of the strongest contributions of biology to mathematics! The approach we take here is based on beginning with nonrandom models and studying their responses to random perturbations. Earlier work in this direction was carried out by R.M. May [135], D. Ludwig [133], and W. Ewens [39]. Since then a large literature has emerged in this area. Most of the results in this chapter are based on random perturbations of problems described by Hoppensteadt [74], [75], [77]. The authors have considered some random perturbations of problems in demographics in [92].

Chapter 12: The area of genetics still represents one of the most outstanding applications of mathematics in the life sciences. As noted above, mathematics has benefited greatly from studies of genetics in providing ideas for the development of random processes. On the other hand, modern genetics and all of its successes would not have been possible without mathematical descriptions. Our work here is based on work in Hoppensteadt [74], [75], [77] and by the authors in [81].

References

[1] Arnold, V.I. (1979) *Mathematical methods in classical mechanics*, 2nd ed., Nauka, Moscow; English transl. of 1st ed., Springer-Verlag.

[2] Arnold, L.; Papanicolaou, G.; and Wihstutz, V. (1986) *Asymptotic analysis of the Lyapunov exponent and rotation number of the random oscillator and applications*, SIAM J. Appl. Math. **46** no. 3, 427–450.

[3] Athreya, K.; Karlin, S. (1971) *Branching processes with random environments*, I: *Extinction probabilities*, Ann. Math. Stat. **42**, 1499–1520.

[4] Athreya, K.; Karlin, S. (1971) *Branching processes with random environments*, II: *Extinction probabilities*, Ann. Math. Stat. **42**, 1843–1858.

[5] Bensoussan, A.; Lions, J.-L.; Papanicolaou, G. (1976) *Perturbations et "au des conditions initiales."* (French) *Singular perturbations and boundary layer theory (Proc. Centrale Lyon, 1976)*, pp. 10–29, Lecture Notes in Math., Vol. 594, Springer, Berlin.

[6] Benzi, R.; Sutera, A.; Vulpiam, A. (1981) The mechanism of stochastic resonance, J. Phys. A14, 1453–1457.

[7] Billingsley, P. (1999) *Convergence of probability measures*, 2nd ed., Wiley Series in Probability and Statistics: Probability and Statistics.

[8] Birkhoff, G. D. (1931) *Proof of the ergodic theorem*, Proc. Nat. Acad. Sci. USA **17**, 656–660.

[9] Blagoveshchenskii, Yu.N. (1962) *Diffusion processes depending on a small parameter*, Teor. Veroyatnost. I Primenen **7**, 135–152; English transl. in Theory Probab. Appl. **7**.

[10] Blagoveshchenskii, Yu.N.; Freidlin, M.I. (1961) *Some properties of diffusion processes depending on a parameter*, Dokl. Akad. Nauk. SSSR, **138**, 508–511; English transl. in Soviet Math. Dokl. **2**.

[11] Bogolyubov, N.N. (1946) *Problems of Dynamical Theory in Statistical Physics*, Gostechizdat (Russian) Moscow.

[12] Borodin, A.N.; Freidlin, M.I. (1995) *Fast oscillating random perturbations of dynamical systems with conservation laws*, Ann. Inst. H. Poincare Probab. Statist. **31**, no. 3, 485–525.

[13] Breiman, L. (1968) *Probability*, Addison-Wesley.

[14] Chung, K. L. (1967) *Markov chains with stationary transition probabilities*, 2nd ed. Die Grundlehren der mathematischen Wissenschaften, Band **104**, Springer-Verlag, Inc., New York.

[15] Chung, K.L.; Williams, R.J. (1990) *Introduction to stochastic integration*, 2nd ed., Birkhäuser.

[16] Coddington, E.A.; Levinson, N. (1955) *Theory of ordinary differential equations*, McGraw-Hill, New York.

[17] Cogburn, R. (1980) *Markov chains in random environments: the case of Markovian environments*, Ann. Probab. **8**, 908–916.

[18] Cogburn, R. (1984) *The ergodic theory of Markov chains in random environments*, Wahr verw. Gebiete **66**, 109–128.

[19] Cogburn, R. (1986) *On products of random stochastic matrices*, Contemporary Math. **50**.

[20] Cogburn, R. (1990) *On direct convergence and periodicity for transition probabilities of Markov chains in random environments*, Ann. Probab. **18**, no. 2, 642–654.

[21] Cohen, J.E. (1976) *Ergodicity of age structure in populations with Markovian vital rates* I: *Countable states*, J. Amer. Statist. Assoc. **71**, 335–339.

[22] Cohen, J.E. (1977) *Ergodicity of age structure in populations with Markovian vital rates* II: *Countable states*, J. Amer. Statist. Assoc. **9**, 18–37.

[23] Cohen, J.E. (1977) *Ergodicity of age structure in populations with Markovian vital rates* III: *Finite-state moments and growth rates, an illustration*, Adv. in Appl. Probab. **9**, 462–475.

[24] Cramer, H.(1938). Sur un nouveau théorème–limite de la théorie des probabilites. 736 *Actual. Sci. Indust.*(Paris) 5–23.

[25] Dawson, D.A.; Kurtz, Thomas G. (1982) *Applications of duality to measure-valued diffusion processes*, Advances in filtering and optimal stochastic control (Cocoyoc, 1982), 91–105, Lecture Notes in Control and Inform. Sci. **42**, Springer, Berlin.

[26] Doeblin, W. (1940) *Elements d'une théorie générale des chaines simple constantes de Markoff*, Ann. Sci. Ecole Norm. Sup. (3) **57**, 61–111.

[27] Donnelly, P.; Kurtz, T.G. (1996) *A countable representation of the Fleming–Viot measure-valued diffusion*, Ann. Probab. **24**, no. 2, 698–742.

[28] Donsker, M.D. (1951) *An invariance principle for certain probability limit theorems*, Mem. Amer. Math. Soc. no. 6.

[29] Donsker, M.D.; Varadhan, S.R.S. (1966) Asymptotic probabilities and differential equations, *Comm. Pure Appl. Math.* 19, 261–286.

[30] Donsker, M.D.; Varadhan, S.R.S. (1975) Asymptotic evaluation of certain Markov process expectations for large time., *Comm. Pure Appl. Math.* 28, 1–47.

[31] Doob, J. L. (1953) *Stochastic Processes*, Wiley, New York.

[32] Doss, H.; Stroock, D. W. (1991) *Nouveaux resultats concernant les petites perturbations de systemes dynamiques* (French) [New results concerning small perturbations of dynamical systems], J. Funct. Anal. **101**, no. 2, 370–391.

[33] Dynkin, E.B. (1963) *Markov processes*, Fitzmatgiz, Moscow; English transl., Vols. I and II, Springer-Verlag, Berlin and Academic Press, New York, 1965.

[34] Ethier, S.N.; Kurtz, T.G. (1981) *The infinitely-many-neutral-alleles diffusion model*, Adv. in Appl. Probab. **13**, no. 3, 429–452.

[35] Ethier, S.N.; Kurtz, T.G. (1985) *The infinitely-many-alleles model with selection as a measure-valued diffusion*, *Stochastic Methods in Biology* (Nagoya), 72–86, Lecture Notes in Biomath. **70**, Springer, Berlin, 1987.

[36] Ethier, S.N.; Kurtz, T.G. (1992) *On the stationary distribution of the neutral diffusion model in population genetics*, Ann. Appl. Probab. **2**, no. 1, 24–35.

[37] Ethier, S.N.; Kurtz, T.G. (1993) *Fleming–Viot processes in population genetics*, SIAM J. Control Optim. **31**, no. 2, 345–386.

[38] Ethier, S.N.; Kurtz, T.G. (1994) *Convergence to Fleming-Viot processes in the weak atomic topology*, Stochastic Process Appl. **54**, no. 1, 1–27.

[39] Ewens, W.J. (1969) Population Genetics, Methuen, London.

[40] Feller, W. (1952) *The parabolic differential equations and the associated semigroups of transformations*, Ann. Math. **55**, 468–519.

[41] Feller, W. (1968) *Introduction to probability theory and its applications*, Wiley, Vol. I, 3rd ed.; Vol. II, 1966.

[42] Figari, R.; Orlandi, E.; Papanicolaou, G. (1982) *Mean field and Gaussian approximation for partial differential equations with random coefficients*, SIAM J. Appl. Math. **42**, no. 5, 1069–1077.

[43] Figari, R.; Papanicolaou, G.; Rubinstein, J. (1985) *The point interaction approximation for diffusion in regions with many small holes*, *Stochastic methods in biology* (Nagoya), 202–209, Lecture Notes in Biomath. **70**, Springer, Berlin, 1987.

[44] Fisher, R.A. (1930) The Genetical Theory of Natural Selection, Republished: Dover Publics., New York.

[45] Freidlin, M.I. (1978) *The averaging principle and theorems on large deviations* (Russian), Uspekhi Mat. Nauk **33**, no. 5 (203), 107–160, 238.

[46] Friedlin, M.I. (2000) Quasideterministic approximation, metastability and stochastic resonance. Phys. D. 137, 333–352.

[47] Freidlin, M.I.; Wentzell, A.D. (1994) *Random perturbations of dynamical systems*, transl. from the 1979 Russian by Joseph Szucs, second edition, *Grundlehren der Mathematischen Wissenschaften* [Fundamental Principles of Mathematical Sciences], 260, Springer-Verlag, New York.

[48] Freidlin, M.I.; Wentzell, A.D. (1994) *Random perturbations of Hamiltonian systems*, Mem. Amer. Math. Soc. **109**, no. 523.

[49] Freidlin, M.I.; Wentzell, A.D. (1994) *Necessary and sufficient conditions for weak convergence of one-dimensional Markov processes*, *The Dynkin Festschrift*, 95–109, *Progr. Probab.* **34**, Birkhäuser, Boston, MA.

[50] Freidlin, M.I.; Wentzell, A.D. (1993) *Diffusion processes on graphs and the averaging principle*, Ann. Probab. **21**, no. 4, 2215–2245.

[51] Friedlin, M.I.; Wentzell, A.D. (1998)*Random Perturbations of Dynamical Systems*,Springer-Verlag, NY.

[52] Friedman, A.; Pinsky, M.A. (1973) *Asymptotic behavior of solutions of linear stochastic differential systems*, Trans. Amer. Math. Soc. **181**, 1–22.

[53] Friedman, A.; Pinsky, M.A. (1974) *Asymptotic stability and spiraling properties for solutions of stochastic equations*, Trans. Amer. Math. Soc. **186**, 331–358.

[54] Gantmacher, F.R. (1959) *Applications of the theory of matrices*, transl., Interscience Publ., Inc., New York.

[55] Gantmacher, F.R. (1998) *The theory of matrices*, Vol. 1, transl. from Russian by K.A. Hirsch, AMS Chelsea Publ., Providence, R.I.

[56] Gikhman, I.I. (1941) *On the effect of a random process on a dynamical system*, Nauchn. Zap. Kiev Univ. Mekh.-Mat. Fak. **5**, 119–132 (Ukrainian).

[57] Gikhman, I.I. (1941) *On passing to the limit in dynamical systems*, Nauchn. Zap. Kiev, Univ. Mekh.-Mat. Fak. **5**, 141–149 (Ukrainian).

[58] Gikhman, I.I. (1947) *On a scheme for formation of random processes*, Dokl. Akad. Nauk SSSR **58**, 961–964 (Russian).

[59] Gikhman, I.I. (1950) *On some differential equations with random functions*, Ukrain. Mat. Zh. **2**, no. 3, 45–69 (Russian).

[60] Gikhman, I.I. (1950) *On the theory of differential equations of random processes*, I and II, Ukrain. Mat. Zh. **2**, no. 4, 37–63; **3** (1951), 317–339; English transl. in Amer. Math. Soc. Transl. (2) **1** (1955).

[61] Gikhman, I.I. (1952) *On a theorem of N.N. Bogolyubov*, Ukrain. Mat. Zh. **4**, 215–219 (Russian).

[62] Gikhman, I.I. (1964) *Differential equations with random functions*, Winter School Theory Probab. and Math. Statist. (Uzhgorod), Izdat. Akad. Nauk Ukrain. SSSR, Kiev, 41–85 (Russian). Stability of solutions of stochastic differential equations, Limit theorems and statistical inference (S.Kh. Sirazhdinov, ed.), "Fan" Tashkent (1966), 14–45; English transl. in *Selected Transl. Math. Statist. and Probab.* **12**, Amer. Math. Soc., Providence, RI, 1973.

[63] Gikhman, I.I.; Dorogovtsev, A.Ya. (1965) *On stability of solutions of stochastic differential equations*, Ukrain. Mat. Zh. **17**, no. 6, 3–21; English transl. in Amer. Math. Soc. Transl. (2) **72** (1968).

[64] Gikhman, I.I.; Skorokhod, A.V. (1969) *Introduction to the theory of random processes*, transl. from Russian by Scripta Technica, Inc., W.B. Saunders Co., Philadelphia, PA.

[65] Gikhman, I.I.; Skorokhod, A.V. (1971), (1973), (1975) *The theory of stochastic processes* I, II, and III, Nauka, Moscow; English transl., Springer-Verlag, 1974, 1975, 1979.

[66] Gikhman, I.I.; Skorokhod, A.V. (1982) *Stochastic differential equations and their applications*, Naukova Dumka, Kiev (Russian).

[67] Guest, P.B. (1991) *Laplace transforms and an introduction to distributions*, Ellis Horward Series: Mathematics and its Applications, New York.

[68] Guo, M.Z.; Papanicolaou, G.; Varadhan, S.R.S. (1988) *Nonlinear diffusion limit for a system with nearest neighbor interactions*, Comm. Math. Phys. **118**, no. 1, 31–59.

[69] Halmos, P. R. (1956) *Lectures on ergodic theory*, Math. Society of Japan, no. 3.

[70] Harris, T.E. (1954/55) *The existence of stationary measures for certain Markov processes*, Proc. Berkeley Sympos. Math. Statist. Probab., Vol. 2, Univ. of California Press, Berkeley, CA, 1956, 113–124.

[71] Hersh, R. (1972) *Random evolutions: a survey of results and problems*, Based on lectures given by Richard Griego, Reuben Hersh, Tom Kurtz and George Papanicolaou. Papers arising from a Conference on Stochastic Differential Equations (Univ. Alberta, Edmonton, Alta.), Rocky Mountain J. Math. **4** (1974), 443–477.

[72] Hopf, E. (1937) *Ergodentheorie*, Springer-Verlag; reprint, Chelsea, NY, 1948.

[73] Hoppensteadt, F. (2000) *Analysis and Simulation of Chaotic Systems* Springer-Verlag, New York.

[74] Hoppensteadt, F. (1975) Mathematics of Population Biology: Demographics, Epidemics and Genetics. CBMS vol. 20, SIAM.

[75] Hoppensteadt, F. (1982) Mathematical Methods of Population Biology, Cambridge U. Press, Cambridge.

[76] Hoppensteadt, F. (1997) Introduction to the Mathematics of Neurons, 2nd ed., Camb. U. Press, Cambridge.

[77] Hoppensteadt, F.; Peskin, C. (2001) Modeling and Simulation in Medicine and the Life Sciences, Springer-Verlag, New York.

[78] Hoppensteadt, F.; Izhikevich, E. (1997) Weakly Connected Neural Networks, Springer-Verlag, New York.

[79] Hoppensteadt, F.; Salehi, H.; Skorokhod, A. (1995) *Randomly perturbed Volterra integral equations and some applications*, Stochastics and Stochastic Reports, **54**, 89–125.

[80] Hoppensteadt, F.; Salehi, H.; Skorokhod, A. (1996) *An averaging principle for dynamical systems in Hilbert space with Markov random perturbations*, Stochastic Processes and their Applications **61**, 85–108.

[81] Hoppensteadt, F.; Salehi, H.; Skorokhod, A. (1996) *Markov chains with small random perturbations with applications to bacterial genetics*, Random Oper. and Stoch. Equ. **4**, 205–227.

[82] Hoppensteadt, F.; Salehi, H.; Skorokhod, A. (1997) *Discrete time semigroups transformations with random perturbations*, J. Dynamics and Differential Equ. **9**, 463–505.

[83] Horowitz, P.; Hill, W. (1989) The Art of Electronics, 2nd ed., Camb. U. Press, Cambridge.

[84] Ibragimov, I.A. (1966) *Some limit theorems for stationary processes*, Theory, Probab. and its Appl. **7**, 349–382.

[85] Ibragimov, I.A.; Linnik, Yu.V. (1971) *Independent and stationary sequences of random variables*, transl. from Russian by J.F.C. Kingman, Wolters-Noordhoff Publ., Groningen.

[86] Ikeda, N.; Watanabe, S. (1981) *Stochastic differential equations and diffusion processes*, North Holland.

[87] Ito, K. (1946) *On a stochastic integral equation*, Proc. Japan Acad. **22**, 32–35.

[88] Ito, K. (1951) *On stochastic differential equations*, Mem. Amer. Math. Soc. **4**.

[89] Jagers, P. (1976) *Branching processes with biological applications*, Wiley, NY.

[90] Karlin, S. (1966) *A first course in stochastic processes*, Academic Press.

[91] Kesten, H.; Papanicolaou, G. (1980/81) *A limit theorem for stochastic processes*, Comm. Math. Phys. **78**, no. 1, 19–63.

[92] Keyfitz, N.; Flieger, W. (1971) Population: Fact and Methods of Demography, W.H. Freeman, San Francisco.

[93] Khasminskii, R.Z. (1963) *On the averaging principle for parabolic and elliptic differential equations and Markov processes with small diffusion*, Teor. Veroyatnost. I Primenen **8**, 3–25; Engl. transl. in Theory Probab. Appl. **8**.

[94] Khasminskii, R.Z. (1966) *On random processes determined by differential equations with a small parameter*, Teor. Veroyatnost. I Primenen **11**, 240–259; English transl. in Theory Probab. Appl. **11** (1966).

[95] Khasminskii, R.Z. (1966) *A limit theorem for solutions of differential equations with a random right-hand side*, Theor. Veroyatnost I Primenen **11**, 444–462; English transl. in Theory Probab. Appl. **11** (1966).

[96] Khasminskii, R.Z. (1968) *On the averaging principle for Ito stochastic differential equations*, Kybernetika (Prague) **4**, 260–279 (Russian).

[97] Khasminskii, R.Z. (1969) *Stability of systems of differential equations under random perturbations of their parameters*, Nauka, Moscow; English transl., Stochastic stability of differential equations, Sijthoff and Noordhoff, Alphen aan den Riin, 1981.

[98] Y. Kifer, *Random Perturbations of Dynamical Systems*, Birkhaeuser, 1988.

[99] Kiguradze, I.T.; Chanturia, T.A. (1985) *Asymptotic properties of solutions of nonautonomous ordinary differential equations*, transl. 1985 Russian original Math. and Its Appl. (Soviet Series), 89, Kluwer Academic Publ. Group, Dordrecht, 1993.

[100] Kolmogorov, A.N. (1937) *Markov chains with a countable number of possible states*, Bull. Math. Univ. Moscou **1**, no. 3, 16 pp. (Russian).

[101] Kolmogorov, A.N. (1937) *A simple proof of the ergodic theorem of Birkhoff and Khinchin*, Uspehi Mat. Nauk. **5**, 52–59.

[102] Korolyuk, V.S. (1997) *Singularly perturbed stochastic systems*, Ukrain. Mat. Zh. **49**, no. 1, 25–34.

[103] Korolyuk, V.S. (1998) *Stability of stochastic systems in diffusion approximation scheme*, Ukrain. Mat. Zh. **50**, no. 1, 36–47; transl. in Ukrain. Mat. J. **50** (1998), no. 1, 40–54.

[104] Korolyuk, V.; Swishchuk, A. (1995) *Evolution of systems in random media*, CRC Press, Boca Raton, FL.

[105] Korolyuk, V.; Swishchuk, A. (1995) *Semi-Markov random evolutions*, transl. from 1992 Russian by V. Zayats and revised by the authors, *Mathematics and its applications* **308**, Kluwer Acad. Publishers, Dordrecht.

[106] Korolyuk, V. S.; Turbin, A.F. (1978) *Mathematical foundations for phase amalgamation of complex systems*, Naukova Dumka, Kiev (Russian).

[107] Korolyuk, V. S.; Turbin, A.F. (1993) *Mathematical foundations of the state lumping of large systems*, transl. from 1978 Russian by V.V. Zayats and Y.A. Atanov and revised by the authors, Mathematics and Its Applications **264**, Kluwer Acad. Publ. Group, Dordrecht.

[108] Krylov, N.V. (1995) *Introduction to the theory of diffusion processes*, transl. from Russian by Valim Khidekel and Gennady Pasechnik, transl. of Mathematical Monographs **142**, American Mathematical Society, Providence, RI.

[109] Krylov, N.M.; Bogolyubov, N.N. (1937) *L'effet de la variation statistique des parametres sur les propriétés ergodiques des systems dynamiques non conservatifs*, Zap. Kafedr. Mat. Fiz. Inst. Budivel. Mat. Akad. Nauk Ukrain. SSR **3**, 154–171 (Ukrainian); French transl., ibid., 172–190.

[110] Krylov, N.M.; Bogolyubov, N.N. (1937) *General measure theory in nonlinear mechanics*, Zap. Kafedr. Mat. Fiz. Inst. Budivel. Mat. Akad. Nauk Ukrain SSR **3**, 55–112 (Ukrainian); French transl., La théorie générale de la mesure dans son application a l'étude des systèmes dynamiques de la mécanique non linéaire, Ann. Math. **38**, no. 2, 65–113.

[111] Krylov, N.M.; Bogolyubov, N.N. (1937) *Les propriétés ergodiques des suites des probabilités en chaine*, C.R. Acad. Sci., Paris, **204** (1937), 1545–1546.

[112] Krylov, N.M.; Bogolyubov, N.N. (1939) *Sur les equations de Fokker–Planck déduites dans la théorie des perturbations a l'aide d'une méthode basée sur les properiétés spectrales de l'hamiltonien perturbateur*, Zap. Kafedr. Mat. Fiz. Inst. Budievel. Mat. Akad. Nauk Ukrain. SSR **4** (1939), 5–80 (Ukrainian); French transl., ibid. 81–157.

[113] Krylov, N.M.; Bogolyubov, N.N. (1939) *On some problems in the ergodic theory of stochastic systems*, Zap. Kafedr. Mat. Fiz. Inst. Budivel. Mat. Akrad. Nauk Ukrain. SSR **4** (1939), 243–287 (Ukrainian).

[114] Kurtz, T.G. (1973) *A limit theorem for perturbed operator semigroups with applications to random evolutions*, J. Functional Analysis **12**, 55–67.

[115] Kurtz, T.G. (1977) *Applications of an abstract perturbation theorem to ordinary differential equations*, Houston J. Math. **3**, no. 1, 67–82.

[116] Kurtz, T.G. (1976) *An abstract averaging theorem*, J. Functional Analysis **23**, no. 2, 135–144.

[117] Kurtz, T.G. (1976) *Diffusion approximations for branching processes, Branching processes* (Conf., Saint Hippolyte, Que., 1976), pp. 269–292, Adv. Probab. Related Topics **5**, Dekker, New York, 1978.

[118] Kurtz, T.G. (1981) *Approximation of population processes*, CBMS-NSF Regional Conference Series in Applied Mathematics **36**, Society for Industrial and Applied Mathematics (SIAM), Philadelphia, PA.

[119] Kurtz, T.G. (1981) *The central limit theorem for Markov chains*, Ann. Probab. **9**, no. 4, 557–560.

[120] Kurtz, T.G. (1991) *Averaging for martingale problems and stochastic approximation, Applied Stochastic Analysis* (New Brunswick, NJ), 186–209, Lecture Notes in Control and Inform. Sci. **177**, Springer, Berlin, 1992.

[121] Kurtz, T.G.; Marchetti, F. (1993) *Averaging stochastically perturbed Hamiltonian systems, Stochastic analysis* (Ithaca, NY), 93–114, Proc. Sympos. Pure Math. **57**, Amer. Math. Soc., Providence, RI, 1995.

[122] Kurtz, T.G.; Protter, P.E. (1995) *Weak convergence of stochastic integrals and differential equations, Probabilistic models for nonlinear partial differential equations* (Montecatini Terme), 1–41, Lecture Notes in Math. **1627**, Springer, Berlin, 1996.

[123] Kurtz, T.G.; Protter, P.E. (1995) *Weak convergence of stochastic integrals and differential equations*, II. *Infinite-dimensional case, Probabilistic models for nonlinear partial differential equations* (Montecatini Terme), 197–285, Lecture Notes in Math. **1627**, Springer, Berlin, 1996.

[124] Kushner, H.J. (1966) *Stability of stochastic dynamical systems, Advances in Control Systems*, Vol. 4, pp. 73–102, Academic Press, New York.

[125] Kushner, H.J. (1977) *Probability methods for approximations in stochastic control and for elliptic equations*, Mathematics in Science and Engineering Vol. 129, Academic Press [Harcourt Brace Jovanovich Publishers], New York.

[126] Kushner, H.J. (1979) *Jump-diffusion approximations for ordinary differential equations with wide-band random right-hand sides*, SIAM J. Control Optim. **17**, no. 6, 729–744.

[127] Kushner, H.J. (1979) *Approximation of solutions to differential equations with random inputs by diffusion processes, Stochastic control theory and stochastic differential systems*, (Proc. Workshop, Deutsch. Forschungsgemeinsch., Univ. Bonn, Bad Honnef) pp. 172–193, Lecture Notes in Control and Information Sci. **16**, Springer, Berlin.

[128] Kushner, H.J. (1980) *A martingale method for the convergence of a sequence of processes to a jump-diffusion process*, Z. Wahrsch. Verw. Gebiete **53**, no. 2, 207–219.

[129] Kushner, H.J. (1981) *Diffusion approximation with discontinuous dynamics and state dependent noise*, J. Math. Anal. Appl. **82**, no. 2, 527–542.

[130] Kushner, H.J. (1984) *Approximation and weak convergence methods for random processes, with applications to stochastic systems theory*, MIT Press

Series in Signal Processing, Optimization, and Control **6**, MIT Press, Cambridge, MA.

[131] Kushner, H.J.; Huang, H. (1980) *Diffusion approximations to output processes of nonlinear systems with wide-band inputs and applications*, IEEE Trans. Inform. Theory **26**, no. 6, 715–725.

[132] Levi, M. (1979) On van der Pol's equation, AMS Memoirs, Providence.

[133] Ludwig, D.A. (1974) Stochastic Population Theories, Lect. Notes, Biomath. vol 3, Springer-Verlag, New York.

[134] Malkin, I.G. (1966) *Theory of stability of motion*, 2nd rev. ed., Nauka, Moscow (Russian).

[135] May, R.M. (1970) Stability and Complexity of Model Ecosystems, Princ. U. Press, Princeton.

[136] McKendrick, A.G. (1926) Applications of mathematics to medical problems, Proc. Edin. Math. Soc, 44:98-130.

[137] Mitropolskii, Yu.A. (1964) *Problems in the asymptotic theory of nonstationary oscillations*, Nauka, Moscow; English transl., Israel Prog. Sci., Jerusalem and Davey, NY, 1965.

[138] Nawrotzki, K. (1976) *Ein zufalliger Ergodensatz fur eine Familie stochastischer Matrizen ohne gemeinsames invariantes*, Verteilungs. Math. Arch. **70**, 17–28.

[139] Nawrotzki, K. (1980) *Stationare Verteilungen endlicher offener Systeme*, EIK **16**, 345–351.

[140] Nawrotzki, K. (1982) *Finite Markov chains in stationary random environments*, Ann. Probab. **20**, 1041–1046.

[141] Oksendal, B. (1992) *Stochastic Differential Equations*, Springer-Verlag, New York.

[142] Orey, S. (1991) *Markov chains with stochastically transition probabilities*, Ann. Probab. **19**, 907–928.

[143] Ornstein, D. (1974) *Ergodic theory, randomness, and dynamical systems*, Yale University Press.

[144] Papanicolaou, G. (1975) *Asymptotic analysis of transport processes*, Bull. Amer. Math. Soc. **81**, 330–392.

[145] Papanicolaou, G. (1975) *Introduction to the asymptotic analysis equations*, Modern modeling of continuum phenomena (Ninth Summer Sem. Appl. Math., Polytech., Inst., Troy, NY), pp. 109–147, Lectures in Appl. Math. **16**, Amer. Providence, R.I., 1977.

[146] Papanicolaou, G. (1980) *Stochastically perturbed bifurcation*, Comm. Methods in Applied Sciences and Engineering (Proc. Fourth Internat. Sympos., Versailles, pp. 659–673), North-Holland, Amsterdam, New York, 1980.

[147] Papanicolaou, G. (1982) *Diffusions with random coefficients*, *Statistics and Probability: Essays in Honor of C.R. Rao*, pp. 547–552, North-Holland, Amsterdam.

[148] Papanicolaou, G. (1983) *Diffusions and random walks in random media*, *The mathematics and physics of disordered media* (Minneapolis, Minn.), 391–399, Lecture Notes in Math. **1035**, Springer, Berlin.

[149] Papanicolaou, G. (1995) *Diffusion in random media, Surveys in applied mathematics*, Vol. 1, 205–253, Surveys Appl. Math. **1**, Plenum, New York.

[150] Papanicolaou, G. Stroock, D., and Varadhan, S.R.S. (1977) *Martingale approach to some limit theorems*, Papers from the Duke Turbulence Conference (Duke Univ., Durham, NC, Paper No. 6).

[151] Papanicolaou, G.; Varadhan, S.R.S. (1973) *A limit theorem with strong mixing in Banach space and two applications to stochastic differential equations*, Comm. Pure Appl. Math. **26**, 497–524.

[152] Papanicolaou, G.; Varadhan, S.R.S. (1978) *Diffusion in regions with many small holes, Stochastic differential systems* (Proc. IFIP-WB 7/1 Working Conf., Vilnius), 190–206, Lecture Notes in Control and Information Sci. **25**, Springer, 1980.

[153] Papoulis, A. (1991) Probability, Random Variables, and Stochastic Processes, 3rd ed., McGraw-Hill, New York.

[154] Pinsky, M.A. (1974) *Multiplicative operator functionals and their asymptotic properties*, Advances in Probability and Related Topics **3**, Dekker, New York, 1–100.

[155] Pinsky, M.A. (1974) *Random evolutions, Probabilistic methods in differential equations* (Proc. Conf., Univ. Victoria, Victoria, B.D.), 89–99, Lecture Notes in Math. **451**, Springer, Berlin, 1975.

[156] Pinsky, M.A. (1978) *Averaging an alternating series*, Math. Mag. **51**, no. 4, 235–237.

[157] Pinsky, M.A. (1986) *Instability of the harmonic oscillator with small noise*, SIAM J. Appl. Math. **46**, no. 3, 451–463.

[158] Pinsky, M.A. (1985) *Lyapunov exponent and rotation number of the stochastic harmonic oscillator*, Random Media (Minneapolis, MN), 229–243, IMA Vol. Math. Appl. **7**, Springer, New York, 1987.

[159] Pinsky, M.A. (1989) *Diffusion processes and related problems in analysis*, Vol. I *Diffusions in analysis and geometry*, Papers from the Intern. Conf. held at Northwestern Univ., Evanston, IL, Progress in Probability **22**, Birkhäuser Boston, Inc., Boston, MA, 1990.

[160] Pinsky, M.A. (1990) *Lyapunov exponent and rotation number of the linear harmonic oscillator, Diffusion processes and related problems in analysis*, Vol. II (Charlotte, NC), 257–267, Progr. Probab. **27**, Birkhäuser, Boston, MA, 1992.

[161] Pinsky, M.A. (1993) *Invariance of the Lyapunov exponent under nonlinear perturbations, Stochastic analysis* (Ithaca, NY), 619–621, Proc. Sympos. Pure Math. **57**, Amer. Math. Soc., Providence, RI, 1995.

[162] Pinsky, M.A. (1995) *A unified approach to stochastic stability*, Nonlinear Dynamics and Stochastic Mechanics, 313–340, CRC Math. Model. Ser., CRC, Boca Raton, FL.

[163] Pinsky, M.A.; Wihstutz, V. (1990) *Diffusion processes and related problems in analysis*, Vol. II *Stochastic flows*, Papers from the Conf. held at the Univ. of NC, Charlotte, NC, 16–18 (edited by Pinsky and Wihstutz).

[164] Pinsky, M.A.; Wihstutz, V. (1992) *Lyapunov exponents and rotation numbers of linear systems with real noise*, Probability Theory, 109–119.

[165] Prohorov, Yu.V. (1956) *Convergence of random processes and limit theorems in probability theory*, Teor. Ver. I ee Prim. **1**, 177–238 (Russian); transl. Theory Prob. Appl. 1, 157-214.

[166] Protter, P. (1978) *Stability of solutions of stochastic differential equations*, Z. Wahrsch. Verw. Gebiete **44**, no. 4, 337–352.

[167] Protter, P. (1990) *Stochastic integration and differential equations. A new approach*, Applications of Mathematics **21**, Springer-Verlag, Berlin.

[168] Revuz, D. (1984) *Markov chains*, 2nd ed., North-Holland Mathematical Library **11**, North-Holland Publ. Co., Amsterdam.

[169] Revuz, D.; Yor, M. (1994) (review) *Continuous martingales and Brownian motion*, 3rd ed., Grundlehren der Mathematischen Wissenschaften [Fundamental Principles of Mathematical Sciences] **293**, Springer-Verlag, Berlin.

[170] Rogers, L.C.G.; Williams, D. (1987) *Diffusions, Markov processes, and Martingales*, Wiley.

[171] Rosenblatt, M. (2000) *Gaussian and non-Gaussian linear time series and random fields*, Springer Series in Statistics, Springer-Verlag, New York.

[172] Sarafyan, V.V.; Skorokhod, A.V. (1987) *On fast-switching dynamical systems*, Teor. Veroyatnost. I Primenen **32**, 658–669; English transl. in Theory Probab. Appl. **32**, 1987.

[173] Shurenkov, V.M. (1981) *Ergodic theory and related equations in the theory of random processes*, Naukova Dumka, Kiev (Russian).

[174] Shurenkov, V.M. (1989) *Ergodic Markov processes*, Nauka, Moscow (Russian).

[175] Skorokhod, A.V. (1965) *Studies in the theory of random processes*, transl. from Russian by Scripta Technica, Inc., Addison-Wesley Publ. Co., Inc., Reading, MA.

[176] Skorokhod, A.V. (1985) *Stochastic equations for ecological systems* (Russian) *Probabilistic methods in biology* 137–146, vi, Adac. Sci. Ukrain. SSR Inst. Math., Kiev.

[177] Skorokhod, A.V. (1988) *Dynamical systems that are subject to the influence of "rapid" dynamical systems with feedback* (Russian) *Asymptotic methods in mathematical physics* (Russian), 241–250, 303, Naukova Dumka, Kiev.

[178] Skorokhod, A.V. (1988) *Stochastic equations for complex systems*, transl. from Russian by L.F. Boron *Mathematics and its Applications* (Soviet Series) **13**, D. Reidel Publ. Co., Dordrecht, Boston, MA.

[179] Skorokhod, A.V. (1989) *Asymptotic methods in the theory of stochastic differential equations*, transl. from Russian by H.H. McFaden. Translations of Mathematical Monographs **78**, American Mathematical Society, Providence, RI.

[180] Skorokhod, A.V. (1989) *Dynamical systems with random influences Functional Analysis*, III (Dubrovnik), 193–234, Various Publ. Ser. **40**, Aarhus Univ., 1992.

[181] Skorokhod, A.V. (1991) *Dynamical systems under the influence of fast random perturbations* (Russian) Ukrain. Mat. Zh. **43**, no. 1.3–21; transl. Ukrainian Math. J. **43** (1991), no. 1, 1–15.

[182] Skorokhod, A.V. (1991) *Random processes with independent increments*, Math. and Its Applications (Soviet Series), Kluwer Academic Publishers.

[183] Skorokhod, A.V. (2000) *On randomly perturbed mechanical systems with two degrees of freedom*, Random Operators and Stochastic Equations, to appear.

[184] Stratonovich, R.P. (1961) *Selected questions of the theory of fluctuations in radio engineering*, Soviet, Radio, Moscow; English transl., Topics in the theory of random noise, Vol. I: General theory of random processes, *Nonlinear transformations of signals and noise*, Gordon and Breach, New York.

[185] Stratonovich, R.P. (1966) *Conditional Markov processes and their application to the theory of optimal control*, Izdat. Moskovm Gos. Univ., Moscow; English transl. Amer. Elsevier, New York, 1968.

[186] Stroock, D.W. (1973/74) *Some stochastic processes which arise from a model of the motion of a bacterium*, Z. Wahrscheinlichkeitstheorie und Verw. Gebiete **28**, 303–315.

[187] Stroock, D.W. (1981) *On the spectrum of Markov semigroups and the existence of invariant measures, Functional analysis in Markov process*, (Katata/Kyoto), 286–307, Lecture Notes in Math. **923**, Springer, Berlin.

[188] Stroock, D.W.; Varadhan, S.R.S. (1969) *Diffusion processes with continuous coefficients* I, Comm. Pure Appl. Math. **22**, 345–400.

[189] Stroock, D.W.; Varadhan, S.R.S. (1969) *Diffusion processes with continuous coefficients* II, Comm. Pure Appl. Math. **22**, 479–530.

[190] Stroock, D.W.; Varadhan, S.R.S. (1971) *Diffusion processes with boundary conditions*, Comm. Pure Appl. Math. **24**, 147–225.

[191] Stroock, D.W.; Varadhan, S.R.S. (1979) *Multidimensional diffusion processes.* Grundlehren der mathematischen Wissenschaften [Fundamental Principles of Mathematical Sciences] **233**, Springer-Verlag, Berlin.

[192] Synge, J.L.; Griffith, B.A. (1959) *Principles of Mechanics*, McGraw-Hill Book Company.

[193] Turbin, A.F. (1972) *An application of the theory of perturbations of linear operators to the solution of some problems connected with Markov chains and semi-Markov processes*, Teor. Veroyatnost. I Mat. Statist. Vyp. **6**, 118–128; English transl. in Theory Probab. Math. Statist. **6**, 1975, 1976.

[194] Varadhan, S.R.S. (1990) *On the derivation of conservation laws for stochastic dynamics, Analysis, et cetera* 677–694, Academic Press, Boston, MA.

[195] Varadhan, S.R.S. (1966) Asymptotic probabilities and differential equations. CPAM, 19, 261–286.

[196] Varadhan, S.R.S. (1984) Large deviations and applications. SIAM, Philadelphia.

[197] Wentzell, A.D. (1976) *Robust limit theorems on large deviations for Markov random processes*, I, II and III, Teor. Veroyatnost. I Primenen **21**, 235–252,

512–526; 24 (1979), 673–691; English transl. in Theory Probab. Appl. **21** (1976); **24** (1979).

[198] Wentzell, A.D. (1990) *Limit theorems on large deviations for Markov stochastic processes*, transl. from Russian Mathematics and Its Applications (Soviet Series) **38**, Kluwer Academic Publishers group, Dordrecht.

[199] Wentzell, A.D.; Freidlin, M.I. (1970) *Small random perturbations of dynamical systems* (Russian), Uspehi Mat. Nauk **25**, no. 1 (151), 3–55.

[200] Widder, D.V. (1941) *The Laplace transform*, Princeton Mathematical Series, Vol. 6, Princeton, NJ.

Index

Applied Mathematical Sciences

(continued from page ii)

60. *Ghil/Childress:* Topics in Geophysical Dynamics: Atmospheric Dynamics, Dynamo Theory and Climate Dynamics.
61. *Sattinger/Weaver:* Lie Groups and Algebras with Applications to Physics, Geometry, and Mechanics.
62. *LaSalle:* The Stability and Control of Discrete Processes.
63. *Grasman:* Asymptotic Methods of Relaxation Oscillations and Applications.
64. *Hsu:* Cell-to-Cell Mapping: A Method of Global Analysis for Nonlinear Systems.
65. *Rand/Armbruster:* Perturbation Methods, Bifurcation Theory and Computer Algebra.
66. *Hlavácek/Haslinger/Necasl/Lovísek:* Solution of Variational Inequalities in Mechanics.
67. *Cercignani:* The Boltzmann Equation and Its Applications.
68. *Temam:* Infinite-Dimensional Dynamical Systems in Mechanics and Physics, 2nd ed.
69. *Golubitsky/Stewart/Schaeffer:* Singularities and Groups in Bifurcation Theory, Vol. II.
70. *Constantin/Foias/Nicolaenko/Temam:* Integral Manifolds and Inertial Manifolds for Dissipative Partial Differential Equations.
71. *Catlin:* Estimation, Control, and the Discrete Kalman Filter.
72. *Lochak/Meunier:* Multiphase Averaging for Classical Systems.
73. *Wiggins:* Global Bifurcations and Chaos.
74. *Mawhin/Willem:* Critical Point Theory and Hamiltonian Systems.
75. *Abraham/Marsden/Ratiu:* Manifolds, Tensor Analysis, and Applications, 2nd ed.
76. *Lagerstrom:* Matched Asymptotic Expansions: Ideas and Techniques.
77. *Aldous:* Probability Approximations via the Poisson Clumping Heuristic.
78. *Dacorogna:* Direct Methods in the Calculus of Variations.
79. *Hernández-Lerma:* Adaptive Markov Processes.
80. *Lawden:* Elliptic Functions and Applications.
81. *Bluman/Kumei:* Symmetries and Differential Equations.
82. *Kress:* Linear Integral Equations, 2nd ed.
83. *Bebernes/Eberly:* Mathematical Problems from Combustion Theory.
84. *Joseph:* Fluid Dynamics of Viscoelastic Fluids.
85. *Yang:* Wave Packets and Their Bifurcations in Geophysical Fluid Dynamics.
86. *Dendrinos/Sonis:* Chaos and Socio-Spatial Dynamics.
87. *Weder:* Spectral and Scattering Theory for Wave Propagation in Perturbed Stratified Media.
88. *Bogaevski/Povzner:* Algebraic Methods in Nonlinear Perturbation Theory.
89. *O'Malley:* Singular Perturbation Methods for Ordinary Differential Equations.
90. *Meyer/Hall:* Introduction to Hamiltonian Dynamical Systems and the N-body Problem.
91. *Straughan:* The Energy Method, Stability, and Nonlinear Convection.
92. *Naber:* The Geometry of Minkowski Spacetime.
93. *Colton/Kress:* Inverse Acoustic and Electromagnetic Scattering Theory, 2nd ed.
94. *Hoppensteadt:* Analysis and Simulation of Chaotic Systems, 2nd ed.
95. *Hackbusch:* Iterative Solution of Large Sparse Systems of Equations.
96. *Marchioro/Pulvirenti:* Mathematical Theory of Incompressible Nonviscous Fluids.
97. *Lasota/Mackey:* Chaos, Fractals, and Noise: Stochastic Aspects of Dynamics, 2nd ed.
98. *de Boor/Höllig/Riemenschneider:* Box Splines.
99. *Hale/Lunel:* Introduction to Functional Differential Equations.
100. *Sirovich (ed):* Trends and Perspectives in Applied Mathematics.
101. *Nusse/Yorke:* Dynamics: Numerical Explorations, 2nd ed.
102. *Chossat/Iooss:* The Couette-Taylor Problem.
103. *Chorin:* Vorticity and Turbulence.
104. *Farkas:* Periodic Motions.
105. *Wiggins:* Normally Hyperbolic Invariant Manifolds in Dynamical Systems.
106. *Cercignani/Illner/Pulvirenti:* The Mathematical Theory of Dilute Gases.
107. *Antman:* Nonlinear Problems of Elasticity.
108. *Zeidler:* Applied Functional Analysis: Applications to Mathematical Physics.
109. *Zeidler:* Applied Functional Analysis: Main Principles and Their Applications.
110. *Diekmann/van Gils/Verduyn Lunel/Walther:* Delay Equations: Functional-, Complex-, and Nonlinear Analysis.
111. *Visintin:* Differential Models of Hysteresis.
112. *Kuznetsov:* Elements of Applied Bifurcation Theory, 2nd ed.
113. *Hislop/Sigal:* Introduction to Spectral Theory: With Applications to Schrödinger Operators.
114. *Kevorkian/Cole:* Multiple Scale and Singular Perturbation Methods.
115. *Taylor:* Partial Differential Equations I, Basic Theory.
116. *Taylor:* Partial Differential Equations II, Qualitative Studies of Linear Equations.

(continued on next page)

Applied Mathematical Sciences

(continued from previous page)

117. *Taylor:* Partial Differential Equations III, Nonlinear Equations.
118. *Godlewski/Raviart:* Numerical Approximation of Hyperbolic Systems of Conservation Laws.
119. *Wu:* Theory and Applications of Partial Functional Differential Equations.
120. *Kirsch:* An Introduction to the Mathematical Theory of Inverse Problems.
121. *Brokate/Sprekels:* Hysteresis and Phase Transitions.
122. *Gliklikh:* Global Analysis in Mathematical Physics: Geometric and Stochastic Methods.
123. *Le/Schmitt:* Global Bifurcation in Variational Inequalities: Applications to Obstacle and Unilateral Problems.
124. *Polak:* Optimization: Algorithms and Consistent Approximations.
125. *Arnold/Khesin:* Topological Methods in Hydrodynamics.
126. *Hoppensteadt/Izhikevich:* Weakly Connected Neural Networks.
127. *Isakov:* Inverse Problems for Partial Differential Equations.
128. *Li/Wiggins:* Invariant Manifolds and Fibrations for Perturbed Nonlinear Schrödinger Equations.
129. *Müller:* Analysis of Spherical Symmetries in Euclidean Spaces.
130. *Feintuch:* Robust Control Theory in Hilbert Space.
131. *Ericksen:* Introduction to the Thermodynamics of Solids, Revised ed.
132. *Ihlenburg:* Finite Element Analysis of Acoustic Scattering.
133. *Vorovich:* Nonlinear Theory of Shallow Shells.
134. *Vein/Dale:* Determinants and Their Applications in Mathematical Physics.
135. *Drew/Passman:* Theory of Multicomponent Fluids.
136. *Cioranescu/Saint Jean Paulin:* Homogenization of Reticulated Structures.
137. *Gurtin:* Configurational Forces as Basic Concepts of Continuum Physics.
138. *Haller:* Chaos Near Resonance.
139. *Sulem/Sulem:* The Nonlinear Schrödinger Equation: Self-Focusing and Wave Collapse.
140. *Cherkaev:* Variational Methods for Structural Optimization.
141. *Naber:* Topology, Geometry, and Gauge Fields: Interactions.
142. *Schmid/Henningson:* Stability and Transition in Shear Flows.
143. *Sell/You:* Dynamics of Evolutionary Equations.
144. *Nédélec:* Acoustic and Electromagnetic Equations: Integral Representations for Harmonic Problems.
145. *Newton:* The *N*-Vortex Problem: Analytical Techniques.
146. *Allaire*: Shape Optimization by the Homogenization Method.
147. *Aubert/Kornprobst:* Mathematical Problems in Image Processing: Partial Differential Equations and the Calculus of Variations.
148. *Peyret:* Spectral Methods for Incompressible Viscous Flow.
149. *Ikeda/Murota:* Imperfect Bifurcation in Structures and Materials.
150. *Skorokhod/Hoppensteadt/Salehi:* Random Perturbation Methods with Applications in Science and Engineering.
151. *Bensoussan/Frehse:* Topics on Nonlinear Partial Differential Equations and Applications.
152. Holden/Risebro: Front Tracking for Hyperbolic Conservation Laws.
153. *Osher/Fedkiw:* Level Sets and Dynamic Implicit Surfaces.